DATE			

ADVANCED ENGINEERING DYNAMICS

ADVANCED ENGINEERING DYNAMICS

Jerry H. Ginsberg

Georgia Institute of Technology

HARPER & ROW, PUBLISHERS, New York
Cambridge, Philadelphia, San Francisco, Washington,
London, Mexico City, São Paulo, Singapore, Sydney

1817

Sponsoring Editor: Cliff Robichaud
Project Editor: Ellen MacElree
Cover Design: Wanda Lubelska
Text Art: RDL Artset Ltd.
Production Manager: Jeanie Berke
Production Assistant: Beth Maglione
Compositor: Waldman Graphics, Inc.
Printer and Binder: R. R. Donnelley & Sons Company
Cover Printer: R. R. Donnelley & Sons Company

ADVANCED ENGINEERING DYNAMICS

Library of Congress Cataloging in Publication Data

Ginsberg, Jerry H., 1944–
    Advanced engineering dynamics.
    Includes index.
    1. Dynamics.   I. Title.
TA352.G55   1987   620.1′04   87–23670
ISBN 0–06–042308–0

88   89   90   91   9   8   7   6   5   4   3   2   1

To my wife
RONA

CONTENTS

CHAPTER 4
Kinematics of Rigid Bodies   113

CHAPTER 5
Newtonian Kinetics of a Rigid Body   153

CHAPTER 6
Introduction to Analytical Mechanics   224

PREFACE

Each opportunity that I have had to teach upper-level courses in the dynamics of physical systems has reinforced my belief that engineering students have specific needs that are not addressed by many texts. Although a classical physics book at the intermediate level might contain many of the topics in an engineering course, those topics must be approached from a different perspective. In addition to being familiar with the various principles, engineers must be able to solve complex problems, involving a multitude of interconnected parts and a variety of constraint conditions. Complicating features, such as friction, may be treated through idealizations, but not ignored. An engineering dynamics text must be concerned with practical applications to current systems, without sacrificing topics that prepare students for developments at the forefront of technology. Consequently, coverage must be extensive, with adequate attention to methodology. Derivations of basic principles must be rigorous, while at the same time emphasizing the significance and suitability of those principles.

I have endeavored to address these needs in a variety of ways. Some instructors say that engineers learn by example. If that statement is true, it is only because an engineer is problem oriented. One of the features of the text that the reader will find to be most prominent is the wealth of examples and homework problems. I have tried to select systems that are recognizable to the student as being relevant. Another aspect of my selection of examples is that they are intended to highlight the many facets of the associated theoretical topics, without being excessively complicated. For example, the method of Lagrangian multipliers is illustrated in some texts by rather simple examples. In contrast, Example 3 in Chapter 7 employs Lagrangian multipliers to obtain the equations of motion for a rolling disk in arbitrary motion. In order to emphasize the continuity of the result with earlier work, this solution is then compared with the case of steady precession of a rolling disk, which is solved in Example 8 of Chapter 5 by using the Newton-Euler equations. This approach is typical, in that the same system is often used to highlight the comparative features of different principles and approaches. The solutions for most examples are discussed in depth, along with qualitative discussions of the results.

A unifying aspect of most engineering applications is that kinematics provides the framework on which the kinetics principles are constructed. It is in this area that some texts have their greatest shortcoming. Thus, a large portion of the present treatment is devoted to a variety of topics in the kinematics of particles and rigid bodies. Particle motion is treated extensively in Chapter 2, from a variety of viewpoints. The fundamental principles for motion relative to a moving reference frame are developed in Chapter 3—after an extensive treatment of rotation transformations and their implications for the movements of points in a rotating body. The ability to formulate the angular velocity and angular acceleration is crucial, yet many texts have not presented a consistent methodology for such an evaluation. The development of this topic in Chapter 3 is drawn from the undergraduate text I coauthored with Dr. Joseph Genin.[1] Our experiences showed this approach to be readily accessible and highly versatile. A large part of the formulation of equations of motion using either Newtonian or Lagrangian concepts must be devoted to constraint equations, so much of Chapter 4 concerns applications involving interconnected systems, such as linkages and rolling systems.

Most engineering students first learn the Newtonian approach to kinetics principles, and that is as it should be. In that way, the student comes to appreciate the relationships among the external force system, the constraint forces, and the changing linear and angular momenta. It is from such study that one seems to develop physical insight. Chapter 5 is devoted to these topics for systems in general spatial motion, as well as the evaluation of moments and products of inertia. The emphasis is on the significance of the angular momentum, and the methodology for the evaluation of its rate of change.

A potential difficulty in analytical mechanics is that students often come to view the techniques as rotelike procedures, which hinders their ability to address new situations. I believe that these difficulties arise because of weaknesses in some presentations of generalized coordinates and virtual displacements, and the related question of the mutual role of constraint forces and constraint conditions. These matters are treated extensively in the beginning of Chapter 6. Such a development greatly assists the student to appreciate the utility of Hamilton's principle and Lagrange's equations, which are derived in the latter part of the chapter. More advanced topics, such as the equations of motion for nonholonomic systems and ignorable coordinates, are developed in Chapter 7 from concepts developed in Chapter 6.

My primary goal in writing this book is to provide technically advanced material for the student whose only prior experience in dynamics is the traditional lower-level undergraduate course. Thus, I have devoted much of the treatment to discussions of the various topics. A person with extensive experience in the subject might find the treatment to be somewhat verbose, but when confronted with the question of whose needs I should address, I chose the student's. I have omitted some advanced topics, such as Hamilton-Jacobi theory, for the sake of brevity, and also because I have not found them to be necessary for the problems most engineers encounter. However, I do offer in Chapter 7 an in-depth development of the Gibbs-Appell equations for quasi-coordinates. The conceptual ease with which the Gibbs-Appell equations treat systems subject to many nonholonomic constraints seems to make them very attractive for applications in the area of robotics. Another reason I present this material is my belief that such a treatment is far more accessible and general than is Kane's approach.[2]

The reader will observe that my primary concern is with the task of modeling systems. Thus, I have not placed great emphasis in most of the text on the evaluation of system response. In the earlier chapters, I generally evaluate the response when it

[1] J. H. Ginsberg and J. Genin, *Dynamics,* 2d ed., John Wiley & Sons, Inc., New York, 1984.

[2] See, for example, T. R. Kane and Levinson, *Dynamics,* McGraw-Hill, New York, 1985. It should be noted that the authors fail to mention the close relationship between their formalism and the Gibbs-Appell formulation. At the time this text went to press, E. A. Desloge, "Relationship between Kane's Equations and the Gibbs-Appell Equations," *Journal of Guidance, Control, and Dynamics,* vol. 10, pp. 120–122, 1987, gave the most recent discussion of this matter.

leads to increased understanding of basic physical phenomena, as in the case of the Foucault pendulum or a rolling coin in steady precession. I consider in-depth evaluations of system response to be the realm of courses on linear and nonlinear vibrations, which are covered by many excellent texts. Also, the availability of versatile differential equation solvers for digital computation places the burden on the engineer to be able to formulate equations of motion, and then to have the physical insight to validate numerical predictions. However, Chapter 8 is devoted to several applications of the basic concepts to gyroscopic systems, wherein detailed response analyses are pursued. Those developments are intended to unify the basic principles and procedures, and also to provide an introduction to the study of inertial guidance systems.

The material in the text originated from notes I developed for a graduate course in the School of Mechanical Engineering at the Georgia Institute of Technology. It is a three-credit, one-trimester course devoted to the bulk of Chapters 1 to 6, as well as the first section in Chapter 7 on constrained generalized coordinates and Lagrangian multipliers. Adequate coverage of all the material in the text can be expected to require at least one semester. An interesting aspect of the course is that a large segment of its enrollment typically consists of students who are performing research in other areas, such as computer-aided design and acoustics. I think that this is in part attributable to the depth of coverage in kinematics, wherein background material, such as the Frenet relations, curvilinear coordinates, and rotation transformations, is developed in a manner that brings out applications in other areas. I also attribute this interest to the fact that the study of dynamics provides an excellent framework for developing an engineering approach to problem solving, in which a variety of concepts must be implemented in a logical manner. The course at Georgia Tech is a prerequisite for the first vibrations course, in order to assure that all students have adequate preparation in the modeling of systems. However, a valid argument can be made that concepts in vibrations provide a useful basis for more general studies in dynamics.

ACKNOWLEDGMENTS

My thanks to the many people who contributed to making this a better book. I owe a large debt to my colleagues at Georgia Tech. Dr. Allan D. Pierce always was supportive. Dr. Aldo Ferri helped me identify troublesome areas when we shared the teaching load in dynamics in the quarter that I completed this manuscript. And a special note of gratitude is due Dr. John G. Papastavridis for sharing his insights in the area of analytical mechanics. Dr. Joseph Genin, my former colleague at Purdue University, and now at New Mexico State University, shared his thoughts on many occasions. Grateful acknowledgment is given to Russell K. Dean, Edward E. Hornsey, and Burton Paul, who have, by their suggestions and criticisms, directly contributed to the preparation of this manuscript. Finally, many students in my courses have assisted me; I especially thank Joseph Vignola, James D. Huggins, and John Ward. I was very fortunate that Clifford Robichaud was my editor at Harper & Row, Inc.; his enthusiasm provided the impetus I needed to complete the book.

My love and appreciation for my wife, Rona A. Ginsberg, go beyond the support she provided in the form of editorial judgment and preparation of the manuscript. Her understanding and devotion were an immeasurable help in my effort to balance family life against a preoccupation with research and writing this text. The patience of our children, Mitchell and Daniel, while I was engrossed in this project, was always reassuring. The debt I owe my parents, Rae and David Ginsberg, for the sacrifices they made during my education is enormous. I regret that my father is not alive to see this book, for he taught me that learning can be a joyful experience.

Jerry H. Ginsberg

Chapter 1

Basic Considerations

1.1 INTRODUCTION

The subject of dynamics is concerned with the relationship between the forces acting on a physical object and the motion that is produced by the force system. Our concern in this text is situations in which the classical laws of physics (i.e., Newtonian mechanics) are applicable. For our purposes, we may consider this to be the case whenever the object of interest is moving much more slowly than the speed of light. In part, this restriction means that we can use the concept of an absolute (i.e., fixed) frame of reference, which will be discussed shortly.

A study of dynamics consists of two phases: kinematics and kinetics. The objective of a kinematical analysis is to describe the motion of the system. It is important to realize that this type of study does not concern itself with what is causing the motion. A kinematical study might be needed to quantify a nontechnical description of the way a system moves, for example, finding the velocity of points on a mechanical linkage. In addition, some features of a kinematical analysis will always arise in a kinetics study, which analyzes the interplay between forces and motion. A primary objective will be the development of procedures for applying kinematics and kinetics principles in a logical and consistent manner, so that one may successfully analyze systems that have novel features. Particular emphasis will be placed on three-dimensional systems, some of which feature phenomena that you might not have encountered in your studies thus far. This is particularly the case if your prior experiences in the area of dynamics were limited to planar motion problems. As we proceed, you might recognize several topics from your earlier courses, both in engineering and in mathematics. Those topics are treated again here because of their importance, and also in order to gain greater understanding and rigor.

1.2 NEWTON'S LAWS

A fundamental aspect of the laws presented by Sir Isaac Newton is the concept of an absolute reference frame, which implies that somewhere in the universe there is an object that is stationary. This concept was discarded in modern physics (relativity theory), but the notion of a fixed reference frame introduces negligible errors for slowly moving objects. The corollary of this concept is the dilemma of what object should be considered to be fixed. Once again, negligible errors are usually produced if one considers the sun to be fixed. However, in most engineering situations it is preferable to use the earth as our reference frame. The primary effect of the earth's motion in most cases is to modify the (in vacuo) free-fall acceleration g resulting from the gravitational attraction between an object and the earth. Other than that effect, it is usually permissible to consider the earth to be an absolute reference frame. (A more careful treatment of the effects of the earth's motion will be part of our study of motion relative to a moving reference frame.)

For the purpose of formulating principles and solving problems, the fixed reference frame will be depicted as a set of coordinate axes, such as *xyz*. It is important to realize that coordinate axes are also often used to represent the directions for the component description of vectorial quantities. The two uses for a coordinate system are not necessarily related. Indeed, we will frequently describe a kinematical quantity relative to a specified frame of reference in terms of its components along the coordinate axes associated with a different frame of reference.

A remarkable feature of Newton's laws is that they address only objects that can be modeled as a single particle. Bodies having finite dimensions are not formally covered by these laws. The three kinematical quantities for a particle with which we are primarily concerned are position, velocity, and acceleration. By definition, a particle occupies only a single point in space. As time evolves, the point occupies a succession of positions. The locus of all positions occupied by the point is its *path*.

The position of a point, as well as the velocity and acceleration, may be described mathematically by giving three independent coordinate values. Such a description is said to be *extrinsic*, because it does not rely on knowledge of the path. This feature is contrasted by an *intrinsic* kinematical description, in which the position, velocity, and acceleration are defined in terms of the properties of the path.

In either case the *position* of the point may be depicted by a vector arrow extending from some reference location, such as the origin of the fixed frame of reference, to the point of interest. We use an overbar to denote any vector, and subscripts denote the point of interest and the reference point. For example $\bar{r}_{P/O}$ denotes the position of point P with respect to point O (the slash, /, may be translated to mean "with respect to"). A typical position vector is shown in Figure 1.1.

The position changes as time goes by, so $\bar{r}_{P/O}$ is a vector function of time. The rules for differentiation of a vector are the same as those for differentiation of a scalar, except that the order of multiplication cannot be changed in treating cross products. The time derivative of the position is called the *velocity*. It is conventional to use one overdot to denote each time derivative. Thus

$$\bar{v} = \frac{d\bar{r}_{P/O}}{dt} = \dot{\bar{r}}_{P/O} \tag{1.1}$$

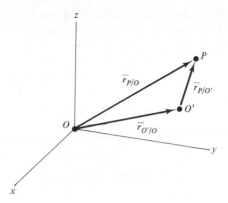

Figure 1.1 Position vectors.

Two aspects of the above are notable. First, no subscripts have been used to denote the velocity vector. If there is any ambiguity as to the point whose velocity is under consideration, the subscript will match that point. It is never necessary to indicate the reference point in the description of the velocity, because the velocity is the same as seen from all locations in an absolute frame of reference. This may be proved from Figure 1.1. If points O and O' are both fixed, then the difference in the position of point P relative to these points is constant, that is, $\bar{r}_{O'/O}$ is constant. The derivative of a constant is obviously zero. In Chapter 3 we will treat cases where the reference frame moves, in which case we will be interested in the motion relative to that reference frame. Equation (1.1) defines the *absolute velocity*, whereas the velocity seen from a moving reference frame is a relative velocity. The same terminology applies to the description of acceleration, whose definition follows. If it is not specified otherwise, the words velocity and acceleration should be understood to mean the absolute quantities.

Because velocity is a vector, it has a magnitude and a direction associated with it. The magnitude is called the *speed*,

$$v = |\bar{v}| \tag{1.2}$$

and the direction of $\bar{v}$ tells us the *heading*. Both of these properties are particularly important for formulations using intrinsic (path-related) variables.

Acceleration needs to be considered because it is the only motion parameter that arises in Newton's laws. The basic relation for this quantity is

$$\bar{a} = \dot{\bar{v}} = \ddot{\bar{r}}_{P/O} \tag{1.3}$$

It might be argued that our senses are accurately attuned to acceleration only when we are experiencing it—it is difficult to judge the acceleration of an object that we are passively observing. Indeed, the time derivative of $\bar{a}$, which is called the *jerk*, primarily occurs in considerations of ride comfort for vehicles.

Newton's laws have been translated in a variety of ways from their original statement in the *Principia* (1687), which was in Latin. We shall use the following version:

First law.   The velocity of a particle can only be changed by the application of a force.

Second law.   The resultant force acting on a particle is proportional to the acceleration of the particle. The factor of proportionality is the mass. In other words,

$$\overline{F} = m\overline{a} \qquad (1.4)$$

Third law.   The forces acting on a body result from an interaction with another body, such that there is a reactive force, that is, reaction, applied to the other body. The action-reaction pair consists of forces having the same magnitude, and acting along the same line of action, but having opposite direction.

We realize that the first law is included in the second, but we retain it primarily because it treats static systems without the need to discuss acceleration. The second law is quite familiar, but it must be emphasized that it is a vector relation. Hence, it can be decomposed into as many as three scalar laws, one for each component. The third law is very important to the modeling of systems. The "models" that are created in a kinetics study are free body diagrams, in which the system is isolated from its surroundings. Careful application of the third law will assist identification of the forces exerted on the body.

The conceptualization of the first and second laws can be traced back to Galileo. Newton's revolutionary idea was the recognition of the third law and its implications for the first and second laws. An interesting aspect of the third law is that it excludes the concept of an inertial force, $-m\overline{a}$, which is usually associated with d'Alembert, because there is no corresponding reactive body. It is for that reason that we shall only employ the inertial force concept in Chapter 6 to develop the principles of analytical mechanics.

It is also worth noting that the class of forces described by the third law is limited— any force obeying this law is said to be a *central force*. An example of a noncentral force arises from the interaction between moving electric charges. Such forces have their origin in relativistic effects. Strictly speaking, the study of *classical mechanics* is concerned only with systems that fully satisfy all of Newton's laws. However, many of the principles and techniques are applicable either directly, or with comparatively minor modifications, to relativity theory.

We should note that the acceleration to be employed in Newton's second law is relative to the hypothetical fixed reference frame. However, the same acceleration can also be observed from a variety of moving reference frames, all of which are translating (that is, the reference directions are not rotating) at a constant velocity relative to the fixed reference. Such reference frames are said to be *inertial*. The fact that Newton's laws are valid in any inertial reference frame is the *principle of Galilean invariance*, or the principle of *Newtonian relativity*.

1.3 SYSTEMS OF UNITS

Newton's second law brings up the question of the units to be used for describing the force and motion variables. Related to that consideration is the dimensionality of a quantity, which refers to the basic measures that are used to form the quantity. In dynamics, the basic measures are time T, length L, mass M, and force F. The law of dimensional homogeneity requires that these four quantities be consistent with the second law. Thus

$$F = \frac{ML}{T^2} \qquad (1.5)$$

It is clear from this relation that only three of the four basic measures are independent. Measures for time and length are easily defined, so this leaves the question of whether mass or force is the third independent quantity. When a system of units is defined such that the unit of mass is fundamental, the units are said to be *absolute*. In contrast, when the force unit is fundamental and the mass unit is given by $M = FT^2/L$, the units are said to be *gravitational*. This terminology stems from the relation among the weight w, the mass m, and the free-fall acceleration.

Newton's law of gravitation† states that the magnitude of the attractive force exerted between the earth and a body of mass m is

$$F = \frac{GMm}{r^2} \qquad (1.6)$$

where r is the distance between the centers of mass, G is the universal gravitational constant, and M is the mass of the earth,

$$G = 6.6732(10^{-11}) \text{ m}^3/\text{kg-s}^2 = 3.4392(10^{-8}) \text{ ft}^4/\text{lb-s}^4$$

$$M = 5.976(10^{24}) \text{ kg} = 4.095(10^{23}) \text{ lb-s}^2/\text{ft}$$

The weight w of a body usually refers to the gravitational attraction of the earth when the body is near the earth's surface. When a body near the earth's surface is falling freely in a vacuum, the gravitational attraction is the mass of the body times the free-fall acceleration, that is,

$$w = mg \qquad (1.7)$$

Matching Eq. (1.6) at the earth's surface to Eq. (1.7) leads to

$$g = \frac{GM}{r_e^2} \qquad (1.8)$$

where r_e is the radius of the earth, $r_e = 6.371(10^6)$ meters.

The relationship between g and the gravitational pull of the earth is actually far more complicated than Eq. (1.8). In fact, g depends on the location along the earth's surface. One reason for such variation is the fact that the earth is not perfectly spherical, which means that r_e is not actually constant. Variation in the value of g also arises because the earth is not homogeneous. In addition to these deviations of the gravitational force, the value of g is influenced by the motion of the earth, which means that it is a

†It is implicit to this development that the inertial mass in Newton's second law is the same as the gravitational mass appearing in the law of gravitation. This fundamental assumption, which is known as the *principle of equivalence*, actually is owed to Galileo, who tested the hypothesis with his experiments on various pendulums. More refined experiments performed subsequently have continued to verify the principle.

measure of acceleration relative to a noninertial reference frame. Consequently, it is not exactly correct to employ Eq. (1.8).

The mass of a particle is constant (assuming no relativistic effects), so defining mass as a fundamental parameter yields a system of units whose definition is not position-dependent. The SI (Standard International) system of units, which is a metric MKS system with standardized prefixes for powers of 10 and standard names for derived units, is an absolute system. The system now known as US Customary is a gravitational system, because its basic unit is force, measured in pounds (lb). The body whose weight is defined as a pound must be at a specified location. If that body were to be moved to a different place, the gravitational force acting on it, and hence the units of force, might be changed.

The relationship $w = mg$ can cause some confusion in either system of units. In the SI system, where mass is basic, any body should be described in terms of its mass in kilograms. Its weight is then found by multiplying by g, which for a standard location on the earth's surface is

$$g = 9.807 \text{ m/s}^2$$

The weight that is found from $w = mg$ has units of newtons,

$$1 \text{ N} = 1 \text{ kg} \times 1 \text{ m/s}^2 = 1 \text{ kg-m/s}^2$$

The situation is not so simple for US Customary units for two reasons. First, many people have used the unit of pound mass in the related absolute system of units. If one also employs a pound force unit, where 1 lbf is the weight of a 1-lbm body at the surface of the earth, then $f = ma$ is not satisfied unless the acceleration is measured as multiples of g. This is an unnecessary complication, so the pound will always be a force unit in our studies. Thus, a body will always be specified in terms of its weight in pounds. Even then, the mass unit is complicated by the fact that two length units, feet and inches, are in common use. We shall use the standard values

$$g = 32.17 \text{ ft/s}^2 \quad \text{or} \quad g = 386.0 \text{ in./s}^2$$

It follows that computing the mass as $m = w/g$ will give a value for m that depends on the length unit in use. The *slug* is a standard name for the mass unit, with

$$1 \text{ slug} = 1 \text{ lb}/(1 \text{ ft/s}^2) = 1 \text{ lb-s}^2/\text{ft}$$

However, this mass unit is not applicable when inches is the length unit. In order to emphasize this matter, we will make it a standard practice in US Customary units to give mass in terms of the basic units. For example, a mass might be listed as 5.2 lb-s^2/ft, or a moment of inertia might be 125 lb-s^2-in.

1.4 VECTOR CALCULUS

It is assumed here that you are familiar with the basic laws and techniques for the algebra of vectors. Specifically, you should be able to represent a vector in terms of its components and perform calculations such as addition, dot products, and cross products using that component representation. If you feel uncertain about your current proficiency, it is strongly recommended that you take some time to review the appropriate topics.

As was mentioned earlier, most of the laws for calculus operations are the same as those for scalar variables. It is only necessary to remember to keep the overbar on vector quantities. In the following, $\overline{A}$ and $\overline{B}$ are time-dependent vector functions, and c and α are scalar functions.

Definition of a Derivative

$$\dot{\overline{A}} = \lim_{\Delta t \to 0} \frac{\overline{A}(t + \Delta t) - \overline{A}(t)}{\Delta t} \tag{1.9}$$

Definite Integration

Let $\overline{B} = \dot{\overline{A}}$. Then

$$\overline{A}(t) = \overline{A}(0) + \int_0^t \overline{B}(\tau)\, d\tau \tag{1.10}$$

Derivative of a Sum

$$\frac{d}{dt}(\overline{A} + \overline{B}) = \dot{\overline{A}} + \dot{\overline{B}} \tag{1.11}$$

Derivative of Products

$$\frac{d}{dt}(c\overline{A}) = \dot{c}\overline{A} + c\dot{\overline{A}} \tag{1.12}$$

$$\frac{d}{dt}(\overline{A} \cdot \overline{B}) = \dot{\overline{A}} \cdot \overline{B} + \overline{A} \cdot \dot{\overline{B}} \tag{1.13}$$

$$\frac{d}{dt}(\overline{A} \times \overline{B}) = \dot{\overline{A}} \times \overline{B} + \overline{A} \times \dot{\overline{B}} \tag{1.14}$$

Chain Rule for Differentiation

Let $\overline{A}$ be a function of some parameter α and let α be a function of time. (This obviously means that $\overline{A}$ is an implicit function of time.) Then

$$\dot{\overline{A}} = \frac{d\overline{A}}{d\alpha}\frac{d\alpha}{dt} = \dot{\alpha}\frac{d\overline{A}}{d\alpha} \tag{1.15}$$

These rules will be used in the next chapter to treat the component representation of vectors with respect to various triads of directions.

1.5 ENERGY AND MOMENTUM

A basic application of the calculus of vectors in dynamics is the derivation of energy and momentum principles, which are integrals of Newton's second law. These integrals represent standard relations between velocity parameters and the properties of the force system. We will derive these laws for particle motion here; the corresponding derivations for a rigid body appear in Chapter 5.

Energy principles are useful when we know how the resultant force varies as a function of the particle's position, in other words when $\Sigma \bar{F}(\bar{r})$ is known. A dot product of Newton's second law with a differential displacement $d\bar{r}$ of the particle yields

$$\Sigma \bar{F} \cdot d\bar{r} = m\bar{a} \cdot d\bar{r} \tag{1.16}$$

Multiplying and dividing the right side by dt, and then using the definition of velocity and acceleration, leads to

$$\Sigma \bar{F} \cdot d\bar{r} = m\bar{a} \cdot \frac{d\bar{r}}{dt}\, dt = m\frac{d\bar{v}}{dt} \cdot \bar{v}\, dt = \frac{1}{2} m \frac{d}{dt}(\bar{v} \cdot \bar{v})\, dt$$

$$= \tfrac{1}{2}m\, d(\bar{v} \cdot \bar{v}) = \tfrac{1}{2}m\, d(v^2) \tag{1.17}$$

The right side is a perfect differential, and the left side is a function of only the position because of the assumed dependence of the force resultant. Hence, we may integrate the differential relation between the two positions. The evaluation of the integral of the left side must account for the variation of the resultant force as the position changes when the particle moves along its path; this is a called a path integral. We therefore find that

$$\oint_1^2 \Sigma \bar{F} \cdot d\bar{r} = \tfrac{1}{2}m(v_2^2 - v_1^2) \tag{1.18}$$

where the subscripts indicate that the speed should be evaluated at either the initial position 1 or the final position 2. The *kinetic energy* is defined as

$$T = \tfrac{1}{2}mv^2 \tag{1.19}$$

and the path integral is the *work done by the force* in moving the particle,

$$W_{1\to2} = \oint_1^2 \Sigma \bar{F} \cdot d\bar{r} \tag{1.20}$$

where the subscript notation for W indicates that the work is done in going from the starting position 1 to the end position 2 along the particle's path. The corresponding form of Eq. (1.18) is

$$T_2 = T_1 + W_{1\to2} \tag{1.21}$$

This is known as the *work-energy principle*.

The operation of evaluating the work is depicted in Figure 1.2. We see there that

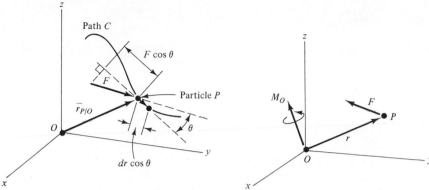

Figure 1.2 Work done by a force.

Figure 1.3 Moment of a force.

a differential amount of work done by the resultant force in moving the particle may be considered in either of two ways. It is the product of the differential distance the particle moves and the component of the resultant force in the direction of movement or, equivalently, the product of the magnitude of the resultant force and the projection of the displacement in the direction of the force. Only in the simple case where the force has a constant component in the direction of the displacement does the work reduce to the simple expression "force multiplied by distance displaced."

The evaluation of the work is a major part of a formulation of the work-energy principle. We will find in Chapter 5 that this task is alleviated by introducing the concept of potential energy.

Two momentum principles follow from Newton's second law. The linear impulse-momentum principle is an immediate result when the resultant force is known as a function of time. Because the acceleration is the time derivative of velocity, multiplying the second law by dt and integrating over an interval $t_1 \leqq t \leqq t_2$ leads to

$$\int_{t_1}^{t_2} \sum \overline{F}\, dt = \int_{t_1}^{t_2} m\overline{a}\, dt = m(\overline{v}_2 - \overline{v}_1) \tag{1.22}$$

The quantity $m\overline{v}$ is the *momentum* of the particle, which we shall denote by the symbol $\overline{P}$. Thus, we have

$$\overline{P} = m\overline{v} \qquad \overline{P}_2 = \overline{P}_1 + \int_{t_1}^{t_2} \sum \overline{F}\, dt \tag{1.23}$$

The time integral of the resultant force is the *impulse*. More precise names for the terms appearing in Eq. (1.23) are the linear momentum and linear impulse, because they are associated with the movement of a particle along a (possibly curved) line. The primary utility of the linear impulse-momentum principle is to treat systems excited by impulsive forces, that is, forces that impart a very large acceleration to a body over a very short time interval. Otherwise, the principle is an obvious consequence of knowing the resultant force as a function of time.

The angular-momentum principle is associated with the moment the resultant force exerts. Let us evaluate the moment $\overline{M}_O$ about origin O of the fixed reference frame in Figure 1.3. Using the position $\overline{r}$ to form the lever arm leads to

$$\overline{M}_O = \overline{r} \times \sum \overline{F} = \overline{r} \times m\overline{a} = \overline{r} \times m\frac{d\overline{v}}{dt} \tag{1.24}$$

We now take the time derivative outside the cross product by compensating the equation with an appropriate term to maintain the identity, specifically,

$$\overline{M}_O = \frac{d}{dt}(\overline{r} \times m\overline{v}) - m\frac{d\overline{r}}{dt} \times \overline{v} = \frac{d}{dt}(\overline{r} \times m\overline{v}) - \overline{v} \times m\overline{v} \tag{1.25}$$

The last term vanishes because the momentum $m\overline{v}$ is parallel to the velocity. The remaining term on the right side of the equation is the time derivative of the moment about origin O of the linear momentum of the particle. We refer to this term as the angular momentum because a moment is associated with a rotational tendency, and use the symbol $\overline{H}_O$ to denote it. Thus,

$$\overline{H}_O = \overline{r} \times m\overline{v} \tag{1.26}$$

Substitution of $\overline{H}_O$ into Eq. (1.25) leads to the derivative form of the *angular impulse-momentum principle*,

$$\overline{M}_O = \frac{d\overline{H}_O}{dt} \tag{1.27}$$

Multiplying the relation by dt and integrating over a time interval $t_1 \leqq t \leqq t_2$ leads to

$$(\overline{H}_O)_2 = (\overline{H}_O)_1 + \int_{t_1}^{t_2} \sum \overline{M}_O \, dt \tag{1.28}$$

where the time integral of the moment is called the *angular impulse* of the resultant force.

Situations where the angular-impulse momentum principle, Eq. (1.28), is needed to study the motion of a particle are few. As is the case for its linear analog, the angular-momentum principle might be useful to treat an impulsive force. Also, when the moment of the resultant force about an axis $\overline{e}$ is zero, the principle yields a conservation principle, $\overline{H}_O \cdot \overline{e}$ is constant. The primary utility of the angular-momentum principle lies in the application of the derivative form, Eq. (1.27), to a rigid body. We will find in Chapter 5 that the angular momentum of a body is related to the rotation of the body.

1.6 BRIEF BIOGRAPHICAL PERSPECTIVE

As we proceed through the various topics, the names of some early scientists and mathematicians will be encountered in a variety of contexts. The magnitude of the contribution of these pioneers cannot be overstated. Indeed, it is a testimonial to their ingenuity that we continue to use so much of their work. A view of the historical relationship between these researchers can greatly enhance our insight. The following is an informal chronological survey of a few individuals who have made key contributions to classical, as opposed to relativistic, physics. More details may be found in the list of references for this chapter.

Galileo, Galilei (1564–1642)

Galileo is best known for experiments on gravity at the leaning tower of Pisa, in his native country, Italy, but there is no conclusive evidence that those experiments actually occurred. From his measurements of the motion of pendulums, which led him to propose the use of a pendulum for a clock, he deduced that gravitational and inertial mass are identical. He refuted Aristotle's ancient statements by observing that the state of motion can only be altered by the presence of other bodies, and that there is no unique inertial reference frame. In astronomy, he developed the astronomical telescope, and used it for many pioneering observations. His last eight years were spent under house arrest for advocating the Copernican view of the solar system, which held that the sun, rather than the earth, is the center of the solar system.

Newton, Sir Isaac (1642–1727)

Newton pursued his studies of physics in England, aware of scientific developments flowering throughout Europe. The foundation for our study of mechanics was laid out by him in *Principia Mathematica Philosophiae Naturalis* (1687). In addition to his basic laws governing the movement of bodies due to forces, Newton developed the classical law of gravitation, thereby providing a scientific basis for Johannes Kepler's empirical laws for the orbits of the planets. Newton is also generally credited with having developed the mathematical calculus.

Euler, Leonhard (1707–1783)

Euler was Swiss by birth, but his scientific work was done first in Russia, then in Berlin, and finally in Russia again. Some of the greatest contributions to the mechanics of rigid bodies are due to Euler, who was also the most prolific mathematician of his century. He derived many new mathematical principles in order to solve physically meaningful problems. In addition to his contributions to the kinematics and kinetics of rigid bodies, which we shall study in later chapters, he developed the theory of functions. He also made important contributions to the solution of ordinary differential equations, to geometry and topology, and to number theory.

Alembert, Jean Le Rond d' (1717–1783)

d'Alembert, who was French, is associated with the notion of an inertial force $-m\bar{a}$. In addition, he introduced the concept of partial differential equations in order to determine the dynamic response of deformable bodies. We will not use the concept of an inertial force as a technique for formulating problems, for it is incorrect in the sense that there is no reaction corresponding to the "inertia force." However, his contribution was crucial to the development by Lagrange of analytical mechanics. Interestingly, d'Alembert and Euler were bitter rivals. d'Alembert was also known as a philosopher and as a music theorist.

Lagrange, Joseph-Louis, Comte de (1736–1813)

The works of Euler and Lagrange are intimately related. Their mutual respect is exemplified by the fact that Lagrange left his native Italy to replace Euler in an important academic position in Berlin upon Euler's recommendation. The fundamental equations of analytical mechanics bear Lagrange's name. These principles, which we will derive in Chapters 6 and 7, employ energy functions and geometrical relations in a viewpoint that emphasizes the way a system behaves as a single entity. Lagrange's equations could be considered to be equivalent to Newton's laws, but they are actually more general. Lagrange made important contributions in celestial mechanics and, like Euler, was active in many areas of mathematics, including calculus of variations, the theory of equations, probability theory, and number theory.

Coriolis, Gustave-Gaspard (1792–1843)

In the list we have assembled here, the Frenchman Coriolis is certainly a minor figure. The identification of an acceleration effect attributable to interaction between rotation of a reference frame and movement relative to that reference frame is due to Coriolis. He also made some important contributions to the study of mechanics of materials and collisions of bodies.

Hamilton, Sir William Rowan (1805–1865)

Hamilton, who was Irish, developed a unified formulation for classical mechanics. "Hamilton's principle," which draws on concepts from the calculus of variations, contains both the Newtonian and the Lagrangian forms of the equations of motion. His principle has even been extended to relativistic and quantum mechanics through appropriate redefinitions of the energy functions. Since Newton's laws are axioms, some researchers have argued that Hamilton's principle, rather than Newton's laws, is the foundation for classical mechanics.

Rayleigh, Lord John William Strutt (1842–1919)

Acclaimed as the last of the great British classical physicists, Rayleigh's name appears in the dynamics of particles and rigid bodies in a relatively minor context. He introduced dissipation effects in Lagrange's equations in the same manner that inertia and conservative forces are described, but we now recognize that technique to be of limited validity. Much of Rayleigh's work was in the field of vibrations, which builds on the concepts we develop here. His contributions in acoustics, optics, and electromagnetism are equally significant. To some he is best known as the person who discovered and isolated argon.

Perhaps the most remarkable aspect of the foregoing survey is the date of Hamilton's death. Although the basic principles were developed more than a century ago, the versatility of the analytical tools and their level of sophistication have continued to be refined. The subject of mechanics is mature only from a philosophical view—many important questions remain to be answered.

REFERENCES

R. B. Lindsay and H. Margenau, *Foundations of Physics*, Dover Publications, Inc., New York, 1957.

E. Mach, *The Science of Mechanics*, 6th ed., Open Court Publishing Co., LaSalle, Ill, 1960.

J. B. Marion, *Classical Dynamics of Particles and Systems*, Academic Press, New York, 1960.

L. Meirovitch, *Methods of Analytical Dynamics*, McGraw-Hill Book Company, New York, 1970.

P. M. Morse and H. Feshbach, *Methods of Theoretical Physics*, McGraw-Hill Book Company, New York, 1953.

K. R. Symon, *Mechanics*, 3d ed., Addison-Wesley, Reading, Mass, 1971.

C. Truesdell, *Essays in the History of Mechanics*, Springer-Verlag, New York, 1960.

Chapter 2

Particle Kinematics

This chapter develops some basic techniques for describing the motion of a particle. Each description is based on a different set of coordinates. Which is best suited to a particular situation depends on a variety of factors, but a primary consideration is whether the coordinates naturally fit the known aspects of the motion. At the end of the chapter, we will examine situations where more than one of these descriptions may be employed beneficially.

2.1 PATH VARIABLES—INTRINSIC COORDINATES

The idea that the motion of a point should be described in terms of the properties of its path may not seem to be obvious. However, that is precisely the way in which one thinks in using a road map and the speedometer and odometer of an automobile. This type of description is known as *path variables*, or less commonly as *intrinsic coordinates*, because the basic parameters that are considered to change are associated with the properties of the path. It is also called *tangent and normal components* because those are the primary directions, as we shall see. We assume that the path is known. The most fundamental variable for a specified path is the arclength s along this curve, measured from some starting point to the point of interest. As shown in Figure 2.1, measurement of s requires statement of positive sense along the path. Thus, negative s means that the point has receded, rather than advancing along its path. It is quite obvious from Figure 2.1 that the position $\bar{r}_{P/O}$ is unambiguously defined by the value of s. Because s changes with time, the position is an implicit function of time, $\bar{r}_{P/O} = \bar{r}(s)$ and $s = s(t)$. It follows that the derivation of formulas for velocity and acceleration will involve chain-rule differentiation. We begin by deriving some basic laws governing the geometry of curves.

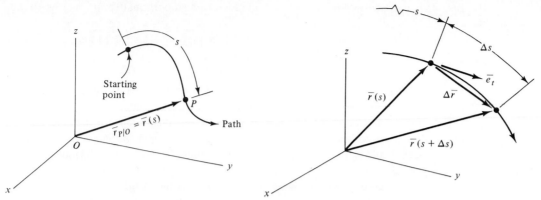

Figure 2.1 Position in path variables.

Figure 2.2 Tangent vector.

2.1.1 Curves in Space—Frenet's Formulas

Figure 2.2 shows the position vector at two locations that are separated by a small arclength Δs, with Δs becoming ds in the limit of infinitesimal differences. The displacement $\Delta \bar{r}$ is the change in the position of the point as it moves from position s to $s + \Delta s$,

$$\Delta \bar{r} = \bar{r}(s + \Delta s) - \bar{r}(s) \tag{2.1}$$

In the limit, the magnitude of $\Delta \bar{r}$ equals ds because a chord progressively approaches the curve. For the same reason, the direction of $\Delta \bar{r}$ approaches tangency to the curve, in the sense of increasing s. This *tangent direction* is defined by the unit tangent vector $\bar{e}_t$, which is the second path variable parameter. A unit vector has the dimensionless value 1 for magnitude, so

$$\bar{e}_t = \frac{d\bar{r}}{ds} \tag{2.2}$$

The tangent vector is one of three unit vectors used for describing vectorial quantities in terms of path variables. The second unit vector in the triad is derived by considering the dependence of $\bar{e}_t$ on s. For this evaluation the dot product giving the magnitude of the unit vector $\bar{e}_t$, that is, $\bar{e}_t \cdot \bar{e}_t = 1$, may be differentiated, with the result that

$$\bar{e}_t \cdot \frac{d\bar{e}_t}{ds} = 0 \tag{2.3}$$

In other words, $d\bar{e}_t/ds$ is always perpendicular to $\bar{e}_t$. The *normal direction*, whose unit vector is $\bar{e}_n$, is defined to be parallel to this derivative. Because parallelism of two vectors corresponds to their proportionality, this definition may be written as

$$\bar{e}_n = \rho \frac{d\bar{e}_t}{ds} \tag{2.4}$$

Because $\bar{e}_n$ is a dimensionless unit vector, the factor of proportionality, ρ, may be found from

$$\frac{1}{\rho} = \left|\frac{d\bar{e}_t}{ds}\right| \tag{2.5}$$

Dimensional consistency of Eq. (2.5) requires that ρ be a length parameter. It is the *radius of curvature*, as we will soon demonstrate for a planar path.

The tangent and normal unit vectors at a selected position form a plane that is tangent to the curve. Although any plane containing $\bar{e}_t$ is tangent to the curve, the plane containing both $\bar{e}_t$ and $\bar{e}_n$ has several interesting features; this plane is referred to as the *osculating plane*.

In order to see why the parameter ρ in Eq. (2.5) is the radius of curvature, let us consider a planar path, in which case the osculating plane is the plane of the curve. Figure 2.3a depicts the tangent and normal vectors associated with two points, A and B, that are separated by an infinitesimal distance ds measured along an arbitrary planar path. Point C, which is the intersection of the normal vectors at the two positions along the curve, is the *center of curvature*. Because ds is infinitesimal, the arc AB seems to be circular. The radius R of this arc is the radius of curvature. The formula for the arc of a circle shows that $d\theta = ds/R$.

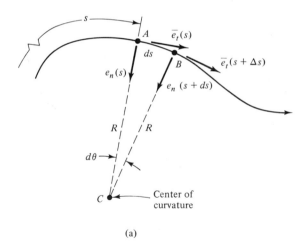

(a)

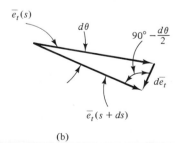

(b)

Figure 2.3 Relation between tangent and normal vectors.

Now consider the increment $d\bar{e}_t = \bar{e}_t(s + ds) - \bar{e}_t(s)$ in Figure 2.3b. The angle $d\theta$ between the normal vectors in Figure 2.3a is also the angle between the tangent vectors. The vector triangle in Figure 2.3b is isosceles because $|\bar{e}_t| = 1$. Hence, the angle between $d\bar{e}_t$ and either tangent vector is $90° - d\theta/2$. Since $d\theta$ is infinitesimal, it must be that $d\bar{e}_t$ is perpendicular to $\bar{e}_t$ in the direction of $\bar{e}_n$. A unit vector has a length of one, so

$$|d\bar{e}_t| = d\theta \, |\bar{e}_t| = \frac{ds}{R}$$

Any vector may be expressed as the product of its magnitude and a unit vector defining the sense of the vector, from which we find that

$$d\bar{e}_t = |d\bar{e}_t| \, \bar{e}_n = \frac{ds}{R} \, \bar{e}_n$$

When this relation is divided by ds, the result agrees with Eq. (2.2), provided that $\rho = R$. Hence, we have proved that the reciprocal of the magnitude of $d\bar{e}_t/ds$ is the radius of curvature of the planar path.

When the path is not planar, the orientation of the osculating plane containing the $\bar{e}_t, \bar{e}_n$ pair will depend on the position along the curve. Nevertheless, ρ is still the radius of curvature. Note that the radius of curvature is generally not a constant, although ρ is obviously the radius of a true circular path.

The development thus far is adequate to obtain formulas for velocity and acceleration. However, additional study of the unit vectors will enhance our understanding of the properties of curves. Because $\bar{e}_t$ and $\bar{e}_n$ are situated in the osculating plane, a third direction is required for the resolution of an arbitrary vector. The direction perpendicular to the osculating plane is called the *binormal*; the corresponding unit vector is $\bar{e}_b$. The cross product of two unit vectors is a unit vector perpendicular to the original two, so we define the binormal direction such that

$$\bar{e}_b = \bar{e}_t \times \bar{e}_n \tag{2.6}$$

An interesting property arises in the derivative of the $\bar{e}_n$ unit vector, which we may represent in terms of its tangent, normal, and binormal components. The component of an arbitrary vector in a specific direction may be obtained from a dot product with a unit vector in that direction. Multiplying each component by the corresponding unit vector and adding the individual contributions then reproduces the original vector.

$$\frac{d\bar{e}_n}{ds} = \left(\bar{e}_t \cdot \frac{d\bar{e}_n}{ds} \right) \bar{e}_t + \left(\bar{e}_n \cdot \frac{d\bar{e}_n}{ds} \right) \bar{e}_n + \left(\bar{e}_b \cdot \frac{d\bar{e}_n}{ds} \right) \bar{e}_b \tag{2.7}$$

We obtain the tangential component in the above from the orthogonality of the unit vectors, which requires that $\bar{e}_t \cdot \bar{e}_n = 0$. Then

$$\bar{e}_t \cdot \frac{d\bar{e}_n}{ds} = -\bar{e}_n \cdot \frac{d\bar{e}_t}{ds} = -\bar{e}_n \cdot \left(\frac{1}{\rho} \bar{e}_n \right) = -\frac{1}{\rho} \tag{2.8}$$

Similarly, because $\bar{e}_n \cdot \bar{e}_n = 1$, we find that

$$\bar{e}_n \cdot \frac{de_n}{ds} = 0 \tag{2.9}$$

The binormal component of the derivative is generally nonzero; its value is defined as the reciprocal of the *torsion* τ,

$$\frac{1}{\tau} = \bar{e}_b \cdot \frac{d\bar{e}_n}{ds} \tag{2.10}$$

The reciprocal is used here for consistency with Eq. (2.5), such that τ has the dimension of length. Substitution of Eqs. (2.8) to (2.10) into Eq. (2.7) results in

$$\frac{d\bar{e}_n}{ds} = -\frac{1}{\rho}\bar{e}_t + \frac{1}{\tau}\bar{e}_b \tag{2.11}$$

The derivative of $\bar{e}_b$ may be obtained by a similar approach. The fact that $\bar{e}_t$, $\bar{e}_n$, and $\bar{e}_b$ are mutually orthogonal, in combination with Eqs. (2.4) and (2.11), yields

$$\bar{e}_t \cdot \bar{e}_b = 0 \Rightarrow \bar{e}_t \cdot \frac{d\bar{e}_b}{ds} = -\frac{d\bar{e}_t}{ds} \cdot \bar{e}_b = -\frac{1}{\rho}\bar{e}_n \cdot \bar{e}_b = 0 \tag{2.12a}$$

$$\bar{e}_n \cdot \bar{e}_b = 0 \Rightarrow \bar{e}_n \cdot \frac{d\bar{e}_b}{ds} = -\frac{d\bar{e}_n}{ds} \cdot \bar{e}_b = -\frac{1}{\tau} \tag{2.12b}$$

$$\bar{e}_b \cdot \bar{e}_b = 1 \Rightarrow \bar{e}_b \cdot \frac{d\bar{e}_b}{ds} = 0 \tag{2.12c}$$

The result is

$$\frac{d\bar{e}_b}{ds} = -\frac{1}{\tau}\bar{e}_n \tag{2.13}$$

Because $\bar{e}_n$ is a unit vector, this relation provides an alternative to Eq. (2.10) for the torsion

$$\frac{1}{\tau} = \left|\frac{d\bar{e}_b}{ds}\right| \tag{2.14}$$

Equations (2.4), (2.11), and (2.13) are *Frenet's formulas* for a spatial curve. The first one shows that the change in the tangent vector due to a small increase in s is primarily in the normal direction. The osculating plane is formed from $\bar{e}_t$ and $\bar{e}_n$. We therefore may consider this plane to be the tangent plane that most closely fits the curve at the position of interest.

Equation (2.13) shows that the vector normal to the osculating plane primarily changes in the direction of $\bar{e}_n$ with small increments in s. This is equivalent to a rotation of the osculating plane about the tangent direction, which is the source of the terminology "torsion." The osculating plane is constant for a planar curve, which corresponds to an infinite value of τ. The greater the degree to which a curve is twisted in space, the smaller will be the value of τ. In a similar vein, ρ measures the amount by which the curve bends in the osculating plane. A small value of ρ corresponds to a highly bent curve, and ρ is infinite for a straight line.

It is possible to specify a path in a variety of ways. Let us suppose that the path is described in *parametric form*. This means that if α is some parameter that can have a range of values, then the x, y, and z coordinates are given in terms of the value of α. The position vector may be written in component form in this case as

$$\bar{r} = x(\alpha)\bar{i} + y(\alpha)\bar{j} + z(\alpha)\bar{k} \tag{2.15}$$

When such a description is available, it is possible to apply Frenet's formulas to evaluate the path variables in terms of the parameter α.

We employ the chain rule in order to determine $\bar{e}_t$ according to Eq. (2.2), from which we find that

$$\bar{e}_t = \frac{d\bar{r}}{d\alpha}\frac{d\alpha}{ds} = \frac{\bar{r}'}{s'}$$
$$\bar{r}' = x'\bar{i} + y'\bar{j} + z'\bar{k} \tag{2.16}$$

where a prime denotes differentiation with respect to α. The fact that $\bar{e}_t$ is a unit vector, $|\bar{e}_t| = 1$, then yields a relation by which the arclength may be computed.

$$s' = (\bar{r}' \cdot \bar{r}')^{1/2} = [(x')^2 + (y')^2 + (z')^2]^{1/2} \tag{2.17a}$$

$$s = \int_{\alpha_0}^{\alpha} [(x')^2 + (y')^2 + (z')^2]^{1/2} \, d\alpha \tag{2.17b}$$

where α_0 is the value at the starting position. The value of s' found from Eq. (2.17a) may be substituted into Eqs. (2.16) in order to evaluate the tangent vector.

The next step is to evaluate $\bar{e}_n$ and ρ, for which the first of Frenet's formulas, Eq. (2.4), is used. From Eq. (2.16) we have

$$\bar{e}_n = \rho\frac{d\bar{e}_t}{ds} = \rho\frac{d\bar{e}_t}{d\alpha}\frac{d\alpha}{ds} = \frac{\rho}{s'}\left(\frac{\bar{r}''}{s'} - \frac{\bar{r}'s''}{(s')^2}\right)$$
$$= \frac{\rho}{(s')^3}(\bar{r}''s' - \bar{r}'s'') \tag{2.18}$$

The value of s' is given by Eq. (2.17a). Differentiating that relation yields

$$s'' = \frac{\bar{r}' \cdot \bar{r}''}{(\bar{r}' \cdot \bar{r}')^{1/2}} = \frac{\bar{r}' \cdot \bar{r}''}{s'} \tag{2.19}$$

The desired expression for the normal vector is obtained by substituting Eq. (2.19) into Eq. (2.18), with the result that

$$\bar{e}_n = \frac{\rho}{(s')^4}[\bar{r}''(s')^2 - \bar{r}'(\bar{r}' \cdot \bar{r}'')] \tag{2.20}$$

Because $\bar{e}_n$ is a unit vector, using a dot product to form the magnitude of this expression leads to the radius of curvature.

$$\frac{1}{\rho} = \frac{1}{(s')^4} \left| [\bar{r}''(s')^2 - \bar{r}'(\bar{r}' \cdot \bar{r}'')] \right|$$

$$= \frac{1}{(s')^4} [\bar{r}'' \cdot \bar{r}''(s')^4 - 2(\bar{r}' \cdot \bar{r}'')^2(s')^2 + (\bar{r}' \cdot \bar{r}')(\bar{r}' \cdot \bar{r}'')^2]^{1/2}$$

which simplifies to

$$\frac{1}{\rho} = \frac{1}{(s')^3} [(\bar{r}'' \cdot \bar{r}'')(s')^2 - (\bar{r}' \cdot \bar{r}'')^2]^{1/2} \qquad (2.21)$$

In the case of a planar curve defined in the form $y = y(x)$, so that $\alpha = x$, this expression reduces to

$$\rho = \frac{[1 + (y')^2]^{3/2}}{|y''|} \qquad (2.22)$$

which is the same as the formula derived in a course on calculus.

After the result of Eq. (2.21) is substituted back into Eq. (2.20), it is a simple matter to evaluate the binormal direction according to Eq. (2.6).

$$\bar{e}_b = \bar{e}_t \times \bar{e}_n = \frac{\bar{r}'}{s'} \times \frac{\rho}{(s')^4} [\bar{r}''(s')^2 - \bar{r}'(\bar{r}' \cdot \bar{r}'')]$$

$$= \frac{\rho}{(s')^3} \bar{r}' \times \bar{r}'' \qquad (2.23)$$

An expression for τ may be obtained by applying the third Frenet formula, Eq. (2.13). The result of differentiating Eq. (2.23) may be written as

$$\frac{d\bar{e}_b}{ds} = \frac{1}{s'} \frac{d\bar{e}_b}{d\alpha} = \frac{1}{s'} \frac{d}{d\alpha} \left[\frac{\rho}{(s')^3} \right] (\bar{r}' \times \bar{r}'') + \frac{\rho}{(s')^4} (\bar{r}' \times \bar{r}''') \qquad (2.24)$$

Next, we substitute this expression and Eq. (2.20) into Eq. (2.13) to find that

$$\frac{1}{\tau} = -\bar{e}_n \cdot \frac{d\bar{e}_b}{ds}$$

$$= -\frac{\rho}{(s')^4} [\bar{r}''(s')^2 - \bar{r}'(\bar{r}' \cdot \bar{r}'')] \cdot \left[\frac{1}{s'} \frac{d}{d\alpha} \left(\frac{\rho}{(s')^3} \right) (\bar{r}' \times \bar{r}'') + \frac{\rho}{(s')^4} (\bar{r}' \times \bar{r}''') \right]$$

We may simplify this equation by recognizing that a cross product is perpendicular to the individual terms in the product. Hence, carrying out the dot product in the above expression, term by term, leads to

$$\frac{1}{\tau} = -\frac{\rho^2}{(s')^6} [\bar{r}'' \cdot (\bar{r}' \times \bar{r}''')] \qquad (2.25)$$

We see from these developments that the parametric description of a curved path enables us to evaluate all the properties of that path. Whether or not path variables is actually suitable for the description of the motion depends on how the movement along the path is specified, as we shall see in the next section.

EXAMPLE 2.1

A particle moves along the hyperbolic paraboloidal surface $z = xy/2$ such that $x = 6 \sin k\xi$, $y = -6 \cos k\xi$, where x, y, and z are meters and ξ is a parameter. Determine path variable unit vectors and the radius of curvature at the position where $\xi = \pi/3k$.

Solution. It is possible to obtain the desired results by direct substitution into Eqs. (2.16) to (2.25). For the sake of increased understanding, we shall instead carry out the sequential operations indicated by the Frenet formulas. The parametric form of the position is given as

$$\bar{r} = 6 \sin k\xi \bar{i} - 6 \cos \xi \bar{j} - 9 \sin 2k\xi \bar{k}$$

The tangent vector is then

$$\bar{e}_t = \frac{d\bar{r}}{ds} = \frac{1}{s'} [6k(\cos k\xi \bar{i} + \sin k\xi \bar{j} - 3 \cos 2k\xi \bar{k})]$$

Setting $|\bar{e}_t| = 1$ yields

$$s' = 6k(1 + 9 \cos^2 2k\xi)^{1/2}$$

As a preliminary to obtaining $\bar{e}_n$, we differentiate s' to find

$$s'' = -\frac{108k^2 \sin 2k\xi \cos 2k\xi}{(1 + 9 \cos^2 2k\xi)^{1/2}}$$

The normal vector is then given by

$$\bar{e}_n = \rho \frac{d\bar{e}_t}{ds} = \frac{\rho}{s'} \frac{d\bar{e}_t}{d\xi} = \frac{\rho}{s'} \left[\frac{\bar{r}''}{s'} - \frac{s''\bar{r}'}{(s')^2} \right]$$

$$= \frac{\rho}{s'} \left[\frac{6k^2}{s'} (-\sin k\xi \bar{i} + \cos k\xi \bar{j} + 6 \sin 2k\xi \bar{k}) \right.$$

$$\left. - \frac{s''(6k)}{(s')^2} (\cos k\xi \bar{i} + \sin k\xi \bar{j} - 3 \cos 2k\xi \bar{k}) \right]$$

If the value of the torsion τ had been requested, it would be necessary to retain $\bar{e}_n$ in functional form, in order to differentiate it. For the present problem we may evaluate all quantities at $\xi = \pi/3k$, with the result that

$$s' = 10.8167k \qquad s'' = 25.9408k^2$$

$$\bar{e}_t = 0.27735\bar{i} + 0.48038\bar{j} + 0.83205\bar{k}$$

$$\bar{e}_n = \rho(-0.105905\bar{i} - 0.080869\bar{j} + 0.081990\bar{k})$$

The requirement that $\bar{e}_n$ be a unit vector yields

$$\rho = 6.392 \text{ m}$$

$$\bar{e}_n = -0.6769\bar{i} - 0.5169\bar{j} + 0.5241\bar{k}$$

Finally, we compute the binormal unit vector from a cross product, according to

$$\bar{e}_b = \bar{e}_t \times \bar{e}_n = 0.6819\bar{i} - 0.7086\bar{j} + 0.1818\bar{k}$$

2.1.2 Kinematical Relations

Situations in which the path variable formulation is useful may be recognized by the fact that some aspect of $s(t)$ is given, for example, its rate, $\dot{s}$, is known. We saw in Eq. (2.1) that $\bar{r}_{P/O}$ is a function of time through the corresponding dependence on s. Using the chain rule to evaluate the derivative then yields

$$\bar{v} = \frac{d\bar{r}}{ds} \dot{s} \tag{2.26}$$

In view of Eq. (2.2), this is equivalent to describing the velocity as

$$\bar{v} = v\bar{e}_t \qquad v = |\bar{v}| = \dot{s} \tag{2.27}$$

These expressions indicate that the speed v is the rate of change of the arclength to the point. They also show that the velocity is always tangent to the path.

A corresponding formula for acceleration may be obtained by differentiating Eq. (2.27) with respect to t. We henceforth shall consider the speed v to be an explicit function of time. Therefore

$$\bar{a} = \frac{d\bar{v}}{dt} = \dot{v}\bar{e}_t + v\frac{d\bar{e}_t}{ds}\dot{s} = \dot{v}\bar{e}_t + v^2\frac{d\bar{e}_t}{ds}$$

Equation (2.4) then leads to

$$\bar{a} = a_t\bar{e}_t + a_n\bar{e}_n$$

$$a_t = \dot{v} \qquad a_n = \frac{v^2}{\rho} \tag{2.28}$$

Several aspects of this relation are important. First, the acceleration will have components in the normal and tangent directions. (There is no binormal component because the curve seems locally to lie in the osculating plane.) Since v^2/ρ is never negative, the normal component of acceleration is always directed toward the center of curvature for that position. In the case of a circular path, ρ is the radius R. Thus, a_n is the *centripetal acceleration*. It must be emphasized that, even though the speed might be constant, there is always a centripetal acceleration, with two exceptions. When a point comes to rest, even momentarily, then $v = 0$. Alternatively, if the path is a straight line or if the point under consideration is an inflection point, then ρ is infinite. In general, the normal acceleration arises because the direction of the velocity changes as the point moves along its path. In contrast, the tangential acceleration arises whenever the speed, which is the magnitude of the velocity, changes.

There are a variety of ways in which the arclength s or speed v might be given. Whenever the speed is described in terms of s, rather than time t, the chain rule may be used to find the tangential acceleration. Specifically

$$\dot{v} = v\frac{dv}{ds}$$

EXAMPLE 2.2

A particle follows a path defined in parametric form by $x = \frac{1}{2}A\pi\xi^2$, $y = A\xi \sin \pi\xi$, $z = A\xi \cos \pi\xi$, where A is a constant. The particle gains speed at the constant rate $\dot{v}$. Determine the speed v that the particle should have at the position where $\xi = \frac{1}{2}$ in order for its x component of acceleration to be zero at that location.

Solution. Because the value of $\dot{v}$ is given and the path is known in parametric form, we form the acceleration using path variables. The required value of v will be found by setting $\bar{a} \cdot \bar{i} = 0$ at $\xi = \frac{1}{2}$.

We begin by using Eqs. (2.16) to (2.21) to obtain the path parameters. Thus

$$\bar{r} = \tfrac{1}{2}A\pi\xi^2\bar{i} + A\xi \sin \pi\xi\bar{j} + A\xi \cos \pi\xi\bar{k}$$
$$\bar{r}' = A\pi\xi\bar{i} + A(\pi\xi \cos \pi\xi + \sin \pi\xi)\bar{j} + A(-\pi\xi \sin \pi\xi + \cos \pi\xi)\bar{k}$$
$$\bar{r}'' = A\pi\bar{i} + A(-\pi^2\xi \sin \pi\xi + 2\pi \cos \pi\xi)\bar{j} + A(-\pi^2\xi \cos \pi\xi - 2\pi \sin \pi\xi)\bar{k}$$

The other parameters depend algebraically on $\bar{r}'$ and $\bar{r}''$, so we may evaluate these quantities for $\xi = \frac{1}{2}$, which gives

$$\bar{r}' = A\left(\frac{\pi}{2}\bar{i} + \bar{j} - \frac{\pi}{2}\bar{k}\right)$$

$$\bar{r}'' = A\pi\left(\bar{i} - \frac{\pi}{2}\bar{j} - 2\bar{k}\right)$$

Sequential application of Eqs. (2.17a), (2.16), and (2.20) then yields

$$s' = A\left(\tfrac{1}{2}\pi^2 + 1\right)^{1/2} = 2.4361A$$

$$\bar{e}_t = \frac{\pi\bar{i} + 2\bar{j} - \pi\bar{k}}{2(2.4361)} = 0.64479\bar{i} + 0.41048\bar{j} - 0.64479\bar{k}$$

$$\bar{e}_n = \frac{\rho}{2.4361^4A}\left[(2.4361)^2\pi\left(\bar{i} - \frac{\pi}{2}\bar{j} - \bar{k}\right) - \frac{\pi^2}{2}\left(\frac{\pi}{2}\bar{i} + \bar{j} - \frac{\ddot{}}{2}\bar{k}\right)\right]$$

$$= \frac{\rho}{A}(0.30927\bar{i} - 0.97161\bar{j} - 0.30927\bar{k})$$

The value of ρ obtained from $|\bar{e}_n| = 1$ is

$$\rho = \frac{A}{[2(0.30927^2) + 0.97161^2]^{1/2}} = 0.93851A$$

which corresponds to

$$\bar{e}_n = 0.29025\bar{i} - 0.91187\bar{j} - 0.29025\bar{k}$$

We may now form the acceleration.

$$\bar{a} = \dot{v}\bar{e}_t + \frac{v^2}{\rho}\bar{e}_n$$

$$= \dot{v}(0.64479\bar{i} + 0.41048\bar{j} - 0.54479\bar{k})$$

$$+ \frac{v^2}{0.93851A}(0.29025\bar{i} - 0.91187\bar{j} - 0.29025\bar{k})$$

Finally, setting $\bar{a} \cdot \bar{i} = 0$ yields

$$0.64479\dot{v} + \frac{0.29025}{0.93851A} v^2 = 0$$

$$v = 1.4439(-A\dot{v})^{1/2}$$

The presence of the negative sign means that $\bar{a} \cdot \bar{i} = 0$ is possible only if the speed is decreasing. This situation is a consequence of the fact that the x components of $\bar{e}_n$ and $\bar{e}_t$ are both positive.

EXAMPLE 2.3

At the instant when a particle is at position A, it has a velocity of 500 m/s directed from point A to point B, and an acceleration of 10 g directed from point A to point O. Determine the corresponding rate of change of the speed, the radius of curvature of the path, and the location of the center of curvature of the path.

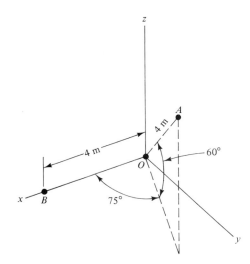

Example 2.3

Solution. The velocity and acceleration are known, so representing them in terms of tangent and normal components should yield relations for the desired parameters. The given vectors are

$$\bar{v} = 500\bar{e}_{B/A} \text{ m/s} \qquad \bar{a} = 10(9.807)\bar{e}_{O/A} \text{ m/s}^2$$

The unit vectors are defined by the positions of the end points, according to

$$\bar{r}_{A/O} = 4\cos 60°(\cos 75°\bar{i} + \sin 75°\bar{j}) + 4\sin 60°\bar{k}$$
$$= 0.5176\bar{i} + 1.9319\bar{j} + 3.464\bar{k} \text{ m}$$
$$\bar{r}_{B/O} = 4\bar{i}$$
$$\bar{e}_{B/A} = \frac{\bar{r}_{B/A}}{|\bar{r}_{B/A}|} = \frac{\bar{r}_{B/O} - \bar{r}_{A/O}}{|\bar{r}_{B/O} - \bar{r}_{A/O}|} = 0.6598\bar{i} - 0.3660\bar{j} - 0.6563\bar{k}$$
$$\bar{e}_{O/A} = -\frac{\bar{r}_{A/O}}{|\bar{r}_{A/O}|} = -0.1294\bar{i} - 0.4830\bar{j} - 0.8660\bar{k}$$

In general, $\bar{v} = v\bar{e}_t$, from which it follows that $\bar{e}_t = \bar{e}_{B/A}$. Then, because $\dot{v}$ is the tangential component of acceleration, we find

$$\dot{v} = a_t = \bar{a} \cdot \bar{e}_t = 98.07\bar{e}_{O/A} \cdot \bar{e}_{B/A} = 64.70 \text{ m/s}^2$$

We may evaluate the normal acceleration by forming the difference between $\bar{a}$ and $\dot{v}\bar{e}_t$, specifically,

$$\frac{v^2}{\rho}\bar{e}_n = \bar{a} - \dot{v}\bar{e}_t = 98.07\bar{e}_{O/A} - 64.70\bar{e}_{B/A}$$

$$= -55.38\bar{i} - 23.69\bar{j} - 42.47\bar{k} \text{ m/s}^2$$

The values of ρ and $\bar{e}_n$ come from the magnitude and direction of this acceleration.

$$\frac{v^2}{\rho} = (55.38^2 + 23.69^2 + 42.47^2)^{1/2} = 73.70 \text{ m/s}^2$$

$$\rho = \frac{v^2}{73.70} = \frac{500^2}{73.70} = 3392 \text{ m}$$

$$\bar{e}_n = \frac{-55.38\bar{i} - 23.69\bar{j} - 42.47\bar{k}}{73.70}$$

$$= -0.7514\bar{i} - 0.3214\bar{j} - 0.5763\bar{k}$$

Finally we locate the center of curvature C by recalling that it is at ρ units in the $\bar{e}_n$ direction relative to the corresponding point on the path. Hence,

$$\bar{r}_{C/O} = \bar{r}_{A/O} + \rho\bar{e}_n = -54.86\bar{i} - 21.76\bar{j} - 39.00\bar{k} \text{ m}$$

2.2 RECTANGULAR CARTESIAN COORDINATES

The path variable formulation that was developed in the previous section describes *intrinsic coordinates* because it relies on knowledge of the path for the definition of the unit vectors and of the position. For the remainder of this chapter we shall consider *extrinsic coordinate systems*, in which these properties are defined in a manner that is independent of the path.

The simplest set of extrinsic coordinates is *rectangular Cartesian coordinates*, which are associated with orthogonal *xyz* axes. Situations where such coordinates might be suitable are recognizable by the fact that vectors (position, velocity, etc.) are described in terms of components with respect to fixed directions, such as left-right and up-down. As shown in Figure 2.4, the components of the position vector are merely the (x, y, z) coordinates projected onto the coordinate axes. These coordinates may all be functions of time. Accordingly, the position is given by

$$\bar{r}_{P/O} = x(t)\bar{i} + y(t)\bar{j} + z(t)\bar{k} \tag{2.30}$$

Differentiating this expression is a simple matter because the unit vectors are constant. Thus the velocity is given by

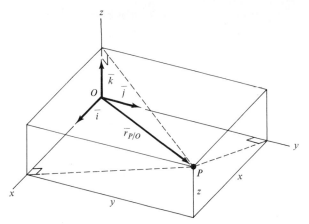

Figure 2.4 Rectangular Cartesian coordinates.

$$\bar{v} = v_x\bar{i} + v_y\bar{j} + v_z\bar{k}$$
$$v_x = \dot{x} \qquad v_y = \dot{y} \qquad v_z = \dot{z} \tag{2.31}$$

from which it follows that the acceleration is

$$\bar{a} = a_x\bar{i} + a_y\bar{j} + a_z\bar{k}$$
$$a_x = \dot{v}_x \qquad a_y = \dot{v}_y \qquad a_z = \dot{v}_z \tag{2.32}$$

A notable feature of these relations is that the motions in the x, y, and z directions are uncoupled. In other words, none of the motion parameters for one direction appear in the other components. One way of regarding this result conceptually is to think of it as a superposition of rectilinear (i.e., straight-line) motions in each of the coordinate directions.

As you might suspect, the simplicity of this formulation limits its usefulness. Practical situations in which the motion is given in terms of fixed directions are not abundant. The most common involves projectile motion near the earth's surface. In that case the force of gravity is considered to be in the downward vertical direction, which means that the acceleration is always downward. Even this case breaks down when one wishes to treat the motion more accurately. For example, if it is desired to account for air resistance, the resistance force is always opposite the velocity. Such a force is readily described in path variables as $-f\bar{e}_t$. Another situation where the description of projectile motion encounters difficulty is when the motion covers a long range, as is the case with missiles. Then the gravitational force is always directed toward a fixed point, rather than having a fixed direction. A kinematical description using curvilinear coordinates is more suitable to this type of consideration.

EXAMPLE 2.4 _____

A 200-gram ball is thrown from the ground with the initial velocity $\bar{v}_0$ shown. In addition to its weight, there is a constant wind force of 8 newtons acting in the easterly direction.

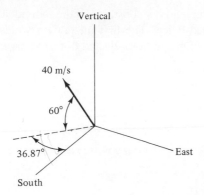

Example 2.4

Find the coordinates of the ball at the instant it returns to the elevation from which it was thrown, and the velocity of the ball at that instant.

Solution. A free body diagram shows the weight and the wind force, both of which act in a fixed direction, labeled z and y, respectively. Forming $\Sigma \, \bar{F} = m \, \bar{a}$ in terms of components relative to these directions yields

$$8\bar{j} - 0.2(9.807)\bar{k} = m\bar{a} = 0.2(\ddot{x}\bar{i} + \ddot{y}\bar{j} + \ddot{z}\bar{k})$$

The initial conditions for the motion are

$$\bar{r}_0 = \bar{0}$$
$$\bar{v}_0 = 40 \cos 60°(\cos 36.87°\bar{i} - \sin 36.87°\bar{j}) + 40 \sin 60°\bar{k}$$
$$= 16\bar{i} - 12\bar{j} + 34.67\bar{k} \text{ m/s}$$

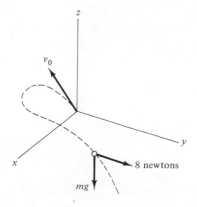

Free body diagram.

We decompose these relations to their individual components.

$$\ddot{x} = 0 \qquad \dot{x}_0 = 16 \text{ m/s} \qquad x_0 = 0 \text{ at } t = 0$$
$$\ddot{y} = 40 \text{ m/s}^2 \qquad \dot{y}_0 = -12 \text{ m/s} \qquad y_0 = 0 \text{ at } t = 0$$
$$\ddot{z} = -9.807 \text{ m/s}^2 \qquad \dot{z}_0 = 34.67 \text{ m/s} \qquad z_0 = 0 \text{ at } t = 0$$

The acceleration in each direction is constant, so the individual acceleration equations may each be integrated twice. The first integration yields the velocity components, with the constants of integration selected to match the initial velocity conditions. Thus

$$\dot{x} = 16 \text{ m/s}$$
$$\dot{y} = 40t - 12 \text{ m/s}$$
$$\dot{z} = -9.807t + 34.67 \text{ m/s}$$

The constants of integration for the second integration are used to satisfy the initial positions, with the result that

$$x = 16t \text{ m}$$
$$y = 20t^2 - 12t \text{ m}$$
$$z = -4.904t^2 + 34.67t \text{ m}$$

Finally, we evaluate the instant when the ball returns to the xy plane by setting $z = 0$ for $t > 0$, which occurs when

$$t = 7.064 \text{ s}$$

Evaluating the position and velocity components for that instant yields

$$\bar{r} = 113.03\bar{i} + 913.3\bar{j} \text{ m}$$
$$\bar{v} = 16\bar{i} + 270.6\bar{j} - 34.64\bar{k} \text{ m/s}$$

EXAMPLE 2.5

A right circular cone is defined by $x^2 + y^2 = 9z^2$ (x, y, and z are inches). The vertical position of a block sliding along the interior of such a cone is observed to be $z = -26 + 8 t^2$ and $x = y^2/3$. Also, $y > 0$ throughout the motion. Determine the velocity and acceleration of the block when $t = 2$ s.

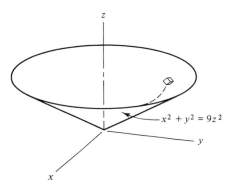

Example 2.5

Solution. The given functions relate z to x and y, z to t, and x to y, respectively. This may be simplified by using the first and third to relate z solely to y. The position equations then become

$$z = -26 + 8t^2 \qquad \frac{1}{9}y^4 + y^2 = 9z^2 \qquad x = \frac{1}{3}y^2 \text{ in.}$$

Differentiation of these expressions yields relations governing $\dot{x}$, $\dot{y}$, and $\dot{z}$.

$$\dot{z} = 16t \qquad \left(\frac{4}{9}y^3 + 2y\right)\dot{y} = 18\,z\dot{z} \qquad \dot{x} = \frac{2}{3}y\dot{y}$$

A second differentiation leads to

$$\ddot{z} = 16$$

$$\left(\frac{4}{9}y^3 + 2y\right)\ddot{y} + \left(\frac{4}{3}y^2 + 2\right)\dot{y}^2 = 18\,(z\ddot{z} + \dot{z}^2)$$

$$\ddot{x} = \frac{2}{3}(y\ddot{y} + \dot{y}^2)$$

We find values for $t = 2$ s by sequential evaluation of the equations. The position coordinates are

$$z = 6 \qquad y = 7.049 \qquad x = 16.562 \text{ in.}$$

These coordinates allow us to evaluate the first derivatives. Thus,

$$\dot{z} = 32 \qquad \dot{y} = 20.36 \qquad \dot{x} = 287.0 \text{ in./s}$$

Next, we use these values to solve the equations for the second derivatives, specifically,

$$\ddot{z} = 16 \qquad \ddot{y} = -47.87 \qquad \ddot{x} = 276.3 \text{ in./s}^2$$

The respective derivatives are the components of the velocity and acceleration, so

$$\bar{v} = 287.0\bar{i} + 20.36\bar{j} + 32\bar{k} \text{ in./s}$$
$$\bar{a} = 276.3\bar{i} - 47.87\bar{j} + 16\bar{k} \text{ in./s}^2$$

The path may be visualized by noting that it lies on a cone; the projection of the path on the xy plane is a parabola having x as the axis of symmetry.

2.3 ORTHOGONAL CURVILINEAR COORDINATES

The description we treat here specifies the position of a point by giving the value of a triad of parameters that form an orthogonal mesh in space. Let α, β, and γ be the three parameters, such that there is a unique transformation between the (x,y,z) rectangular Cartesian coordinates and the (α,β,γ) values.

$$x = x(\alpha,\beta,\gamma) \qquad y = y(\alpha,\beta,\gamma) \qquad z = z(\alpha,\beta,\gamma)$$
$$\alpha = \alpha(x,y,z) \qquad \beta = \beta(x,y,z) \qquad \gamma = \gamma(x,y,z) \qquad (2.33)$$

Typical are formulations in terms of cylindrical or spherical coordinates.

When two of the parameters (α, β, γ) are held constant, and the third is given a range of values, the first group of Eqs. (2.33) specify a curve in space in parametric form. When the constant parameter pair is given an assortment of values, the result is a family of curves. Repeating this procedure with each of the other pairs of parameters held constant produces two more families of curves. This is the aforementioned mesh. The families of curves are mutually orthogonal in the cases that we shall treat. For this reason, (α, β, γ) are said to be *orthogonal curvilinear coordinates*. The name for each set of coordinates usually corresponds to one of the types of surfaces on which one of the curvilinear coordinates is constant. This is illustrated in Figure 2.5 for cylindrical and spherical coordinates.

It is difficult to depict a three-dimensional situation, so Figure 2.6 shows a two-dimensional grid. Neighboring curves for each family are separated by values of the other coordinate that differ by an infititesimal value. The distance between intersection points on the grid is not the same as the value of the increment in that coordinate. The ratio between the differential arclength along a coordinate curve between intersections and the increment in the coordinate corresponding to the intersections is the *stretch ratio* for that coordinate, denoted h_λ, with $\lambda = \alpha$, β, or γ. The corresponding arclength along a coordinate curve is s_λ, so

$$ds_\lambda = h_\lambda \, d\lambda \qquad \lambda = \alpha, \beta, \text{ or } \gamma \tag{2.34}$$

The relationship between the curvilinear coordinate transformation, Eqs. (2.33), and the stretch ratios will be established shortly.

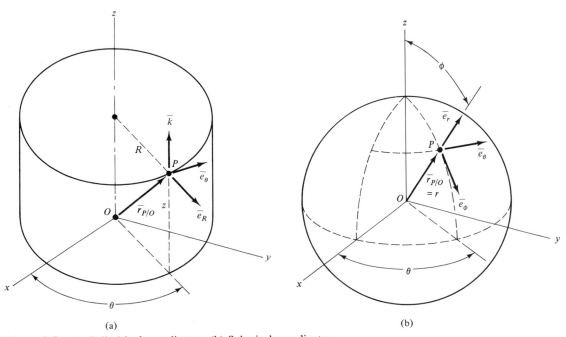

(a)   (b)

Figure 2.5   (a) Cylindrical coordinates. (b) Spherical coordinates.

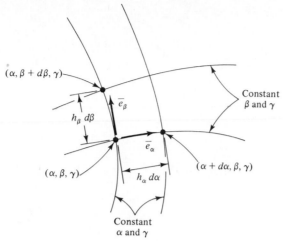

Figure 2.6 Curvilinear coordinate mesh.

2.3.1 Coordinates and Unit Vectors

Moving along any of the coordinate arcs for a curvilinear coordinate mesh is very much like the situation in path variables. Hence, there are three unit vectors $\bar{e}_\lambda$, consisting of the tangent to each of the coordinate curves intersecting at any point. Incrementing one coordinate with the other two fixed is a process of partial differentiation, so the unit vectors may be obtained from

$$\bar{e}_\lambda = \frac{\partial \bar{r}}{\partial s_\lambda}$$

$$\bar{e}_\lambda = \frac{1}{h_\lambda} \frac{\partial \bar{r}}{\partial \lambda} \qquad \lambda = \alpha, \beta, \text{ or } \gamma \tag{2.35}$$

Note that the unit vectors may depend on the value of each curvilinear coordinate. However, in many cases the unit vectors might be independent of one or more of the coordinates. Two such cases are cylindrical and spherical coordinates. The unit vectors obtained for each were depicted in Figure 2.5. In cylindrical coordinates $\bar{e}_R$ and $\bar{e}_\theta$ depend only on the azimuthal (transverse) angle θ, and $\bar{e}_z = \bar{k}$ is constant. In spherical coordinates all the unit vectors depend only on the polar angle ϕ and azimuthal angle θ.

It is conventional to employ a right-handed coordinate system in order to avoid sign errors in the evaluation of cross products. Consistency with this convention is obtained in curvilinear coordinates by ordering (α, β, γ) such that

$$\bar{e}_\alpha \times \bar{e}_\beta = \bar{e}_\gamma \tag{2.36}$$

Explicit expressions for the stretch ratios may be obtained in geometrically simple cases such as cylindrical and spherical coordinates by drawing diagrams resembling Figure 2.6 for each coordinate line. More difficult cases are treated by using the fact that $\bar{e}_\lambda$ is a unit vector, so that

$$h_\lambda = \left| \frac{\partial \bar{r}}{\partial \lambda} \right| \tag{2.37}$$

The derivation of the acceleration equation will require differentiation of the unit vectors. Rather than differentiating Eq. (2.35) directly, we shall follow a more circuitous approach that will yield explicit expressions in terms of the stretch ratios. The approach here is similar to the way in which some of the Frenet formulas were derived for path variables. First, the derivative of any unit vector $\bar{e}_\lambda$ with respect to any coordinate μ is resolved into components as

$$\frac{\partial \bar{e}_\lambda}{\partial \mu} = \left(\bar{e}_\alpha \cdot \frac{\partial \bar{e}_\lambda}{\partial \mu} \right) \bar{e}_\alpha + \left(\bar{e}_\beta \cdot \frac{\partial \bar{e}_\lambda}{\partial \mu} \right) \bar{e}_\beta + \left(\bar{e}_\gamma \cdot \frac{\partial \bar{e}_\lambda}{\partial \mu} \right) \bar{e}_\gamma \tag{2.38}$$

Both λ and μ in the foregoing correspond to α, β, or γ, so we must consider permutations of the general term $\bar{e}_\nu \cdot (\partial \bar{e}_\lambda / \partial \mu)$. All cases where the unit vectors $\bar{e}_\lambda$ and $\bar{e}_\nu$ match are covered by

$$\bar{e}_\lambda \cdot \bar{e}_\lambda = 1 \Rightarrow \bar{e}_\lambda \cdot \frac{\partial \bar{e}_\lambda}{\partial \mu} = 0 \tag{2.39}$$

Other cases will be evaluated with the aid of a sequence of identities. When ν is different from λ, the orthogonality of the unit vectors yields

$$\bar{e}_\nu \cdot \bar{e}_\lambda = 0 \Rightarrow \bar{e}_\nu \cdot \frac{\partial \bar{e}_\lambda}{\partial \mu} = - \bar{e}_\lambda \cdot \frac{\partial \bar{e}_\nu}{\partial \mu} \qquad \nu \neq \lambda \tag{2.40}$$

The following relation originates from Eq. (2.35).

$$\frac{\partial}{\partial \mu} (h_\lambda \bar{e}_\lambda) = \frac{\partial}{\partial \lambda} (h_\mu \bar{e}_\mu)$$

which leads to

$$h_\lambda \frac{\partial \bar{e}_\lambda}{\partial \mu} + \frac{\partial h_\lambda}{\partial \mu} \bar{e}_\lambda = h_\mu \frac{\partial \bar{e}_\mu}{\partial \lambda} + \frac{\partial h_\mu}{\partial \lambda} \bar{e}_\mu \tag{2.41}$$

We may now consider the various combinations of the general term $\bar{e}_\nu \cdot (\partial \bar{e}_\lambda / \partial \mu)$ in which $\bar{e}_\lambda$ and $\bar{e}_\nu$ are different. In the first case the coordinate for the derivative matches the component, that is, $\nu = \mu$. Such terms are obtained from the dot product of Eq. (2.41) with $\bar{e}_\mu$. The properties in Eqs. (2.39) and (2.40) lead to

$$\bar{e}_\mu \cdot \frac{\partial \bar{e}_\lambda}{\partial \mu} = - \bar{e}_\lambda \cdot \frac{\partial \bar{e}_\mu}{\partial \mu} = \frac{1}{h_\lambda} \frac{\partial h_\mu}{\partial \lambda} \qquad \mu \neq \lambda \tag{2.42}$$

The last case requiring treatment is that in which λ, μ, and ν are different from each other. A dot product of Eq. (2.41) with $\bar{e}_\nu$ yields

$$h_\lambda \bar{e}_\nu \cdot \frac{\partial \bar{e}_\lambda}{\partial \mu} = h_\mu \bar{e}_\nu \cdot \frac{\partial \bar{e}_\mu}{\partial \lambda} \qquad \lambda, \nu, \mu \text{ distinct} \tag{2.43}$$

The next steps involve alternate application of permutations of the properties in Eqs. (2.40) and (2.43) to the *right side* of Eq. (2.43). This gives

$$h_\lambda \bar{e}_\nu \cdot \frac{\partial \bar{e}_\lambda}{\partial \mu} = -h_\mu \bar{e}_\mu \cdot \frac{\partial \bar{e}_\mu}{\partial \lambda} = -\frac{h_\mu h_\lambda}{h_\nu} \bar{e}_\mu \cdot \frac{\partial \bar{e}_\lambda}{\partial \nu}$$

$$= \frac{h_\mu h_\lambda}{h_\nu} \bar{e}_\lambda \cdot \frac{\partial \bar{e}_\mu}{\partial \nu} = h_\lambda \bar{e}_\lambda \cdot \frac{\partial \bar{e}_\nu}{\partial \mu} = -h_\lambda \bar{e}_\nu \cdot \frac{\partial \bar{e}_\lambda}{\partial \mu}$$

The foregoing is a contradiction unless

$$\bar{e}_\nu \cdot \frac{\partial \bar{e}_\lambda}{\partial \mu} = 0 \qquad \lambda, \nu, \mu \text{ distinct} \tag{2.44}$$

Equations (2.39), (2.42), and (2.44) are the identities we seek.† With these relations, we may express the components of $\partial \bar{e}_\lambda / \partial \mu$ in Eq. (2.38) in terms of the stretch ratios. There are nine combinations of λ and μ values, whose individual components are evaluated by selecting the appropriate case from the identities. We will list only the results for $\lambda = \alpha$. The others follow by permutation of the symbols.

$$\frac{\partial \bar{e}_\alpha}{\partial \alpha} = -\frac{1}{h_\beta} \frac{\partial h_\alpha}{\partial \beta} \bar{e}_\beta - \frac{1}{h_\gamma} \frac{\partial h_\alpha}{\partial \gamma} \bar{e}_\gamma$$

$$\frac{\partial \bar{e}_\alpha}{\partial \beta} = \frac{1}{h_\alpha} \frac{\partial h_\beta}{\partial \alpha} \bar{e}_\beta \qquad \frac{\partial \bar{e}_\alpha}{\partial \gamma} = \frac{1}{h_\alpha} \frac{\partial h_\gamma}{\partial \alpha} \bar{e}_\gamma \tag{2.45}$$

EXAMPLE 2.6 _____

The two-dimensional hyperbolic-elliptic coordinate system is defined by

$$x = a \cosh \alpha \sin \beta \qquad y = a \sinh \alpha \cos \beta$$

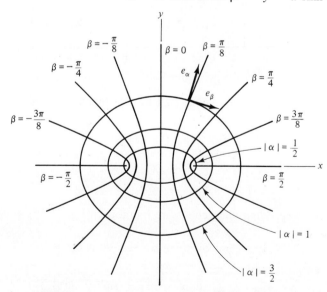

Example 2.5

---

† A common operation in tensor analysis is covariant differentiation of a quantity that is defined in terms of basis vectors for an arbitrary curvilinear coordinate system. The derivative may be expressed in terms of *Christoffel symbols*. Equations (2.39), (2.42), and (2.44) define some of the Christoffel symbols for the case of an orthogonal coordinate system.

where a is a constant. Evaluate the unit vectors of this system in terms of components relative to the x and y axes; then describe the derivatives of the unit vectors.

Solution.   The name of this set of coordinates stems, in part, from the fact that lines of constant α are ellipses,

$$\left(\frac{x}{a \cosh \alpha}\right)^2 + \left(\frac{y}{\alpha \sinh \alpha}\right)^2 = 1$$

where $2a \cosh \alpha$ and $2a \sinh \alpha$ are the lengths of the major and minor diameters. Also, lines of constant β are hyperbolas,

$$\left(\frac{x}{a \sin \beta}\right)^2 = \left(\frac{y}{a \cos \beta}\right)^2 + 1$$

where $x = \pm y \tan \beta$ are the asymptotes and $\pm a \sin \beta$ are the intercepts on the x axis.

The stretch ratios and unit vectors come from the partial derivatives of the position vector.

$$\frac{\partial \bar{r}}{\partial \alpha} = \frac{\partial}{\partial \alpha}(x\bar{i} + y\bar{j}) = a \sinh \alpha \sin \beta \bar{i} + a \cosh \alpha \cos \beta \bar{j}$$

$$\frac{\partial \bar{r}}{\partial \beta} = \frac{\partial}{\partial \beta}(x\bar{i} + y\bar{j}) = a \cosh \alpha \cos \beta \bar{i} - a \sinh \alpha \sin \beta \bar{j}$$

We find from Eq. (2.37) that

$$h_\alpha = \left|\frac{\partial \bar{r}}{\partial \alpha}\right| = a (\sinh^2 \alpha \sin^2 \beta + \cosh^2 \alpha \cos^2 \beta)^{1/2}$$

$$= a [(\cosh^2 \alpha - 1) \sin^2 \beta + \cosh^2 \alpha \cos^2 \beta]^{1/2}$$

$$= a (\cosh^2 \alpha - \sin^2 \beta)^{1/2}$$

Steps comparable to the foregoing lead to the stretch ratio $h_\beta = |\partial \bar{r}/\partial \beta|$. It is convenient to define

$$h = (\cosh^2 \alpha - \sin^2 \beta)^{1/2}$$

The result is that

$$h_\beta = h_\alpha = ah$$

The corresponding unit vectors are found from Eq. (2.35) to be

$$\bar{e}_\alpha = \frac{1}{h}(\sinh \alpha \sin \beta \bar{i} + \cosh \alpha \cos \beta \bar{j})$$

$$\bar{e}_\beta = \frac{1}{h}(\cosh \alpha \cos \beta \bar{i} - \sinh \alpha \sin \beta \bar{j})$$

The orthogonality of the mesh is confirmed by these unit vectors, because they show that $\bar{e}_\alpha \cdot \bar{e}_\beta = 0$.

The derivatives of the unit vectors involve partial derivatives of the stretch ratios, which in the present case are obtained from

$$\frac{\partial h}{\partial \alpha} = \frac{1}{h} \cosh \alpha \sinh \alpha$$

$$\frac{\partial h}{\partial \beta} = -\frac{1}{h} \sin \beta \cos \beta$$

The corresponding expressions resulting from Eq. (2.45) are

$$\frac{\partial \bar{e}_\alpha}{\partial \alpha} = -\frac{1}{h}\frac{\partial h}{\partial \beta}\bar{e}_\beta = \frac{\sin \beta \cos \beta}{h^2}\bar{e}_\beta$$

$$\frac{\partial \bar{e}_\alpha}{\partial \beta} = \frac{1}{h}\frac{\partial h}{\partial \alpha}\bar{e}_\beta = \frac{\cosh \alpha \sinh \alpha}{h^2}\bar{e}_\beta$$

$$\frac{\partial \bar{e}_\beta}{\partial \alpha} = \frac{1}{h}\frac{\partial h}{\partial \beta}\bar{e}_\alpha = -\frac{\sin \beta \cos \beta}{h^2}\bar{e}_\alpha$$

$$\frac{\partial \bar{e}_\beta}{\partial \beta} = -\frac{1}{h}\frac{\partial h}{\partial \alpha}\bar{e}_\alpha = -\frac{\cosh \alpha \sinh \alpha}{h^2}\bar{e}_\alpha$$

2.3.2 Kinematical Formulas

Our task in this section is to express the velocity and acceleration in terms of the parameters of a curvilinear coordinate system. For this development we consider the motion to be specified through the dependence of the curvilinear coordinates on time. The velocity is the time derivative of the position vector, which is a function of the curvilinear coordinates. The definition of the unit vectors in Eq. (2.35) then results in

$$\bar{v} = \dot{\alpha}\frac{\partial \bar{r}}{\partial \alpha} + \dot{\beta}\frac{\partial \bar{r}}{\partial \beta} + \dot{\gamma}\frac{\partial \bar{r}}{\partial \gamma} = h_\alpha \dot{\alpha}\bar{e}_\alpha + h_\beta \dot{\beta}\bar{e}_\beta + h_\gamma \dot{\gamma}\bar{e}_\gamma$$

This expression may be written in summation form as

$$\bar{v} = \sum_{\lambda = \alpha,\beta,\gamma} h_\lambda \dot{\lambda}\bar{e}_\lambda \tag{2.46}$$

The acceleration is derived by differentiating Eq. (2.46) with respect to time. For this, only the curvilinear coordinates λ are considered to be explicit functions of time. The unit vectors $\bar{e}_\lambda$ and the stretch ratios h_λ are functions of the coordinates, so their time derivative is obtained through application of the chain rule.

$$\bar{a} = \sum_{\lambda = \alpha,\beta,\gamma}\left[h_\lambda \ddot{\lambda}\bar{e}_\lambda + \sum_{\mu = \alpha,\beta,\gamma}\left(\frac{\partial h_\lambda}{\partial \mu}\bar{e}_\lambda + h_\lambda \frac{\partial \bar{e}_\lambda}{\partial \mu}\right)\dot{\lambda}\dot{\mu}\right] \tag{2.47}$$

Explicit expressions for a specific set of curvilinear coordinates may be obtained from the above by evaluating the stretch ratios and the derivatives of the unit vectors according to Eqs. (2.37) and (2.45), respectively.

It is apparent that each acceleration component might consist of several terms in the most general case. The situation for many common sets of curvilinear coordinates is simplified by the fact that the stretch ratios do not depend on all the coordinate values. For the coordinates defined in Figure 2.5, we have

Cylindrical coordinates (R, θ, z)

$$
\begin{aligned}
x &= R \cos \theta \qquad y = R \sin \theta \qquad z = z \\
h_R &= 1 \qquad h_\theta = R \qquad h_z = 1 \\
\bar{r}_{P/O} &= R \, \bar{e}_R + z\bar{k} \qquad \bar{e}_z = \bar{k} \\
\bar{v} &= \dot{R}\bar{e}_R + R\dot{\theta}\bar{e}_\theta + \dot{z}\bar{k} \\
\bar{a} &= (\ddot{R} - R\dot{\theta}^2)\bar{e}_R + (R\ddot{\theta} + 2\dot{R}\dot{\theta})\bar{e}_\theta + \ddot{z}\bar{k}
\end{aligned}
\tag{2.48}
$$

Spherical Coordinates (r, ϕ, θ)

$$
\begin{aligned}
x &= r \sin \phi \cos \theta \qquad y = r \sin \phi \sin \theta \qquad z = r \cos \phi \\
h_r &= 1 \qquad h_\phi = r \qquad h_\theta = r \sin \phi \\
\bar{r}_{P/O} &= r\bar{e}_r \\
\bar{v} &= \dot{r}\bar{e}_r + r\dot{\phi}\,\bar{e}_\phi + r\dot{\theta} \sin \phi \,\bar{e}_\theta \\
\bar{a} &= (\ddot{r} - r\dot{\phi}^2 - r\dot{\theta}^2 \sin^2\phi)\bar{e}_r + (r\ddot{\phi} + 2\dot{r}\dot{\phi} - r\dot{\theta}^2 \sin \phi \cos \phi)\bar{e}_\phi \\
&\quad + (r\ddot{\theta} \sin \phi + 2\dot{r}\dot{\theta} \sin \phi + 2r\dot{\phi}\,\dot{\theta} \cos \phi)\bar{e}_\theta
\end{aligned}
\tag{2.49}
$$

2.3.3 Interpretation

The cylindrical and spherical coordinate expressions for acceleration have terms that contain products of the rate of change of two different coordinates, for example, $\dot{R}\,\dot{\theta}$. Such terms are preceded by a factor of 2. In order to understand this situation, consider Eq. (2.47). Mixed product terms are associated with two effects. One term arises when $\partial h_\lambda / \partial \mu$ is nonzero, so that constant rates of change in the coordinates do not produce constant rates of movement along the coordinate curves.

The other contribution corresponds to a nonzero value of $\partial \bar{e}_\lambda / \partial \mu$. Recall that $\bar{e}_\lambda$ is parallel to the λ coordinate curve, along which the coordinates other than λ are constant. A point moves from one such coordinate curve to a neighbor when the value of μ changes. As shown in Figure 2.7, this leads to a change in the orientation of the $\bar{e}_\lambda$ component of velocity if the λ curves are not parallel.

The mixed product terms are usually called Coriolis accelerations after G. Coriolis, who successfully explained the phenomenon. It is more instructive to regard them as effects that arise from an interaction of motion along more than one coordinate curve. Only in the special case of Cartesian coordinates can the motion be regarded as a superposition of motions along the three coordinate curves.

Explanations for the other acceleration terms are simpler. Any term containing a second derivative of a coordinate represents the tangential acceleration along the corre-

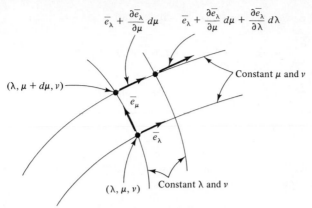

$$\bar{e}_\lambda + \frac{\partial \bar{e}_\lambda}{\partial \mu}\, d\mu \qquad \bar{e}_\lambda + \frac{\partial \bar{e}_\lambda}{\partial \mu}\, d\mu + \frac{\partial \bar{e}_\lambda}{\partial \lambda}\, d\lambda$$

Constant μ and ν

$(\lambda, \mu + d\mu, \nu)$

$\bar{e}_\mu$

$\bar{e}_\lambda$

(λ, μ, ν) Constant λ and ν

Figure 2.7   Derivatives of a curvilinear coordinate system unit vector.

sponding coordinate curve. Similarly, a term containing a square of the first derivative of a coordinate arises as a centripetal acceleration effect associated with motion along the corresponding coordinate curve.

EXAMPLE 2.7

Derive the expressions for velocity and acceleration in terms of cylindrical coordinates.

Solution.   The first step is to evaluate the unit vectors and stretch ratios. For the coordinate system in Figure 2.5a, we have

$$\bar{r} = R \cos \theta \bar{i} + R \sin \theta \bar{j} + z\bar{k}$$

Then

$$h_R \bar{e}_R = \frac{\partial \bar{r}}{\partial R} = \cos \theta \bar{i} + \sin \theta \bar{j}$$

$$h_\theta \bar{e}_\theta = \frac{\partial \bar{r}}{\partial \theta} = -R \sin \theta \bar{i} + R \cos \theta \bar{j}$$

$$h_z \bar{e}_z = \frac{\partial \bar{r}}{\partial z} = \bar{k}$$

Setting the magnitude of each unit vector to unity yields the stretch ratios. Thus

$$h_R = 1 \qquad h_\theta = R \qquad h_z = 1$$

which corresponds to

$$\bar{e}_R = \cos \theta \bar{i} + \sin \theta \bar{j}$$
$$\bar{e}_\theta = -\sin \theta \bar{i} + \cos \theta \bar{j}$$

The derivatives of the unit vectors are required to form Eq. (2.47) for $\bar{a}$. Although Eqs. (2.45) may be applied for this purpose, we may employ direct differentiation with equal ease in the present case. Specifically,

$$\frac{\partial \bar{e}_R}{\partial R} = \bar{0} \qquad \frac{\partial \bar{e}_R}{\partial \theta} = -\sin\theta\,\bar{i} + \cos\theta\,\bar{j} = \bar{e}_\theta \qquad \frac{\partial \bar{e}_R}{\partial z} = \bar{0}$$

$$\frac{\partial \bar{e}_\theta}{\partial R} = \bar{0} \qquad \frac{\partial \bar{e}_\theta}{\partial \theta} = -\cos\theta\,\bar{i} - \sin\theta\,\bar{j} = -\bar{e}_R \qquad \frac{\partial \bar{e}_\theta}{\partial z} = \bar{0}$$

$$\frac{\partial \bar{e}_z}{\partial R} = \frac{\partial \bar{e}_z}{\partial \theta} = \frac{\partial \bar{e}_z}{\partial z} = \bar{0}$$

We obtain an expression for the velocity by expanding Eq. (2.46) and substituting the various terms. Thus,

$$\bar{v} = h_R \dot{R} \bar{e}_R + h_\theta \dot{\theta} \bar{e}_\theta + h_z \dot{z} \bar{e}_z = \dot{R} \bar{e}_R + R\dot{\theta} \bar{e}_\theta + \dot{z} \bar{k}$$

Applying the same procedure to Eq. (2.47) yields the acceleration. Specifically

$$\bar{a} = h_R \ddot{R} \bar{e}_R + \left(\frac{\partial h_R}{\partial R} \bar{e}_R + h_R \frac{\partial \bar{e}_R}{\partial R} \right) \dot{R}^2 + \left(\frac{\partial h_R}{\partial \theta} \bar{e}_R + h_R \frac{\partial \bar{e}_R}{\partial \theta} \right) \dot{R}\dot{\theta}$$

$$+ \left(\frac{\partial h_R}{\partial z} \bar{e}_R + h_R \frac{\partial \bar{e}_R}{\partial z} \right) \dot{R}\dot{z} + h_\theta \ddot{\theta} \bar{e}_\theta + \left(\frac{\partial h_\theta}{\partial R} \bar{e}_\theta + h_\theta \frac{\partial \bar{e}_\theta}{\partial R} \right) \dot{R}\dot{\theta}$$

$$+ \left(\frac{\partial h_\theta}{\partial \theta} \bar{e}_\theta + h_\theta \frac{\partial \bar{e}_\theta}{\partial \theta} \right) \dot{\theta}^2 + \left(\frac{\partial h_\theta}{\partial z} \bar{e}_\theta + h_\theta \frac{\partial \bar{e}_\theta}{\partial z} \right) \dot{\theta}\dot{z}$$

$$+ h_z \ddot{z} \bar{e}_z + \left(\frac{\partial h_z}{\partial R} \bar{e}_z + h_z \frac{\partial \bar{e}_z}{\partial R} \right) \dot{R}\dot{z} + \left(\frac{\partial h_z}{\partial \theta} \bar{e}_z + h_z \frac{\partial \bar{e}_z}{\partial \theta} \right) \dot{\theta}\dot{z}$$

$$+ \left(\frac{\partial h_z}{\partial z} \bar{e}_z + h_z \frac{\partial \bar{e}_z}{\partial z} \right) \dot{z}^2$$

$$\bar{a} = \ddot{R} \bar{e}_R + \dot{R}\dot{\theta} \bar{e}_\theta + R\ddot{\theta} \bar{e}_\theta + \dot{R}\dot{\theta} \bar{e}_\theta - R\dot{\theta}^2 \bar{e}_R + \ddot{z} \bar{k}$$

Collecting like components in this expression leads to the same result as Eq. (2.48). Note that the second and fourth terms in the present derivation, which combine to form the Coriolis acceleration, originated from different sources. One term is associated with $h_R \, (\bar{e}_R / \partial \theta) \, \dot{R}\dot{\theta}$, so it is a consequence of the dependence of the radial unit vector on the azimuthal angle θ. The other term is associated with $(\partial h_\theta / \partial R) \, \bar{e}_\theta \dot{R}\dot{\theta}$, which arises because the movement in the azimuthal direction accelerates owing to dependence of the azimuthal stretch ratio on the radial distance.

EXAMPLE 2.8

An airplane climbs at a constant speed v at a constant climb angle β. The airplane is being tracked by a radar station at point A on the ground. Determine the radial velocity $\dot{R}$ and the angular velocity $\dot{\theta}$ as functions of the tracking angle θ.

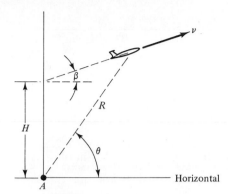

Example 2.8

Solution. We shall employ a trigonometric approach here, in which the desired parameters are obtained from differentiation of geometrical relations. (A simpler solution to this problem may be found in Example 2.10 in the next section, which matches the given velocity to the cylindrical coordinate formulas.) First, we construct the distance vt the airplane has traveled after passing point B above the radar station. This forms one side of a triangle whose other sides are R and H. Then the law of sines yields

$$\frac{R}{\sin(\pi/2 + \beta)} = \frac{vt}{\sin(\pi/2 - \theta)} = \frac{H}{\sin(\theta - \beta)}$$

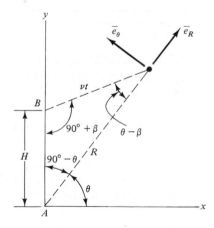

Polar coordinate system.

Thus

$$R \sin(\theta - \beta) = H \sin(\pi/2 + \beta) = H \cos \beta$$
$$vt \sin(\theta - \beta) = H \sin(\pi/2 - \theta) = H \cos \theta$$

Differentiating each expression leads to

$$\dot{R} \sin(\theta - \beta) + R\,\dot{\theta} \cos(\theta - \beta) = 0$$
$$v \sin(\theta - \beta) + vt\,\dot{\theta} \cos(\theta - \beta) = -H\,\dot{\theta} \sin \theta$$

These are simultaneous equations for $\dot{R}$ and $\dot{\theta}$, whose solution is

$$\dot{\theta} = -\frac{v \sin (\theta - \beta)}{vt \cos (\theta - \beta) + H \sin \theta}$$

$$\dot{R} = \frac{Rv \cos (\theta - \beta)}{vt \cos (\theta - \beta) + H \sin \theta}$$

These expressions are not in the desired form because they depend on t and R. The equations obtained from the law of sines indicate that

$$R = H \frac{\cos \beta}{\sin (\theta - \beta)} \qquad vt = H \frac{\cos \theta}{\sin (\theta - \beta)}$$

We use these relations to eliminate R and vt from the expressions for $\dot{R}$ and $\dot{\theta}$, which leads to

$$\dot{\theta} = -\frac{v}{H} \frac{\sin (\theta - \beta)}{\cos \theta \cot (\theta - \beta) + \sin \theta}$$

$$\dot{R} = v \frac{\cot (\theta - \beta) \cos \beta}{\cos \theta \cot (\theta - \beta) + \sin \theta}$$

In order to simplify these relations, we multiply numerator and denominator of each by $\sin (\theta - \beta)$, and then employ the identity for the cosine of the difference of angles. This yields

$$\dot{\theta} = -\frac{v}{H} \frac{\sin^2 (\theta - \beta)}{\cos \beta} \qquad \dot{R} = v \cos (\theta - \beta)$$

EXAMPLE 2.9 _____

The flyball governor rotates about the vertical axis at the rate of 1800 rev/min, and the angle ϕ locating the arms relative to the vertical is known to vary as $\phi = (\pi/3) \sin (120\pi t)$ rad, where t is in units of seconds. Determine the velocity and the acceleration of sphere A as a function of time. Then evaluate these expressions for the instants when the elevation of the sphere is a maximum and a minimum.

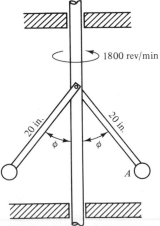

1800 rev/min

20 in.

20 in.

ϕ ϕ

A

Example 2.9

Solution.   The motion of each sphere is described in terms of a distance from a fixed point, a rotation about a fixed axis through that point, and an angle relative to the axis, each of which matches a spherical coordinate description. The unit vectors in the sketch are in the sense of increasing coordinate values. The radial distance and rotation rate about the fixed axis are both constant, so

$$r = 20 \text{ in.} \qquad \dot{r} = \ddot{r} = 0$$

$$\dot{\theta} = 1800 \left(\frac{2\pi}{60}\right) = 60\pi \text{ rad/s} \qquad \ddot{\theta} = 0$$

$$\phi = \frac{\pi}{3} \sin (120\pi t) \text{ rad} \qquad \dot{\phi} = 40\pi^2 \cos (120\pi t) \text{ rad/s}$$

$$\ddot{\phi} = -4800\pi^3 \sin (120\pi t) \text{ rad/s}^2$$

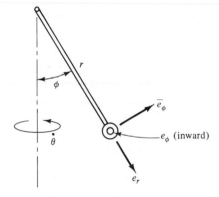

Spherical coordinate system.

The spherical coordinate formulas for these variables become

$$\bar{v} = 800\pi^2 \cos (120\pi t) \, \bar{e}_\phi + 1200\pi \sin \left[\frac{\pi}{3} \sin (120\pi t)\right] \bar{e}_\theta \text{ in./s}$$

$$\bar{a} = -\left\{32{,}000\pi^4 \cos^2 (120\pi t) + 72{,}000\pi^2 \sin^2 \left[\frac{\pi}{3} \sin (120\pi t)\right]\right\} \bar{e}_r$$

$$- \left\{96{,}000\pi^3 \sin (120\pi t) + 36{,}000\pi^2 \sin \left[\frac{2\pi}{3} \sin (120\pi t)\right]\right\} \bar{e}_\phi$$

$$+ 96{,}000\pi^3 \cos (120\pi t) \cos \left[\frac{\pi}{3} \sin (120\pi t)\right] \bar{e}_\theta \text{ in./s}^2$$

These general results may now be evaluated at the desired instants. The highest elevation of a sphere corresponds to the maximum value $\phi = \pi/3$, at which time $\sin (120\pi t) = 1$. Correspondingly, $\cos (120\pi t) = 0$ at this instant, which yields

$$\bar{v} = 3264 \bar{e}_\theta \text{ in./s}$$

$$\bar{a} = -5.330(10^5)\bar{e}_r - 3.2843(10^6)\bar{e}_\phi \text{ in./s}^2$$

The lowest elevation occurs at $\phi = 0$, $\cos(120\pi t) = \pm 1$, which yields

$$\bar{v} = \pm 7896\bar{e}_\theta \text{ in./s}$$
$$\bar{a} = -3.1171(10^6)\bar{e}_r \pm 2.9766(10^6)\bar{e}_\theta \text{ in./s}^2$$

Note that the $\pm$ sign arises because the sphere may be swinging in either direction at $\phi = 0$.

2.4 MIXED KINEMATICAL DESCRIPTIONS

The degree to which the system parameters match those of the coordinate system is a key factor affecting the selection of a kinematical description. A specific concern is whether the quantities that are either given or to be determined are like those for the chosen description. For example, suppose that the path of a particle is known to be as shown in Figure 2.8. If the rate of movement along that path is specified in terms of the speed v, we would certainly want to employ a path variable description. On the other hand, specification of the rate of motion in terms of the angle θ measured from the x axis would certainly suggest that cylindrical coordinates be employed.

We could consider the kinematical description that best matches the parameters of the actual system to be the "natural" one. We shall investigate here situations in which no one formulation is entirely natural but more than one has elements that are suitable. Such a situation arises for the path in Figure 2.8 when the rate of movement is given in terms of the speed, but it is desired to evaluate $\dot{R}$ and $\ddot{R}$. It is almost axiomatic that if one of the kinematical descriptions (path variables, Cartesian coordinates, or one of the curvilinear coordinate systems) has some aspect that suits a problem, it should be employed. Thus, the task that confronts us here is to establish how to implement two different descriptions simultaneously.

The general concept is to match the velocities and accelerations obtained from each of the formulations of interest. This matching depends on the fact that the unit vectors for one formulation may be resolved into components relative to the other. For simplicity, let us begin by considering planar motion. Let $\bar{e}_\alpha$, $\bar{e}_\mu$ be the unit vectors for one kinematical description (for example, $\bar{e}_\alpha$ and $\bar{e}_\beta$ are the tangent and normal directions), and let $\bar{e}_\lambda$, $\bar{e}_\mu$ be the unit vectors for the other description. These unit vectors are depicted in Figure 2.9.

As shown in the figure, the orientation of one set of unit vectors relative to the other is defined by the angle ψ. (The definition of this angle as that between $\bar{e}_\alpha$ and $\bar{e}_\lambda$ is arbitrary.) The components of $\bar{e}_\lambda$ and $\bar{e}_\mu$ relative to $\bar{e}_\alpha$ and $\bar{e}_\beta$ are found from this figure to be

$$\bar{e}_\lambda = \cos \psi \bar{e}_\alpha + \sin \psi \bar{e}_\beta$$
$$\bar{e}_\mu = -\sin \psi \bar{e}_\alpha + \cos \psi \bar{e}_\beta \tag{2.50}$$

The velocity may be expressed in terms of components relative to either set of unit vectors. Thus,

$$\bar{v} = v_\alpha \bar{e}_\alpha + v_\beta \bar{e}_\beta = v_\lambda \bar{e}_\lambda + v_\mu \bar{e}_\mu \tag{2.51}$$

We are assuming that at this stage, the first set of components v_α, v_β have been related

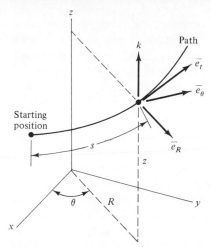

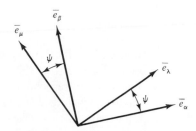

Figure 2.8 Joint usage of path variables and cylindrical coordinates.

Figure 2.9 Transformation of unit vectors in a plane.

to the parameters associated with (α, β) through the corresponding velocity formula, such as $\bar{v} = \dot{R}\bar{e}_R + R\dot{\theta}\bar{e}_\theta$ for polar coordinates. The same operation is also assumed to have been performed for the second set of components. Each velocity component might contain unknown kinematical parameters.

The next step is to convert the (λ, μ) components to (α, β) components. This is achieved by substituting Eqs. (2.50) into Eq. (2.51), with the result that

$$\bar{v} = v_\alpha \bar{e}_\alpha + v_\beta \bar{e}_\beta$$
$$= v_\lambda(\cos \psi \bar{e}_\alpha + \sin \psi \bar{e}_\beta) + v_\mu(-\sin \psi \bar{e}_\alpha + \cos \psi \bar{e}_\beta)$$

These are two descriptions of the velocity in terms of the same set of components. Equality of vectors requires that their corresponding components be equal, which leads to the following algebraic relations:

$$v_\alpha = v_\lambda \cos \psi - v_\mu \sin \psi$$
$$v_\beta = v_\lambda \sin \psi + v_\mu \cos \psi \tag{2.52}$$

These relations may be used to solve for two unknown parameters in the velocity components.

As an illustration of this procedure, suppose that (α, β) represents path variables and (λ, μ) represents polar coordinates. Substitution of the respective velocity components into Eqs. (2.52) then yields

$$v = \dot{R} \cos \psi - R\dot{\theta} \sin \psi$$
$$0 = \dot{R} \sin \psi + R\dot{\theta} \cos \psi$$

The values of the radial distance R and the angle of orientation θ are known if the position is specified. Thus, the above represents two relations among the three rate variables v, $\dot{R}$, and $\dot{\theta}$.

Relations such as those listed in Eqs. (2.52) could be used in either of two general situations. It might be that the velocity is already known in terms of components in the (λ, μ) system. In that case, Eqs. (2.52) provide the conversion to components and parameters associated with the (α, β) system. The more interesting situation is that of a mixed kinematical description, that is, one in which the velocity is only partially known in terms of either of the two descriptions. In that case, Eqs. (2.52) provide the means to ascertain the unknown parameters in each system, and thereby the velocity itself.

The same approach may be applied to treat acceleration. Specifically, the individual formulas for acceleration may be matched by employing the unit vector transformation, Eq. (2.50). However, doing so requires that the velocity parameters, such as $|\bar{v}|$ or $\dot{R}$, be evaluated first because they occur in the acceleration components. In other words, the velocity relations must be solved before accelerations can be addressed.

The discussion treated the case of planar motion, but the same procedure also applies to three-dimensional motion. The kinematical formulas in that case have three components, so matching corresponding components will lead to three simultaneous equations. The primary difficulty that arises in this extension is the evaluation of the transformation of the unit vectors. The component representation in Eq. (2.50) was achieved by visual projections of a unit vector onto the other directions. The same procedure may be performed in a three-dimensional case. An alternative approach for complicated geometries is to use rotation transformation properties established in the next chapter.

EXAMPLE 2.10 _____

Use the concept of a mixed kinematical description to determine $\dot{R}$ and $\dot{\theta}$ for the airplane in Example 2.8.

Solution. The path and speed of the airplane are given, both of which are path variable parameters. We must determine the rates of change of polar coordinates, which are cylindrical coordinates with $z = 0$. Thus, we draw a sketch that depicts the unit vectors for both formulations at an arbitrary θ.

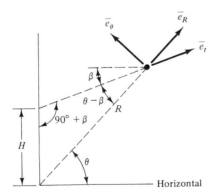

Unit vectors.

The velocity in terms of each set of unit vectors is

$$\bar{v} = v\bar{e}_t = \dot{R}\bar{e}_R + R\dot{\theta}\bar{e}_\theta$$

Resolving $\bar{e}_t$ into polar coordinate components yields

$$\bar{e}_t = \cos(\theta - \beta)\bar{e}_R - \sin(\theta - \beta)\bar{e}_\theta$$

so that

$$\bar{v} = v \cos (\theta - \beta)\bar{e}_R - v \sin (\theta - \beta)\bar{e}_\theta = \dot{R}\bar{e}_R + R \dot{\theta}\bar{e}_\theta$$

The result of matching like components is

$$\dot{R} = v \cos (\theta - \beta) \qquad \dot{\theta} = -\frac{v}{R} \sin (\theta - \beta)$$

All that remains is to express R in terms of θ, which we find from the law of sines.

$$\frac{R}{\sin (\pi/2 + \beta)} = \frac{H}{\sin (\theta - \beta)} \Rightarrow R = H \frac{\cos \beta}{\sin (\theta - \beta)}$$

Thus

$$\dot{R} = v \cos (\theta - \beta) \qquad \dot{\theta} = -\frac{v}{H} \frac{\sin^2 (\theta - \beta)}{\cos \beta}$$

There is no doubt that this solution is easier than the one in Example 2.8. In essence, the mixed kinematical description avoids the need to differentiate functions, because the various kinematical formulas are themselves derivatives of the position.

EXAMPLE 2.11

Arm AB rotates clockwise at the constant rate of 40 rad/s as it pushes the slider along guide CD, which is described by $y = x^2/200$ (x and y are in millimeters). Determine the velocity and acceleration of the collar when it is at the position $x = 200$ mm.

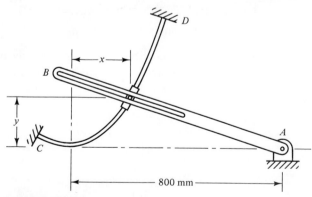

Example 2.11

Solution. The planar motion is specified by a rotation rate, but the path is not described in terms of polar coordinates. Hence, we shall follow an approach that employs path variables and polar coordinates. A sketch shows both sets of unit vectors at $x = 200$ mm, which corresponds to $y = x^2/200 = 200$ mm.

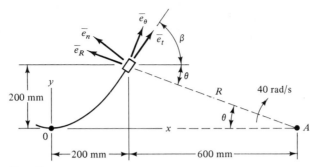

Coordinate systems.

The polar coordinates are found from a right triangle to be

$$R = [(600)^2 + (200)^2]^{1/2} = 632.5 \text{ mm} \qquad \theta = \tan^{-1}\left(\tfrac{200}{600}\right) = 18.435°$$

The slope of the guide bar at this location yields the angle of orientation of the tangent vector.

$$\beta = \tan^{-1}\left(\frac{dy}{dx}\right) = \tan^{-1}\left(\frac{x}{100}\right) = 63.435°$$

Matching like velocity components in each formulation is the next step. We find that

$$\bar{v} = \dot{R}\bar{e}_R + R\dot{\theta}\bar{e}_\theta = v\bar{e}_t = v[-\cos(\theta + \beta)\,\bar{e}_R + \sin(\theta + \beta)\,\bar{e}_\theta]$$
$$\bar{v} \cdot \bar{e}_R = \dot{R} = -v\cos(\theta + \beta) \qquad \bar{v} \cdot \bar{e}_\theta = R\dot{\theta} = v\sin(\theta + \beta)$$

The value of $\dot{\theta}$ is given to be 40 rad/s, and R, θ, and β have been evaluated. The corresponding results obtained from these relations are

$$v = 25{,}557 \text{ mm/s} = 25.56 \text{ m/s}$$
$$\dot{R} = -3614 \text{ mm/s} = -3.614 \text{ m/s}$$

Because we have evaluated all velocity parameters, we may now follow a similar procedure for acceleration. In order to resolve $\bar{e}_n$, we note that it is oriented toward the center of curvature, which is up and to the left for the parabolic curve.

$$\bar{a} = (\ddot{R} - R\dot{\theta}^2)\bar{e}_R + (R\ddot{\theta} + 2\dot{R}\dot{\theta})\bar{e}_\theta = \dot{v}\bar{e}_t + \frac{v^2}{\rho}\bar{e}_n$$

$$= \dot{v}[-\cos(\theta + \beta)\bar{e}_R + \sin(\theta + \beta)\bar{e}_\theta]$$

$$+ \frac{v^2}{\rho}[\sin(\theta + \beta)\bar{e}_R + \cos(\theta + \beta)\bar{e}_\theta]$$

The unknowns in these equations are $\dot{v}$, $\ddot{R}$, and ρ. We compute the latter from Eq. (2.22), which gives, for $x = 200$ mm,

$$\rho = \frac{[1 + (x/100)^2]^{3/2}}{\frac{1}{100}} = 1118.0 \text{ mm}$$

The result of matching like acceleration components is

$$\bar{a} \cdot \bar{e}_R = \ddot{R} - R\dot{\theta}^2 = -\dot{v} \cos(\theta + \beta) + \frac{v^2}{\rho} \sin(\theta + \beta)$$

$$\bar{a} \cdot \bar{e}_\theta = R\ddot{\theta} + 2\dot{R}\dot{\theta} = \dot{v} \sin(\theta + \beta) + \frac{v^2}{\rho} \cos(\theta + \beta)$$

The value of $\dot{\theta}$ is constant, and we found $\dot{R}$ and v earlier. Hence, we solve the a_θ equation for $\dot{v}$, then substitute that result into the equation for a_R.

$$\dot{v} = \frac{2\dot{R}\dot{\theta} - (v^2/\rho) \cos(\theta + \beta)}{\sin(\theta + \beta)} = -3.755(10^5) \text{ mm/s}^2$$

$$\ddot{R} = R\dot{\theta}^2 - \dot{v} \cos(\theta + \beta) + \frac{v^2}{\rho} \sin(\theta + \beta) = 1.6434(10^6) \text{ mm/s}^2$$

The parameters that were requested are the velocity and acceleration, which we may form in terms of either kinematical formulation. Thus

$$\bar{v} = 25.56\bar{e}_t \text{ m/s} = -3.614\bar{e}_R + 25.30\bar{e}_\theta \text{ m/s}$$

$$\bar{a} = -375.5\bar{e}_t + 584.2\bar{e}_n \text{ m/s}^2 = 631.4\bar{e}_R - 289.1\bar{e}_\theta \text{ m/s}^2$$

EXAMPLE 2.12 _____

The cord suspending a spherical pendulum is pulled in at a constant rate of 5 m/s. At the instant when the pendulum is 2 m long the azimuth angle $\theta = 0°$ and the angle of inclination of the cable is $\phi = 30°$. At this instant, $\dot{\theta} = 2$ rad/s, $\ddot{\theta} = 0$, $\dot{\phi} = -5$ rad/s, and $\ddot{\phi} = -10$ rad/s^2. Determine the speed and the rate of change of the speed of the small body at the end of the cable at this instant. Also, determine the corresponding radius of curvature.

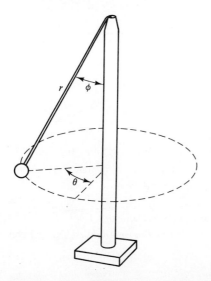

Example 2.12

Solution. In this situation the motion is fully specified in terms of spherical coordinates, whereas the desired parameters are path variables. Hence, the procedure is to construct the velocity and acceleration using spherical coordinates, and then to relate those expressions to the path variable formulas.

Because the cable is being pulled in at a constant rate, we set $\dot{r} = -5$ m/s and $\ddot{r} = 0$ when $r = 2$ m. Substitution of these values and the stated rotation rates into Eqs. (2.49) gives

$$\bar{v} = -5\bar{e}_r - 10\bar{e}_\phi + 2\bar{e}_\theta \text{ m/s}$$
$$\bar{a} = -52\bar{e}_r + 26.54\bar{e}_\phi - 44.64\bar{e}_\theta \text{ m/s}^2$$

The speed is the magnitude of the velocity, so

$$v = (v_r^2 + v_\phi^2 + v_\theta^2)^{1/2} = 11.358 \text{ m/s}$$

Because $\bar{v} = v\bar{e}_t$, we find that the tangent vector is

$$\bar{e}_t = \frac{\bar{v}}{v} = -0.4402\bar{e}_r - 0.8805\bar{e}_\phi + 0.17609\bar{e}_\theta$$

We know that $\dot{v}$ is the tangential component of acceleration, which we find by using a dot product.

$$\dot{v} = \bar{a} \cdot \bar{e}_t = -8.336 \text{ m/s}^2$$

Finally, in order to evaluate ρ, we form the difference between the total acceleration and the tangential acceleration. Specifically,

$$\frac{v^2}{\rho} \bar{e}_n = \bar{a} - \dot{v}\bar{e}_t = -55.67\bar{e}_r + 19.20\bar{e}_\phi - 43.17\bar{e}_\theta \text{ m/s}^2$$

We evaluate the magnitude of this acceleration and use the value of v to find

$$\rho = \frac{v^2}{(55.67^2 + 19.20^2 + 43.17^2)^{1/2}} = 1.7668 \text{ m}$$

REFERENCES

J. H. Ginsberg and J. Genin, *Dynamics*, 2d ed., John Wiley & Sons, Inc., New York, 1984.

J. B. Marion, *Classical Dynamics of Particles and Systems*, Academic Press, New York, 1960.

P. M. Morse and H. Feshbach, *Methods of Theoretical Physics*, McGraw-Hill Book Company, New York, 1953.

I. H. Shames, *Engineering Mechanics*, 2d ed., Prentice-Hall, Inc., Englewood Cliffs, N.J., 1967.

A. P. Willis, *Vector Analysis with an Introduction to Tensor Analysis*, Dover Publications, Inc., New York, 1958.

HOMEWORK PROBLEMS

2.1 A particle follows a planar path defined by $x = (A/k\zeta) \sin k\zeta$, $y = (A/k\zeta) \cos k\zeta$, such that its speed is $v = \beta\zeta$, where A, k, and β are constants. Determine the velocity and acceleration at $\zeta = 3\pi/4k$.

2.2 A small block slides in the interior of a smooth semicircular cylinder. Because friction is negligible, the speed of the block is given by $v^2 = 2gh$, where h is the vertical distance the block has fallen. Determine the velocity and acceleration of the block as a function of the distance the block travels in a case where the block is released at position A, at the horizontal diameter.

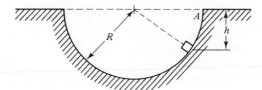

Problem 2.2

2.3 The collar slides over the stationary guide defined by $x = ky^2$ in the vertical plane. The speed of the collar is the constant value v. This motion is implemented by applying a force $\overline{F}$ of variable magnitude parallel to the x axis. Derive an expression for the magnitude of $\overline{F}$ and of the normal reaction as functions of the y coordinate of the collar.

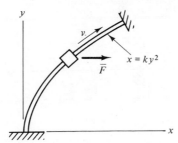

Problem 2.3

2.4 A slider moves over a curved guide whose shape in the vertical plane is given by $y = \cosh x$. Starting from $x = 0$, the speed is observed to vary as $v = v_0 (1 - ks)$, where s is the distance traveled and k is a constant. Derive expressions for the velocity and acceleration of the slider as a function of x.

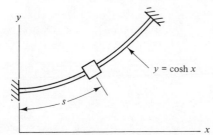

Problem 2.4

2.5 A curve is defined by $x = ec\psi$, $y = e \sin c\psi$, $z = -e \cos c\psi$. Determine the path variable unit vectors, the radius of curvature, and the torsion of this curve as a function of ψ.

2.6 A particle moves along the paraboloid of revolution $y = (x^2 + z^2)/3$, such that $x = -3k\zeta \sinh k\zeta$, $z = 3k\zeta \cosh k\zeta$, where x, y, and z are meters and ζ is a parameter. At the position where $\zeta = 1/k$, its speed is $5k$ and its speed is decreasing at the rate $2k^2$. Determine the velocity and acceleration at this position.

2.7 Determine the radius of curvature and the torsion of the path in Problem 2.6 at the given position.

2.8 A particle moves along the paraboloid of revolution $y = (x^2 + z^2)/k$ such that $x = k\zeta \sin \omega\zeta$, $y = k\zeta^2$, where k and ω are constants and ζ is a parameter. Consider the case where the parameter $\zeta = t^2$, where t is time in units of seconds. Derive expressions for the velocity and acceleration.

2.9 Pin P is pushed by arm ABC through the groove, $y = 2(1 - 4x^2)$, where x and y are in feet. The velocity of arm ABC is constant at 10 ft/s to the right. Determine the velocity and acceleration of the collar at the position $x = 2$ ft.

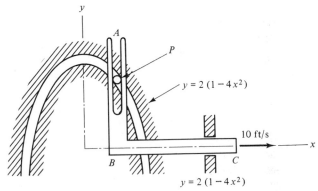

Problem 2.9

2.10 A ball is thrown down an incline whose angle of elevation is θ. The initial velocity is u at an angle of elevation β. Derive an expression for the distance D measured along the incline at which the ball will return to the incline. Also determine the maximum height H (measured perpendicularly to the incline) of the ball, and the corresponding velocity of the ball at that position.

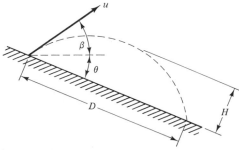

Problem 2.10

2.11 A 10-mg dust particle is injected into an electrostatic precipitator with an initial velocity of 10 m/s, as shown. The z axis is vertical and the electrostatic force is 0.2 mN, acting in the positive y direction. Determine the location at which the dust particle will strike a collector plate that is situated in the vertical plane, $y = 400$ mm.

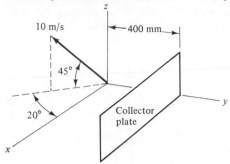

Problem 2.11

2.12 For laminar flow at low Reynolds number, the air resistance on an object is $-c\bar{v}$, where c is a constant and $\bar{v}$ is the velocity of the object. A sphere of mass m is thrown from the ground with an initial speed v_0 at an angle of elevation β in the (vertical) xy plane. Determine the position and velocity of the sphere as a function of time.

2.13 Derive the expressions for velocity and acceleration in terms of spherical coordinates.

2.14 A ball rolls on the interior of a paraboloid of revolution given by $x^2 + y^2 = cz$. The angle of rotation about the z axis is $\theta = 4\pi \sin^2 \omega t$, and the elevation of the ball is $z = b\theta$, where b, c, and ω are constants. Determine the velocity and acceleration when $t = 4\pi/3\omega$.

2.15 The polar velocity components of fluid particles in a certain flow are known to be $v_R = (A \cos \phi)/R^2$, $v_\phi = (A \sin \phi)/R^2$, where R, ϕ are the polar coordinates. Determine the corre ponding expressions for the acceleration.

2.16 A particle follows a planar path defined in polar coordinates by $R = R(\theta)$ such that $\dot{\theta}$ is constant. Derive expressions for the velocity and acceleration of the particle. Then use those results to derive an expression for the radius of curvature of a path in polar coordinates.

2.17 A cable that passes through a hole at point A is pulled inward at the constant rate of 8 ft/s, thereby causing the $\frac{1}{2}$-lb collar to move along the circular guide bar. The system is situated

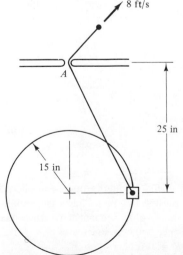

Problem 2.17

in the vertical plane. Determine the speed and the rate of change of the speed of the slider at the instant shown in the sketch. Also evaluate the corresponding tension in the cable.

2.18 A radar station at the origin measures the azimuth angle θ, the elevation angle ϕ, and the radial distance r to a target. At the instant when a missile passes position A, its velocity is 500 m/s directed from point A to point B. Its acceleration at this instant is 10 g, directed toward the origin. Determine the first and second derivatives of these position parameters at this instant.

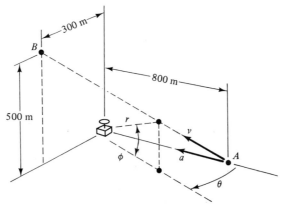

Problem 2.18

2.19 A block is pushed along the stationary hemisphere such that it completes one revolution about the vertical centerline in 2 s. The angle of inclination ϕ relative to the vertical axis is $\phi = 8t^2$ rad. Determine the velocity and acceleration of the block as a function of time.

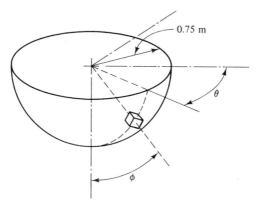

Problem 2.19

2.20 A 250-g block is pushed over the smooth interior of a hemispherical shell whose interior radius is 0.75 m. The force $\overline{F}$ causing this motion acts tangentially to the hemispherical surface. The block completes one revolution about the vertical centerline in 2 s, and it moves outward such that the distance to the axis increases as $R = 0.24 + 4t^2$ m. Determine the force $\overline{F}$ at the instant when $R = 0.6$ m.

2.21 Toroidal coordinates (ρ, θ, ψ) are useful for magnetohydrodynamic studies in tokamaks. Such coordinates reference position to a circle of radius r, such that the transformation to Cartesian coordinates is $x = (r + \rho \cos \psi) \cos \theta$, $y = (r + \rho \cos \psi) \sin \theta$, $z = \rho \sin \psi$. Derive expressions for the unit vectors for this coordinate system, and then describe the derivatives of the unit vectors with respect to the toroidal coordinates.

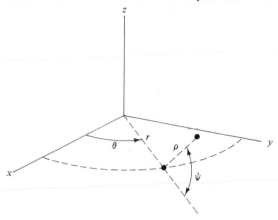

Problem 2.21

2.22 Obtain expressions for velocity and acceleration in terms of the toroidal coordinates in Problem 2.21.

2.23 The instantaneous velocity of a point is $\bar{v} = 10\bar{i} - 4\bar{j} + 6\bar{k}$ m/s, and the acceleration is $\bar{a} = -30\bar{i} - 25\bar{j} + 15\bar{k}$ m/s². Determine the corresponding speed, rate of change of the speed, and the radius of curvature of the path.

2.24 An airplane heading eastward is observed to be in a 20° climb at a true airspeed of 1500 ft/s. At this instant its acceleration components are $2g$ eastward, $5g$ northward, and $1.5g$ downward. Determine the rate of change of the speed, as well as the radius of curvature and the location relative to the airplane of the center of curvature of the path.

2.25 Pin P slides inside the 500-mm-radius groove at a constant rate of 1 m/s. Determine the values of $\dot{\phi}$ and $\ddot{\phi}$ at the instant when $\theta = 90°$.

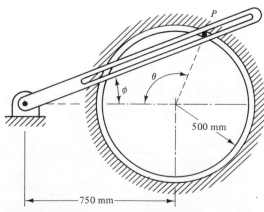

Problem 2.25

2.26 Solve Problem 2.25 for the instant when $\theta = 135°$.

2.27 The elevation of the center of mass of a rowboat in a storm is observed to be $x = 2 \sin (\pi x/3)$ (x and y in feet). Its speed at $x = 1$ ft is measured as 20 ft/s, and the speed at that position is decreasing at 100 ft/s^2. Determine the horizontal and vertical components of the acceleration at that instant.

Chapter 3

Relative Motion

A moving body, such as an automobile, frequently provides a useful reference frame for our observations of motion. Even when we are not moving, it is often easier to describe the motion of a point by referring it to a moving object. This is the case for many common machines, such as linkages. In this chapter we develop the ability to correlate observations of position, velocity, and acceleration from fixed and moving reference frames. Figure 3.1 depicts a general situation in which point P is being observed from a moving reference frame xyz, whereas XYZ is a fixed reference frame. In order to use xyz as a reference frame, we must know its location. Thus, the position $\bar{r}_{O'/O}$ of the origin O' must be specified. Also, because the motion is being described in terms of the xyz reference frame, we locate point P by giving its (x, y, z) coordinates. The corresponding relative position is

$$\bar{r}_{P/O'} = x\bar{i} + y\bar{j} + z\bar{k} \tag{3.1}$$

It is apparent from Figure 3.1 that the absolute position $\bar{r}_{P/O}$ is

$$\bar{r}_{P/O} = \bar{r}_{O'/O} + \bar{r}_{P/O'} \tag{3.2}$$

This relation would be straightforward were it not for the fact that one aspect of the relative position has not been discussed. We must be able to relate the unit vectors of the coordinate systems associated with the fixed and moving reference frames. Otherwise, we will be unable to add like components in the evaluation of Eq. (3.2).

3.1 ROTATION TRANSFORMATIONS

Let us consider a general situation in which two coordinate systems, xyz and $x'y'z'$, are employed to represent the components of a vector. Only the orientation of the axes is of interest here, so the origins of the coordinate systems may be made to coincide. Figure

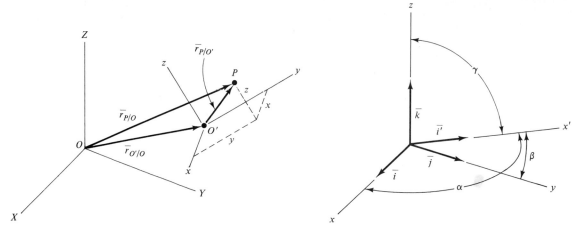

Figure 3.1   Absolute and relative position.   **Figure 3.2**   Direction angles.

3.2 depicts the *direction angles* α, β, γ between the x' axis and each of the *xyz* axes. An examination of Figure 3.2 shows that the values of the direction angles should be limited to the range $0 < \alpha, \beta, \gamma < \pi$ in order to avoid ambiguity. The components of $\bar{i}'$ are the projections of the vector onto the axes of *xyz*, which, in turn, are determined from the direction angles according to

$$\begin{aligned} \bar{i}' &= (\bar{i}' \cdot \bar{i})\bar{i} + (\bar{i}' \cdot \bar{j})\bar{j} + (\bar{i}' \cdot \bar{k})\,\bar{k} \\ &= \cos\alpha\,\bar{i} + \cos\beta\,\bar{j} + \cos\gamma\,\bar{k} \end{aligned} \tag{3.3}$$

This expression indicates that the cosines of the direction angles are more significant to our investigation: they are the *direction cosines*. We obviously are equally interested in all the unit vectors. Thus,

Define $l_{p'q} = l_{qp'}$ to be the cosine of the angle between axis p' and axis q, with p and q representing x, y, or z.

Extending Eq. (3.3) to the other unit vectors then yields

$$\begin{aligned} \bar{i}' &= l_{x'x}\bar{i} + l_{x'y}\bar{j} + l_{x'z}\bar{k} \\ \bar{j}' &= l_{y'x}\bar{i} + l_{y'y}\bar{j} + l_{y'z}\bar{k} \\ \bar{k}' &= l_{z'x}\bar{i} + l_{z'y}\bar{j} + l_{z'z}\bar{k} \end{aligned} \tag{3.4}$$

It is convenient to rewrite these equations in matrix form as

$$\begin{Bmatrix} \bar{i}' \\ \bar{j}' \\ \bar{k}' \end{Bmatrix} = [R] \begin{Bmatrix} \bar{i} \\ \bar{j} \\ \bar{k} \end{Bmatrix} \tag{3.5}$$

where

$$[R] = \begin{bmatrix} l_{x'x} & l_{x'y} & l_{x'z} \\ l_{y'x} & l_{y'y} & l_{y'z} \\ l_{z'x} & l_{z'y} & l_{z'z} \end{bmatrix} \tag{3.6}$$

The matrix $[R]$ is the *rotation transformation matrix*. It is a generalization of the conversion between systems of unit vectors that we employed to discuss joint kinematical descriptions in Section 2.4. An important property of $[R]$ results from the fact that $\bar{i}, \bar{j}, \bar{k}$ are an orthogonal set of unit vectors, as are $\bar{i}', \bar{j}', \bar{k}'$. The dot product of any pair of vectors in Eq. (3.4) can be written in a general form as

$$l_{p'x}l_{q'x} + l_{p'y}l_{q'y} + l_{p'z}l_{q'z} = \delta_{pq} \qquad p, q = x, y, z \qquad (3.7)$$

where δ_{pq} denotes the Kronecker delta, $\delta_{pq} = 1$ if $p = q$ and $\delta_{pq} = 0$ otherwise.

Equation (3.7) is symmetric in p and q. Hence, it represents six equations (one for each p,q pair) relating the nine direction cosines. It follows that there are only three independent direction angles. The selection of which angles are arbitrary is not entirely free. For example, the values of α, β, and γ in Figure 3.2 are not independent because $\cos^2 \alpha + \cos^2 \beta + \cos^2 \gamma = 1$. This restriction arises because these three angles only locate one axis.

The left side of Eq. (3.7) corresponds to the product of any row of $[R]$ and a column formed from any row of $[R]$. The latter is equivalent to the transpose of a row of $[R]$. It follows that Eq. (3.7) details the elements of the following matrix product:

$$[R][R]^T = [U] \qquad (3.8)$$

where $[U]$ is the identity (i.e., *u*nit) matrix. Premultiplication of Eq. (3.8) by $[R]^{-1}$ leads to the conclusion that

$$[R]^{-1} = [R]^T \qquad (3.9)$$

Because $[R]$ represents the rotation transformation between two sets of rectangular Cartesian coordinate systems, any matrix satisfying Eq. (3.9) is said to be *orthonormal*.

An important aspect of an orthonormal matrix is obtained by reversing the transformation such that $\bar{i}, \bar{j}, \bar{k}$ are represented in terms of their $x'y'z'$ components. Let $[R']$ denote the transformation matrix for this representation, so

$$\left\{ \begin{matrix} \bar{i} \\ \bar{j} \\ \bar{k} \end{matrix} \right\} = [R'] \left\{ \begin{matrix} \bar{i}' \\ \bar{j}' \\ \bar{k}' \end{matrix} \right\} \qquad (3.10)$$

Now solve Eq. (3.5) for $\bar{i}, \bar{j}, \bar{k}$ and compare the result with this equation. The comparison and Eq. (3.9) reveal that

The matrix $[R']$ representing the inverse transformation is the inverse of the original transformation matrix $[R]$, which is identical to the transpose of $[R]$.

$$[R'] = [R]^{-1} = [R]^T \qquad (3.11)$$

The foregoing relationship could have been derived differently by evaluating $[R']$ according to the approach in Eq. (3.4). That would have shown that the elements of $[R']$ are $l_{pq'}$, which are identically equal to $l_{q'p}$.

The importance of the transformation matrix stems from the fact that it relates the components of arbitrary vectors with respect to two coordinate systems, not just the unit vectors. In order to demonstrate this feature, we recall that any vector $\bar{A}$ is independent of the coordinate system used to describe its components, so

$$\overline{A} = A_x \overline{i} + A_y \overline{j} + A_z \overline{k} = A_{x'} \overline{i}' + A_{y'} \overline{j}' + A_{z'} \overline{k}'$$

This expression may be written in matrix form as

$$[\overline{i}' \quad \overline{j}' \quad \overline{k}'] \begin{Bmatrix} A_{x'} \\ A_{y'} \\ A_{z'} \end{Bmatrix} = [\overline{i} \quad \overline{j} \quad \overline{k}] \begin{Bmatrix} A_x \\ A_y \\ A_z \end{Bmatrix} \tag{3.12}$$

In order to eliminate the unit vectors, substitute the transpose of Eq. (3.10) into the above. The transpose of a product is the product of the transposes, so

$$[\overline{i}' \quad \overline{j}' \quad \overline{k}'] \begin{Bmatrix} A_{x'} \\ A_{y'} \\ A_{z'} \end{Bmatrix} = [\overline{i}' \quad \overline{j}' \quad \overline{k}'][R']^{\mathrm{T}} \begin{Bmatrix} A_x \\ A_y \\ A_z \end{Bmatrix} \tag{3.13}$$

In view of the inverse property in Eq. (3.11), the foregoing reduces to

$$\begin{Bmatrix} A_{x'} \\ A_{y'} \\ A_{z'} \end{Bmatrix} = [R] \begin{Bmatrix} A_x \\ A_y \\ A_z \end{Bmatrix} \tag{3.14}$$

A particularly simple type of transformation arises when the $x'y'z'$ system may be pictured as being the result of a rotation about one of the axes of the xyz coordinate system. The three possibilities, involving rotation about either the x, y, or z axis, are depicted in Figure 3.3. Denote the corresponding transformation matrices by a subscript that corresponds to the rotation axis. Then Eq. (3.6) leads to

$$[R_x] = \begin{bmatrix} 1 & 0 & 0 \\ 0 & \cos\theta_x & \sin\theta_x \\ 0 & -\sin\theta_x & \cos\theta_x \end{bmatrix} \tag{3.15a}$$

$$[R_y] = \begin{bmatrix} \cos\theta_y & 0 & -\sin\theta_y \\ 0 & 1 & 0 \\ \sin\theta_y & 0 & \cos\theta_y \end{bmatrix} \tag{3.15b}$$

$$[R_z] = \begin{bmatrix} \cos\theta_z & \sin\theta_z & 0 \\ -\sin\theta_z & \cos\theta_z & 0 \\ 0 & 0 & 1 \end{bmatrix} \tag{3.15c}$$

One of the most common types of vectors to be involved in a rotation transformation is the position of a point with respect to the origin. In such cases the vector components appearing in Eq. (3.14) are the position coordinates of the point relative to the respective coordinate systems.

The transformation of position coordinates may be used in two ways. Sometimes the position of the point with respect to the fixed reference frame XYZ is known. The position coordinates relative to a reference frame that has moved away from XYZ may be found directly from Eq. (3.14). A less obvious situation arises when it is stated that a point remains fixed relative to a moving reference frame. In that case the coordinates of the point *relative to the moving reference frame* do not change from the values they

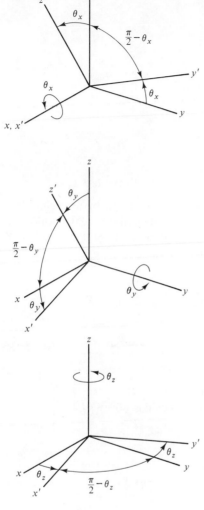

Figure 3.3 Rotations about coordinate axes.

had prior to any rotation. The subsequent coordinates of the point *relative to the fixed reference frame* may then be found by inverting Eq. (3.14), for which the orthonormal property, Eq. (3.11), is suitable.

EXAMPLE 3.1 _____

A force $\overline{F}$ may be described in terms of its components with respect to either the *XYZ* or *xyz* reference frames shown in the sketch.

   (a) If $\overline{F} = 100\overline{I} - 50\overline{J} + 150\overline{K}$ N, determine the components of the force relative to the *xyz* coordinate system.

   (b) If $\overline{F} = 100\overline{i} - 50\overline{j} + 150\overline{k}$ N, determine the components of the force relative to the *XYZ* coordinate system.

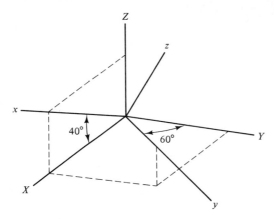

Example 3.1

Solution. We will find the transformation matrix by resolving the unit vectors of *xyz* into components relative to *XYZ*, and then imposing the orthogonality condition. Since the *x* axis lies in the *XZ* plane at an angle of 40° from the *X* axis, and the direction angle between the *Y* and *y* axes is 60°, we have

$$\bar{i} = l_{xX}\bar{I} + l_{xY}\bar{J} + l_{xZ}\bar{K}$$
$$\bar{j} = l_{yX}\bar{I} + l_{yY}\bar{J} + l_{yZ}\bar{K}$$

where we know that

$$l_{xX} = \cos 40° \qquad l_{xY} = 0 \qquad l_{xZ} = \sin 40° \qquad l_{yY} = \cos 60°$$

These expressions indicate that

$$[R] = \begin{bmatrix} \cos 40° & 0 & \sin 40° \\ l_{yX} & \cos 60° & l_{yZ} \\ l_{zX} & l_{zY} & l_{zZ} \end{bmatrix}$$

where

$$\begin{Bmatrix} x \\ y \\ z \end{Bmatrix} = [R] \begin{Bmatrix} X \\ Y \\ Z \end{Bmatrix}$$

Because $[R]$ is an orthogonal transformation, $[R][R]^T = [U]$, which is

$$\begin{bmatrix} .7660 & 0 & 0.6428 \\ l_{yX} & 0.50 & l_{yZ} \\ l_{zX} & l_{zY} & l_{zZ} \end{bmatrix} \begin{bmatrix} 0.7660 & l_{yX} & l_{zX} \\ 0 & 0.50 & l_{zY} \\ 0.6428 & l_{yZ} & l_{zZ} \end{bmatrix} = [U]$$

The corresponding equations derived from each element of the product are

(1,2): $0.7660\, l_{yX} + 0.6428\, l_{yZ} = 0$

(1,3): $0.7660\, l_{zX} + 0.6428\, l_{zZ} = 0$

(2,2): $l_{yX}^2 + 0.25 + l_{yZ}^2 = 1$

(2,3): $l_{yX}\, l_{zX} + 0.5\, l_{zY} + l_{yZ}\, l_{zZ} = 0$

(3,3): $l_{zX}^2 + l_{zY}^2 + l_{zZ}^2 = 1$

We solve Eqs. (1,2) and (2,2) first, and then use that result to determine the other direction cosines.

(1,2): $\qquad l_{yZ} = -1.1918 l_{yX}$

(2,2): $\qquad (1.1918^2 + 1)l_{yX}^2 = 0.75$

$\qquad\qquad l_{yX} = 0.5567 \qquad l_{yZ} = -0.6634$

(1,3): $\qquad l_{zZ} = -1.1918 l_{zX}$

(2,3): $\qquad 0.5567 l_{zX} + 0.50 l_{zY} - 0.6634(-1.1918 l_{zX}) = 0$

$\qquad\qquad l_{zY} = -2.695 l_{zX}$

(3,3): $\qquad (1 + 2.695^2 + 1.1918^2)l_{zX}^2 = 1$

$\qquad\qquad l_{zX} = -0.3218 \qquad l_{zY} = 0.8660 \qquad l_{zZ} = 0.3830$

Note that the signs of l_{yX} and l_{zX} were selected according to whether the given sketch indicates that the angles between the respective axes are acute or obtuse.

Now that $[R]$ is known, we may transform the vectors. In case (a), we know the XYZ components, so

$$\begin{Bmatrix} F_x \\ F_y \\ F_z \end{Bmatrix} = [R] \begin{Bmatrix} 100 \\ -50 \\ 150 \end{Bmatrix} = \begin{Bmatrix} 173.02 \\ -68.84 \\ -18.03 \end{Bmatrix}$$

$$\bar{F} = 173.02\,\bar{i} - 68.84\,\bar{j} - 18.03\,\bar{k} \text{ N}$$

In case (b), we use the inverse transformation because we know the xyz components. Specifically,

$$\begin{Bmatrix} F_X \\ F_Y \\ F_Z \end{Bmatrix} = [R]^{\text{T}} \begin{Bmatrix} 100 \\ -50 \\ 150 \end{Bmatrix} = \begin{Bmatrix} 0.05 \\ 104.90 \\ 154.90 \end{Bmatrix}$$

$$\bar{F} = 0.05\,\bar{i} + 104.90\,\bar{j} + 154.90\,\bar{k} \text{ N}$$

3.2 FINITE ROTATIONS

A *spatial rotation* features rotation about two or more nonparallel axes. The orientation of a reference frame that undergoes such a rotation clearly will depend on the orientation of each axis of rotation and the amount of rotation about each axis. It is less apparent that the final alignment of the reference frame also is dependent on the sequence in which the individual rotations occur. Two situations commonly arise. The conceptually simpler case involves *space-fixed axes*, by which we mean that the individual axes have fixed orientations in space. The contrasting situation is that of *body-fixed axes*. In a body-fixed rotation sequence, each rotation is about one of the axes of the coordinate system at the preceding step in the sequence. For example, a body-fixed sequence θ_y, θ_z, θ_x occurs first about the initial position of the y axis, then about the orientation of the z axis after the first rotation, then finally about the x axis after the second rotation. We shall see that although a body-fixed rotation is more difficult than a space-fixed rotation to describe in words, the transformation matrix for a body-fixed rotation is easier to obtain.

3.2.1 Body-Fixed Rotations

We begin by following a specific sequence of body-fixed rotations. The first rotation occurs about the original orientation of the x axis, and the second rotation occurs about the new orientation of the y axis. After we have derived the transformation for this case, we will generalize the result to an arbitrary sequence of rotations.

As shown in Figure 3.4a, we choose the fixed XYZ system such that it coincides with the initial orientation of xyz. We mark the orientation of xyz after the θ_x rotation as $x_1y_1z_1$. The transformation matrix for a single-axis rotation about the x axis was given in Eq. (3.15a). Adpating that matrix to the current notation leads to

$$\begin{Bmatrix} x_1 \\ y_1 \\ z_1 \end{Bmatrix} = [R_1] \begin{Bmatrix} X \\ Y \\ Z \end{Bmatrix} \qquad [R_1] = [R_x] \tag{3.16}$$

where $[R_1]$ is used to denote the transformation, rather than $[R_x]$, in order to emphasize that it corresponds to the first rotation.

The result of the second rotation is depicted in Figure 3.4b. The θ_y rotation moves xyz from $x_1y_1z_1$ to its final orientation. Since this corresponds to a single-axis rotation

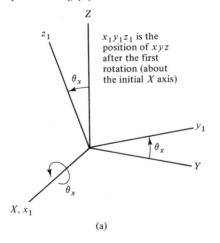

(a)

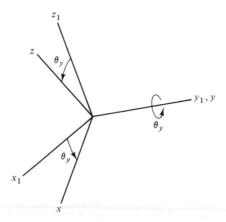

(b)

Figure 3.4 Body-fixed rotations.

about the y_1 axis, we may apply Eq. (3.15b) directly. Thus,

$$\begin{Bmatrix} x \\ y \\ z \end{Bmatrix} = [R_2] \begin{Bmatrix} x_1 \\ y_1 \\ z_1 \end{Bmatrix} = [R_2][R_1] \begin{Bmatrix} X \\ Y \\ Z \end{Bmatrix} \qquad [R_2] = [R_y] \qquad (3.17)$$

As an alternative to Eq. (3.17), the (x,y,z) values could have been expressed directly in terms of the XYZ components by using the overall transformation matrix $[R]$, so that

$$\begin{Bmatrix} x \\ y \\ z \end{Bmatrix} = [R] \begin{Bmatrix} X \\ Y \\ Z \end{Bmatrix} \qquad [R] = [R_2][R_1] \qquad (3.18)$$

The virtue of the notation we have employed comes from the recognition that each transformation $[R_i]$ is the result of the ith rotation. Thus, we may conclude that Eq. (3.18) is valid for any set of body-fixed rotations. In other words,

Let xyz be a reference frame that undergoes a sequence of rotations about its own axes, and let XYZ mark the initial orientation of xyz. The transformation from XYZ to the final xyz components is obtained by premultiplying (from right to left) the sequence of transformation matrices for the individual single-axis rotations. For n rotations:

$$[R] = [R_n] \cdots [R_2][R_1] \qquad (3.19)$$

EXAMPLE 3.2 _____

An xyz coordinate system, which initially coincided with the XYZ coordinate system, first undergoes a rotation $\theta_1 = 65°$ about its y axis, followed by $\theta_2 = -145°$ about its z axis. For this rotation determine:

(a) The coordinates relative to xyz in its final orientation of a stationary point at $X = 2, Y = -3, Z = 4$ m.

(b) The final coordinates relative to XYZ of a point that is situated at $x = 2, y = -3, z = 4$ m relative to the moving reference frame.

Solution. We begin by sketching the orientation of xyz after the first rotation, which we mark as $x_1y_1z_1$.

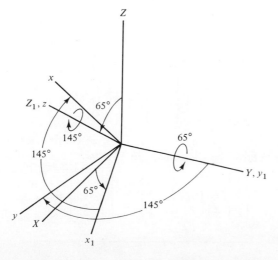

Coordinate systems.

The rotation is about the y axis, so the first transformation is

$$\begin{Bmatrix} x_1 \\ y_1 \\ z_1 \end{Bmatrix} = [R_1] \begin{Bmatrix} X \\ Y \\ Z \end{Bmatrix}$$

$$[R_1] = \begin{bmatrix} \cos\theta_1 & 0 & \cos(\theta_1 + 90°) \\ 0 & 1 & 0 \\ \cos(90° - \theta_1) & 0 & \cos\theta_1 \end{bmatrix}$$

$$= \begin{bmatrix} 0.4226 & 0 & -0.9063 \\ 0 & 1 & 0 \\ 0.9063 & 0 & 0.4226 \end{bmatrix}$$

The second rotation is about the axis marked z_1. The second transformation, for $\theta_2 = -145°$, is

$$\begin{Bmatrix} x \\ y \\ z \end{Bmatrix} = [R_2] \begin{Bmatrix} x_1 \\ y_1 \\ z_1 \end{Bmatrix}$$

$$[R_2] = \begin{bmatrix} \cos\theta_2 & \cos(90° - \theta_2) & 0 \\ \cos(90° + \theta_2) & \cos\theta_2 & 0 \\ 0 & 0 & 1 \end{bmatrix}$$

$$= \begin{bmatrix} -0.8192 & -0.5736 & 0 \\ 0.5736 & -0.8192 & 0 \\ 0 & 0 & 1 \end{bmatrix}$$

The combination of the two transformations is

$$\begin{Bmatrix} x \\ y \\ z \end{Bmatrix} = [R] \begin{Bmatrix} X \\ Y \\ Z \end{Bmatrix}$$

$$[R] = [R_2][R_1] = \begin{bmatrix} -0.3462 & -0.5736 & 0.7424 \\ 0.2424 & -0.8192 & -0.5199 \\ 0.9063 & 0 & 0.4226 \end{bmatrix}$$

In case (a), the coordinates relative to XYZ are known, so we employ $[R]$ directly to obtain

$$\begin{Bmatrix} x \\ y \\ z \end{Bmatrix} = [R] \begin{Bmatrix} 2 \\ -3 \\ 4 \end{Bmatrix} = \begin{Bmatrix} 3.998 \\ 0.863 \\ 3.503 \end{Bmatrix} \text{ m}$$

The situation in case (b) is one in which the final xyz coordinates are known. The inverse of $[R]$ then returns the coordinates to XYZ, so

$$\begin{Bmatrix} X \\ Y \\ Z \end{Bmatrix} = [R]^T \begin{Bmatrix} 2 \\ -3 \\ 4 \end{Bmatrix} = \begin{Bmatrix} 2.206 \\ 1.310 \\ 4.735 \end{Bmatrix} \text{ m}$$

3.2.2 Space-Fixed Rotations

We shall develop here the transformation matrix for a sequence of rotations about axes that are fixed in space. The method will follow a course that parallels the development in the previous section. Thus, we begin by considering a set of rotations θ_x about the fixed X axis, followed by θ_y about the fixed Y axis.

In addition to xyz, which undergoes both rotations, and XYZ, which remains fixed, we introduce two other reference frames. Figure 3.5a shows $x_1y_1z_1$, which is the orientation of xyz after the first rotation, θ_x. This reference frame does not undergo the second rotation. The other reference frame we need is $x_2y_2z_2$, which is defined in Figure 3.5b to be the reference frame that coincided with XYZ before the second rotation.

The transformation from XYZ to $x_1y_1z_1$ is straightforward to obtain, because the X and x_1 axes are coincident. The corresponding transformation is $[R_x]$ in Eq. (3.15a). We therefore have

$$\begin{Bmatrix} x_1 \\ y_1 \\ z_1 \end{Bmatrix} = [R_1] \begin{Bmatrix} X \\ Y \\ Z \end{Bmatrix} \qquad [R_1] = [R_x] \tag{3.20}$$

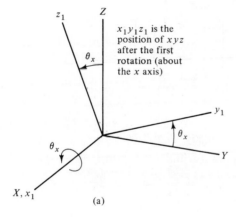

$x_1y_1z_1$ is the position of xyz after the first rotation (about the x axis)

(a)

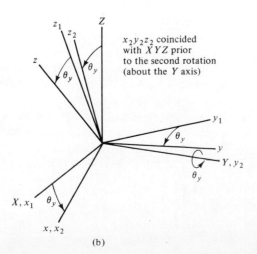

$x_2y_2z_2$ coincided with XYZ prior to the second rotation (about the Y axis)

(b)

Figure 3.5 Space-fixed rotations.

Similarly, the transformation from XYZ to $x_2y_2z_2$ is given by $[R_y]$ in Eq. (3.15b) because the Y and y_2 axes are coincident. Thus, we have

$$\begin{Bmatrix} x_2 \\ y_2 \\ z_2 \end{Bmatrix} = [R_2] \begin{Bmatrix} X \\ Y \\ Z \end{Bmatrix} \qquad [R_2] = [R_y] \qquad (3.21)$$

It still remains to relate the final xyz reference frame to any of the others. This is the reason for the introduction of $x_2y_2z_2$. In the second rotation, xyz goes from $x_1y_1z_1$ to its final orientation, while $x_2y_2z_2$ goes from XYZ to its final position. Since both reference frames undergo the same rotation, their relative position is not altered. Hence, the relation between xyz and $x_2y_2z_2$ is the same as that between $x_1y_1z_1$ and XYZ. It follows from Eq. (3.20) that

$$\begin{Bmatrix} x \\ y \\ z \end{Bmatrix} = [R_1] \begin{Bmatrix} x_2 \\ y_2 \\ z_2 \end{Bmatrix} \qquad (3.22)$$

It is a simple matter to eliminate the intermediate coordinate values by substituting Eq. (3.21) into Eq. (3.22). The result is that

$$\begin{Bmatrix} x \\ y \\ z \end{Bmatrix} = [R_1][R_2] \begin{Bmatrix} X \\ Y \\ Z \end{Bmatrix} = [R] \begin{Bmatrix} X \\ Y \\ Z \end{Bmatrix} \qquad [R] = [R_1][R_2] \qquad (3.23)$$

As we did for body-fixed axes, we may conclude that Eqs. (3.23) are generally valid. Specifically,

Let xyz be a reference frame that undergoes a sequence of rotations about space-fixed axes XYZ with which it initially coincided. The transformation from XYZ to the final xyz components is obtained by postmultiplying (from left to right) the sequence of transformation matrices for the individual single-axis rotations. For n rotations:

$$[R] = [R_1][R_2] \cdots [R_n] \qquad (3.24)$$

The similarity of Eqs. (3.19) and (3.24) is significant. We see that the result of a sequence of body-fixed rotations matches that obtained from the reverse sequence of space-fixed rotations, and vice versa. This similarity is also a source of errors. The single-axis transformations in each case might be identical. However, for a given sequence of rotations, the order of multiplication of the individual matrices must be consistent with the type of rotation: right to left for body-fixed rotations and left to right for space-fixed rotations.

We have seen in this section and the preceding one that the sequence in which individual rotations occur must be considered when the overall transformation is formed. Suppose we were to reverse the sequence in which two space-fixed rotations occur. That would reverse the order of multiplication in Eq. (3.23), such that

$$\begin{Bmatrix} x \\ y \\ z \end{Bmatrix} = [\hat{R}] \begin{Bmatrix} X \\ Y \\ Z \end{Bmatrix} \qquad [\hat{R}] = [R_2][R_1] \qquad (3.25)$$

Because $[R_2][R_1]$ does not, in general, equal $[R_1][R_2]$, the transformation $[\hat{R}]$ will generally be different from the transformation $[R]$ in Eq. (3.23). The same observation would arise if we were to reverse the sequence in which the individual body-fixed rotations forming Eq. (3.19) occur. We must conclude that in either type of rotation,

The final orientation of a coordinate system depends on the sequence in which rotations occur, as well as the magnitude of the individual rotations and the orientation of their respective axes.

An important corollary is that finite spatial rotations cannot be represented as vectors, because vector addition is independent of the order of addition.

EXAMPLE 3.3

Consider Example 3.2 in the case where the rotations of an xyz coordinate system are $\theta_1 = 65°$ about the Y axis, followed by $\theta_2 = -145°$ about the fixed Z axis, where xyz coincides with XYZ prior to any rotation. As was requested in that problem, determine for this set of rotations

(a) The coordinates relative to xyz in its final orientation of a stationary point at $X = 2$, $Y = -3$, $Z = 4$ m.

(b) The final coordinates relative to XYZ of a point that is situated at $x = 2$, $y -3$, $z = 4$ m relative to the moving reference frame.

Solution. The transformation matrices for the individual rotations are the same as in the previous example, but they combine differently because the rotations are about space-fixed axes. Here, the overall transformation results from the sequential product from left to right, that is,

$$\begin{Bmatrix} x \\ y \\ z \end{Bmatrix} = [R] \begin{Bmatrix} X \\ Y \\ Z \end{Bmatrix} \qquad [R] = [R_1][R_2]$$

The values of $[R_1]$ and $[R_2]$ are detailed in the solution to Example 3.2; the corresponding product specified above is

$$[R] = \begin{bmatrix} -0.3462 & -0.2424 & -0.9063 \\ 0.5736 & -0.8192 & 0 \\ -0.7424 & -0.5199 & 0.4226 \end{bmatrix}$$

Direct application of $[R]$ for case (a) yields

$$\begin{Bmatrix} x \\ y \\ z \end{Bmatrix} = [R] \begin{Bmatrix} 2 \\ -3 \\ 4 \end{Bmatrix} = \begin{Bmatrix} -3.590 \\ 3.605 \\ 1.765 \end{Bmatrix}$$

while using $[R]^{-1}$ in case (b) gives

$$\begin{Bmatrix} X \\ Y \\ Z \end{Bmatrix} = [R]^{\mathrm{T}} \begin{Bmatrix} 2 \\ -3 \\ 4 \end{Bmatrix} = \begin{Bmatrix} -5.383 \\ -0.107 \\ -0.122 \end{Bmatrix}$$

EXAMPLE 3.4 ──────────────────────────────────────

A reference frame *xyz*, initially coincident with fixed reference frame *XYZ*, is rotated through an angle β about axis *OA*, counterclockwise when viewed from *A* to *O*. The orientation of this fixed rotation axis is specified by the azimuthal angle ψ and angle of elevation θ relative to the *XY* plane. Point *P*, which is fixed in *xyz*, had coordinates (χ, η, ξ) relative to *XYZ* prior to the rotation. Determine the coordinates of point *P* relative to *XYZ* after the rotation. Describe the resulting transformation as a sequence of single-axis rotations.

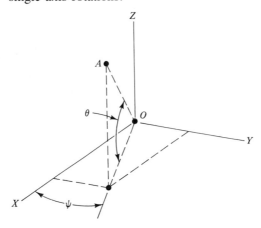

Example 3.4

Solution. Thus far, we have only addressed individual rotations about axes that coincide with a coordinate axis. We will find in the solution that follows that the overall effect may be treated as a sequence of body-fixed rotations. Our primary reason for placing the example in the section on space-fixed axes is that the solution requires the introduction of intermediate coordinate systems, much like the technique that was used to develop Eq. (3.24).

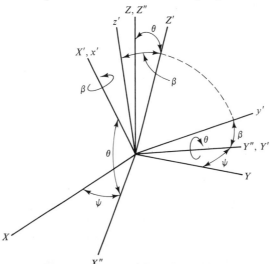

Coordinate systems.

In addition to XYZ, we will need another fixed reference frame $X'Y'Z'$ whose X' axis coincides with the rotation axis OA. We wish to describe the overall result as a sequence of single-axis rotations. Correspondingly, we introduce the intermediate fixed reference frame $X''Y''Z''$, which is rotated relative to XYZ by the angle ψ about the Z axis. Then $X'Y'Z'$ is rotated relative to $X''Y''Z''$ by the angle of inclination θ about the *negative Y''* axis. The transformations between these coordinate axes are

$$\begin{Bmatrix} X'' \\ Y'' \\ Z'' \end{Bmatrix} = [R_\psi] \begin{Bmatrix} X \\ Y \\ Z \end{Bmatrix} \qquad \begin{Bmatrix} X' \\ Y' \\ Z' \end{Bmatrix} = [R_\theta] \begin{Bmatrix} X'' \\ Y'' \\ Z'' \end{Bmatrix}$$

The individual transformations, which may be constructed directly or obtained from Eqs. (3.15), are

$$[R_\psi] = \begin{bmatrix} \cos\psi & \sin\psi & 0 \\ -\sin\psi & \cos\psi & 0 \\ 0 & 0 & 1 \end{bmatrix} \qquad [R_\theta] = \begin{bmatrix} \cos\theta & 0 & \sin\theta \\ 0 & 1 & 0 \\ -\sin\theta & 0 & \cos\theta \end{bmatrix}$$

The combined transformation is therefore

$$\begin{Bmatrix} X' \\ Y' \\ Z' \end{Bmatrix} = [R_\theta][R_\psi] \begin{Bmatrix} X \\ Y \\ Z \end{Bmatrix} \tag{1}$$

Thus, the $X'Y'Z'$ coordinates of the arbitrary point P prior to any rotation are given by

$$\begin{Bmatrix} X'_0 \\ Y'_0 \\ Z'_0 \end{Bmatrix} = [R_\theta][R_\psi] \begin{Bmatrix} \chi \\ \eta \\ \xi \end{Bmatrix} \tag{2}$$

In order to describe the rotation about axis OA, we introduce another reference frame, $x'y'z'$, which coincided with $X'Y'Z'$ prior to the rotation β about the positive X' axis. The transformation from $X'Y'Z'$ to $x'y'z'$ is

$$\begin{Bmatrix} x' \\ y' \\ z' \end{Bmatrix} = [R_\beta] \begin{Bmatrix} X' \\ Y' \\ Z' \end{Bmatrix} \qquad [R_\beta] = \begin{bmatrix} 1 & 0 & 0 \\ 0 & \cos\beta & \sin\beta \\ 0 & -\sin\beta & \cos\beta \end{bmatrix} \tag{3}$$

Since point P is fixed relative to any reference frame that undergoes the rotation about axis OA, the coordinates of point P relative to $x'y'z'$ remain (X'_0, Y'_0, Z'_0). The corresponding coordinates of point P relative to $X'Y'Z'$ after the β rotation may be found by using the inverse of the transformation in Eq. (3), which gives

$$\begin{Bmatrix} X'_f \\ Y'_f \\ Z'_f \end{Bmatrix} = [R_\beta]^{\mathrm{T}} \begin{Bmatrix} X'_0 \\ Y'_0 \\ Z'_0 \end{Bmatrix}$$

Substituting Eq. (2) into this relation yields

$$\begin{Bmatrix} X'_f \\ Y'_f \\ Z'_f \end{Bmatrix} = [R_\beta]^{\mathrm{T}} [R_\theta][R_\psi] \begin{Bmatrix} \chi \\ \eta \\ \xi \end{Bmatrix}$$

The above describes the final coordinates of point P relative to the fixed reference frame $X'Y'Z'$. We may obtain the corresponding coordinates relative to XYZ by applying the inverse of the transformation in Eq. (1), which leads to

$$\begin{Bmatrix} X_f \\ Y_f \\ Z_f \end{Bmatrix} = [R_\psi]^{\mathrm{T}} [R_\theta]^{\mathrm{T}} \begin{Bmatrix} X'_f \\ Y'_f \\ Z'_f \end{Bmatrix} = [R_\psi]^{\mathrm{T}} [R_\theta]^{\mathrm{T}} [R_\beta]^{\mathrm{T}} [R_\theta] [R_\psi] \begin{Bmatrix} \chi \\ \eta \\ \xi \end{Bmatrix}$$

An interesting interpretation of this result comes from recognizing that the coordinates are equivalent to arbitrary vector components relative to the respective reference frames. Let $[R]$ denote the transformation from XYZ to xyz. The inverse property then shows that

$$[R] = [[R_\psi]^{\mathrm{T}} [R_\theta]^{\mathrm{T}} [R_\beta]^{\mathrm{T}} [R_\theta] [R_\psi]]^{\mathrm{T}}$$

$$= [R_\psi]^{\mathrm{T}} [R_\theta]^{\mathrm{T}} [R_\beta] [R_\theta] [R_\psi]$$

Recall that the inverse of a rotation about a coordinate axis is a rotation by the same amount in the opposite sense. Therefore, this expression for $[R]$ suggests that xyz may be rotated away from XYZ by a sequence of rotations. An appropriate sequence of either the space-fixed or body-fixed type may be constructed. For example, we may follow Eq. (3.19) to represent the result in terms of body-fixed rotations. Proceeding from right to left, the products in $[R]$ indicate that to obtain xyz starting from XYZ,

1. Rotate through an angle ψ about the z axis: $[R_\psi]$.
2. Rotate through an angle $-\theta$ about the y axis: $[R_\theta]$.
3. Rotate through an angle β about the x axis: $[R_\beta]$.
4. Rotate through an angle θ about the y axis: $[R_\theta]^{\mathrm{T}}$.
5. Rotate through an angle $-\psi$ about the x axis: $[R_\psi]^{\mathrm{T}}$.

From a conceptual viewpoint, we perform the first two rotations in order to bring a coordinate axis into coincidence with the axis of rotation, and then apply the rotation about the specified axis. Finally, we perform the opposite of the first two rotations in order to return to the reference frame that started at the original orientation.

3.3 ANGULAR VELOCITY AND DERIVATIVES OF ROTATING VECTORS

Discrepancies between spatial rotations that differ only in their sequence become less significant as the magnitude of each rotation decreases. Indeed, the limiting case of infinitesimal rotations yields a result that is independent of the sequence. In order to demonstrate this fact, let us evaluate the transformation matrix associated with a set of space-fixed rotations $d\theta_X$, $d\theta_Y$, $d\theta_Z$, in that order, about the fixed reference frame axes.

The transformation matrices for the individual rotations are given by Eqs. (3.15). Second-order differential quantities are negligible compared with first-order terms, so we set $\cos d\theta = 1$ and $\sin d\theta = d\theta$. The limiting forms of the individual transformation matrices are therefore

$$[R_1] = \begin{bmatrix} 1 & 0 & 0 \\ 0 & 1 & d\theta_X \\ 0 & -d\theta_X & 1 \end{bmatrix} \quad [R_2] = \begin{bmatrix} 1 & 0 & -d\theta_Y \\ 0 & 1 & 0 \\ d\theta_Y & 0 & 1 \end{bmatrix}$$

$$[R_3] = \begin{bmatrix} 1 & d\theta_Z & 0 \\ -d\theta_Z & 1 & 0 \\ 0 & 0 & 1 \end{bmatrix} \tag{3.26}$$

Second-order differentials are also negligible when these matrices are multiplied. The resulting transformation is found from Eqs. (3.24) to be

$$\begin{Bmatrix} x \\ y \\ z \end{Bmatrix} = [R_1] [R_2] [R_3] \begin{Bmatrix} X \\ Y \\ Z \end{Bmatrix}$$

$$= \begin{bmatrix} 1 & d\theta_Z & -d\theta_Y \\ -d\theta_Z & 1 & d\theta_X \\ d\theta_Y & -d\theta_X & 1 \end{bmatrix} \begin{Bmatrix} X \\ Y \\ Z \end{Bmatrix} \tag{3.27}$$

The foregoing is the general transformation between the fixed and moving coordinate systems. Let us apply it to the situation where we wish to follow an arbitrary point P whose coordinates relative to the fixed reference frame prior to any rotation are known. Denote these initial coordinates as (X_0, Y_0, Z_0). If point P is fixed relative to the moving reference frame, then these are always the coordinates of point P relative to xyz. Let (X_f, Y_f, Z_f) be the fixed reference frame coordinates after the rotation. Solving Eq. (3.27) with the aid of the orthonormal property yields

$$\begin{Bmatrix} X_f \\ Y_f \\ Z_f \end{Bmatrix} = \begin{bmatrix} 1 & -d\theta_Z & d\theta_Y \\ d\theta_Z & 1 & -d\theta_X \\ -d\theta_Y & d\theta_X & 1 \end{bmatrix} \begin{Bmatrix} X_0 \\ Y_0 \\ Z_0 \end{Bmatrix} \tag{3.28}$$

Consider changing the order in which $[R_1]^T$, $[R_2]^T$, and $[R_3]^T$ are multiplied to form Eq. (3.28). The final result will be the same because second-order differentials are negligible with respect to first-order differentials. Since the right to left order of multiplication in Eq. (3.24) matches the rotation sequence, we may conclude that

The final orientation of a coordinate system is unaffected by the sequence in which a set of infinitesimal rotations are performed.

An important corollary follows from the fact that the transformations for space-fixed and body-fixed axes only differ by the sequence in which the individual rotation transformations are multiplied. Consequently, we observe from the foregoing statement that the same transformation is obtained if a set of infinitesimal rotations are imparted about body-fixed or space-fixed axes.

Equation (3.28) relates the final coordinates of a point to the initial values. The differential change in the (fixed frame) coordinates is found to be

$$\begin{Bmatrix} dX \\ dY \\ dZ \end{Bmatrix} = \begin{Bmatrix} X_f - X_0 \\ Y_f - Y_0 \\ Z_f - Z_0 \end{Bmatrix} = \begin{bmatrix} 0 & -d\theta_Z & d\theta_Y \\ d\theta_Z & 0 & -d\theta_X \\ -d\theta_Y & d\theta_X & 0 \end{bmatrix} \begin{Bmatrix} X_0 \\ Y_0 \\ Z_0 \end{Bmatrix} \quad (3.29)$$

The increments described by Eq. (3.29) are the components of the infinitesimal displacement $d\bar{r}$. The vector form of $d\bar{r}$ is

$$\begin{aligned} d\bar{r} &= dX\bar{I} + dY\bar{J} + dZ\bar{K} \\ &= (d\theta_Y Z_0 - d\theta_Z Y_0)\bar{I} + (d\theta_Z X_0 - d\theta_X Z_0)\bar{J} + (d\theta_X Y_0 - d\theta_Y X_0)\bar{K} \end{aligned}$$
$$(3.30)$$

A simpler representation of the foregoing is obtained by recognizing that it is the cross product of a differential rotation vector $\overline{d\theta}$ and the relative position $\bar{r}_{P/O'}$. Define

$$\overline{d\theta} = d\theta_X\bar{I} + d\theta_Y\bar{J} + d\theta_Z\bar{K} \quad (3.31)$$

and describe the position vector in terms of components with respect to the fixed reference frame as

$$\bar{r}_{P/O'} = X_0\bar{I} + Y_0\bar{J} + Z_0\bar{K} \quad (3.32)$$

Then the resulting expression for the infinitesimal displacement is

$$d\bar{r} = \overline{d\theta} \times \bar{r}_{P/O'} \quad (3.33)$$

Several aspects of Eqs. (3.31) and (3.33) are noteworthy. The individual rotations were defined in Figure 3.5a and b to be positive according to the right-hand rule. Thus the component representation of $\overline{d\theta}$ in Eq. (3.31) is equivalent to a vectorial superposition of the infinitesimal rotations, as was anticipated earlier. The overbar is placed above the entire symbol $\overline{d\theta}$ in order to emphasize that there is no finite rotation vector from which the differential is formed. A principal advantage of Eq. (3.33) over Eq. (3.29) is that the vector form does not rely on a coordinate system to represent components. Specifically, Eq. (3.33) remains valid if its vectors are represented in terms of components relative to the axes of the moving reference frame. (We often shall use such a description.)

Finally, observe that the position vector and its differential occur in Eq. (3.33) only because the derivation began by considering the position coordinates. Suppose we had considered an arbitrary vector $\bar{A}$ that is attached to a moving reference frame, so that its components relative to that frame are constant. The result would have been an expression for the infinitesimal change in $\bar{A}$, that is,

$$d\bar{A} = \overline{d\theta} \times \bar{A} \quad (3.34)$$

There is a simple explanation for this relation. Figure 3.6a shows a typical vector $\bar{A}$ before and after an infinitesimal rotation $\overline{d\theta}$. The direction of the rotation vector is parallel to the axis of rotation, and the angle of rotation is $|\overline{d\theta}|$. The change in $\bar{A}$ is found as the difference between the new and previous vectors. This difference is depicted in Figure 3.6b, where the tails of the vectors have been brought to the axis represented by $\overline{d\theta}$.

The sketch shows that only the component of $\bar{A}$ perpendicular to the axis changes; call this component $\bar{A}_\perp$. The line in Figure 3.6b representing $\bar{A}_\perp$ rotates through the

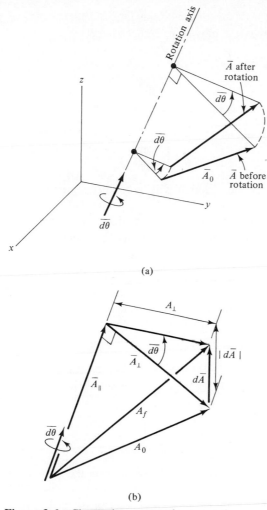

(a)

(b)

Figure 3.6   Change in a vector due to rotation.

angle $|d\theta|$. Hence, the arc that represents dA has a length $A_\perp |d\theta|$, and the direction of $d\overline{A}$ is perpendicular to both $\overline{A}$ and the $\overline{d\theta}$ vector. The magnitude of a cross product is defined to be the product of the magnitude of one vector and the perpendicular component of the other vector, and the sense of the product is perpendicular to the individual vectors. It follows that the pictorial analysis fully agrees with Eq. (3.34). In other words, the change in $\overline{A}$ results from the movement of its tip perpendicular to the plane formed by $\overline{d\theta}$ and $\overline{A}$, in the sense of the rotation. We will often recall this interpretation when we treat the change of a vector due to a rotation.

It was mentioned earlier that the properties of a differential change in a variable are much like those for the rate of change of that variable. Hence, we define the angular velocity to be

$$\overline{\omega} = \frac{\overline{d\theta}}{dt} = \dot{\theta}_X \overline{I} + \dot{\theta}_Y \overline{J} + \dot{\theta}_Z \overline{K} \qquad (3.35)$$

Similarly, dividing Eq. (3.34) by dt yields the rate of change of any vector having constant components relative to a moving reference frame.

$$\dot{\overline{A}} = \overline{\omega} \times \overline{A} \qquad (3.36)$$

This theorem may be interpreted in words as:

Let $\overline{A}$ be a vector whose components relative to a moving reference frame are constant, and let $\overline{\omega}$ be the angular velocity of that reference frame. Then the rate of change of $\overline{A}$ is the cross product of $\overline{\omega}$ and $\overline{A}$.

Our primary application for this theorem will be to differentiate the unit vectors of a moving reference frame.

Another generalization results from the fact that $\overline{\omega}$, like $\overline{d\theta}$, is a vector quantity. This permits us to represent the angular velocity as the vector sum of rotations about arbitrary axes, rather than only the axes of the fixed reference frame. Specifically,

Let xyz be a coordinate system that is undergoing a spatial rotation. Let $\overline{e}_i$ be a unit vector parallel to the ith axis of rotation in the sense of the rotation according to the right-hand rule, and let ω_i be the corresponding rate of rotation in radians per second. Then the angular velocity of xyz is given by

$$\overline{\omega} = \sum_i \omega_i \overline{e}_i \qquad (3.37)$$

Equations (3.36) and (3.37) are powerful tools that we shall employ frequently in our further studies of kinematics, and they will also play a vital role for the kinetics of rigid bodies.

EXAMPLE 3.5 _____

The disk is rotating about shaft AB at 3600 rev/min as the system rotates about the vertical axis at 20 rad/s. Determine the angular velocity of the disk. Then determine the approximate displacement of point C on the perimeter of the disk 2 μs after the instant depicted in the sketch.

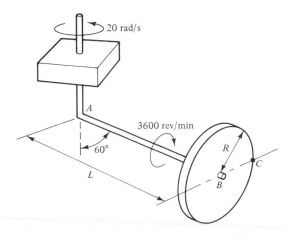

Example 3.5

Solution. In Chapter 4, we will study Eulerian angles, which will enable us to determine precisely the position of points resulting from rotations about various axes. However, the time interval in the present case is very small. As a result, we may approximate increments over the interval as differentials. This leads to

$$\overline{\Delta\theta} \approx \overline{\omega} \, \Delta t \qquad \Delta\overline{r}_C \approx \overline{\Delta\theta} \times \overline{r}_{C/A}$$

where the reference point for the position is selected as point A because that is the point that is fixed in the rotation.

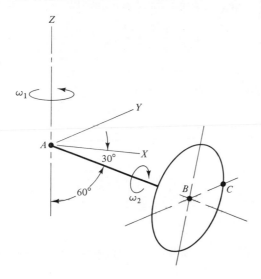

Coordinate system.

We obtain the angular velocity by vectorially adding the rotation rates. Thus,

$$\overline{\omega} = \omega_1 \overline{K} + \omega_2 \overline{e}_{A/B}$$

where, according to the right-hand rule, the sense of the rotation ω_2 about shaft AB is directed from point B to point A. For the instant under consideration, the resolution of $\overline{e}_{A/B}$ into components relative to the fixed reference frame is

$$\overline{e}_{A/B} = -\sin 60°\overline{I} + \cos 60°\overline{K}$$

Substituting ω_1, ω_2, and $\overline{e}_{A/B}$ into the expression for $\overline{\omega}$ yields

$$\overline{\omega} = 20\overline{K} + 3600\left(\frac{2\pi}{60}\right)(-0.866\overline{I} + 0.50\overline{K})$$

$$= -326.5\overline{I} + 208.5\overline{K} \text{ rad/s}$$

At this instant, the position is

$$\overline{r}_{C/A} = L(\sin 60°\overline{I} - \cos 60°\overline{K}) + R\overline{J}$$

Thus, for $\Delta t = 2(10^{-6})$ s, we find

$$\Delta\overline{r}_{C/A} \approx (-326.5\overline{I} + 208.5\overline{K}) \times (0.866L\overline{I} + R\overline{J} - 0.50L\overline{K})[2(10^{-6})]$$

$$\Delta\overline{r}_{C/A} = (-4.170\,R\overline{I} + 0.346L\overline{J} - 6.530R\overline{K}) \times 10^{-4}$$

EXAMPLE 3.6

The Frenet formulas give the derivatives of the path variable unit vectors with respect to the arclength s along an arbitrary curve. Since $\dot{s} = v$, the chain rule may be used to determine the time rates of change of the unit vectors. Furthermore, since these unit vectors form an orthonormal set, they define a moving reference frame. Use the results for the time derivatives of the unit vectors to determine the angular velocity of the $\bar{e}_t$-$\bar{e}_n$-$\bar{e}_b$ reference frame in terms of the path variable parameters.

Solution. It is useful to begin by recalling the Frenet formulas, which are

$$\frac{d\bar{e}_t}{ds} = \frac{1}{\rho} \bar{e}_n \qquad \frac{d\bar{e}_n}{ds} = -\frac{1}{\rho} \bar{e}_t + \frac{1}{\tau} \bar{e}_b \qquad \frac{d\bar{e}_b}{ds} = -\frac{1}{\tau} \bar{e}_n$$

In order to convert these to time derivatives, we observe that if $\bar{e}$ is a vector that depends on the arclength s locating a point, and $s = s(t)$, then

$$\dot{\bar{e}} = \frac{d\bar{e}}{ds} \dot{s} = v \frac{d\bar{e}}{ds}$$

Hence, we have

$$\dot{\bar{e}}_t = v \frac{d\bar{e}_t}{ds} = \frac{v}{\rho} \bar{e}_n \qquad \dot{\bar{e}}_n = v \frac{d\bar{e}_n}{ds} = -\frac{v}{\rho} \bar{e}_t + \frac{v}{\tau} \bar{e}_b$$

$$\dot{\bar{e}}_b = v \frac{d\bar{e}_b}{ds} = -\frac{v}{\tau} \bar{e}_n$$

Let $\bar{\omega}$ be the angular velocity of $\bar{e}_t$-$\bar{e}_n$-$\bar{e}_b$, which may be written in component form as

$$\bar{\omega} = \omega_t \bar{e}_t + \omega_n \bar{e}_n + \omega_b \bar{e}_b$$

In general, $\dot{\bar{e}} = \bar{\omega} \times \bar{e}$, so that

$$\dot{\bar{e}}_t = \bar{\omega} \times \bar{e}_t = \omega_b \bar{e}_n - \omega_n \bar{e}_b = \frac{v}{\rho} \bar{e}_n$$

$$\dot{\bar{e}}_n = \bar{\omega} \times \bar{e}_n = -\omega_b \bar{e}_t + \omega_t \bar{e}_b = -\frac{v}{\rho} \bar{e}_t + \frac{v}{\tau} \bar{e}_b$$

$$\dot{\bar{e}}_b = \bar{\omega} \times \bar{e}_b = \omega_n \bar{e}_t - \omega_b \bar{e}_n = -\frac{v}{\tau} \bar{e}_n$$

Matching like components in each equation leads to

$$\omega_t = \frac{v}{\tau} \qquad \omega_n = 0 \qquad \omega_b = \frac{v}{\rho} \qquad \bar{\omega} = \frac{v}{\tau} \bar{e}_t + \frac{v}{\rho} \bar{e}_b$$

3.4 ANGULAR ACCELERATION

One of the primary properties of the motion of a reference frame is its angular acceleration, which is defined as the time derivative of the angular velocity. It is conventional to denote the angular acceleration as $\bar{\alpha}$, so

$$\bar{\alpha} = \dot{\bar{\omega}} \tag{3.38}$$

We saw in the preceding section that $\bar{\omega}$ is the sum of rotation rates about various axes. Even if the rotation rates are constants, there will be an angular acceleration whenever any of the axes do not have a fixed orientation.

Let us consider the manner in which $\bar{\omega}$ may be differentiated. The only general statement regarding $\bar{\omega}$ is Eq. (3.37), which calls for a summation of rotation-rate vectors that are formed according to the right-hand rule. Each unit vector $\bar{e}_i$ has been defined to be aligned with an axis of rotation. In order to expedite the description of each $\bar{e}_i$, we:

Define a moving reference frame $x_i y_i z_i$ for each rotation, such that one of the axes of $x_i y_i z_i$ always coincides with that rotation axis. Hence, one of the axes of $x_i y_i z_i$ coincides with $\bar{e}_i$. Let $\overline{\Omega}_i$ be the angular velocity of $x_i y_i z_i$.

Note that each of these reference frames may be, but is not necessarily, the same as the xyz frame whose angular acceleration is being evaluated. By definition, the unit vector $\bar{e}_i$ is fixed relative to $x_i y_i z_i$. Hence, the derivative of $\bar{e}_i$ is known from Eq. (3.36) to be $\overline{\Omega}_i \times \bar{e}_i$. The rules for differentiating a sum and a product then lead to

$$\bar{\alpha} = \sum_i [\dot{\omega}_i \bar{e}_i + \omega_i (\overline{\Omega}_i \times \bar{e}_i)] \tag{3.39}$$

After Eq. (3.39) has been formed using the unit vectors of the various reference frames, all terms in $\bar{\omega}$ and $\bar{\alpha}$ should be transformed to a common coordinate system. The principles governing the kinetics of a rigid body that will be derived in Chapter 5 require that the components of $\bar{\omega}$ and $\bar{\alpha}$ be expressed relative to a body-fixed reference frame. Hence, we shall employ the same representation here.

The task of finding the angular acceleration is a key aspect of evaluations based on relative-motion concepts. Hence, it is imperative to understand the meaning of the terms in Eqs. (3.37) and (3.39). In particular, each $\overline{\Omega}_i$ corresponds to the angular velocity of a reference frame in which the ith axis of rotation seems to be fixed. The term in Eq. (3.39) containing $\overline{\Omega}_i \times \bar{e}_i$ represents the effect of changing the sense of the ith axis of rotation; the effect is perpendicular to that axis. In contrast, the term $\dot{\omega}_i \bar{e}_i$ is parallel to the ith axis. It arises whenever the rotation rate is changed from its current value. Note that a rotation about a fixed axis, such as that in a planar motion, only produces the latter effect. It is for this reason that intuitive judgments obtained from experiences with planar motion are often incorrect.

EXAMPLE 3.7 ───

Determine the angular acceleration of the disk in Example 3.5.

Solution. The procedure here differs from the solution to Example 3.5 owing to the need to define reference frames associated with each rotation. As shown in the sketch, the Z axis of the fixed XYZ frame coincides with the vertical axis of rotation, so $\bar{e}_1 = \bar{K}$. For the other rotation, we attach the xyz reference frame to the flywheel.

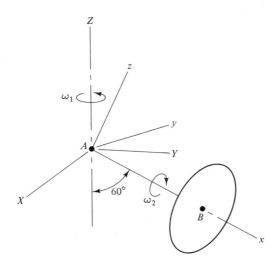

Coordinate systems.

(The location of the origin of xyz is unimportant for these operations.) We align the x axis with shaft AB, so that $\bar{e}_2 = \bar{i}$ throughout the motion.

The corresponding general description of the angular velocity is

$$\bar{\omega} = \omega_1 \bar{K} - \omega_2 \bar{i}$$

where the sign for the ω_2 rotation is a consequence of the right-hand rule. Both rotation rates are constant. Since $\bar{K}$ is constant, $\dot{\bar{K}} = \bar{0}$. In contrast, the angular velocity of xyz is $\bar{\omega}$, so $\dot{\bar{i}} = \bar{\omega} \times \bar{i}$. The derivative of the general expression for $\bar{\omega}$ therefore is

$$\bar{\alpha} = -\omega_2(\bar{\omega} \times \bar{i})$$

Employing $\bar{\omega}$ and $\bar{\alpha}$ in subsequent computations would require their resolution into components relative to a single coordinate system. As was mentioned in the discussion following Eq. (3.39), it is standard practice to employ the axes of the reference frame associated with $\bar{\omega}$, that is, the axes of xyz. We could consider xyz to have undergone an arbitrary rotation about shaft AB. However, it is equally valid, and much more convenient, to define xyz such that, at the instant of interest, its z axis lies in the vertical plane formed by the two rotation axes. This is the orientation depicted in the sketch. Thus,

$$\bar{K} = -\cos 60°\bar{i} + \sin 60°\bar{k}$$

An interesting corollary of the foregoing procedure is that the orientation of the fixed-reference-frame axes is unimportant if they are not rotation axes.

Substituting for $\bar{K}$, ω_1, and ω_2 in the general expressions yields

$$\bar{\omega} = 20(-0.50\bar{i} + 0.866\bar{k}) - 120\pi\bar{i}$$
$$= -387.0\bar{i} + 17.32\bar{k} \text{ rad/s}$$
$$\bar{\alpha} = -(120\pi)(-387.0\bar{i} + 17.32\bar{k}) \times \bar{i}$$
$$= -6530\bar{j} \text{ rad/s}^2$$

A simple verification of this result for $\bar{\alpha}$ is the observation that, as the system rotates about the fixed vertical axis, the tip of the angular velocity term $-\omega_2 \bar{i}$ moves in the negative y direction. This agrees with the direction of the computed value of $\bar{\alpha}$.

EXAMPLE 3.8 _____

The gyroscopic turn indicator consists of a flywheel that spins about its axis of symmetry at the constant rate ω_1, as the assembly rotates about the fixed horizontal shaft at the variable rate ω_2. The angle β locating the axis of the flywheel relative to the horizontal shaft is an arbitrary function of time. Determine the angular acceleration of the flywheel at an arbitrary instant.

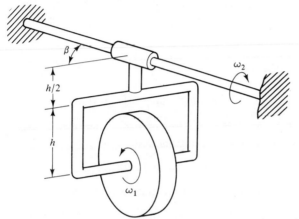

Example 3.8

Solution. We define xyz to be attached to the flywheel, so that the angular motion of xyz is identical to that of the flywheel. The ω_2 rotation is about the fixed horizontal shaft. Correspondingly, we define the fixed reference frame such that the unit vector parallel to that rotation axis is $\bar{e}_2 = -\bar{I}$. (There is no need to specify the other fixed axes, because we will resolve all terms into components relative to the moving reference frame.) The ω_1 rotation is about the axis of the flywheel. Since this direction is fixed relative to the flywheel, it is logical to define xyz such that the unit vector along this rotation axis is $\bar{e}_1 = \bar{i}$. The third rotation is about an axis that is always perpendicular to the horizontal shaft. In order to describe it, we attach an $x'y'z'$ reference frame to the gimbal supporting the flywheel. We define the z' axis to coincide with the $\dot{\beta}$ rotation, so $\bar{e}_3 = \bar{k}'$. (Note that we defined the x' axis to coincide with the axis of the flywheel, so we could have alternatively defined $\bar{e}_1 = \bar{i}'$. However, it would be incorrect to assign $\bar{e}_3$ to $\bar{k}$, because the z axis rotates relative to the gimbal.)

The angular velocity of the flywheel is the sum of the individual effects, so

$$\bar{\omega} = -\omega_2 \bar{I} + \omega_1 \bar{i} + \dot{\beta} \bar{k}'$$

We associated $\bar{e}_1$ with xyz, and $\bar{e}_2$ with XYZ. The corresponding angular velocities to be used in differentiating the unit vectors of those reference frames are

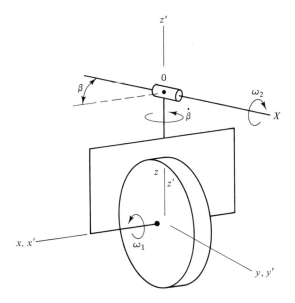

Coordinate systems.

$$\overline{\Omega}_1 = \overline{\omega} \quad \text{and} \quad \overline{\Omega}_2 = \overline{0}$$

The angular motion of $x'y'z'$ is like that of xyz, except that it does not include the spinning rotation ω_1. Hence, the angular velocity of $x'y'z'$ is

$$\overline{\Omega}_3 = -\omega_2\overline{I} + \dot{\beta}\overline{k}'$$

Using these angular velocities to differentiate the general expression for $\overline{\omega}$ yields

$$\overline{\alpha} = -\dot{\omega}_2\overline{I} + \ddot{\beta}\overline{k}' + \omega_1(\overline{\omega} \times \overline{i}) + \dot{\beta}(\overline{\Omega}_3 \times \overline{k}')$$

where we have set $\dot{\omega}_1 = 0$, as specified.

Both $\overline{\omega}$ and $\overline{\alpha}$ must be represented in terms of a single set of directions, for which we select xyz. The resolution is simplified significantly if we define xyz such that the z and z' axes coincide at the instant of interest—this is the configuration in the sketch. Then,

$$\overline{I} = -\cos\beta\overline{i} + \sin\beta\overline{j} \qquad \overline{k}' = \overline{k}$$

Substitution of these unit vectors into the earlier relations yields

$$\overline{\omega} = (\omega_1 + \omega_2 \cos\beta)\overline{i} - \omega_2 \sin\beta\overline{j} + \dot{\beta}\overline{k}$$

$$\overline{\Omega}_3 = \omega_2 \cos\beta\overline{i} - \omega_2 \sin\beta\overline{j} + \dot{\beta}\overline{k}$$

$$\overline{\alpha} = -\dot{\omega}_2(-\cos\beta\overline{i} + \sin\beta\overline{j}) + \ddot{\beta}\overline{k}$$
$$\qquad + \omega_1[(\omega_1 + \omega_2 \cos\beta)\overline{i} - \omega_2 \sin\beta\overline{j} + \dot{\beta}\overline{k}] \times \overline{i}$$
$$\qquad + \dot{\beta}[\omega_2 \cos\beta\overline{i} - \omega_2 \sin\beta\overline{j} + \dot{\beta}\overline{k}] \times \overline{k}$$

$$\overline{\alpha} = (\dot{\omega}_2 \cos\beta - \omega_2 \dot{\beta} \sin\beta)\overline{i} + (-\dot{\omega}_2 \sin\beta + \omega_1\dot{\beta} - \omega_2\dot{\beta} \cos\beta)\overline{j}$$
$$\qquad\qquad\qquad\qquad\qquad\qquad + (\ddot{\beta} + \omega_1\omega_2 \sin\beta)\overline{k}$$

3.5 DERIVATIVE OF AN ARBITRARY VECTOR

We saw in Section 3.3 that the derivative of a vector having constant components relative to a moving reference frame is determined by the angular velocity of that reference frame. Then in Section 3.4 we treated a much more general situation in which the vector to be differentiated (i.e., $\overline{\omega}$), had variable components. In that case the individual contributions were described in terms of unit vectors $\overline{e}_i$ that were oriented arbitrarily. These unit vectors were not necessarily associated with the same reference frame. Here, we shall address the conventional representation of a vector quantity, in which all unit vectors are associated with the axes of the (moving) xyz reference frame.

Let $\overline{\omega}$ denote the angular velocity of xyz, and let $\overline{A}$ be a vector whose components A_x, A_y, A_z relative to the moving reference frame are functions of time. The component representation of $\overline{A}$ is

$$\overline{A} = A_x \overline{i} + A_y \overline{j} + A_z \overline{k} \tag{3.40}$$

The unit vectors, as well as the components, in this expression are functions of time. Equation (3.34) describes the derivatives of $\overline{i}$, $\overline{j}$, and $\overline{k}$. The angular velocity to be used for the differentiation in each case is $\overline{\omega}$ because the unit vectors are those for the moving axes. The rules for the derivative of a sum and a product therefore yield

$$\dot{\overline{A}} = \dot{A}_x \overline{i} + \dot{A}_y \overline{j} + \dot{A}_z \overline{k} + A_x(\overline{\omega} \times \overline{i}) + A_y(\overline{\omega} \times \overline{j}) + A_z(\overline{\omega} \times \overline{k})$$

$$\dot{\overline{A}} = \frac{\delta \overline{A}}{\delta t} + \overline{\omega} \times \overline{A} \tag{3.41}$$

where $\delta \overline{A}/\delta t$ describes the rate of change of $\overline{A}$ due to the time dependency of the components.

$$\frac{\delta \overline{A}}{\delta t} = \dot{A}_x \overline{i} + \dot{A}_y \overline{j} + \dot{A}_z \overline{k} \tag{3.42}$$

These relations have a simple explanation. Suppose that you were to observe any vector quantity, such as a position $\overline{r}$, while you were situated on a moving reference frame. If you were to describe the position in terms of coordinates measured relative to the axes of the moving frame, the term $\delta \overline{r}/\delta t$ is the only effect you would observe. In general, $\delta \overline{A}/\delta t$ is a time derivative as seen from a moving reference frame. We know from the previous section that the term $\overline{\omega} \times \overline{A}$ gives the portion of the change of $\overline{A}$ that is attributable to rotation of the reference frame. Hence, Eq. (3.41) is merely a statement that these two effects superpose, as you might have expected.

3.6 VELOCITY AND ACCELERATION USING A MOVING REFERENCE FRAME

Equations (3.41) and (3.42) will be employed in a variety of ways. Here we shall use those expressions to derive formulas that relate observations of velocity and acceleration

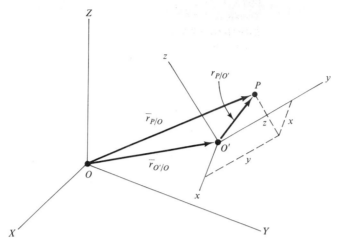

Figure 3.7   Position relative to a moving reference frame.

relative to moving and fixed reference frames. The general situation confronting us is depicted in Figure 3.7, which is the same as the diagram that introduced this chapter. We have already seen that the observations of the position of point P from the fixed and moving reference frame are related by

$$\bar{r}_{P/O} = \bar{r}_{O'/O} + \bar{r}_{P/O'} \tag{3.43}$$

By definition, the absolute velocity is the time derivative of the position with respect to the fixed reference frame. Hence, differentiating Eq. (3.43) yields

$$\bar{v}_P = \bar{v}_{O'} + \frac{d}{dt}\bar{r}_{P/O'} \tag{3.44}$$

Presumably the (moving) xyz reference frame has been chosen for its convenience in describing the position of point P. We therefore describe $\bar{r}_{P/O'}$ in terms of the coordinates of point P with respect to the axes of this reference frame. Thus,

$$\bar{r}_{P/O'} = x\bar{i} + y\bar{j} + z\bar{k} \tag{3.45}$$

Differentiating Eq. (3.45) is not difficult, because it matches the situation addressed in the previous section. We apply Eq. (3.41), which superposes the effects of the changing coordinate values and the rotation of the vectors. This gives

$$\frac{d}{dt}\bar{r}_{P/O'} = \dot{x}\bar{i} + \dot{y}\bar{j} + \dot{z}\bar{k} + \bar{\omega} \times \bar{r}_{P/O'} \tag{3.46}$$

Suppose you were an observer moving with the xyz reference frame. You would only see the (x,y,z) coordinates change. Thus, the first three terms on the right side of Eq. (3.46) describe the velocity of point P as seen from the moving reference frame. Let $(\bar{v}_P)_{xyz}$ denote this *relative velocity*, where the subscript P details the point under consideration and the trailing group of subscripts describes the reference frame from which the motion is viewed. Equations (3.44) and (3.46) may be written in this notation as

$$\bar{v}_P = \bar{v}_{O'} + (\bar{v}_P)_{xyz} + \bar{\omega} + \bar{r}_{P/O'} \tag{3.47}$$

where the relative velocity is

$$(\bar{v}_P)_{xyz} = \frac{\delta}{\delta t} \bar{r}_{P/O'} = \dot{x}\bar{i} + \dot{y}\bar{j} + \dot{z}\bar{k} \tag{3.48}$$

Note that it is a matter of convenience to omit the specification of the reference frame when indicating an absolute velocity (and acceleration).

Equation (3.47) confirms a superposition of effects that could have been anticipated. Since $\bar{\omega} \times \bar{r}_{P/O'}$ is perpendicular to the plane formed by $\bar{\omega}$ and $\bar{r}_{P/O'}$, the rotation of the reference frame contributes a transverse velocity. This term combines with the velocity of the origin O', and the relative velocity, to form $\bar{v}_P$.

Our intuition is not as correct when it is applied to acceleration. A relation for the acceleration of a point is obtained by differentiating Eq. (3.47). The derivative of the velocity of origin O' is its acceleration. Because Eq. (3.48) gives the relative velocity in terms of its components relative to the moving frame, we employ Eq. (3.41) to differentiate $(\bar{v}_P)_{xyz}$.

$$\frac{d}{dt} (\bar{v}_P)_{xyz} = (\bar{a}_P)_{xyz} + \bar{\omega} \times (\bar{v}_P)_{xyz} \tag{3.49}$$

where the *relative acceleration* is

$$(\bar{a}_P)_{xyz} = \frac{\delta}{\delta t} (\bar{v}_P)_{xyz} = \ddot{x}\bar{i} + \ddot{y}\bar{j} + \ddot{z}\bar{k} \tag{3.50}$$

The third term in Eq. (3.47) is the product $\bar{\omega} \times \bar{r}_{P/O'}$. Equation (3.46) gives the derivative of $\bar{r}_{P/O'}$. Since the angular acceleration is the derivative of the angular velocity, the total derivative of the third term is

$$\frac{d}{dt} (\bar{\omega} \times \bar{r}_{P/O'}) = \bar{\alpha} \times \bar{r}_{P/O'} + \bar{\omega} \times [(\bar{v}_P)_{xyz} + \bar{\omega} \times \bar{r}_{P/O'}] \tag{3.51}$$

The resulting acceleration relation is thereby found to be

$$\bar{a}_P = \bar{a}_{O'} + (\bar{a}_P)_{xyz} + \bar{\alpha} \times \bar{r}_{P/O'} + \bar{\omega} \times (\bar{\omega} \times \bar{r}_{P/O'}) + 2\bar{\omega} \times (\bar{v}_P)_{xyz} \tag{3.52}$$

The term in the foregoing preceded by the 2 factor is called the *Coriolis acceleration*. In order to understand this term, consider the second derivative of a product, specifically

$$\frac{d^2}{dt^2} (uv) = \ddot{u}\,v + 2\,\dot{u}\,\dot{v} + u\,\ddot{v}$$

Since acceleration is the second derivative of position, the occurrence of a factor 2 should not be surprising. For further insight, recall the analysis in Chapter 2 of curvilinear coordinates for particle motion. We saw there that the Coriolis acceleration originates from two distinct effects. The same is true here, since Eqs. (3.49) and (3.51) both contributed to the overall effect. The term in Eq. (3.49) is associated with the change in the direction of the relative velocity resulting from rotation of the reference frame. In contrast, the term in Eq. (3.51) corresponds to the change in the velocity component $\bar{\omega} \times \bar{r}_{P/O'}$ that results from changing the components of $\bar{r}_{P/O'}$. Comparable effects were found for curvilinear coordinates.

Experienced dynamicists recognize Coriolis acceleration as a combination of two explainable effects resulting from movement relative to a rotating reference frame. It is, to a certain extent, a misnomer to use a single name to describe the corresponding term in Eq. (3.52).

The other terms in Eq. (3.52) could have been predicted in advance. The additive nature of the accelerations of the origin O' of the xyz reference frame and of point P relative to xyz requires no explanation. The term $\bar{\alpha} \times \bar{r}_{P/O'}$ is the angular acceleration contribution, analogous to the velocity term $\bar{\omega} \times \bar{r}_{P/O'}$ resulting from an angular velocity. In spatial motion, the angular acceleration is usually not parallel to the angular velocity. As a result, the direction of the corresponding acceleration might occasionally be different from your expectation.

Now consider the term $\bar{\omega} \times (\bar{\omega} \times \bar{r}_{P/O'})$. Figure 3.8 shows the construction of this acceleration. The magnitude of $(\bar{\omega} \times \bar{r}_{P/O'})$ is $r_{\perp} |\bar{\omega}|$, and the corresponding direction is perpendicular to the plane formed by $\bar{\omega}$ and $\bar{r}_{P/O'}$. Then $\bar{\omega} \times (\bar{\omega} \times \bar{r}_{P/O'})$ is perpendicular to the rotation axis, pointing inward; its magnitude is $|\bar{\omega}|^2 r_{\perp}$. In other words, this term describes the centripetal acceleration that would be found if the reference frame was rotating at rate $|\bar{\omega}|$ about an axis that intersects the origin O' and is parallel to $\bar{\omega}$.

One aspect of the relative velocity and relative acceleration greatly facilitates their evaluation. These terms may be regarded as the effects that would be present if the reference frame was held stationary but the relative motion was unchanged. They were described in Eqs. (3.48) and (3.50), respectively, in terms of a Cartesian coordinate

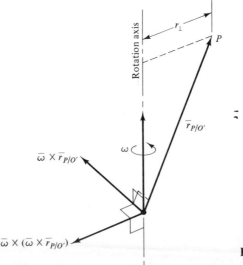

Figure 3.8 Centripetal acceleration.

description. However, it is apparent that other kinematical descriptions, such as path variables and curvilinear coordinates, might be more appropriate in some situations. If such an approach is employed, it will be necessary to convert the corresponding unit vectors to the set of components selected for the other vectors in the velocity and acceleration relations.

It is instructive to close this discussion by considering two special cases. The situation where the *xyz* frame translates corresponds to $\bar{\omega}$ identically equal to zero. Hence, $\bar{\alpha}$ also is zero. The relative motion equations then reduce to

Translating reference frame *xyz*:

$$\bar{v}_P = \bar{v}_{O'} + (\bar{v}_P)_{xyz}$$
$$\bar{a}_P = \bar{a}_{O'} + (\bar{a}_P)_{xyz}$$

$$(3.53)$$

The motion of the origin and of the point relative to the moving reference frame are additive—there are no corrections for direction changes due to rotation.

Let us use Eqs. (3.53) to reexamine fixed reference frames. Suppose that *xyz*, as well as *XYZ*, is fixed. Then the above relations show that the velocity and acceleration are the same, regardless of which reference frame is selected. This verifies the earlier statement that there is no need to indicate the origin of the reference frame in the notation for velocity and acceleration.

Another interesting situation arises if the reference frame is translating at a constant velocity. This means that $\bar{a}_{O'} = \bar{O}$. The second of Eqs. (3.53) shows that the accelerations viewed from the fixed and moving references are identical. (Note that origin O' must be following a straight path in order for it to have no acceleration.) A reference frame having this type of motion is said to be an *inertial* or *Galilean reference frame*. The terminology arises from the fact that the absolute acceleration is observable from the reference frame, so the frame may be employed to formulate Newton's laws.

The second special case arises when point *P* is fixed with respect to the moving reference frame. This is the situation treated in Section 3.2 to study position changes in finite rotation. The velocity and acceleration relations simplify substantially, because $(\bar{v}_P)_{xyz}$ and $(\bar{a}_P)_{xyz}$ are both identically zero. Then

Fixed position relative to *xyz*:

$$\bar{v}_P = \bar{v}_{O'} + \bar{\omega} \times \bar{r}_{P/O'}$$
$$\bar{a}_P = \bar{a}_{O'} + \bar{\alpha} \times \bar{r}_{P/O'} + \bar{\omega} \times (\bar{\omega} \times \bar{r}_{P/O'})$$

$$(3.54)$$

A primary reason for highlighting this situation is that it is descriptive of the motion of a rigid body. If the reference frame is attached to the body, then the position vectors between points in the body have constant components relative to the moving reference frame. Also, the angular motions of the body and of the reference frame are synonymous in this case. The motion of rigid bodies is the focus of the next chapter.

EXAMPLE 3.9

Bar *BC* is pinned to the T-bar, which is rotating about the vertical axis at the constant rate Ω. The angle θ is an arbitrary function of time. Determine the velocity and acceleration of point *C* using an *xyz* reference frame that is attached to the T-bar with its *x* axis aligned with segment *AB*.

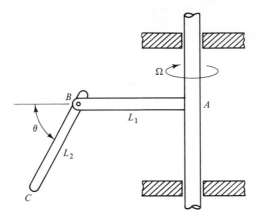

Example 3.9

Solution.   We place the origin of *xyz* at joint *A* in order to avoid evaluating the motion of the origin. The given *xyz* reference frame only rotates about the vertical axis, which is fixed, so we designate $\bar{e}_1 = -\overline{K} = -\bar{k}$. Hence, the motion of the reference frame is

$$\bar{v}_A = \bar{a}_A = \overline{0} \qquad \bar{\omega} = -\Omega\overline{K} = -\Omega\bar{k} \qquad \bar{\alpha} = \overline{0}$$

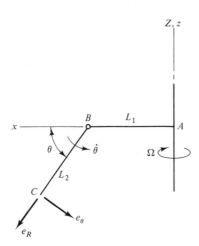

Coordinate systems.

Relative to *xyz*, point *C* follows a circular path centered at point *B*. Using the polar coordinate defined in the sketch to represent the relative motion yields

$$(\bar{v}_C)_{xyz} = L_2\,\dot{\theta}\bar{e}_\theta$$

$$(\bar{a}_C)_{xyz} = -L_2\,\dot{\theta}^2\bar{e}_R + L_2\,\ddot{\theta}\bar{e}_\theta$$

We now transform these expressions to *xyz* components in order to employ them in the relative-motion equation. First, we form

$$\bar{e}_R = \cos\theta\bar{i} - \sin\theta\bar{k}$$

$$\bar{e}_\theta = -\sin\theta\bar{i} - \cos\theta\bar{k}$$

Substitution of these unit vectors into the relative velocity and acceleration yields

$$(\bar{v}_C)_{xyz} = L_2 \,\dot{\theta}\,(-\sin\theta\,\bar{i} - \cos\theta\,\bar{k})$$

$$(\bar{a}_C)_{xyz} = L_2\,[(-\dot{\theta}^2\cos\theta - \ddot{\theta}\sin\theta)\bar{i} + (\dot{\theta}^2\sin\theta - \ddot{\theta}\cos\theta)\bar{k}]$$

Also, the vector from the origin of xyz to point C is

$$\bar{r}_{C/A} = (L_1 + L_2\cos\theta)\bar{i} - L_2\sin\theta\,\bar{k}$$

We are now ready to form the absolute motion. The result of substituting the individual terms into Eq. (3.47) is

$$\bar{v}_C = \bar{v}_A + (\bar{v}_C)_{xyz} + \bar{\omega} \times \bar{r}_{C/A}$$

$$= -L_2\,\dot{\theta}\sin\theta\,\bar{i} - (L_1 + L_2\cos\theta)\,\Omega\bar{j} - L_2\,\dot{\theta}\cos\theta\,\bar{k}$$

Similar steps for Eq. (3.52) yield

$$\bar{a}_C = \bar{a}_A + (\bar{a}_C)_{xyz} + \bar{\alpha}\times\bar{r}_{C/A} + \bar{\omega}\times(\bar{\omega}\times\bar{r}_{C/A}) + 2\bar{\omega}\times(\bar{v}_C)_{xyz}$$

$$= -[L_2\,\ddot{\theta}\sin\theta + L_2\,\dot{\theta}^2\cos\theta + (L_1 + L_2\cos\theta)\,\Omega^2]\bar{i}$$

$$+ 2L_2\Omega\,\dot{\theta}\sin\theta\,\bar{j} - L_2(\ddot{\theta}\cos\theta - \dot{\theta}^2\sin\theta)\bar{k}$$

EXAMPLE 3.10

Determine the acceleration of point C in Example 3.9 using an xyz reference frame that is attached to bar BC with its x axis aligned with that bar.

Solution. Since xyz must be fixed to bar BC, we cannot place the origin of xyz at the stationary point A. Point B is suitable as this origin, because we know that it follows a circular path centered at point A. The radial direction for this motion is $\bar{e}_R$ in the sketch, and we define the azimuthal direction to be inward.

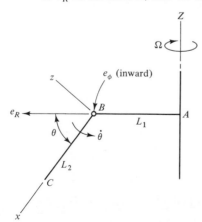

Coordinate systems.

Hence, the motion of the origin is

$$\bar{v}_B = L_1\Omega\bar{e}_\phi \qquad \bar{a}_B = -L_1\Omega^2\bar{e}_R$$

We are required to align the x axis with bar BC, and it is convenient to select the z axis to lie in the vertical plane, as shown. One rotation of xyz occurs about the vertical axis at rate Ω, so $\bar{e}_1 = -\bar{K}$. The other rotation is $\dot{\theta}$ about the axis of the pin, outward as viewed in the sketch according to the right-hand rule. The y axis always coincides with this axis of rotation, and y must be outward in the sketch because xyz must be a right-handed set of coordinates. Hence, we set $\bar{e}_2 = \bar{j}$, which leads to

$$\bar{\omega} = -\Omega\bar{K} + \dot{\theta}\bar{j}$$

Differentiating this expression in accordance with the general relation, Eq. (3.39), gives, for constant Ω,

$$\bar{\alpha} = \ddot{\theta}\bar{j} + \dot{\theta}(\bar{\omega} \times \bar{j})$$

One asset of fixing xyz to bar BC is that point C is stationary relative to xyz, so

$$(\bar{v}_C)_{xyz} = (\bar{a}_C)_{xyz} = \bar{0}$$

As the last step prior to forming Eqs. (3.47) and (3.52), we transform all vectors to components relative to xyz. The various unit vectors are found by referring to our sketch, which gives

$$\bar{e}_R = \cos\theta\bar{i} + \sin\theta\bar{k} \qquad \bar{e}_\phi = -\bar{j} \qquad \bar{K} = -\sin\theta\bar{i} + \cos\theta\bar{k}$$

which then leads to

$$\bar{v}_B = -L_1\Omega\bar{j} \qquad \bar{a}_B = -L_1\Omega^2(\cos\theta\bar{i} + \sin\theta\bar{k})$$
$$\bar{\omega} = \Omega\sin\theta\bar{i} + \dot{\theta}\bar{j} - \Omega\cos\theta\bar{k}$$
$$\bar{\alpha} = \Omega\dot{\theta}\cos\theta\bar{i} + \ddot{\theta}\bar{j} + \Omega\dot{\theta}\sin\theta\bar{k}$$

A verification of the correctness of this expression is that the two components containing the product $\Omega\dot{\theta}$ form a vector in the direction of $\bar{e}_R$. This is the direction in which the tip of the angular velocity $\dot{\theta}\bar{j}$ moves owing to the rotation about the vertical axis.

The velocity and acceleration relations, Eqs. (3.47) and (3.52), now give, for $\bar{r}_{C/B} = L_2\bar{i}$,

$$\bar{v}_C = \bar{v}_B + \bar{\omega} \times \bar{r}_{C/B}$$
$$= -(L_1 + L_2\cos\theta)\Omega\bar{j} - L_2\dot{\theta}\bar{k}$$
$$\bar{a}_C = \bar{a}_B + \bar{\alpha} \times \bar{r}_{C/B} + \bar{\omega} \times (\bar{\omega} \times \bar{r}_{C/B})$$
$$= -[L_2\dot{\theta}^2 + (L_1 + L_2\cos\theta)\Omega^2\cos\theta]\bar{i}$$
$$+ 2L_2\Omega\dot{\theta}\sin\theta\bar{j} - [L_2\ddot{\theta} + (L_1 + L_2\cos\theta)\Omega^2\sin\theta]\bar{k}$$

If we were to transform the present components to those used in Example 3.9, we would find that the results represent identical vectors.

EXAMPLE 3.11 _____

Disk B rotates at 900 rev/min relative to the turntable, which is rotating about a fixed axis at a constant rate of 300 rev/min. Determine the acceleration of point C on the perimeter of the disk at the instant shown using

(a) A moving coordinate system that is attached to the turntable

(b) A moving coordinate system that is attached to the disk

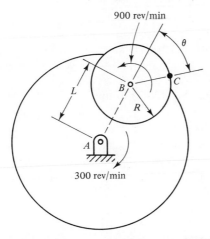

Example 3.11

Solution. This is a case of planar motion, so we orient the z axis normal to the plane in both formulations. The reference frame $x_1y_1z_1$ in the first case is attached to the turntable, so center point A is a convenient origin. Placing the x axis along line AB leads to the following description of the motion of $x_1y_1z_1$:

$$\bar{v}_A = \bar{a}_A = \bar{0} \qquad \bar{\omega} = 300\left(\frac{2\pi}{60}\right)\bar{k} \qquad \bar{\alpha} = \bar{0}$$

We could use polar coordinates to formulate the relative velocity and acceleration. For the sake of variety, we shall employ Eqs. (3.54). We visualize the

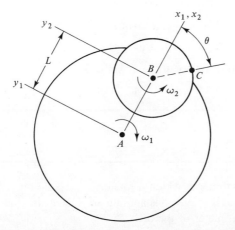

Coordinate systems.

motion that would remain if the turntable were stationary. The angular motion of the disk relative to the turntable is then seen to be

$$\overline{\omega}_{rel} = -900 \left(\frac{2\pi}{60}\right) \overline{k} \text{ rad/s} \qquad \overline{\alpha}_{rel} = \overline{0}$$

Since point C is fixed to the disk and point B is fixed relative to xyz, we find

$$(\overline{v}_C)_{xyz} = \overline{\omega}_{rel} \times \overline{r}_{C/B} = (-30\pi\overline{k}) \times (R\cos\theta\overline{i} - R\sin\theta\overline{j})$$
$$= -30\pi R (\sin\theta\overline{i} + \cos\theta\overline{j})$$
$$(\overline{a}_C)_{xyz} = \overline{\omega}_{rel} \times (\overline{\omega}_{rel} \times \overline{r}_{C/B}) = -900\pi^2(R\cos\theta\overline{i} - R\sin\theta\overline{j})$$

We substitute these quantities into Eq. (3.47) to find

$$\overline{v}_C = \overline{v}_A + (\overline{v}_C)_{xyz} + \overline{\omega} \times \overline{r}_{C/A}$$
$$= -30\pi R(\sin\theta\overline{i} + \cos\theta\overline{j}) + 10\pi\overline{k} \times [(L + R\cos\theta)\overline{i} - R\sin\theta\overline{j}]$$
$$= -20\pi R \sin\theta\overline{i} + 10\pi(L - 2R\cos\theta)\overline{j}$$
$$\overline{a}_C = \overline{a}_A + (\overline{a}_C)_{xyz} + \overline{\alpha} \times \overline{r}_{C/A} + \overline{\omega} \times (\overline{\omega} \times \overline{r}_{C/A}) + 2\overline{\omega} \times (\overline{v}_C)_{xyz}$$
$$= -100\,\pi^2(L + 4R\cos\theta)\overline{i} + 400\,\pi^2 R \sin\theta\overline{j}$$

A formulation based on fixing reference frame $x_2y_2z_2$ to the disk is considerably easier. We place the origin of $x_2y_2z_2$ at center point B, because it is the only point on the disk that follows a simple path. The rotation of $x_2y_2z_2$ consists of a superposition of $\omega_1 = 10\pi$ rad/s in the sense of $\overline{e}_1 = \overline{k}$, and $\omega_2 = 30\pi$ rad/s in the sense of $\overline{e}_2 = -\overline{k}$. Thus,

$$\overline{\omega} = \omega_1\overline{k} - \omega_2\overline{k} = -20\pi\overline{k} \text{ rad/s} \qquad \overline{\alpha} = \overline{0}$$
$$\overline{v}_B = 10\pi L\overline{j} \qquad \overline{a}_B = -100\pi^2 L\overline{i}$$

There is no relative motion to evaluate in this case, so we have

$$\overline{v}_C = \overline{v}_B + \overline{\omega} \times \overline{r}_{C/B} = 10\pi L\overline{j} + (-20\pi\overline{k}) \times (R\cos\theta\overline{i} - R\sin\theta\overline{j})$$
$$= -20\pi R \sin\theta\overline{i} + 10\pi(L - 2R\cos\theta)\overline{j}$$
$$\overline{a}_C = \overline{a}_B + \overline{\omega} \times (\overline{\omega} \times \overline{r}_{C/B})$$
$$= -100\pi^2 L\overline{i} - 400\pi^2(R\cos\theta\overline{i} - R\sin\theta\overline{j})$$
$$= -100\pi^2(L + 4R\cos\theta)\overline{i} + 400\,\pi^2 R \sin\theta\overline{j}$$

EXAMPLE 3.12 ——

The disk is mounted on a bent shaft about which it rotates at variable rate ω_1, while that shaft rotates about the horizontal axis at the constant rate ω_2. Determine the velocity and acceleration of point D on the perimeter of the disk.

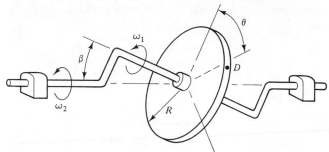

Example 3.12

Solution. The center C is a convenient origin. We select the xyz reference frame to be attached to the disk, in order to eliminate the relative motion,

$$(\bar{v}_D)_{xyz} = (\bar{a}_D)_{xyz} = \bar{0}$$

In the sketch, the x axis has been aligned with the bent shaft, so that $\bar{e}_1 = -\bar{i}$ for the ω_1 rotation. The ω_2 rotation is described by $\bar{e}_2 = -\bar{I}$ for the fixed horizontal axis, so it is convenient to define the y axis such that the x, y, and x axes are coplanar.

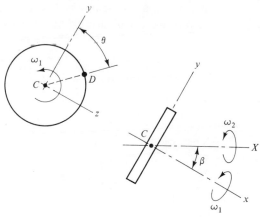

Coordinate systems.

Since ω_2 is constant, the motion of xyz is described by

$$\bar{v}_C = \bar{a}_C = \bar{0}$$

$$\bar{\omega} = -\omega_1\bar{i} - \omega_2\bar{I} \qquad \bar{\alpha} = -\dot{\omega}_1\bar{i} - \omega_1(\bar{\omega} \times \bar{i})$$

In view of the definition of the y axis, the fixed rotation axis is

$$\bar{I} = \cos\beta\bar{i} + \sin\beta\bar{j}$$

so

$$\bar{\omega} = -(\omega_1 + \omega_2\cos\beta)\bar{i} - \omega_2\sin\beta\bar{j}$$

$$\bar{\alpha} = -\dot{\omega}_1\bar{i} - \omega_1\omega_2\sin\beta\bar{k}$$

As noted earlier, there is no relative motion and the origin is fixed, so Eqs. (3.54) yield

$$\bar{v}_D = \bar{\omega} \times \bar{r}_{D/C} = \bar{\omega} \times [R \cos \theta \bar{j} + R \sin \theta \bar{k}]$$
$$= -\omega_2 R \sin \beta \sin \theta \bar{i} + (\omega_1 + \omega_2 \cos \beta) R \sin \theta \bar{j}$$
$$- (\omega_1 + \omega_2 \cos \beta) R \cos \theta \bar{k}$$

$$\bar{a}_D = \bar{\alpha} \times \bar{r}_{D/C} + \bar{\omega} \times (\bar{\omega} \times \bar{r}_{D/C})$$
$$= \omega_2(2\omega_1 + \omega_2 \cos \beta) R \sin \beta \cos \theta \bar{i}$$
$$+ [\dot{\omega}_1 R \sin \theta - (\omega_1 + \omega_2 \cos \beta)^2 R \cos \theta] \bar{j}$$
$$- [\dot{\omega}_1 R \cos \theta + (\omega_1^2 + \omega_2^2 + 2\omega_1 \omega_2 \cos \beta) R \sin \theta] \bar{k}$$

EXAMPLE 3.13

Let ω_x, ω_y, and ω_z denote the pitch, roll, and yaw rates, respectively, of a ship about xyz axes that are attached to the ship with the orientations shown. All these rotation rates are variable quantities. Consider an elevator car whose path perpendicularly intersects the centerline at a distance L forward from the center of mass G. At the instant when the elevator car passes the centerline, it has a constant speed u relative to the ship. Determine the velocity and acceleration of the car relative to the center point G at this instant.

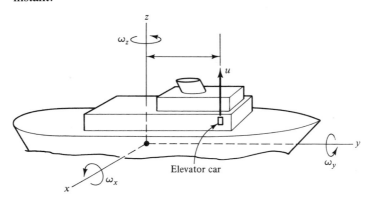

Example 3.13

Solution. The elevator follows a straight path relative to the ship, so it is convenient to attach xyz to the ship. The given rotations are about body-fixed axes, so we have $\bar{e}_1 = \bar{i}$, $\bar{e}_2 = \bar{j}$, and $\bar{e}_3 = \bar{k}$, corresponding to the rates ω_x, ω_y, and ω_z, respectively. Thus, the rotation of xyz is

$$\bar{\omega} = \omega_x \bar{i} + \omega_y \bar{j} + \omega_z \bar{k}$$
$$\bar{\alpha} = \dot{\omega}_x \bar{i} + \dot{\omega}_y \bar{j} + \dot{\omega}_z \bar{k} + \omega_x(\bar{\omega} \times \bar{i}) + \omega_y(\bar{\omega} \times \bar{j}) + \omega_z(\bar{\omega} \times \bar{k})$$
$$= \dot{\omega}_x \bar{i} + \dot{\omega}_y \bar{j} + \dot{\omega}_z \bar{k} + \bar{\omega} \times (\omega_x \bar{i} + \omega_y \bar{j} + \omega_z \bar{k})$$
$$= \dot{\omega}_x \bar{i} + \dot{\omega}_y \bar{j} + \dot{\omega}_z \bar{k}$$

It should be noted that these expressions for $\bar{\omega}$ and $\bar{\alpha}$ are generally true. They indicate that the angular-acceleration components are always the derivatives of the angular-velocity components, provided that all components are relative to body-fixed axes. This observation is a key aspect to the development of kinetics principles in Chapter 5.

The motion of the center of mass is not specified, so let $\bar{v}_B$ and $\bar{a}_B$ represent the velocity and acceleration of the origin. The motion of the elevator relative to the ship is defined by

$$(\bar{v}_E)_{xyz} = u\bar{k} \qquad (\bar{a}_E)_{xyz} = \bar{0}$$

and the position of this instant is $\bar{r}_{A/E} = L\bar{j}$. Correspondingly, we find

$$\bar{v}_E = \bar{v}_G + (\bar{v}_E)_{xyz} + \bar{\omega} \times \bar{r}_{E/G}$$

$$= \bar{v}_G - \omega_z L\bar{i} + (u + \omega_x L)\bar{k}$$

$$\bar{a}_E = \bar{a}_G + (\bar{a}_E)_{xyx} + \bar{\alpha} \times \bar{r}_{E/G} + \bar{\omega} \times (\bar{\omega} \times \bar{r}_{E/G}) + 2\bar{\omega} \times (\bar{v}_E)_{xyz}$$

$$= \bar{a}_G + (-\dot{\omega}_z L + \omega_x \omega_y L + 2\omega_y u)\bar{i}$$

$$+ (-\omega_z^2 L - \omega_x^2 L - 2\omega_x u)\bar{j} + (\dot{\omega}_x L + \omega_y \omega_z L)\bar{k}$$

3.7 OBSERVATIONS FROM A MOVING REFERENCE SYSTEM

The treatment in the previous section implicitly assumed that the motion of some point could be more readily described in terms of a moving reference frame rather than a fixed one. Such is not always the case, since there are situations where the absolute motion is known and the relative motion must be evaluated. For example, it might be necessary to ensure that one part of a machine merges with another part in a smooth manner, as in the case of gears. The influence of the earth's motion on the dynamic behavior of a system is an important case where aspects of the absolute motion are known.

One approach is to interchange the absolute and relative reference frames, based on the fact that the kinematical relationships do not actually require that one of the reference frames be stationary. Thus, in this viewpoint, if the angular velocity of xyz relative to XYZ is $\bar{\omega}$, then the angular velocity of XYZ as viewed from xyz is $-\bar{\omega}$. The difficulty with this approach is that it is prone to errors, particularly in signs, because of the need to change the observer's viewpoint for the formulation. The simpler approach, which does not require redefinitions of the basic quantities, manipulates the earlier relations.

The basic concept is quite straightforward. When the absolute velocity $\bar{v}_P$ and absolute acceleration $\bar{a}_P$ are known, Eqs. (3.47) and (3.52) may be solved for the relative-motion parameters. Specifically,

$$(\bar{v}_P)_{xyz} = \bar{v}_P - \bar{v}_{O'} - \bar{\omega} \times \bar{r}_{P/O'} \tag{3.55}$$

$$(\bar{a}_P)_{xyz} = \bar{a}_P - \bar{a}_{O'} - \bar{\alpha} \times \bar{r}_{P/O'} - \bar{\omega} \times (\bar{\omega} \times \bar{r}_{P/O'}) \tag{3.56}$$

$$- 2\bar{\omega} \times (\bar{v}_P)_{xyz}$$

If it is appropriate, the relative velocity may be removed from the acceleration relation by substitution of Eq. (3.55). The result is

$$(\bar{a}_P)_{xyz} = \bar{a}_P - \bar{a}_{O'} - \bar{\alpha} \times \bar{r}_{P/O'} + \bar{\omega} \times (\bar{\omega} \times \bar{r}_{P/O'}) - 2\bar{\omega} \times (\bar{v}_P - \bar{v}_{O'})$$

(3.57)

The steps required to apply these relations are like those already established, since the angular velocity and angular acceleration still describe the rotation of the moving reference frame as seen from the fixed one.

One of the most common applications of these relations is to situations where the rotation of the earth must be considered. Our observations are usually in terms of earth-based instruments. However, Newton's second law relates the forces acting on a body to the motion relative to some hypothetical inertial reference frame, that is, $\bar{a}_P = \Sigma \bar{F}/m$ for a particle.

Consider an observer at point O' on the earth's surface. A natural definition for the reference frame employed by this observer is east-west and north-south for position along the surface, and vertical for measurements off the surface. Such a reference frame is depicted in Figure 3.9, where the $\bar{i}$ vector is northward and the $\bar{j}$ vector is westward. The observation point O' in the figure is located by the latitude angle λ measured from the equator and the longitude angle ϕ measured from some reference location, such as the prime meridian (the longitude of Greenwich, England).

For the present purposes, it is adequate to employ an approximate model of the earth. The earth rotates about its polar axis at $\omega_e = 2\pi$ rad/23.934 h = $7.292(10^{-5})$ rad/s. The orbital rate of rotation of the earth about the sun, ω_0, is smaller by an approximate factor of 365, since one such revolution requires a full year. The centripetal acceleration of a point at the equator due to the spin about the polar axis is $\omega_e^2 R_e$, where R_e is the earth's diameter, $R_e \approx 6370$ km. For comparison, the centripetal acceleration due to the orbital motion is $\omega_0^2 R_0$, where the mean orbital radius is $R_0 \approx 149.6(10^6)$ km.

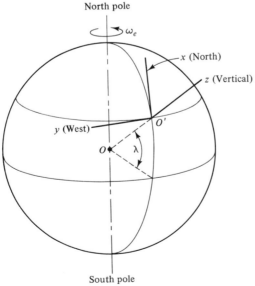

Figure 3.9   Reference frame fixed to the earth.

Since $\omega_0^2 R_0 \approx 0.176\omega_e^2 R_e$, and $\omega_e^2 R_e$ is itself quite feeble (≈ 0.034 m/s^2), it is reasonable to consider the center of the earth to be stationary. Furthermore, if the relatively minor wobble of the polar axis is neglected, the model of the earth reduces to a sphere that rotates about the polar axis at the constant rate ω_e. The corresponding expression for the acceleration relative to the earth is therefore

$$(\bar{a}_P)_{xyz} = \frac{\Sigma \bar{F}_a}{m} + \frac{\bar{F}_g}{m} - \bar{a}_{O'} - \bar{\omega}_e \times (\bar{\omega}_e \times \bar{r}_{P/O'}) - 2\bar{\omega}_e \times (\bar{v}_P)_{xyz} \quad (3.58)$$

where the term $\bar{F}_g$ represents the gravitational force acting on the body and $\Sigma \bar{F}_a$ represents the applied loads and reactions.

Now consider a particle near the earth's surface at the instant after it has begun to fall freely. Let the point O' be close to the particle, so that $\bar{r}_{P/O'} \approx \bar{0}$. If air resistance is negligible, there are no applied forces, $\Sigma \bar{F}_a = \bar{0}$. By definition, the acceleration observed from the earth is g vertically downward. Recall that the definition of the z axis in Figure 3.9 was that it is the upward vertical, which now means that the observed free-fall acceleration is $-g\bar{k}$. Then Eq. (3.58) gives

$$-g\bar{k} = \frac{\bar{F}_g}{m} - \bar{a}_{O'} \quad (3.59)$$

This relation may be quantified by using the inverse-square law for gravity, as well as the rigid-body acceleration equation to describe $\bar{a}_{O'}$. The result is

$$g\bar{k} = \frac{G M_e}{R_e^3}\bar{r}_{O'/O} + \bar{\omega} \times (\bar{\omega} \times \bar{r}_{O'/O}) \quad (3.60)$$

There are two primary aspects of interest in this relation. The centripetal acceleration term is not parallel to the position $\bar{r}_{O'/O'}$ except at the poles, where $\bar{\omega}$ is parallel to $\bar{r}_{O'/O'}$ and at the equator, where $\bar{\omega}$ is perpendicular to $\bar{r}_{O'/O}$. Hence, the $\bar{k}$ direction, which people perceive as vertical, generally does not intersect the center of the earth. (It should be noted that $\bar{k}$ does coincide with a meridional plane, which is any plane formed by $\bar{\omega}$ and $\bar{r}_{O'/O'}$.) At a latitude of 45°, $\bar{k}$ deviates from $\bar{r}_{O/O'}$ by approximately 0.1°. Equally significant is the effect of the centripetal acceleration on the weight mg that is measured at the earth's surface. This effect is most noticeable at the equator, where the centripetal acceleration is $\omega_e^2 R_e$ parallel to $\bar{r}_{O'/O}$. The value $g = 9.807$ m/s^2 represents a reasonable average value when the latitude is not specified.

In view of Eq. (3.59), the acceleration relative to the earth given by Eq. (3.58) becomes

$$(\bar{a}_P)_{xyz} = \frac{\Sigma \bar{F}_a}{m} - g\bar{k} - \bar{\omega}_e \times (\bar{\omega}_e \times \bar{r}_{P/O'}) - 2\bar{\omega}_e \times (\bar{v}_P)_{xyz} \quad (3.61)$$

Usually, the term $\bar{\omega}_e \times (\bar{\omega}_e \times \bar{r}_{P/O'})$ may be neglected unless $\bar{r}_{P/O'}$ is a large fraction of the earth's radius. Therefore, the primary difference between Eq. (3.61) and the form of Newton's second law that ignores the motion of the earth is the Coriolis term.

It is a straightforward matter to describe each of the terms in Eq. (3.61) in terms of components relative to the earth-based xyz reference frame. The position, velocity, and acceleration relative to this system will be described by the Cartesian coordinates

(x, y, z). Because the angular velocity of the earth is parallel to the polar axis and the deviation of the z axis from the line to the center of the earth is small, the angular motion is essentially

$$\overline{\omega}_e = \omega_e(\cos \lambda \overline{i} + \sin \lambda \overline{k}) \qquad \overline{\alpha} = \overline{0} \qquad (3.62)$$

Correspondingly, Eq. (3.61) becomes

$$\ddot{x} - 2 \omega_e \dot{y} \sin \lambda = \frac{F_x}{m}$$

$$\ddot{y} + 2\omega_e(\dot{x} \sin \lambda - \dot{z} \cos \lambda) = \frac{F_y}{m} \qquad (3.63)$$

$$\ddot{z} + 2 \omega_e \dot{y} \cos \lambda = \frac{F_z}{m} - g$$

where (F_x, F_y, F_z) are the components of the applied forces. These equations may be solved for the forces required to have a specified motion relative to the earth. Alternatively, they may be regarded as a set of coupled differential equations for the relative position in situations where the forces are specified.

The fact that the Coriolis term is perpendicular to the velocity as seen by an observer on the earth leads to some interesting anomalies. In the Northern Hemisphere, the component of $\overline{\omega}_e$ perpendicular to the earth's surface is outward. If a particle is constrained to follow a horizontal path relative to the earth in the Northern Hemisphere, the Coriolis term $2\overline{\omega}_e \times (\overline{v}_P)_{xyz}$ is as shown in Figure 3.10. It follows that a horizontal force acting to the left of the direction of motion is required if that direction is to be maintained.

A story that has been passed down from professor to student over the years, without substantiation, states that a railroad line had two sets of north-south tracks along which trains ran in only one direction. For the track along which trains ran northward, the inner surface of the east rail was supposedly more shiny, because of the westward Coriolis force it had to exert on the flange of the wheels. Correspondingly, the track for trains running south was more shiny on the inner surface of the west rail. The veracity of this statement is questionable, owing to the smallness of the effect in comparison with other forces, such as the wind.

If a transverse force is not present to maintain a particle in a straight relative path, as required by Eq. (3.61), the particle will deviate to the right. This observation leads to a qualitative explanation of the fact that a liquid being drained through the center of a cylindrical tank will exhibit a counterclockwise spiraling flow. (The flow will be

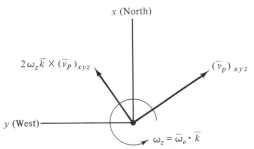

Figure 3.10 Coriolis acceleration due to the earth's rotation.

clockwise in the Southern Hemisphere.) The same phenomenon acts on a much larger scale to set up the flow patterns in hurricanes and typhoons. Meteorological models used to predict general weather patterns must account for the Coriolis acceleration effect.

EXAMPLE 3.14 _____

A child standing on a merry-go-round rotating about the vertical axis at the constant rate ω attempts to catch a ball traveling in the radial direction horizontally at speed v. Determine the velocity and acceleration of the ball as seen by the child.

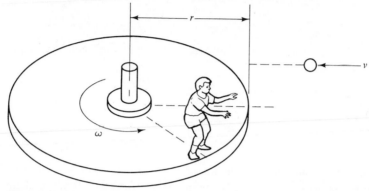

Example 3.14

Solution.   We will evaluate the relative motion by using the kinematical formulas. However, it is useful first to develop a contrasting solution that differentiates the relative position directly. Let $t = 0$ be the instant when the child catches the ball, so $t < 0$ characterizes an arbitrary instant before the ball is caught. Correspondingly, the angle θ locating the child is $\theta = \omega(-t)$ and the distance R to the ball is $R = r + v(-t)$.

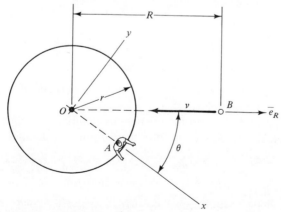

Coordinate systems.

The child is assumed to be stationary with respect to the turntable, so we attach xyz to that body. In order to expedite constructing $\bar{r}_{B/A}$, we place the origin of *xyz* at the center of the turntable and align the *x* axis with the radial line to the child. Correspondingly, we find

$$\bar{r}_{B/A} = (R \cos \theta - r)\bar{i} + R \sin \theta \bar{j}$$
$$= [(r - vt) \cos \omega t - r]\bar{i} - (r - vt) \sin \omega t \bar{j}$$

By definition, the relative velocity is obtained by differentiating the components of the relative position, that is,

$$(\bar{v}_B)_{xyz} = \frac{\delta}{\delta t}(\bar{r}_{B/A}) = [-v \cos \omega t - (r - vt) \omega \sin \omega t]\bar{i}$$
$$+ [v \sin \omega t - (r - vt) \omega \cos \omega t]\bar{j}$$

Similarly, we obtain the relative acceleration by differentiating the relative velocity components.

$$(\bar{a}_B)_{xyz} = \frac{\delta}{\delta t}(\bar{v}_B)_{xyz}$$
$$= [2v\omega \sin \omega t - (r - vt)\omega^2 \cos \omega t]\bar{i}$$
$$+ [2 v \omega \cos \omega t + (r - vt)\omega^2 \sin \omega t]\bar{j}$$

The solution using the kinematical formulas barely resembles the operations in the previous solution. The motion of the reference frame is defined by

$$\bar{v}_A = r\omega \bar{j} \qquad \bar{a}_A = -r\omega^2 \bar{i} \qquad \bar{\omega} = \omega \bar{k} \qquad \bar{\alpha} = \bar{0}$$

We also know the absolute motion of the ball to be

$$\bar{v}_B = -v\bar{e}_R = -v(\cos \theta \bar{i} + \sin \theta \bar{j}) \qquad \bar{a}_B = \bar{0}$$

We write the position as

$$\bar{r}_{B/A} = (R \cos \theta - r)\bar{i} + R \sin \theta \bar{j}$$

so the relative-velocity equation yields

$$(\bar{v}_B)_{xyz} = v_B - v_A - \bar{\omega} \times \bar{r}_{B/A}$$
$$= -v(\cos \theta \bar{i} + \sin \theta \bar{j}) - r\omega \bar{j} - (\omega \bar{k}) \times [(R \cos \theta - r)\bar{i} + R \sin \theta \bar{j}]$$
$$= (-v \cos \theta + \omega R \sin \theta)\bar{i} + (-v \sin \theta - \omega R \cos \theta)\bar{j}$$

Substitution of $R = r - vt$ and $\theta = -\omega t$ into this expression would yield the same result as the previous one.

The same approach leads to the relative acceleration, except that the result for the relative velocity is required to form the Coriolis acceleration. Hence,

$$(\bar{a}_B)_{xyz} = \bar{a}_B - \bar{a}_A - \bar{\alpha} \times \bar{r}_{B/A} - \bar{\omega} \times (\bar{\omega} \times \bar{r}_{B/A}) - 2\bar{\omega} \times (\bar{v}_B)_{xyz}$$
$$= r\omega^2 \bar{i} + \omega^2[(R \cos \theta - r)\bar{i} + R \sin \theta \bar{j}]$$
$$+ 2\omega[(-v \sin \theta - \omega R \cos \theta)\bar{i} + (v \cos \theta - \omega R \sin \theta)\bar{j}]$$
$$= (-\omega^2 R \cos \theta - 2\omega v \sin \theta)\bar{i} + (-\omega^2 R \sin \theta + 2\omega v \cos \theta)\bar{j}$$

Once again, this expression is equivalent to the acceleration in the first solution to this problem.

We could employ either approach with equal ease in this problem because it treated a case of planar motion. The relative-motion equations become increasingly advantageous as the rotation of the reference frame becomes more complicated.

EXAMPLE 3.15 _____

When a small ball is suspended by a cable from an ideal swivel joint that permits three-dimensional motion, the system is called a *spherical pendulum*. Suppose such a pendulum, whose cable length is l, is released from rest with the ball at a distance $b \ll l$ north of the point below the pivot. Analyze the effect of the earth's rotation on the motion. It may be assumed that the angle between the suspending cable and the vertical is always very small.

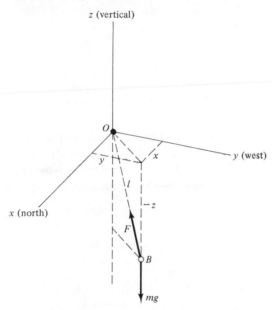

Free body diagram.

Solution. We may apply Eqs. (3.63) directly to this system. A free body diagram of the ball shows the weight mg and the tensile force F exerted by the cable, which may be described in terms of xyz components as

$$\bar{F} = F\bar{e}_{O/B} = F\frac{\bar{r}_{O/B}}{|\bar{r}_{O/B}|} = F\frac{-x\bar{i} - y\bar{j} - z\bar{k}}{l}$$

Because the cable length l is constant, the z coordinate, which is negative, must satisfy

$$z = -[l^2 - x^2 - y^2]^{1/2}$$

We obtain the corresponding velocity and acceleration components by successive differentiation.

$$\dot{z} = +(x\,\dot{x} + y\,\dot{y})\,(l^2 - x^2 - y^2)^{-1/2}$$

$$\ddot{z} = +(x\,\ddot{x} + \dot{x}^2 + y\,\ddot{y} + \dot{y}^2)\,(l^2 - x^2 - y^2)^{-1/2}$$
$$+ (x\,\dot{x} + y\,\dot{y})^2\,(l^2 - x^2 - y^2)^{-3/2}$$

The specification that the angle with the vertical is small means that $x \ll l$ and $y \ll l$. Introducing this approximation above yields

$$\dot{z} \approx +\frac{1}{l}(x\dot{x} + y\dot{y})$$

$$\ddot{z} \approx +\frac{1}{l}(\dot{x}^2 + \dot{y}^2)$$

It follows that for motions in which the angle with the vertical remains small, we may use the approximations that $\dot{z} \approx 0$ and $\ddot{z} \approx 0$. In other words, we may consider the ball to move in the horizontal plane.

The corresponding form of Eqs. (3.63) is

$$\ddot{x} - 2\omega_e \dot{y} \sin \lambda = -\frac{F}{m}\frac{x}{l}$$

$$\ddot{y} + 2\omega_e \dot{x} \sin \lambda = -\frac{F}{m}\frac{y}{l}$$

$$2\omega_e \dot{y} \cos \lambda = -\frac{F}{m}\frac{z}{l} - g$$

Now observe that $z \approx -l$ because x and y are small, and that the Coriolis acceleration in motion relative to the earth is much weaker than the free-fall acceleration. The last of the equations of motion above therefore leads to the approximation that $F/m \approx g$. We introduce this approximation in the first two equations of motion, which become

$$\ddot{x} + \Omega^2 x - 2p\dot{y} = 0$$
$$\ddot{y} + \Omega^2 y + 2p\dot{x} = 0$$

where the coefficients are

$$\Omega^2 = \frac{g}{l} \qquad p = \omega_e \sin \lambda$$

Evaluating the motion requires that we solve this pair of linear, coupled, ordinary differential equations. It is convenient to begin by assuming that the characteristic exponents are imaginary, based on the fact that $x(t)$ and $y(t)$ are oscillatory solutions when Coriolis acceleration is ignored. Thus, we seek solutions having the form

$$x = Ae^{i\mu t} \qquad y = Be^{i\mu t} \tag{1}$$

where μ is real but A and B may be complex constants. These are solutions of the equations of motion if

$$(-\mu^2 + \Omega^2)A - 2ip\mu B = 0 \tag{2}$$
$$2ip\mu A + (-\mu^2 + \Omega^2)B = 0$$

In order for A and B to be nonzero, the determinant of this pair of homogeneous equations for A and B must vanish. This leads to the characteristic equation

$$(-\mu^2 + \Omega^2)^2 - 4p^2\mu^2 = 0$$

This characteristic equation factorizes into

$$\Omega^2 - \mu^2 = +2p\mu \qquad \text{or} \qquad \Omega^2 - \mu^2 = -2p\mu \qquad (3)$$

Solving each of these quadratic equations leads to the four characteristic exponents

$$\mu_{11} = \mu_1 \qquad \mu_{21} = \mu_2 \qquad \mu_{12} = -\mu_2 \qquad \mu_{22} = -\mu_1 \qquad (4)$$

where the first subscript refers to which of Eqs. (3) is satisfied, and

$$\mu_1 = (p^2 + \Omega^2)^{1/2} - p \qquad \mu_2 = (p^2 + \Omega^2)^{1/2} + p \qquad (5)$$

For each root μ_{jk}, there is a corresponding set of coefficients A_{jk} and B_{jk}. The determinant of Eqs. (2) vanishes for each root, so only one of the equations is independent. We solve the second of Eqs. (2) for B_{jk} in terms of A_{jk}, with the result that

$$B_{jk} = -\frac{2ip\mu_{jk}}{\Omega^2 - \mu_{jk}^2} A_{jk}$$

This expression may be simplified by invoking the first of Eqs. (3) for μ_{1k} and the second for μ_{2k}, which leads to

$$B_{1k} = -iA_{1k} \qquad B_{2k} = iA_{2k} \qquad (6)$$

The most general solution for x and y is a sum of solutions in the form of Eqs. (1) for each characteristic exponent. We therefore successively substitute each of Eqs. (4) and the corresponding Eqs. (6) into Eqs. (1), and add the individual contributions, which yields

$$
\begin{aligned}
x &= A_{11}e^{i\mu_1 t} + A_{21}e^{i\mu_2 t} + A_{22}e^{-i\mu_1 t} + A_{12}e^{-i\mu_2 t} \\
y &= -iA_{11}e^{i\mu_1 t} + iA_{21}e^{i\mu_2 t} + iA_{22}e^{-i\mu_1 t} - iA_{12}e^{-i\mu_2 t}
\end{aligned}
\qquad (7)
$$

In order for x and y to be real functions, it must be that A_{22} is the complex conjugate of A_{11} and A_{12} is the complex conjugate of A_{21}. In Eqs. (7), we represent A_{11} and A_{21} in polar form as

$$A_{11} = \tfrac{1}{2}C_1 e^{i\phi_1} \qquad A_{21} = \tfrac{1}{2}C_2 e^{i\phi_2}$$

and apply Euler's identity in order to express x and y as real functions. The result is

$$
\begin{aligned}
x &= C_1 \cos(\mu_1 t + \phi_1) + C_2 \cos(\mu_2 t + \phi_2) \\
y &= C_1 \sin(\mu_1 t + \phi_1) - C_2 \sin(\mu_2 t + \phi_2)
\end{aligned}
\qquad (8)
$$

The coefficients C_1, C_2, ϕ_1, and ϕ_2 must satisfy initial conditions. It was given in the problem statement that the ball was released from rest at a distance b to the north of the pivot. Thus, the initial conditions are

$$x = b \qquad y = 0 \qquad \dot{x} = \dot{y} = 0 \qquad \text{when } t = 0$$

The corresponding solution obtained from Eqs. (8) is

$$x = \frac{b}{\mu_1 + \mu_2} [\mu_2 \cos (\mu_1 t) + \mu_1 \cos (\mu_2 t)] \tag{9}$$

$$y = \frac{b}{\mu_1 + \mu_2} [\mu_2 \sin (\mu_1 t) - \mu_1 \sin (\mu_2 t)]$$

It is convenient at this stage to simplify the characteristic roots by making use of the fact that $p \ll \Omega$ because of the smallness of ω_e. We therefore expand each characteristic exponent in powers, and drop higher-order terms. This leads to

$$\mu_1 = \Omega - p \qquad \mu_2 = \Omega + p$$

The values of μ_1 and μ_2 are very close, $\mu_1 \approx \mu_2 \approx \Omega$. Minor errors will be introduced if we use this approximation in the coefficients, but not the arguments of the sinusoidal terms. Combining this approximation with the trigonometric identities for the sum of sines or cosines leads to

$$x = \tfrac{1}{2}b[\cos (\mu_1 t) + \cos (\mu_2 t)] = b \cos \left(\frac{\mu_1 - \mu_2}{2} t \right) \cos \left(\frac{\mu_1 + \mu_2}{2} t \right)$$

$$= b \cos (pt) \cos (\Omega t)$$

$$y = \tfrac{1}{2}b [\sin (\mu_1 t) - \sin (\mu_2 t)] = b \sin \left(\frac{\mu_1 - \mu_2}{2} t \right) \cos \left(\frac{\mu_2 + \mu_1}{2} t \right)$$

$$= - b \sin (pt) \cos (\Omega t)$$

The nature of the path becomes obvious when we observe that $\sin (pt)$ and $\cos (pt)$ vary much more slowly than $\sin (\Omega t)$ because $p \ll \Omega$. The above solutions satisfy

$$y = -x \tan(pt)$$

which is the equation of a straight line whose slope is $-\tan(pt)$, if we neglect the variation in the value of pt. As shown in the diagram, the path at each instant seems to be at an angle pt relative to the north (i.e., the x axis), measured clockwise

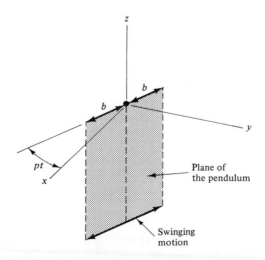

Motion of the spherical pendulum.

when viewed downward. The vertical plane in which the cable lies therefore seems to an observer on the earth to rotate about the vertical axis at $\omega_z = -p = -\omega_e \sin \lambda$. This is exactly opposite the angular-velocity component of the earth in the vertical direction—an observer viewing the pendulum from a fixed reference frame in outer space sees the plane of the pendulum as being fixed.

The movement of the plane of a spherical pendulum relative to the earth was used by the French physicist, Jean Louis Foucault (1819–1869), in 1851 to demonstrate the earth's rotation. What is perhaps the most famous Foucault pendulum in current usage may be found in the General Assembly building at United Nations headquarters in New York City.

In closing, we should note that the spherical pendulum for arbitrary, small initial values would seem to follow an elliptical path. The major and minor axes of the ellipse would rotate relative to the earth at angular speed $\omega_z = -\omega_e \sin \lambda$.

REFERENCES

R. A. Frazer, W. J. Duncan, and A. R. Collar, *Elementary Matrices*, Cambridge University Press, New York, 1960.

J. H. Ginsberg and J. Genin, *Dynamics*, 2d ed., John Wiley & Sons, Inc., New York, 1984.

H. Goldstein, *Classical Mechanics*, 2d ed., Addison-Wesley Publishing Company, Reading, Mass., 1980.

D. T. Greenwood, *Principles of Dynamics*, Prentice-Hall, Inc., Englewood Cliffs, N.J., 1965.

J. B. Marion, *Classical Dynamics of Particles and Systems,* Academic Press, New York, 1960.

L. Meirovitch, *Methods of Analytical Dynamics*, McGraw-Hill Book Company, New York, 1970.

I. H. Shames, *Engineering Mechanics,* 2d ed., Prentice-Hall, Inc., Englewood Cliffs, N.J., 1967.

C. E. Smith, *Applied Mechanics: Dynamics*, John Wiley & Sons, Inc., New York, 1982.

HOMEWORK PROBLEMS

3.1 The x axis coincides with the main diagonal for the box and the y axis coincides with the right face. Determine the coordinates relative to xyz of corners D and E.

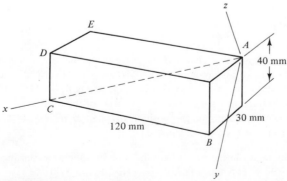

Problem 3.1

3.2 The rectangular box is supported from the ceiling by a ball-and-socket joint at corner A. Commencing from the position shown, the box is rotated about edge AB by 75° counterclockwise as viewed from A to B. Then the box is rotated about edge AC by 50° clockwise as viewed from A to C. Determine the displacement of corner D due to these rotations.

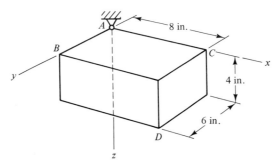

Problems 3.2 and 3.3

3.3 The rectangular box is supported from the ceiling by a ball-and-socket joint. Commencing from the position shown, the box is given a rotation about the fixed X axis of 120°, followed by a rotation about the fixed Z axis of $-40°$. Determine the displacement of corner D due to these rotations.

3.4 Starting from the position shown, the box is rotated by 40° about diagonal AB, clockwise as viewed from corner B toward corner A. Determine the coordinates of corner C relative to the fixed reference frame XYZ after this rotation.

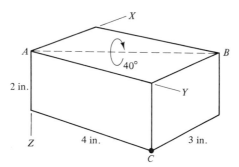

Problem 3.4

3.5 Disk A may rotate about shaft AB by angle θ_3, while angle θ_1 is the rotation about the vertical axis and θ_2 is the rotation of shaft AB away from vertical. Starting from the position where $\theta_1 = \theta_2 = \theta_3 = 0°$, the rotation sequence is $\theta_1 = 115°$, $\theta_2 = -55°$, $\theta_3 = 40°$. Prior to these rotations, point D on the diameter of the disk was at $X = 100$ mm, $Y = 0$, $Z = -200$ mm. Determine the position of point D relative to the fixed XYZ reference frame after the rotations are completed.

3.6 Solve Problem 3.5 in the case where the rotation sequence, starting from the given initial position, is $\theta_2 = 30°$, $\theta_1 = 90°$, $\theta_3 = -40°$, $\theta_1 = -90°$.

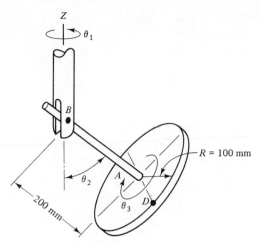

Problems 3.5 and 3.6

3.7 Bar AB is welded to collar B which pivots about segment BC of the triangular arm BCD. The rotation of arm BCD about the horizontal x axis is the angle θ, and the rotation of bar AB relative to the triangular bar is the angle ψ. Consider the case where $\theta = 110°$ and $\psi = 20°$. Determine the displacement of end A from its position when $\theta = \psi = 0°$.

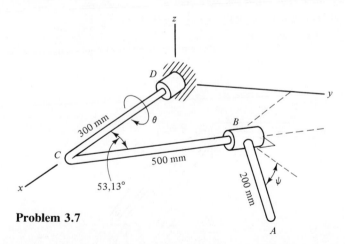

Problem 3.7

3.8 The radiator fan of an automobile engine whose crankshaft is aligned with the automobile axis is rotating at 1000 rev/min, clockwise as viewed from the front of the vehicle. The automobile is following a 100-ft-radius left turn at a constant speed of 45 mi/h. Determine the angular velocity and angular acceleration of the fan.

3.9 The flywheel of the gyroscope rotates about its own axis at $\omega_2 = 6000$ rev/min, and the outer gimbal support is rotating about the horizontal axis at the rate $\omega_1 = 10$ rad/s, $\dot{\omega}_1 = 100$ rad/s^2. Determine the angular velocity and angular acceleration of the flywheel if θ is constant at 75°.

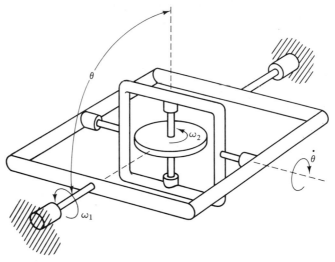

Problems 3.9 and 3.10

3.10 The flywheel of the gyroscope rotates about its own axis at ω_2 = 6000 rev/min. At the instant when θ = 120°, the inner gimbal support is rotating relative to the outer gimbal at $\dot{\theta}$ = 6 rad/s and $\ddot{\theta}$ = −90 rad/s². The corresponding rotation of the outer gimbal about the horizontal axis is ω_1 = 10 rad/s, $\dot{\omega}_1$ = 100 rad/s². Determine the angular velocity and angular acceleration of the flywheel at this instant.

3.11 The disk spins about its own axis at 1200 rev/min as the system rotates about the vertical axis at 20 rev/min. Determine the angular velocity and angular acceleration of the disk if β is constant at 30°.

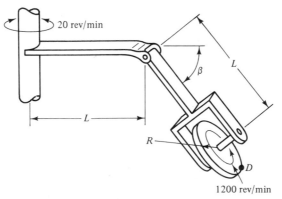

Problems 3.11 to 3.13

3.12 Solve Problem 3.11 for the case where $\dot{\beta}$ = 10 rad/s and $\ddot{\beta}$ = −50 rad/s² when β = 36.87°.

3.13 The disk spins about its own axis at 1200 rev/min as the system rotates about the vertical axis at 20 rev/min. The angle of inclination is constant at β = 30°. Determine the velocity and acceleration of point D on the perimeter of the disk when it is at its lowest position, as shown.

3.14 Collar C moves at the constant speed u relative to the curved bar, which rotates in the horizontal plane at the constant rate ω. Derive expressions for the velocity and acceleration of the collar as a function of the angle θ.

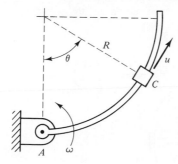

Problem 3.14

3.15 A servomotor maintains the angle ϕ of bar BC relative to bar AB at $\phi = 2\theta$, where θ is the angle of inclination of bar AB. Determine the acceleration of end C if $\dot{\theta} = 50$ rad/s and $\ddot{\theta} = 0$ when $\theta = 15°$. Then determine the corresponding rate at which the speed of end C is increasing.

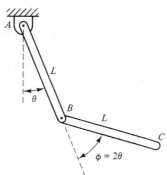

Problem 3.15

3.16 The disk rotates at ω_1 about its axis, and the rotation rate of the forked shaft is ω_2. Both rates are constant. Use two different approaches to determine the velocity and acceleration

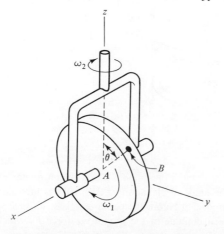

Problem 3.16

of an arbitrarily selected point B on the perimeter. Describe the results in terms of components relative to the xyz axes in the sketch.

3.17 Collar c slides relative to the curved rod at a constant speed u, while the rod rotates about the horizontal axis at the constant rate Ω. Determine the acceleration of the collar in terms of θ.

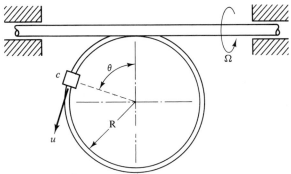

Problem 3.17

3.18 The rotation rates of the bars in Problem 3.7 are constant at $\dot{\theta} = 20$ rad/s, $\dot{\psi} = 10$ rad/s. For the instant when $\theta = 30°$ and $\psi = 50°$, determine the velocity and acceleration of end A of the bar.

3.19 Use the concepts of relative motion to derive the formulas for velocity and acceleration of a point in terms of a set of spherical coordinates.

3.20 The telescope rotates about the fixed vertical axis at 4 rev/h as the angle θ oscillates at $\theta = (\pi/3) \sin (\pi t/7200)$ rad, where t has units of seconds. Determine the velocity and acceleration of points C and D as functions of time.

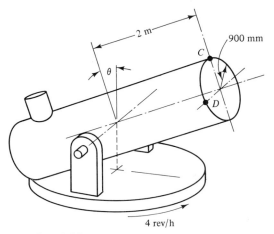

Problem 3.20

3.21 A pellet of mass m moves through the smooth barrel which rotates about the vertical axis at angular speed Ω as the angle of elevation of the barrel is increased at the rate $\dot{\theta}$. Both rates are constant. At the instant before the pellet emerges, its speed relative to the barrel is u. At that instant, the magnitude of the propulsive force $\overline{F}$, which acts parallel to the barrel, is a factor of 50 times greater than the weight of the pellet. Derive expressions for the acceleration term $\ddot{u}$ and the force the pellet exerts on the walls of the barrel at this instant.

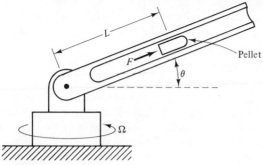

Problem 3.21

3.22 In the position shown, the turret is rotating about the vertical axis at the constant rate of 1.2 rad/s. At this instant the barrel is being raised at the rate of 0.6 rad/s, which is decreasing at 100 rad/s². The tank is at rest. Immediately preceding its emergence, a 30-lb cannon shell is traveling at a speed of 1500 ft/s relative to the barrel, and the internal propulsive pressure within the barrel has been dissipated. Determine the acceleration of the cannon shell at this instant, and the corresponding forces exerted by the cannon shell on the barrel.

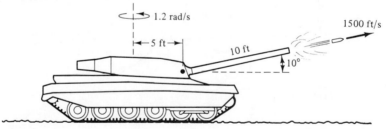

Problem 3.22

3.23. A servomotor at joint A rotates arm AB about its axis through an angle ψ, and another servomotor at joint B controls the angle θ of arm BC relative to arm AB. When $\psi = 0°$,

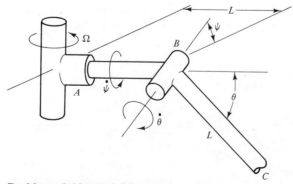

Problems 3.23 and 3.24

arm BC lies in the vertical plane. Consider a situation where $\dot{\theta}$ is constant at 2 rad/s, while $\dot{\psi} = 4$ rad/s and $\ddot{\psi} = 12$ rad/s². Concurrently with these rotations, the entire assembly is rotating about the vertical axis at the constant rate $\Omega = 8$ rad/s. Determine the angular velocity and angular acceleration of link BC if $\psi = 0°$ and $\theta = 30°$ at this instant.

3.24 Consider the system in Problem 3.23 in a situation where $\dot{\theta}$ is constant at 2 rad/s, while $\dot{\psi}$ is constant at 4 rad/s. The entire assembly is rotating about the vertical axis at the constant rate $\Omega = 8$ rad/s. Determine the velocity and acceleration of end C at the instant when $\psi = 53.13°$ and $\theta = 20°$.

3.25 Consider a roller coaster that is constructed such that a car follows a specified curve Γ while the axles of the car are always parallel to the binormal to Γ. In order to achieve this, each point on the tracks is located by measuring equal distances $h/2$ (where h is the distance between wheels) in the binormal direction relative to the point on Γ. As a result, the longitudinal axis of a car is always parallel to the tangential direction. Derive an expression for the acceleration of an arbitrary point P in the car in terms of the speed v of a car, the rate of increase $\dot{v}$, and the properties of curve Γ. The coordinates of point P are (x, y, z) relative to a body-fixed coordinate system whose origin follows curve Γ and whose orientation is selected such that $\bar{i} = \bar{e}_t$ and $\bar{j} = \bar{e}_n$.

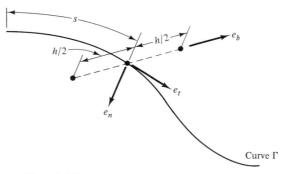

Problem 3.25

3.26 An airplane is executing a manuever in which its roll, pitch, and yaw rates are 2 rad/s, 0.5 rad/s, and -0.2 rad/s, respectively. All rates are constant. Determine the velocity and acceleration of the nose relative to the center of mass G, as seen by an observer on the ground. The point of interest at the nose is 5 m forward and 0.2 m below point G.

3.27 A submarine is following a horizontal 500-m circle at 20 knots (1 knot = 1.852 km/h). At the instant when it is aligned with the horizontal and vertical directions, it is rolling at 0.5 rad/s and pitching at 0.1 rad/s, as shown. Both rates are maxima at this instant.

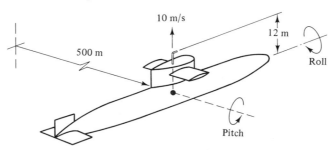

Problem 3.27

Determine the velocity and acceleration of the tip of the periscope, which is being extended at the constant rate of 10 m/s.

3.28 A test chamber for astronauts rotates about axis AB at a constant angular speed ω_1 as the entire assembly rotates about the horizontal axis at angular speed ω_2, which also is constant. An astronaut is seated securely in the chamber at center point O, which is collinear with both axes of rotation. Object C has a constant absolute velocity $\bar{v}_C$ parallel to the horizontal axis. Determine the acceleration of this object as seen by the astronaut.

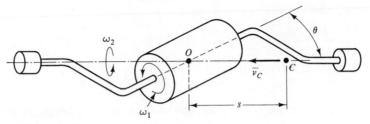

Problem 3.28

3.29 Airplane B has a velocity of 600 mi/h eastward. Airplane A, whose speed is constant at 1200 mi/h, is executing a 5-mile-radius turn. Radar equipment on aircraft A can measure the separation distance R and the angle ϕ relative to the longitudinal axis, as well as the rates of change of these parameters. For the instant shown, determine $\dot{R}$, $\ddot{R}$, $\dot{\phi}$, and $\ddot{\phi}$.

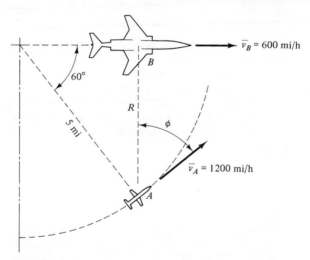

Problem 3.29

3.30 A small disk slides with negligible friction on a horizontal sheet of ice. The initial velocity of the disk was u in the southerly direction. Determine the distance and sense of the shift s in the position after the disk has traveled distance d southward. How would this result have changed if the initial velocity was northward or eastward?

3.31 A ball is thrown vertically from the ground at speed v. Assuming that air resistance is negligible, derive an expression for the shift due to the Coriolis effect in the position where it returns to the ground. Evaluate the result for $v = 40$ m/s at a latitude of 45°.

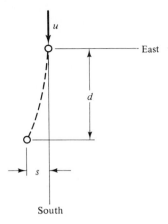

Problem 3.30

3.32 An object falls in a vacuum after being released at a distance H above the surface of the earth. The line extending from the center of the earth to this object is at latitude λ, and point O' on the earth's surface is concurrent with this line. Determine the location (east-west and north-south relative to point O') at which the object strikes the ground in each of the following cases:

(a) The block is initially at rest relative to the earth.

(b) The block was initially at rest relative to a reference frame that translates with the center of the earth but does not execute the earth's spinning rotation.

For the sake of simplicity, the gravitational attraction may be considered to be constant at mg. Explain the difference between the results in cases (a) and (b).

3.33 A small block of mass m is attached to a horizontal turntable by two pairs of opposing springs whose stiffnesses are k_x and k_y. The springs are unstretched when the block coincides with the axis of the turntable, and the (x, y) coordinates of the block relative to the turntable are much less than the radius of the turntable. Derive linearized differential equations for x and y for the case where the turntable rotates at the constant rate Ω. Then solve those equations for initial conditions in which the block is released from rest relative to the turntable at $x = b$, $y = 0$. Discuss how this system may be used as an analog for the Foucault pendulum.

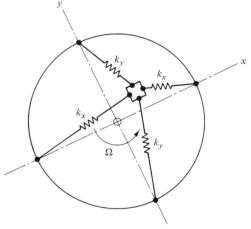

Problem 3.33

Chapter 4

Kinematics of Rigid Bodies

The concept of a rigid body is an artificial one, in that all materials deform when forces are applied to them. Nevertheless, this artifice is very useful when we are concerned with an object whose movement due to deformation is only a minor part of its motion. In addition, formulations of the motion of deformable bodies often find it convenient to decompose the overall motion into rigid body and deformational contributions.

4.1 GENERAL EQUATIONS

A rigid body is defined to be a collection of particles whose distance of separation is invariant. In this circumstance, any set of coordinate axes *xyz* that is scribed in the body will maintain its orientation relative to the body. Such a coordinate system forms a *body-fixed reference frame*. The orientation of *xyz* relative to the body and the location of its origin are arbitrary. A typical situation is depicted in Figure 4.1.

Because all points in the rigid body maintain their relative position, their velocity and acceleration relative to *xyz* is zero. Thus the velocity and acceleration of point A in Figure 4.1 are given by

$$\bar{v}_A = \bar{v}_{O'} + \overline{\omega} \times \bar{r}_{A/O'}$$
$$\bar{a}_A = \bar{a}_{O'} + \overline{\alpha} \times \bar{r}_{A/O'} + \overline{\omega} \times (\overline{\omega} \times \bar{r}_{A/O'}) \qquad (4.1)$$

Similar relations apply for another point B.

$$\bar{v}_B = \bar{v}_{O'} + \overline{\omega} \times \bar{r}_{B/O'}$$
$$\bar{a}_B = \bar{a}_{O'} + \overline{\alpha} \times \bar{r}_{B/O'} + \overline{\omega} \times (\overline{\omega} \times \bar{r}_{B/O'}) \qquad (4.2)$$

The foregoing yield relations between the motion of A and B that do not require

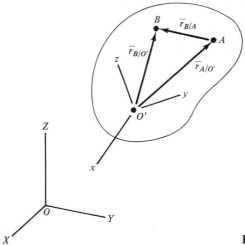

Figure 4.1 Position of points in a rigid body.

that either point be the origin. Subtracting each of Eqs. (4.1) from the corresponding Eqs. (4.2) yields

$$\bar{v}_B - \bar{v}_A = \omega \times (\bar{r}_{B/O'} - \bar{r}_{A/O'})$$
$$\bar{a}_B - \bar{a}_A = \bar{\alpha} \times (\bar{r}_{B/O'} - \bar{r}_{A/O'}) + \bar{\omega} \times [\bar{\omega} \times (\bar{r}_{B/O'} - \bar{r}_{A/O'})] \qquad (4.3)$$

Because $\bar{r}_{B/A} = \bar{r}_{B/O'} - \bar{r}_{A/O'}$, we find that

$$\bar{v}_B = \bar{v}_A + \bar{\omega} \times \bar{r}_{B/A}$$
$$\bar{a}_B = \bar{a}_A + \bar{\alpha} \times \bar{r}_{B/A} + \bar{\omega} \times (\bar{\omega} \times \bar{r}_{B/A}) \qquad (4.4)$$

The arbitrariness of the origin O' could have been used as an argument for deriving Eqs. (4.4) directly. However, the approach used here leads to an important observation.

Given a set of n points in a rigid body, there are $n - 1$ independent equations in the form of Eqs. (4.4) between their velocities or accelerations. These equations may be obtained by relating one point to each of the other $n - 1$.

As was demonstrated in Eqs. (4.1), (4.2), and (4.4), other relations that we might try to write are linear combinations of the independent ones. This interdependence between the motion of points in a body is a consequence of the rigidity, which maintains points at fixed relative positions.

Equations (4.4) describe the motion of a rigid body as the combination of the movement of point A and a rotational effect about point A. This observation is a manifestation of *Chasle's theorem*, which states that

The *general motion* of a rigid body is a superposition of a *translation* and a *pure rotation*. In the translation, all points follow the movement of an arbitrary point A in the body and the orientation remains constant. The rotational motion is such that the arbitrary point A remains at rest.

Note the arbitrariness of the point selected for the translation. This means that the only unique property of the kinematics of a body is the rotation, as described by the orientation, the angular velocity, and the angular acceleration. Various methods for locating a point via intrinsic and extrinsic coordinates were discussed in Chapter 2. The next section will present a standardized way for describing orientation.

A basic tool in the analysis of velocity for a body in planar motion is the instant-center method. In essence, this technique is based on considering a body in general motion (translation plus rotation) to be rotating about a rest point, which is called the *instantaneous center of zero velocity* or, more briefly, the *instant center*. In general, the instant center has an acceleration, so the point on the body that has no velocity changes as the motion evolves. For this reason the method is not suitable for the evaluation of acceleration. The instant-center method also is not useful for evaluating the velocity of bodies in arbitrary spatial motion, for reasons that will become apparent.

In order to explore the instant-center method, consider the velocities of two points A and B in a rigid body. If point A is at rest, then $\bar{v}_B = \bar{\omega} \times \bar{r}_{B/A}$, where $\bar{\omega}$ is the angular velocity of the body. According to this relationship, the speed of point B is proportional to the distance from that point to the axis parallel to $\bar{\omega}$ that intersects point A. Also, the sense of $\bar{v}_B$ will be perpendicular to the radial vector $\bar{r}_{B/A}$ in the sense of the rotation according to the right-hand rule. The instant center may be located by using these properties for two points whose motion is known. The velocities of other points may then be computed by using the same relations for circular motion.

The difficulty is that, in general, there is no point for which $\bar{v}_A = \bar{0}$. This is readily proved by taking a dot product of $\bar{v}_B$ in Eq. (4.4) with $\bar{\omega}$, which yields

$$\bar{\omega} \cdot \bar{v}_B = \bar{\omega} \cdot \bar{v}_A$$

This relation states that all points in a body have the same velocity in the sense of $\bar{\omega}$. Therefore, if there is a situation where $\bar{v}_A = \bar{0}$, all points in the body must have velocities that are perpendicular to $\bar{\omega}$. The conditions imposed on many systems do not satisfy such a restriction, although it is identically satisfied for planar motion.

Pure spatial rotation is one exception, because it is, by definition, the case where some point in the body actually is fixed. General motion occurs when a rotating body has no fixed point. Chasle's theorem could be used to analyze a general motion as the superposition of a translation in the direction of $\bar{\omega}$ that follows a selected point A, and a rotation about an axis parallel to $\bar{\omega}$ through point A. We shall not pursue such a representation because it does little to impove our ability to formulate problems. However, some people do find it to be a useful way to visualize spatial motion.

EXAMPLE 4.1 _____

Observation of the motion of the block reveals that at a certain instant the velocity of corner A is parallel to the diagonal AE. At this instant components relative to the body-fixed xyz coordinate system of the velocities of the other corners are known to be $(v_B)_x = 10$, $(v_C)_z = 20$, $(v_D)_x = 10$, $(v_E)_y = 5$, where all values are in units of meters per second. Determine whether these values are possible, and if so, evaluate the velocity of corner F.

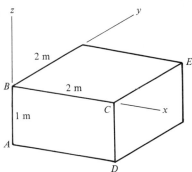

Example 4.1

Solution. We know that $\bar{v}_A = v_A \bar{e}_{E/A}$, with the sign of v_A unspecified, and also that $\bar{v}_B \cdot \bar{i} = 10$, $\bar{v}_C \cdot \bar{k} = 20$, $\bar{v}_D \cdot \bar{i} = 10$, and $\bar{v}_E \cdot \bar{j} = 5$ m/s. If these values are possible, there will be values of $\bar{\omega}$ and the velocity of any point that satisfy the relative-velocity equations. We select point A for this purpose, because the only unknown aspect of its velocity is the speed, that is,

$$\bar{v}_A = v_A \bar{e}_{E/A} = v_A \frac{\bar{r}_{E/A}}{|\bar{r}_{E/A}|} = v_A \frac{2\bar{i} + 2\bar{j} + \bar{k}}{(2^2 + 2^2 + 1)^{1/2}}$$

$$= \tfrac{1}{3} v_A (2\bar{i} + 2\bar{j} + \bar{k})$$

The angular velocity is unknown, so

$$\bar{\omega} = \omega_x \bar{i} + \omega_y \bar{j} + \omega_z \bar{k}$$

The velocity equations relating point A to the other points are

$$\begin{aligned}
\bar{v}_B &= \bar{v}_A + \bar{\omega} \times \bar{r}_{B/A} & \bar{r}_{B/A} &= \bar{k} \\
\bar{v}_C &= \bar{v}_A + \bar{\omega} \times \bar{r}_{C/A} & \bar{r}_{C/A} &= 2\bar{i} + \bar{k} \\
\bar{v}_D &= \bar{v}_A + \bar{\omega} \times \bar{r}_{D/A} & \bar{r}_{D/A} &= 2\bar{i} \\
\bar{v}_E &= \bar{v}_A + \bar{\omega} \times \bar{r}_{E/A} & \bar{r}_{E/A} &= 2\bar{i} + 2\bar{j} + \bar{k}
\end{aligned}$$

We substitute the expressions for $\bar{v}_A$ and $\bar{\omega}$ into these equations, and then evaluate the given velocity components by taking the appropriate dot products. This leads to

$$\begin{aligned}
\bar{v}_B \cdot \bar{i} &= \tfrac{2}{3} v_A + \omega_y = 10 \\
\bar{v}_C \cdot \bar{k} &= \tfrac{1}{3} v_A - 2\omega_y = 20 \\
\bar{v}_D \cdot \bar{i} &= \tfrac{2}{3} v_A = 20 \\
\bar{v}_E \cdot \bar{j} &= \tfrac{2}{3} v_A + 2\omega_x - 2\omega_y = 5
\end{aligned}$$

Although these are four equations for the four unknown parameters, the equations are not solvable. The third equation yields $v_A = 30$ m/s, from which the first equation gives $\omega_y = 0$, while the second gives $\omega_y = -5.0$ rad/s. Therefore, the motion is not possible. Note that if the first two equations were compatible, we still would not be able to determine ω_z because that angular velocity component does not appear in any of the equations.

4.2 EULERIAN ANGLES

Three independent direction angles define the orientation of a set of *xyz* axes. Because there are a total of nine direction angles locating *xyz* with respect to a fixed reference frame *XYZ*, an independent set may be selected in a variety of ways. Eulerian angles treat this matter as a specific sequence of rotations.

Let us follow the intermediate orientations of a moving reference frame as it is rotated away from its initial alignment with *XYZ*. The first rotation, called the *precession*, is about the fixed *Z* axis. The angle of rotation in the precession is denoted ψ, as depicted in Figure 4.2. The orientation of the moving reference frame after it has undergone only the precession is denoted as $x'y'z'$. The transformation from *XYZ* to $x'y'z'$ may be found from Figure 4.2 to be

$$\left\{\begin{matrix} x' \\ y' \\ z' \end{matrix}\right\} = [R_\psi] \left\{\begin{matrix} X \\ Y \\ Z \end{matrix}\right\} \tag{4.5}$$

where

$$[R_\psi] = \begin{bmatrix} \cos\psi & \sin\psi & 0 \\ -\sin\psi & \cos\psi & 0 \\ 0 & 0 & 1 \end{bmatrix} \tag{4.6}$$

The second rotation is about the y' axis. The orientation of the moving reference frame after this rotation is denoted $x''y''z''$ in Figure 4.3. This is the *nutation*, and θ is the angle of nutation. The second transformation is given by

$$\left\{\begin{matrix} x'' \\ y'' \\ z'' \end{matrix}\right\} = [R_\theta] \left\{\begin{matrix} x' \\ y' \\ z' \end{matrix}\right\} \tag{4.7}$$

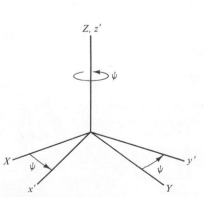

Figure 4.2 Precession.

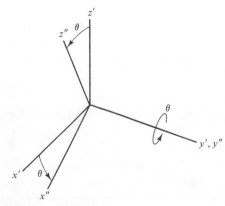

Figure 4.3 Nutation.

where

$$[R_\theta] = \begin{bmatrix} \cos\theta & 0 & -\sin\theta \\ 0 & 1 & 0 \\ \sin\theta & 0 & \cos\theta \end{bmatrix} \qquad (4.8)$$

The last rotation is the *spin*, in which the reference frame moves from $x''y''z''$ to its final orientation. The z'' axis is the axis for the spin, and the angle of spin is denoted as ϕ. The transformation from $x''y''z''$ to the final xyz system is found from Figure 4.4 to be

$$\begin{Bmatrix} x \\ y \\ z \end{Bmatrix} = [R_\phi] \begin{Bmatrix} y'' \\ y'' \\ z'' \end{Bmatrix} \qquad (4.9)$$

where

$$[R_\phi] = \begin{bmatrix} \cos\phi & \sin\phi & 0 \\ -\sin\phi & \cos\phi & 0 \\ 0 & 0 & 1 \end{bmatrix} \qquad (4.10)$$

The overall transformation is found by combining Eqs. (4.5), (4.7), and (4.9), with the result that

$$\begin{Bmatrix} x \\ y \\ z \end{Bmatrix} = [R] \begin{Bmatrix} X \\ Y \\ Z \end{Bmatrix} \qquad [R] = [R_\phi][R_\theta][R_\psi] \qquad (4.11)$$

The full sequence of rotations is depicted in Figure 4.5 by following the tips of the axes.

The angular velocity and angular acceleration are readily expressed in terms of the angles of precession, nutation, and spin by adding the rotation rates about the respective axes. For this task, we note that the precession axis is defined to be the (fixed) Z axis, so the precessional portion of the angular velocity is always $\dot{\psi}\overline{K}$. The nutation occurs about the y' axis. (The term *line of nodes* is sometimes used to refer to the y' axis because points on this axis do not move in the nutation.) A general description of the nutational angular velocity is therefore $\dot{\theta}\bar{j}'$. Finally, the spin is about the z'' axis. Since this rotation features a coincidence of the z'' and z axes, the spin angular velocity may be written as $\dot{\phi}\overline{k}$. The angular velocity is the (vector) sum of the individual rotation rates, so

$$\overline{\omega} = \dot{\psi}\overline{K} + \dot{\theta}\bar{j}' + \dot{\phi}\overline{k} \qquad (4.12)$$

A general expression for angular acceleration is obtained from the foregoing by noting that $\overline{K}$ is fixed in space, whereas $\bar{j}'$ and $\overline{k}$ are unit vectors for $x'y'z'$ and xyz, respectively. Thus, $\dot{\overline{k}} = \overline{\omega} \times \overline{k}$ and $\dot{\bar{j}}' = \overline{\omega}' \times \bar{j}'$, where $\overline{\omega}'$ is the angular velocity of $x'y'z'$. Because $x'y'z'$ only undergoes precession, this term is

$$\overline{\omega}' = \dot{\psi}\overline{K} \qquad (4.13)$$

Using $\overline{\omega}$ and $\overline{\omega}'$ to differentiate the unit vectors then leads to

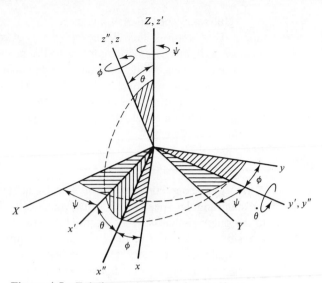

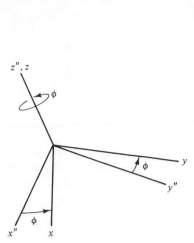

Figure 4.4 Spin.

Figure 4.5 Eulerian angles and reference frames.

$$\overline{\alpha} = \ddot{\psi}\overline{K} + \ddot{\theta}\overline{j}' + \dot{\theta}\dot{\overline{j}}' + \ddot{\phi}\overline{k} + \dot{\phi}\dot{\overline{k}}$$
$$= \ddot{\psi}\overline{K} + \ddot{\theta}\overline{j}' + \dot{\theta}(\overline{\omega}' \times \overline{j}') + \ddot{\phi}\overline{k} + \dot{\phi}(\overline{\omega} \times \overline{k}) \qquad (4.14)$$

If these expressions are to be used in computations, they must be transformed to a common set of components. Many situations require *xyz* components. From Figure 4.5 we find that the unit vectors $\overline{K}$ and $\overline{j}'$ are

$$\overline{K} = \sin \theta(-\cos \phi \overline{i} + \sin \phi \overline{j}) + \cos \theta \overline{k}$$
$$\overline{j}' = \sin \phi \overline{i} + \cos \theta \overline{j} \qquad (4.15)$$

Thus, the angular velocity and angular acceleration are

$$\overline{\omega} = (-\dot{\psi} \sin \theta \cos \phi + \dot{\theta} \sin \phi)\overline{i} + (\dot{\psi} \sin \theta \sin \phi + \dot{\theta} \cos \phi)\overline{j} + (\dot{\psi} \cos \theta + \dot{\phi})\overline{k} \qquad (4.16)$$

$$\overline{\alpha} = (-\ddot{\psi} \sin \theta \cos \phi + \ddot{\theta} \sin \phi - \dot{\psi}\dot{\theta} \cos \theta \cos \phi + \dot{\psi}\dot{\phi} \sin \theta \sin \phi + \dot{\phi}\dot{\theta} \cos \phi)\overline{i}$$
$$+ (\ddot{\psi} \sin \theta \sin \phi + \ddot{\theta} \cos \phi + \dot{\psi}\dot{\theta} \cos \theta \sin \phi + \dot{\psi}\dot{\phi} \sin \theta \cos \phi - \dot{\theta}\dot{\phi} \sin \phi)\overline{j}$$
$$+ (\ddot{\psi} \cos \theta + \ddot{\phi} - \dot{\psi}\dot{\theta} \sin \theta)\overline{k} \qquad (4.17)$$

These expressions, particularly the one for $\overline{\alpha}$, are quite complicated. For that reason, the $x''y''z''$ axes, which do not undergo the spin, are sometimes selected for the representation. Then

$$\overline{K} = -\sin \theta \overline{i}'' + \cos \theta \overline{k}'' \qquad \overline{j}' = \overline{j}'' \qquad \overline{k} = \overline{k}'' \qquad (4.18)$$

Substitution of Eqs. (4.18) into Eqs. (4.12) and (4.14) results in substantially simpler expressions. They will be equivalent to Eqs. (4.16) and (4.17) for the instant when $\phi = 0$, corresponding to $\overline{i}'' = \overline{i}$ and $\overline{j}'' = \overline{j}$.

Utilization of Eulerian angles requires recognition of the appropriate axes of rotation. This involves identifying a fixed axis of rotation as the precession axis. Then the nutation axis precesses orthogonally to the precession axis. Finally, the spin axis precesses and nutates, while it remains perpendicular to the nutation axis. In many cases, the nutation or spin rates may be zero, in which case either of the respective angles is constant. This results in a degree of arbitrariness in the selection of the axes. Indeed, the case of rotation about a single axis can be considered to be solely precession, nutation, or spin, as one wishes.

EXAMPLE 4.2 _____

A free gyroscope consists of a flywheel that rotates relative to the inner gimbal at the constant angular speed of 8000 rev/min, while the rotation of the inner gimbal relative to the outer gimbal is $\gamma = 0.2 \sin (100\pi t)$ rad. The rotation of the outer gimbal is $\beta = 0.5 \sin (50\pi t)$ rad. Use the Eulerian angle formulas to determine the angular velocity and angular acceleration of the flywheel at $t = 4$ ms. Express the results in terms of components relative to the body-fixed xyz and space-fixed XYZ reference frames, where the z axis was parallel to the Z axis at $t = 0$.

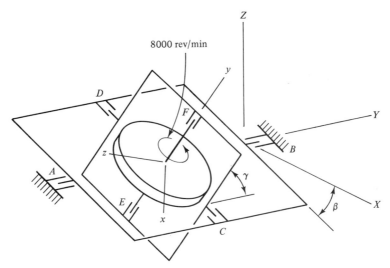

Example 4.2

Solution.   The primary task in applying the Eulerian angle formulas is identification of the precession, nutation, and spin in terms of the given rotations. The angle β is the rotation about the fixed Y axis, so we shall replace ψ in the formulas by β, and $\bar{K}$ by $\bar{J}$. The line of nodes, which is the nutation axis, must be perpendicular to the precession axis, and the spin axis must be perpendicular to the line of nodes. Therefore, we identify the y axis as the spin axis and axis CD as the line of nodes, which we designate as x'. Correspondingly, θ in the formulas becomes γ, since that is the angle between the precession and spin axes, and $\bar{k}$ will be replaced by $\bar{j}$. Finally, we identify the spin angle by using the given information that the z and Z axes were coincident at $t = 0$, at which time $\beta = \gamma = 0$ also.

Because y' was the line of nodes in the derivation, we replace in the formulas $\bar{j}'$ by $\bar{i}'$, $\bar{j}$ by $\bar{i}$, and $\bar{i}$ by $\bar{k}$. Also, the given constant spin rate leads to the spin angle $\phi = 8000(2\pi/60)t$ rad.

In terms of the present notation, Eqs. (4.16) and (4.17) become

$$\bar{\omega} = (-\dot{\beta} \sin\gamma \cos\phi + \dot{\gamma} \sin\phi)\bar{k}$$
$$+ (\dot{\beta} \sin\gamma \sin\phi + \dot{\gamma} \cos\phi)\bar{i} + (\dot{\beta} \cos\gamma + \dot{\phi})\bar{j}$$

$$\bar{\alpha} = (-\ddot{\beta} \sin\gamma \cos\phi + \ddot{\gamma} \sin\phi - \dot{\beta}\dot{\gamma} \cos\gamma \cos\phi + \dot{\beta}\dot{\phi} \sin\gamma \sin\phi$$
$$+ \dot{\phi}\dot{\gamma} \cos\phi)\bar{k} + (\ddot{\beta} \sin\gamma \sin\phi + \ddot{\gamma} \cos\phi + \dot{\beta}\dot{\gamma} \cos\gamma \sin\phi$$
$$+ \dot{\beta}\dot{\phi} \sin\gamma \cos\phi - \dot{\gamma}\dot{\phi} \sin\phi)\bar{i} + (\ddot{\beta} \cos\gamma + \ddot{\phi} - \dot{\beta}\dot{\gamma} \sin\gamma)\bar{j}$$

The Eulerian angles and their derivatives at $t = 4$ s are

$\beta = 0.2939$ rad	$\dot{\beta} = 63.54$ rad/s	$\ddot{\beta} = -7252$ rad/s^2
$\gamma = 0.19021$ rad	$\dot{\gamma} = 19.416$ rad/s	$\ddot{\gamma} = -18{,}773$ rad/s^2
$\phi = 3.351$ rad	$\dot{\phi} = 837.8$ rad/s	$\ddot{\phi} = 0$

The corresponding angular velocity and acceleration are

$$\bar{\omega} = 21.49\bar{i} + 900.2\bar{j} + 7.71\bar{k} \text{ rad/s}$$
$$\bar{\alpha} = 11{,}934\bar{i} - 7354\bar{j} - 14{,}256\bar{k} \text{ rad/s}^2$$

These are the component representatives in terms of the body-fixed xyz axes. We may employ the rotation transformation in Eq. (4.11) to convert the result to the space-fixed XYZ axes, provided that we preserve the permutation of the axis labels in the present application. Thus, we write

$$\begin{Bmatrix} z \\ x \\ y \end{Bmatrix} = [R] \begin{Bmatrix} Z \\ X \\ Y \end{Bmatrix}$$

where

$$[R] = [R_\phi][R_\gamma][R_\beta]$$

$$= \begin{bmatrix} \cos\phi & \sin\phi & 0 \\ -\sin\phi & \cos\phi & 0 \\ 0 & 0 & 1 \end{bmatrix} \begin{bmatrix} \cos\gamma & 0 & -\sin\gamma \\ 0 & 1 & 0 \\ \sin\gamma & 0 & \cos\gamma \end{bmatrix} \begin{bmatrix} \cos\beta & \sin\beta & 0 \\ -\sin\beta & \cos\beta & 0 \\ 0 & 0 & 1 \end{bmatrix}$$

$$= \begin{bmatrix} -0.85909 & -0.47723 & 0.18494 \\ 0.47876 & -0.87707 & -0.03931 \\ 0.18096 & 0.05477 & 0.98196 \end{bmatrix}$$

Applying $[R]^{-1} = [R]^T$ to evaluate the XYZ components of $\bar{\omega}$ and $\bar{\alpha}$ yields

$$\begin{Bmatrix} \omega_Z \\ \omega_X \\ \omega_Y \end{Bmatrix} = [R]^T \begin{Bmatrix} 7.71 \\ 21.49 \\ 900.2 \end{Bmatrix} = \begin{Bmatrix} 166.57 \\ 26.78 \\ 884.54 \end{Bmatrix}$$

$$\begin{Bmatrix} \alpha_Z \\ \alpha_X \\ \alpha_Y \end{Bmatrix} = [R]^T \begin{Bmatrix} -14{,}256 \\ 11{,}934 \\ -7354 \end{Bmatrix} = \begin{Bmatrix} 16{,}630 \\ -4066 \\ -10{,}327 \end{Bmatrix}$$

Thus

$$\overline{\omega} = 26.8\overline{I} + 884.5\overline{J} + 166.6\overline{K} \text{ rad/s}$$
$$\overline{\alpha} = -4066\overline{I} - 10{,}327\overline{J} + 16{,}630\overline{K} \text{ rad/s}^2$$

4.3 INTERCONNECTIONS

Interesting kinematical questions arise when the motion of a body is restricted by other objects. Such restrictions are associated with pin or slider connections between bodies, as well as with a variety of other methods for constructing mechanical systems. The kinematical manifestations of these connections are *constraint equations*, which are mathematical statements of conditions that the connection imposes on the motion of a point or on the angular motion of a body. The kinematical constraints are imposed by *constraint forces* (and couples), which are more commonly known as the *reactions*. The role of constraint forces will be treated in the chapters on kinetics.

A simple, though common, constraint condition arises when a body is only permitted to execute a *planar motion*. By definition, planar motion means that all points in the body follow parallel planes, which can only happen if the angular velocity is always perpendicular to these planes. Let XY of the fixed reference frame and xy of the body-fixed reference frame be coincident planes of motion. Points that differ only in their z coordinate execute the same motion in this case, so they may be considered to be situated in the xy plane. Hence, the kinematical equations for planar motion are

$$\overline{\omega} = \omega\overline{K} = \omega\overline{k} \qquad \overline{\alpha} = \dot{\omega}\overline{K} = \dot{\omega}\overline{k}$$
$$\overline{r}_{B/A} = X\overline{I} + Y\overline{J} = x\overline{i} + y\overline{j}$$
$$\overline{v}_B = \overline{v}_A + \overline{\omega} \times \overline{r}_{B/A} \tag{4.19}$$
$$\overline{a}_B = \overline{a}_A + \overline{\alpha} \times \overline{r}_{B/A} - \omega^2\overline{r}_{B/A}$$

Note that the centripetal acceleration term above is simplified from $\overline{\omega} \times (\overline{\omega} \times \overline{r}_{B/A})$ to $-\omega^2\overline{r}_{B/A}$ by an identity that is valid only when $\overline{r}_{B/A}$ is perpendicular to $\overline{\omega}$. These relations are depicted in Figure 4.6.

Constraint conditions arising from connections may be determined by examining the characteristics of the connection. For example, the *ball-and-socket joint* connecting bodies 1 and 2 in Figure 4.7 allows the bar to have arbitrary orientations. However, point B_1 at the center of the ball moves in unison with point B_2 at the center of the socket. Thus the constraint conditions are

$$\overline{v}_{B1} = \overline{v}_{B2} \tag{4.20}$$

From the above, we may consider the joint to occupy a single point B that belongs to either rigid body.

The case of a *pin connection* between bodies has some elements in common with a ball-and-socket joint. Figure 4.8 depicts such a connection, with the z axis aligned along the axis of the pin. As is true of the ball-and-socket joint, the motion of the point of commonality between the bodies is the same. Consequently, Eq. (4.20) must be satisfied. However, the pin also introduces restrictions on the rotations.

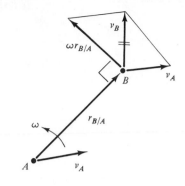

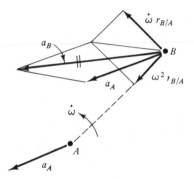

Figure 4.6 Velocity and accleration in planar motion.

In order to develop the constraint equations for angular motion let us define xyz in Figure 4.8 to be fixed to the pin. Each body can only rotate relative to the pin about the z axis, which has been selected to align with the axis of the pin. Let the rate of rotation of each body about the pin be $\dot{\phi}_i \bar{k}$ ($i = 1, 2$). The angular velocity of each body is the sum of the angular velocity of the pin and the angular velocity of that body relative to

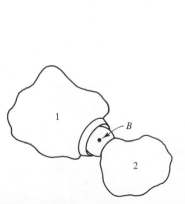

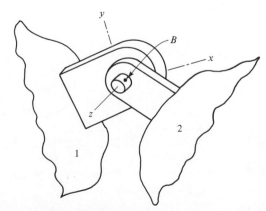

Figure 4.7 Ball-and-socket joint.   **Figure 4.8** Pin connection.

the pin, so that

$$\overline{\omega}_i = \overline{\omega}_{\text{pin}} + \dot{\phi}_i \overline{k} \qquad i = 1, 2 \qquad (4.21)$$

Since the pin can have any rotation rate about the z axis without affecting the rotation of the bodies it joins, let that rate be zero. Thus, without loss of generality,

$$\overline{\omega}_{\text{pin}} = \omega_x \overline{i} + \omega_y \overline{j} \qquad (4.22)$$

Then, because $\overline{i}$, $\overline{j}$, and $\overline{k}$ are fixed to the pin, the angular motion of the bodies is given by

$$\overline{\omega}_i = \omega_x \overline{i} + \omega_y \overline{j} + \dot{\phi}_i \overline{k} \qquad (4.23)$$
$$\overline{\alpha}_i = \dot{\omega}_x \overline{i} + \dot{\omega}_y \overline{j} + \ddot{\phi}_i \overline{k} + \omega_x(\overline{\omega}_{\text{pin}} \times \overline{i}) + \omega_y(\overline{\omega}_{\text{pin}} \times \overline{j}) + \dot{\phi}_i(\overline{\omega}_{\text{pin}} \times \overline{k}) \quad (4.24)$$

These are constraint equations on the angular motion that must be satisfied in addition to Eq. (4.20) for the connecting points. Note that situations where it is more convenient to employ a coordinate system that does not have an axis aligned with the pin may be treated by transforming the components of Eqs. (4.23) and (4.24) to the desired directions.

Another common method for connecting bodies consists of a *collar* that slides over a bar, as depicted in Figure 4.9. (This connection is also known as a *slider*.) Similar to the treatment of a ball-and-socket joint, point B_1 in the figure denotes the pivot point on the collar. The collar is free to slide over bar CD. Let B_2 denote the point on bar CD that coincides at this instant with the projection of point B_1.

The constraint condition in this case is obtained by attaching a (moving) reference frame to bar CD. With respect to this reference frame, the collar only seems to be moving in or out along line CD. (Either sense may be assumed if it is not specified.) Let u denote the speed of the collar relative to bar CD, and let subscript CD denote a motion term that is relative to bar CD. Then

$$(\overline{v}_{B1})_{CD} = u\overline{e}_{D/C} \qquad (\overline{a}_{B1})_{CD} = \dot{u}\overline{e}_{D/C} \qquad (4.25)$$

Note that curvature of bar CD would introduce a centripetal acceleration term in the above, but it would not require any other modification.

The collar is usually small in comparison with the bodies it joins, in which case the distance between points B_1 and B_2 is negligible. Setting $\overline{r}_{B1/B2} = \overline{0}$ reduces the rigid-

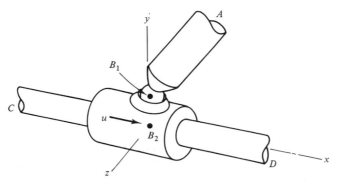

Figure 4.9 Collar connection.

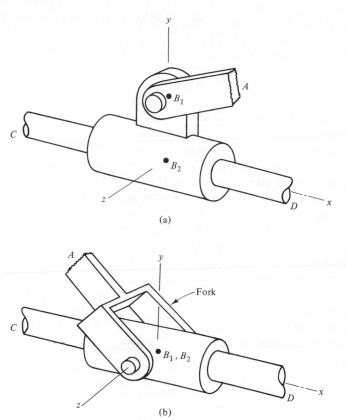

Figure 4.10 Collar connections.

body equations relating points B_1 and B_2 to

$$\bar{v}_{B1} = \bar{v}_{B2} + u\bar{e}_{D/C}$$
$$\bar{a}_{B1} = \bar{a}_{B2} + \dot{u}\bar{e}_{D/C} + 2\bar{\omega}_{CD} \times u\bar{e}_{D/C} \tag{4.26}$$

These are the constraint equations imposed by the collar connection on the motion of points B_1 and B_2.

The collar introduces no restriction on the angular motion if it is fastened to body 1 by a ball-and-socket joint. However, a common connection method is a pin (Figure 4.10a) or a fork-and-clevis joint (Figure 4.10b). In such cases, the rotation constraint produced by the connecting pin and the collar is that both bars must have the same rotation about the x and y axes, both of which are perpendicular to the axis of the pin. In other words, the constraint on angular motion is the same as that described by Eqs. (4.23) and (4.24).

It should be obvious from the discussion thus far that the connections need to be examined in detail to identify all constraints on the motion. If all the permutations and novel features of various types of connections were to be tabulated, it would not aid our understanding. It is preferable to consider each connection on a case-by-case basis (if it is not familiar), and then to employ the type of reasoning developed thus far to identify the constraint equations.

Describing the constraint conditions for a system in mathematical terms is one aspect of an overall kinematical study. It also is necessary to relate the velocity and acceleration of constrained points in each body through Eqs. (4.4). When such relations are broken down into components, one obtains algebraic equations for the kinematical variables describing the motion of each body. In combination with the constraint equations, the result will be a system of simultaneous equations.

If the motion of the system is *fully constrained*, this system of equations will be solvable such that, for each body, the linear motion of a point (velocity or acceleration) and the angular motion may be evaluated. If the system is only *partially constrained*, there will be more kinematical variables than the number of kinematical equations. The simultaneous equations may then be solved for a set of excess variables in terms of the other. The excess variables in this case depend on the nature of the force system, so their evaluation requires a kinetics study. Another possibility is that the kinematical equations are not solvable. In that case, there are too many constraints on the motion of the system. This means that no motion is possible—such a system is *rigid*.

In order to demonstrate these matters, consider bar AB in Figure 4.11, which is constrained by collars to follow nonplanar guide bars. The connection at collar A is a ball-and-socket joint, whereas the one at collar B is a pin.

The collars are constrained to follow the guide bars. Since the guides are fixed, the velocity constraint equations (4.26) reduce to

$$\bar{v}_A = u_A \bar{e}_{F/E} \qquad \bar{v}_B = u_B \bar{e}_{D/C} \qquad (4.27)$$

where the sense of these velocities will be determined by the sign of u_A and u_B. Let us for the moment ignore the constraint on the rotation of bar AB that is introduced by the

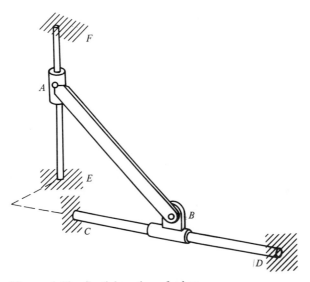

Figure 4.11   Spatial motion of a bar.

pin on collar B. Then the angular velocity of the bar is an unknown $\bar{\omega}_{A/B}$ having three components. Because points A and B are in the same rigid body, their velocities are related by

$$\bar{v}_A = \bar{v}_B + \bar{\omega}_{AB} \times \bar{r}_{A/B} \qquad (4.28)$$

This kinematical relation reduces to three scalar equations for five unknowns: u_A, u_B, and the three components of $\bar{\omega}_{AB}$. In order to further characterize the system, we need to express the constraint on $\bar{\omega}_{AB}$ introduced by the pin on collar B. The round cross-sectional shape of guide bar CD permits collar B and bar AB to execute an arbitrary rotation about the guide bar. Since the guide bar is fixed, it serves as the precession axis for the overall rotation of bar AB. Thus, one component of the angular velocity of bar AB is $\dot{\psi}\bar{e}_{D/C}$. In addition, bar AB may rotate relative to the collar about the axis of the pin in the collar. This axis is always perpendicular to the guide bar, which is the precession axis. It follows that the pin defines the axis of nutation. The corresponding contribution to the angular velocity is $\dot{\theta}\bar{e}_\theta$, where the orientation $\bar{e}_\theta$ of the nutation axis is determined from the fact that the pin is oriented perpendicular to the plane formed by bars CD and AB,

$$\bar{e}_\theta = \frac{\bar{e}_{B/A} \times \bar{e}_{D/C}}{|\bar{e}_{B/A} \times \bar{e}_{D/C}|} \qquad (4.29)$$

The precession and nutation are the only rotations permitted by the sliding collar connection, so the angular velocity of bar AB is given by

$$\bar{\omega}_{AB} = \dot{\psi}\bar{e}_{D/C} + \dot{\theta}\frac{\bar{e}_{B/A} \times \bar{e}_{D/C}}{|\bar{e}_{B/A} \times \bar{e}_{D/C}|} \qquad (4.30)$$

This construction reduces the number of velocity unknowns in the system to four, u_A, u_B, $\dot{\psi}$, and $\dot{\theta}$. Two possibilities arise now. In a fully constrained situation, the overall motion will be defined through some kinematical input such as a specified motion for either collar. This removes the corresponding velocity parameter from the list of unknowns, thereby making Eqs. (4.27), (4.28), and (4.30) into a solvable set. In contrast, if the motion is induced by a given set of forces, the system is partially constrained. In that case, the foregoing kinematical equations lead to characterization of the motion in terms of a single kinematical unknown. Kinetics principles would relate this unknown to the force system.

Let us consider other possibilities. Suppose that bar AB was connected to both collars by pins. That would introduce another constraint on $\bar{\omega}_{AB}$ like Eq. (4.30). In most cases, it would not be possible to satisfy simultaneously both angular motion constraints, unless $\bar{\omega}_{AB} = \bar{0}$. This is the case of a rigid system. One notable exception where two pin connections would be acceptable is planar motion. This occurs when the guide bars are coplanar, and the axis of each pin is perpendicular to the plane containing the guide bars. Then $\bar{\omega}_{AB}$ has only a single component perpendicular to the plane, and the velocity relation, Eq. (4.28), only has components in the plane. In other words, the planar system with pin connections to the collars is not rigid.

A situation of partial constraint is obtained when bar AB is connected to both collars by ball-and-socket joints, because Eq. (4.30) then does not apply. The lack of constraint in such a case is associated with the ability of bar AB to spin about its own axis. Since such a rotation does not affect the motion of either collar, the kinematical equations can be solved for a relation between u_A and u_B, although there will be no unique solution for $\overline{\omega}_{AB}$. If it is desired that the number of equations and unknowns match, such a problem could be remedied by considering one of the joints to be a pin. Another approach in this case is to obtain an additional relation by setting $\overline{\omega}_{AB} = \omega_x \overline{i} + \omega_y \overline{j} + \omega_x \overline{i}$ and then requiring that $\overline{\omega}_{AB} \cdot \overline{e}_{B/A} = 0$.

Thus far the discussion has only addressed the analysis of velocities. The treatment of acceleration follows a parallel development. However, it is necessary to evaluate the velocities first. One reason for this is that the angular velocities occur in the acceleration relation, Eq. (4.4). Also, velocity parameters are required to describe the acceleration of a point that follows a curved path. For example, if we were to employ path variables to formulate the relative acceleration of a collar as it slides over a curved bar, we would require knowledge of the speed of the collar in order to express the centripetal acceleration. In all other respects the analysis of acceleration is essentially a retracing of the procedure whereby accelerations are obtained.

An area of special interest in kinematics is concerned with *linkages*, in which bars are interconnected in order to convert an input motion to a different output motion. From the standpoint of our general approach to rigid-body motion, the treatment of linkages presents no special problems. The constrained points in the system are the ends of the linkage, the connection points, and any point whose motion is specified. The constraint conditions on velocity (and then acceleration) of the constrained points are expressed according to the type of connection, e.g., pin or collar. All constraint conditions on the rotation of each member of the linkage resulting from the method of interconnection are used to express the angular velocity (and then the angular acceleration). An example of such a rotation constraint is given by Eqs. (4.23) and (4.24) for a slider connected to a member by means of a pin.

After the constraint equations have been formulated, the basic approach is to relate the velocity (and then acceleration) of the constrained points to that of the other constrained points using the kinematical relation between the motion of two points in a rigid body. This may be achieved by starting at each end and working inward toward a selected connection in a progressive manner from one link to its neighbor. The constraint equations are used to describe the various terms that arise in the basic kinematical equations for each member.

The ultimate result of the procedure will be two vector expressions for the velocity (or acceleration) of the selected connection, each relation corresponding to different paths through the linkage. Since the two expressions describe the same point, they must be equal. Equating the corresponding components of each expression leads to a set of algebraic equations that should be solvable for all unknown parameters (assuming that the linkage is fully constrained, so that its motion is defined by the input).

There is one complication that arises in treating linkages. The position of all constrained points in the linkage at any given instant must be described. This is not an overwhelming difficulty for planar linkages, but the geometrical relations for an arbitrary spatial linkage require careful consideration in their formulation. Rotation transformation matrices sometimes prove to be useful for this task.

EXAMPLE 4.3

Collar B is pinned to arm AB as it slides over a circular guide bar. The guide bar translates to the left at a constant speed v, such that the distance from pivot A to the center C is vt. Derive expressions for the angular velocity and angular acceleration of arm AB.

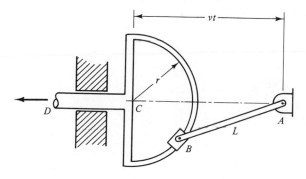

Example 4.3

Solution. In general, the task of treating linkages is expedited by defining a space-fixed set of axes XYZ. This coordinate system, which is depicted in the figure, is sometimes referred to as a *global coordinate system*. It is used to relate vectors that are associated with different bodies.

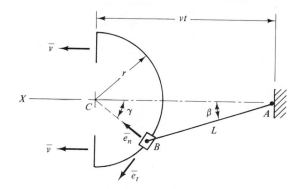

Coordinate system.

First, we relate the angle of orientation of each bar to time by using the law of cosines, which gives

$$\cos \beta = \frac{v^2t^2 + L^2 - r^2}{2\,Lvt}$$

$$\cos \gamma = \frac{v^2t^2 + r^2 - L^2}{2rvt}$$

Hence, we may consider β and γ to be known at any instant t. We could differentiate these expressions in order to determine the time derivatives of these angles. However, the derivatives are cumbersome, particularly for the second derivative. Also, an approach based on differentiation of analytical expressions is inherently limited to situations where the geometrical relations are relatively uncomplicated.

The vectors $\bar{e}_t$ and $\bar{e}_n$ in the sketch are the path variable unit vectors for the motion of the collar relative to guide bar CD. We also know that the collar is pinned to bar AB. Since body CD is translating at speed v to the left, we have

$$\bar{v}_B = v\bar{I} + u\bar{e}_t = \bar{\omega}_{AB} \times \bar{r}_{B/A}$$

where u is the relative speed of the collar.

Resolving the vectors into components relative to XYZ yields

$$\bar{v}_B = v\bar{I} + u(\sin\gamma\,\bar{I} - \cos\gamma\,\bar{J}) = (-\dot{\beta}\bar{K}) \times (L\cos\beta\,\bar{I} - L\sin\beta\,\bar{J})$$

$$\bar{v}_B \cdot \bar{I} = v + u\sin\gamma = -\dot{\beta}\,L\sin\beta$$
$$\bar{v}_B \cdot \bar{J} = -u\cos\gamma = -\dot{\beta}\,L\cos\beta$$

The solution of the component equations is

$$u = -\frac{v}{\sin\gamma + \cos\gamma\,\tan\beta} = -\frac{v\cos\beta}{\sin(\gamma+\beta)}$$

$$\dot{\beta}L = -\frac{v}{\cos\beta\,\tan\gamma + \sin\beta} = -\frac{v\cos\gamma}{\sin(\gamma+\beta)}$$

The corresponding velocity of collar B is

$$\bar{v}_B = (\bar{v}_B \cdot \bar{I})\bar{I} + (\bar{v}_B \cdot \bar{J})\bar{J}$$

$$\bar{v}_B = \frac{v\cos\gamma}{\sin(\gamma+\beta)}(\sin\beta\,\bar{I} + \cos\beta\,\bar{J})$$

The same approach is valid for acceleration. Since the translational velocity of guide bar CD is constant, we have

$$\bar{a}_B = \dot{u}\bar{e}_t + \frac{u^2}{r}\bar{e}_n = \bar{\alpha}_{AB} \times \bar{r}_{B/A} - \omega_{AB}^2\bar{r}_{B/A}$$

In component form, this yields

$$\bar{a}_B = \dot{u}(\sin\gamma\,\bar{I} - \cos\gamma\,\bar{J}) + \frac{u^2}{r}(\cos\gamma\,\bar{I} + \sin\gamma\,\bar{J})$$

$$= (-\ddot{\beta}\bar{K}) \times (L\cos\beta\,\bar{I} - L\sin\beta\,\bar{J}) - \dot{\beta}^2(L\cos\beta\,\bar{I} - L\sin\beta\,\bar{J})$$

$$\bar{a}_B \cdot \bar{I} = \dot{u}\sin\gamma + \frac{u^2}{r}\cos\gamma = -\ddot{\beta}\,L\sin\beta - \dot{\beta}^2\,L\cos\beta$$

$$\bar{a}_B \cdot \bar{J} = -\dot{u}\cos\gamma + \frac{u^2}{r}\sin\gamma = -\ddot{\beta}\,L\cos\beta + \dot{\beta}^2\,L\sin\beta$$

The identities for the sine and cosine of the sum of two angles leads to the following solutions of the component equations:

$$\dot{u} = -\frac{\dot{\beta}^2 L}{\sin(\gamma+\beta)} - \frac{u^2}{r}\cot(\gamma+\beta)$$

$$\ddot{\beta}\,L = -\dot{\beta}^2 L\cot(\beta+\gamma) - \frac{u^2}{r\sin(\gamma+\beta)}$$

The corresponding expression for the acceleration in component form is

$$\bar{a}_B = (\bar{a}_B \cdot \bar{I})\bar{I} + (\bar{a}_B \cdot \bar{J})\bar{J}$$

$$\bar{a}_B = -\frac{\dot{\beta}^2 Lr \sin \gamma - u^2 \sin \beta}{r \sin (\gamma + \beta)}\bar{I} + \frac{\dot{\beta}^2 Lr \cos \gamma + u^2 \cos \beta}{r \sin (\gamma + \beta)}\bar{J}$$

These results for $\bar{v}_B$ and $\bar{a}_B$ are implicit functions of time, because we first expressed β and γ in terms of t, and then found $\dot{\beta}$ and u in terms of β and γ.

EXAMPLE 4.4 _____

Collar A moves downward and to the right at a constant speed of 100 ft/s. The connection of link AB to collar A is a ball-and-socket joint, whereas that at collar B is a pin. Determine the velocity and acceleration of collar B, and the angular velocity and angular acceleration of bar AB, for the position shown.

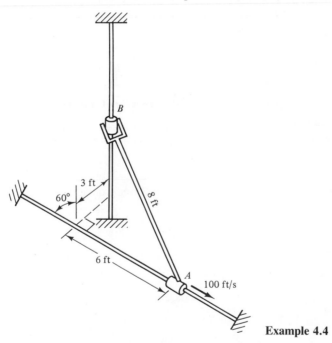

Example 4.4

Solution. We begin by expressing the constraints on the motion of each collar in terms of components relative to the fixed XYZ axes. Since each follows a straight path, we have

$$\bar{v}_A = 100(\cos 30°\bar{I} - \sin 30°\bar{K}) = 86.60\bar{I} - 50\bar{K} \text{ ft/s} \qquad \bar{a}_A = \bar{0}$$
$$\bar{v}_B = v_B(-\bar{K}) \qquad \bar{a}_B = \dot{v}_B(-\bar{K})$$

Next, we describe the constraint on the angular motion of bar AB imposed by the pin at end B. The angular velocity of bar AB consists of a precession about the

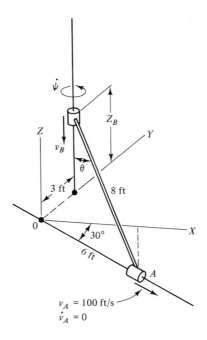

Coordinate system.

guide bar at end B and a nutation about the axis of the pin, which is orthogonal to the vertical guide bar and bar AB. Thus

$$\overline{\omega}_{AB} = \dot{\psi}\overline{K} + \dot{\theta}\overline{e}_\theta \qquad \overline{e}_\theta = \frac{\overline{r}_{A/B} \times \overline{K}}{|\overline{r}_{A/B} \times \overline{K}|}$$

Since the pin can only undergo the precessional rotation, the angular acceleration is

$$\overline{\alpha}_{AB} = \ddot{\psi}\overline{K} + \ddot{\theta}\overline{e}_\theta + \dot{\theta}(\dot{\psi}\overline{K} \times \overline{e}_\theta)$$

In order to express these vectors in component form, we evaluate the Z coordinate of end B at this instant,

$$Z_B = [8^2 - 3^2 - (6 \cos 30°)^2]^{1/2} - 6 \sin 30° = 2.292 \text{ ft}$$

Then

$$\overline{r}_{A/B} = 6 \cos 30°\overline{I} - 3\overline{J} - (Z_B + 6 \sin 30°)\overline{K}$$
$$= 5.196\overline{I} - 3\overline{J} - 5.292\overline{K} \text{ ft}$$
$$\overline{e}_\theta = \frac{-3\overline{I} - 5.196\overline{J}}{(3^2 + 5.196^2)^{1/2}} = -0.50\overline{I} - 0.8660\overline{J}$$
$$\overline{\omega}_{AB} = \dot{\theta}(-0.50\overline{I} - 0.8660\overline{J}) + \dot{\psi}\overline{K}$$
$$\overline{\alpha}_{AB} = \ddot{\psi}\overline{K} + \ddot{\theta}(-0.50\overline{I} - 0.8660\overline{J}) + \dot{\psi}\dot{\theta}(0.8660\overline{I} - 0.50\overline{J})$$

Now that the effects of the constraints have been described, it only remains to relate the motion of ends A and B. The velocity relation gives

$$\bar{v}_A = \bar{v}_B + \bar{\omega}_{AB} \times \bar{r}_{AB}$$

$$86.60\bar{I} - 50\bar{K} = -v_B\bar{K} + (-0.5\dot\theta\bar{I} - 0.8660\dot\theta\bar{J} + \dot\psi\bar{K})$$
$$\times (5.196\bar{I} - 3\bar{J} - 5.292\bar{K})$$

$$\bar{v}_A \cdot \bar{I} = 86.60 = 4.582\dot\theta + 3\dot\psi$$
$$\bar{v}_A \cdot \bar{J} = 0 = -2.646\dot\theta + 5.196\dot\psi$$
$$\bar{v}_A \cdot \bar{K} = -50 = -v_B + 6.0\dot\theta$$

$$\dot\theta = 14.174 \text{ rad/s} \qquad \dot\psi = 7.218 \text{ rad/s} \qquad v_B = 135.04 \text{ ft/s}$$

Substitution of these results into the earlier expressions yields

$$\bar{\omega}_{AB} = -7.087\bar{I} - 12.275\bar{J} + 7.218\bar{K} \text{ rad/s}$$
$$\bar{v}_B = -135.04\bar{K} \text{ ft/s}$$

We may now relate the acceleration of the collars using

$$\bar{a}_A = \bar{a}_B + \bar{\alpha}_{AB} \times \bar{r}_{A/B} + \bar{\omega}_{AB} \times (\bar{\omega}_{AB} \times \bar{r}_{AB})$$

When we substitute the results for $\dot\theta$ and $\dot\psi$ into the general expression for $\bar{\alpha}_{AB}$, we obtain

$$\bar{\alpha}_{AB} = (-0.5\ddot\theta + 88.60)\bar{I} + (-0.8660\ddot\theta - 51.15)\bar{J} + \ddot\psi\bar{K}$$

The constraints on the accelerations of the collars are that $\bar{a}_A = \bar{0}$ and $\bar{a}_B = -\dot{v}_B\bar{K}$, which leads to

$$\bar{a}_A = \bar{0} = -\dot{v}_B\bar{K} + [(-0.5\ddot\theta + 88.60)\bar{I} + (-0.8660\ddot\theta - 51.15)\bar{J} + \ddot\psi\bar{K}]$$
$$\times (5.196\bar{I} - 3\bar{J} - 5.292\bar{K}) + (-7.087\bar{I} - 12.275\bar{J} + 7.218\bar{K})$$
$$\times [(-7.087\bar{I} - 12.275\bar{J} + 7.218\bar{K}) \times (5.196\bar{I} - 3\bar{J} - 5.292\bar{K})]$$

$$\bar{a}_A \cdot \bar{I} = 0 = -5.292(-0.8660\ddot\theta - 51.15) + 3\ddot\psi - 1043.9$$
$$\bar{a}_A \cdot \bar{J} = 0 = 5.292(-0.50\ddot\theta + 88.60) + 5.196\ddot\psi + 1227.9$$
$$\bar{a}_A \cdot \bar{K} = 0$$
$$= -\dot{v}_B - 3(-0.5\ddot\theta + 88.60) - 5.196(-0.8660\ddot\theta - 51.15) + 1063.2$$

The solution of these simultaneous equations is

$$\ddot\theta = 286.9 \text{ rad/s}^2 \qquad \ddot\psi = -180.5 \text{ rad/s}^2 \qquad \dot{v}_B = 2784.3 \text{ ft/s}^2$$

from which we obtain

$$\bar{\alpha}_{AB} = -54.8\bar{I} - 299.6\bar{J} - 180.5\bar{K} \text{ rad/s}^2$$
$$\bar{a}_B = -1721.1\bar{K} \text{ ft/s}^2$$

EXAMPLE 4.5

Two shafts lying in a common horizontal plane at a skew angle β are connected by a cross-link universal joint that is called a *cardan joint*. Derive an expression for the rotation rate ω_2 in terms of ω_1 and the instantaneous angle of rotation ϕ_1, where cross link AB is horizontal when $\phi_1 = 0$.

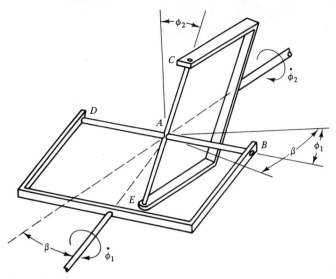

Example 4.5

Solution.   There are a variety of approaches to this problem, but they all require that the angles of rotation ϕ_1 and ϕ_2 of the respective shafts be related. Once such a relation is developed, it is a simple matter to obtain the angular rates by differentiation.

Arm AC is perpendicular to arm AB, so $\bar{e}_{C/A} \cdot \bar{e}_{B/A} = 0$. In order to describe these unit vectors, we define coordinate axes $X_1 Y_1 Z_1$ and $X_2 Y_2 Z_2$ such that each X axis is parallel to the corresponding axis of rotation and each Z axis is perpendicular to the plane formed by the two rotation axes. Because arm AB rotates about the X_1 axis, while arm AC rotates about the X_2 axis, we have

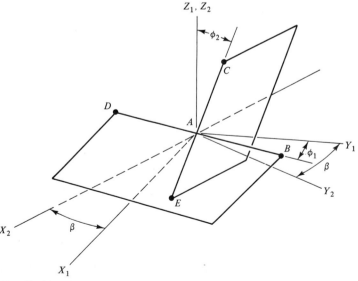

Coordinate systems.

$$\bar{e}_{B/A} = \cos \theta_1 \bar{J}_1 - \sin \phi_1 \bar{K}_1 \qquad \bar{e}_{C/A} = \sin \phi_2 \bar{J}_2 + \cos \phi_2 \bar{K}_2$$

Because the Z axes coincide and the angle between the Y axes is β, the orthogonality condition becomes

$$\bar{e}_{B/A} \cdot \bar{e}_{C/A} = \cos \phi_1 \sin \phi_2 \cos \beta - \sin \phi_1 \cos \phi_2 = 0$$

$$\tan \phi_2 = \frac{1}{\cos \beta} \tan \phi_1$$

This is a general relation between the angles, so it may be differentiated. Thus

$$\frac{1}{\cos^2 \phi_2} \dot{\phi}_2 = \frac{1}{\cos \beta \cos^2 \phi_1} \dot{\phi}_1$$

In order to remove the dependence on ϕ_2, we employ a trigonometric identity to find

$$\cos^2 \phi_2 = \frac{1}{1 + \tan^2 \phi_2} = \frac{1}{1 + \tan^2 \phi_1 / \cos^2 \beta}$$

which leads to

$$\dot{\phi}_2 = \frac{\cos \beta}{\sin^2 \phi_1 + \cos^2 \beta \cos^2 \phi_1} \dot{\phi}_1$$

It is interesting to observe that the maximum and minimum values of $\dot{\phi}_2$ are $\dot{\phi}_1 / \cos \beta$ at $\phi_1 = 0$ and π, and $\dot{\phi}_1 \cos \beta$ at $\phi_1 = \pi/2$ and $3\pi/2$, respectively. This oscillation relative to the input speed $\dot{\phi}_1$ makes the cardan joint by itself unsuitable as a constant-velocity joint. In conventional front-engine, rear-wheel-drive automobiles, two cardan joints are employed in the drive train in opposition. The reciprocal arrangement produces a final speed that matches the input.

4.4 ROLLING

A common constraint condition arises when bodies rotate as they move over each other. The fact that the contacting surfaces cannot penetrate each other imposes a restriction on the velocity components perpendicular to the plane of contact (i.e., the tangent plane). Figure 4.12 shows two surfaces in contact, as viewed edgewise along their plane of contact. The z axis in the figure is defined to be normal to the plane of contact. Because the surface of each body is impenetrable, the velocity components normal to the contact plane must match. Let C_1 and C_2 be contacting points on each body. Then

$$\bar{v}_{C1} \cdot \bar{k} = \bar{v}_{C2} \cdot \bar{k} \tag{4.31}$$

The special case of *rolling without slipping* imposes an additional constraint. A variety of viewpoints are available to treat this type of motion. Consider two pairs of points on the perimeter of bodies that roll over each other without slipping such as B_i and C_i in Figure 4.13. These points are selected such that B_1 and B_2 were the points of

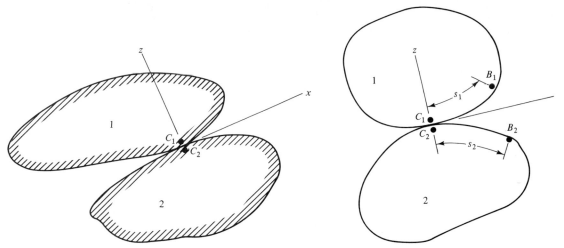

Figure 4.12 Rolling contact.　　　　　　**Figure 4.13** No slip condition.

contact at an earlier instant, and C_1 and C_2 are the current points of contact. (The figure considers a planar situation, in order to readily depict all points of contact.) The elimination of slipping means that the arclength s_1 along the perimeter of body 1 between points B_1 and C_1 is the same as the arclength s_2 along body 2 between points B_2 and C_2. The restriction on arclengths is one way of describing the constraint imposed by the condition of no slipping.

The most common application of this type of description is for a wheel rolling along the ground. The path of a point on the circumference of the wheel is a *cycloid*. The geometrical parameters needed to characterize this path are depicted in Figure 4.14, where the origin of the fixed reference frame has been placed at the starting position of the center of the wheel. At that position point A on the cylinder contacted the ground, whereas point B is the current contact point. When there is no slippage, the arclength between points A and B is the same as the distance x that the center of the wheel has displaced. Thus,

$$x = R\theta \tag{4.32}$$

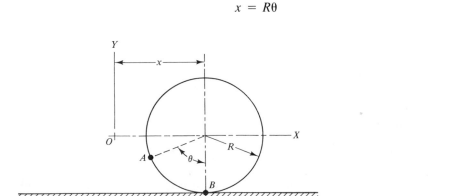

Figure 4.14 Rolling wheel.

From this relation the position of point A may be described in parametric form as a function of θ as

$$\bar{r}_{A/O} = R(\theta - \sin\theta)\bar{i} - R\cos\theta\bar{j} \qquad (4.33)$$

The cycloidal path is depicted in Figure 4.15.

Expressions for the velocity and acceleration of point A may be found by differentiation of the foregoing relationship, in conjunction with the fact that the speed of the center is $v = \dot{x} = R\dot{\theta}$. The results are

$$\bar{v}_A = v(1 - \cos\theta)\bar{i} + v\sin\theta\bar{j}$$
$$\bar{a}_A = [\dot{v}(1 - \cos\theta) + \frac{v^2}{R}\sin\theta]\bar{i} + [\dot{v}\sin\theta + \frac{v^2}{R}\cos\theta]\bar{j} \qquad (4.34)$$

An aspect of the velocity and acceleration of particular relevance to further developments arises at $\theta = 0$, at which position $\bar{v}_A = \bar{0}$ and $\bar{a}_A = (v^2/R)\bar{j}$. In other words, the contact point comes to rest, and the acceleration is upward. This corresponds to the cusp in the cycloidal path.

One difficulty with a formulation in terms of arclengths is that it becomes increasingly difficult to use as the complexity of the motion increases. The alternative method to be developed below uses constraint conditions on velocity and acceleration that are derived from the "equal arclength" rule.

Consider the limiting situation where the points of contact B_i and C_i in Figure 4.13 correspond to instants that are very close. The points of contact on each body then seem to have the same motion along the contact plane at the instant they contact. Hence, the velocities of contacting points of bodies that roll over each other without slipping must match in all directions.

$$\bar{v}_{C1} = \bar{v}_{C2} \qquad \text{for no slipping} \qquad (4.35)$$

Acceleration is more complicated because the contacting points on each body come together and then separate. This means that they have different accelerations in the normal direction. A common misconception arises from the case of the rolling wheel in Figure 4.14, as well as other planar situations. Evaluation of the expression for acceleration of point A on the circumference of the wheel showed that $\bar{a}_A$ is upward when the point contacts the ground. This is often taken (incorrectly) to be generally valid. Specifically, the result is interpreted to mean that, in the absence of slipping, the contact points may

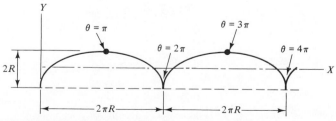

Figure 4.15 Cycloidal path.

only accelerate perpendicularly to the contact plane. However, the statement is not true in many cases of spatial motion, where the normal to the plane of contact does not necessarily coincide with the osculating plane of the path of a contacting point.

As a way of identifying the difficulty, consider the two spheres in Figure 4.16 that rotate at constant rates ω_1 and ω_2 about fixed parallel axes, such that there is no slipping between the contacting points A and B. The perpendicular distances R_1 and R_2 from the points of contact to the respective axes of rotation may be found from similar triangles, and the ratio of the rotation rates is such that the condition $v_A = v_B$ must be satisfied. Because points on each sphere follow circular paths, their (centripetal) acceleration is perpendicular to the rotation axes, as shown in the figure. As a result, each acceleration has a component parallel to the plane of contact.

The lack of a direct constraint condition for the acceleration presents a dilemma, whose resolution lies in the existence of another constraint. The shape of each rolling body is constant. That shape is usually expressed by a functional relationship for the distance from a reference point on the body to points on the perimeter. Round objects are of primary concern for engineering applications; the point of reference in that case is the center. The constraint that there is a constant distance from the contact point to the center of a round body must be satisfied. In effect, this means that the center is subject to a kinematical constraint.

In order to explore this idea, consider the planar situation of a planetary gear rolling over a fixed inner gear (Figure 4.17). Because the distance from the center A of the planetary gear to the point of contact C is constant, point A follows a circular path of radius $r_1 + r_2$. Thus, for the xyz coordinate system depicted in the figure, the velocity and acceleration of point A are described according to path variables as

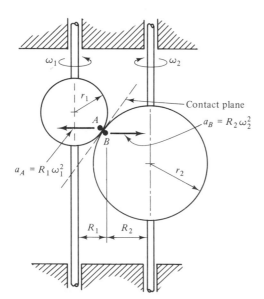

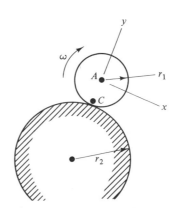

Figure 4.16 Contact between rotating spheres.

Figure 4.17 Rolling of circular shapes.

$$\bar{v}_A = v_A \bar{i} \qquad \bar{a}_A = \dot{v}_A \bar{i} - \frac{v_A^2}{r_1 + r_2}\bar{j} \qquad (4.36)$$

Because there is no slipping at point C, the velocity constraint $\bar{v}_C = \bar{0}$ yields

$$\bar{v}_A = \bar{\omega} \times \bar{r}_{A/C} = (-\omega\bar{k}) \times r_1\bar{j} = \omega r_1\bar{i} \qquad (4.37)$$

Thus, the constraint that the center A move tangent to its circular path is satisfied by the expression for $\bar{v}_A$ obtained independently from the second expression. Comparing the two expressions for $\bar{v}_A$ yields the familiar relation

$$\bar{v}_A = \omega r_1 \qquad (4.38)$$

Now consider acceleration. Because of the round shape and the restriction to planar motion, the algebraic relation between the speed of each geometric center and the angular speeds of the contacting bodies will not change. Such a relation may be differentiated with respect to t, so that

$$\dot{v}_A = \dot{\omega} r_1 \qquad (4.39)$$

Substitution of Eqs. (4.38) and (4.39) into Eq. (4.36) yields a complete description of $\bar{a}_A$ without consideration of $\bar{a}_C$.

The same approach may be extended directly to treat cases of spatial motion in which the orientation of the contacting bodies relative to each other does not change as the bodies roll. An example of such a situation is a cone rolling over a plane. Example 4.7 illustrates another situation of this type.

In some spatial motions the relationships for rotation rates might be position-dependent. Nevertheless, the basic concept of differentiating a general expression remains unchanged. A common system in which the geometrical properties are not constant is a rolling coin. Figure 4.18 depicts a disk that is rolling without slipping over a flat surface in a wobbly manner. Let $\dot{\psi}$ be the precession rate (about the fixed Z axis), and let $\dot{\theta}$ be

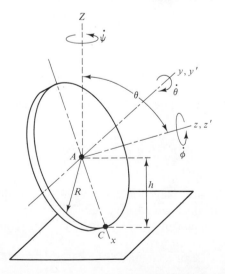

Figure 4.18 Rolling disk in arbitrary motion.

the nutation rate (about the horizontal axis in the plane of the disk). Let $x'y'z'$ be a reference frame whose origin is at the center of the disk, with the z' axis coincident with the axis of the disk and the y' axis coincident with the line of nodes for the Eulerian angles. The body-fixed xyz axes spin at $\dot{\phi}$ relative to $x'y'z'$.

The importance of the round shape in this case is that, if all other quantities are held fixed, the motion of the system will not be altered by changing the value of the spin angle ϕ. As a consequence of this invariance, the $x'y'z'$ axes may be used to obtain a general relation for the velocity of point A. The expressions for the angular velocity $\bar{\omega}'$ of the $x'y'z'$ reference frame and for $\bar{\omega}$ of the disk in terms of the Eulerian angles are

$$\begin{aligned}
\bar{\omega}' &= -\dot{\psi}\sin\theta\,\bar{i}' + \dot{\theta}\bar{j}' + \dot{\psi}\cos\theta\,\bar{k}' \\
\bar{\omega} &= -\dot{\psi}\sin\theta\,\bar{i}' + \dot{\theta}\bar{j}' + (\dot{\phi} + \dot{\psi}\cos\theta)\bar{k}'
\end{aligned} \tag{4.40}$$

The condition that point C not slip relative to the ground leads to relations between the motion of the center A and the Eulerian angles. It follows from $\bar{v}_C = \bar{0}$ that

$$\bar{v}_A = \bar{\omega} \times \bar{r}_{A/C} = -R(\dot{\phi} + \dot{\psi}\cos\theta)\bar{j}' + R\dot{\theta}\bar{k}' \tag{4.41}$$

This is a general relation for $\bar{v}_A$ in terms of the Eulerian angles. It therefore may be differentiated to determine $\bar{a}_A$. Toward this end, remember that $\bar{\omega}'$ must be employed to express the derivatives of the unit vectors in Eq. (4.41). Thus,

$$\begin{aligned}
\bar{a}_A &= \frac{\delta\bar{v}_A}{\delta t} + \bar{\omega}' \times \bar{v}_A \\
&= R[\dot{\theta}^2 + (\dot{\phi} + \dot{\psi}\cos\theta)\dot{\psi}\cos\theta]\bar{i}' + R(-\ddot{\phi} - \ddot{\psi}\cos\theta + 2\dot{\psi}\dot{\theta}\sin\theta)\bar{j}' \\
&\quad + R[\ddot{\theta} + (\dot{\phi} + \dot{\psi}\cos\theta)\dot{\psi}\sin\theta]\bar{k}' \tag{4.42}
\end{aligned}$$

Equations (4.41) and (4.42) provide general relations between the motion of the center and the Eulerian angles when there is no slipping. Although they contain a variety of effects, one is readily identifiable. The components of $\bar{v}_A$ and $\bar{a}_A$ perpendicular to the contact plane, that is, in the direction of the Z axis, are found with the aid of Figure 4.18 to be

$$\bar{v}_A \cdot \bar{K} = R\dot{\theta}\cos\theta \qquad \bar{a}_A \cdot \bar{K} = R(\ddot{\theta}\cos\theta - \dot{\theta}^2\sin\theta)$$

These expressions are readily explained. Figure 4.18 indicates that the elevation of point A above the ground is

$$h = R\sin\theta$$

Successive differentiation of h shows that $\dot{h} = \bar{v}_A \cdot \bar{K}$, $\ddot{h} = \bar{a}_A \cdot \bar{K}$.

Although the roundness of the disk played a less obvious role in this motion, it was crucial. If the disk were elliptical, it would have been necessary to describe the velocity of the center point as a function of the spin angle and the properties of the ellipse. Differentiating such a representation would have been substantially more difficult than the corresponding tasks in the case of a circular disk.

EXAMPLE 4.6

The cylinder of radius R rolls without slipping inside a semicylindrical cavity. The angle ϕ locating point P on the disk is zero when the center of the cylinder coincides with the vertical centerline. Derive expressions for the velocity and acceleration of point P in terms of the variable speed v of the center C.

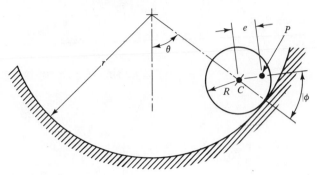

Example 4.6

Solution. It is convenient to orient the body-fixed reference frame such that, at the instant of interest, the x axis is tangent to the path of the center and the y axis intersects the center of curvature. Then movement of the wheel to the right corresponds to

$$\bar{v}_C = v\bar{i} \qquad \bar{a}_C = \dot{v}\bar{i} + \frac{v^2}{r-R}\bar{j} \qquad \bar{\omega} = -\omega\bar{k} \qquad \bar{\alpha} = -\dot{\omega}\bar{k}$$

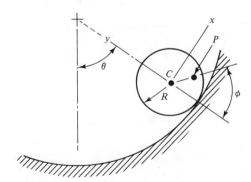

Coordinate system.

However, because there is no slipping at the contact point A, it also must be that

$$\bar{v}_C = \bar{\omega} \times \bar{r}_{C/A}$$

from which it follows that

$$\omega = \frac{v}{R} \qquad \dot{\omega} = \frac{\dot{v}}{R}$$

Now that $\bar{\omega}$, $\bar{\alpha}$, $\bar{v}_C$, and $\bar{a}_C$ have been characterized, the velocity and acceleration of point P may be evaluated.

$$\bar{v}_P = \bar{v}_C + \bar{\omega} \times \bar{r}_{P/C} = v\bar{i} + \left(-\frac{v}{R}\bar{k}\right) \times (e \sin \phi \bar{i} - e \cos \phi \bar{j})$$

$$= v\left(1 - \frac{e}{R}\cos\phi\right)\bar{i} - v\frac{e}{R}\sin\phi\bar{j}$$

$$\bar{a}_P = \bar{a}_C + \bar{\alpha} \times \bar{r}_{P/C} - \omega^2\bar{r}_{P/C}$$

$$= \dot{v}\bar{i} + \frac{v^2}{r - R}\bar{j} + \left(-\frac{\dot{v}}{R}\bar{k}\right) \times (e \sin \phi \bar{i} - e \cos \phi \bar{j})$$

$$- \left(\frac{v}{R}\right)^2 (e \sin \phi \bar{i} - e \cos \phi \bar{j})$$

$$= \left[\dot{v}\left(1 - \frac{e}{R}\cos\phi\right) - \frac{v^2 e}{R^2}\sin\phi\right]\bar{i}$$

$$+ \left[-\dot{v}\frac{e}{R}\sin\phi + v^2\left(\frac{1}{r - R} + \frac{e}{R^2}\cos\phi\right)\right]\bar{j}$$

EXAMPLE 4.7

The shaft of disk A rotates about the vertical axis at the constant rate Ω as the disk rolls without slipping over the inner surface of the cylinder. Determine the angular velocity and angular acceleration of the disk and the acceleration of the point on the disk that contacts the cylinder.

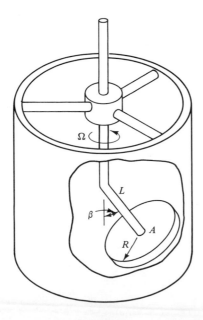

Example 4.7

Solution. Let $\dot{\phi}$ denote the spin rate about the axis of the disk. The z axis of the body-fixed reference frame is defined in the sketch to coincide with the spin axis, so

$$\bar{\omega} = -\Omega \bar{K} - \dot{\phi}\bar{k}$$

Since Ω is constant, $\dot{\phi}$ is also constant. Thus

$$\bar{\alpha} = -\dot{\phi}(\bar{\omega} \times \bar{k})$$

Resolving all vectors into components relative to *xyz* leads to

$$\bar{\omega} = \Omega \sin\beta\bar{i} - (\Omega \cos\beta + \dot{\phi})\bar{k}$$
$$\bar{\alpha} = \Omega\dot{\phi} \sin\beta\bar{j}$$

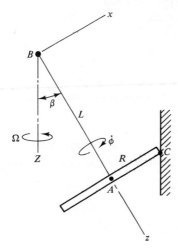

Coordinate system.

It still remains to relate $\dot{\phi}$ to Ω, for which we use the condition that there is no slipping at contact point C. Because point B occupies a fixed position relative to the disk, we could base the formulation on the fact that points B and C are two points on the disk that are at rest at this instant. For the sake of greater generality, we shall relate the velocity of point C to the center A, which follows a horizontal circle of radius $L \sin\beta$. Hence,

$$\bar{v}_A = -\Omega L \sin\beta\bar{j}$$

Then

$$\bar{v}_C = \bar{0} = \bar{v}_A + \bar{\omega} \times \bar{r}_{C/A} = -\Omega L \sin\beta\bar{j} + \bar{\omega} \times R\bar{i}$$
$$= [-\Omega L \sin\beta - (\Omega \cos\beta + \dot{\phi})R]\bar{j}$$

from which we obtain

$$\dot{\phi} = -\Omega \left(\frac{L}{R} \sin\beta + \cos\beta\right)$$

$$\bar{\omega} = \Omega \sin\beta \left(\bar{i} + \frac{L}{R}\bar{k}\right)$$

$$\bar{\alpha} = -\Omega^2 \left(\frac{L}{R} \sin^2\beta + \sin\beta \cos\beta\right)\bar{j}$$

We already identified $\bar{v}_C = \bar{0}$. The acceleration of the contact point is obtained from the kinematical relation based on

$$\bar{a}_A = -L\,\Omega^2 \sin\beta\,\bar{e}_n = -L\,\Omega^2 \sin\beta(-\cos\beta\bar{i} - \sin\beta\bar{j})$$

from which we obtain

$$\bar{a}_C = \bar{a}_A + \bar{\alpha} \times \bar{r}_{C/A} + \bar{\omega} \times (\bar{\omega} \times \bar{r}_{C/A})$$

$$= -\Omega^2 \left(\frac{L}{R}\sin^2\beta + \sin\beta\cos\beta\right)(L\bar{i} - R\bar{k})$$

In order to interpret these results, we recall that $\bar{r}_{C/B} = Ri + Lj$, which leads to the observation that $\bar{\omega} \times \bar{r}_{C/B} = \bar{0}$ and $\bar{a}_C \cdot \bar{r}_{C/B} = 0$. In other words, $\bar{\omega}$ is parallel to $\bar{r}_{C/B}$, while $\bar{a}_C$ is perpendicular to that direction. Both features have a straightforward explanation. First, note that the locus of lines connecting pivot B and contact point C is a cone. This is called the *space cone*. On the other hand, the line connecting point B to a specific point on the rim of the disk lies on a cone relative to the xyz reference frame, which is fixed to the disk; this is the *body cone*. The last sketch depicts both cones. The motion of the disk is equivalent to the body cone rolling without slipping over the space cone. The instantaneous axis of rotation is the line of contact between the cones. The acceleration of any point on the body cone that is on this line of contact is normal to the rotation axis. The concept of space and body cones is particularly useful for the treatment of gyroscopic effects in Chapter 8.

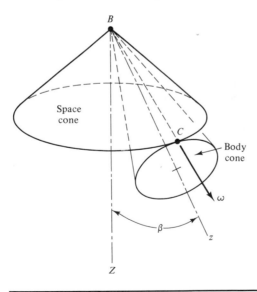

Space and body cones.

EXAMPLE 4.8 —————————————————————————————

A disk rolls without slipping on the XY plane. At the instant shown, the horizontal diameter ACB is parallel to the X axis. Also, at this instant, the horizontal components of the velocity of the center C are known to be 5 ft/s in the X direction and 3 ft/s in the Y direction, while the Y component of the velocity of point B is 6 ft/s. Determine the precession, nutation, and spin rates for the Eulerian angles in Figure 4.18.

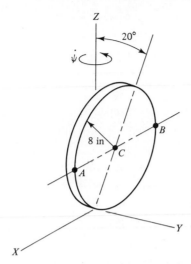

Example 4.8

Solution. For convenience in describing the given components of $\bar{v}_B$ and $\bar{v}_C$ we define the reference frame $x'y'z'$, whose y' axis is the line of nodes and whose z' axis is the axis of the disk. The body-fixed xyz reference frame spins relative to $x'y'z'$, so the coincidence of the reference frames is an instantaneous occurrence.

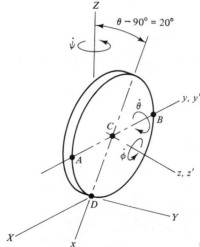

Coordinate systems.

Because we have labeled the axes for the Eulerian angles consistently with Figure 4.18, we may employ Eqs. (4.40) directly. Note that those expressions use components relative to $x'y'z'$, which is especially suitable here because the y' axis is parallel to the XY plane. Since θ is the angle between the Z and z axes, we set

$\theta = 110°$ for the instant of interest, which yields

$$\bar{\omega} = -0.9397\dot{\psi}\bar{i}' + \dot{\theta}\bar{j}' + (\dot{\phi} - 0.3420\dot{\psi})\bar{k}'$$

Since there is no slipping at the contact point A, the velocities of points C and B are

$$\bar{v}_C = \bar{\omega} \times (-\tfrac{8}{12}\bar{i}') = -\tfrac{2}{3}(\dot{\phi} - 0.3420\dot{\psi})\bar{j}' + \tfrac{2}{3}\dot{\theta}\bar{k}'$$

$$\bar{v}_B = \bar{\omega} \times (-\tfrac{8}{12}\bar{i}' + \tfrac{8}{12}\bar{j}') = \tfrac{2}{3}(\dot{\phi} - 0.3420\dot{\psi})(-\bar{i}' - \bar{j}') + \tfrac{2}{3}(\dot{\theta} - 0.9397\dot{\psi})\bar{k}'$$

These velocities must match the given components. The fact that $\bar{j}' = -\bar{I}$ at this instant substantially expedites the evaluation of dot products, which we find to be

$$\bar{v}_C \cdot \bar{I} = 5 = \tfrac{2}{3}[-(\dot{\phi} - 0.3420\dot{\psi})\bar{j}' \cdot \bar{I} + \dot{\theta}\bar{k}' \cdot \bar{I}]$$

$$= \tfrac{2}{3}(\dot{\phi} - 0.3420\dot{\psi}) \text{ ft/s}$$

$$\bar{v}_C \cdot \bar{J} = 3 = \tfrac{2}{3}[-(\dot{\phi} - 0.3420\dot{\psi})\bar{j}' \cdot \bar{J} + \dot{\theta}\bar{k}' \cdot \bar{J}] = \tfrac{2}{3}\dot{\theta}\cos 20°$$

$$\bar{v}_B \cdot \bar{J} = 6 = \tfrac{2}{3}[-(\dot{\phi} - 0.3420\dot{\psi})(\bar{i}' \cdot \bar{J} + \bar{j}' \cdot \bar{J}) + (\dot{\theta} - 0.9397\dot{\psi})\bar{k}' \cdot \bar{J}]$$

$$= \tfrac{2}{3}[-(\dot{\phi} - 0.3420\dot{\psi})(-\sin 20°) + (\dot{\theta} - 0.9397\dot{\psi})(\cos 20°)]$$

The solution of these simultaneous equations is

$$\dot{\psi} = -2.191 \qquad \dot{\theta} = 4.789 \qquad \dot{\phi} = 6.751 \text{ rad/s}$$

REFERENCES

R. N. Arnold and M. Maunder, *Gyrodynamics and Its Engineering Applications*, Academic Press, New York, 1961.

J. H. Ginsberg and J. Genin, *Dynamics*, 2d ed., John Wiley & Sons, New York, 1984.

H. Goldstein, *Classical Mechanics*, 2d ed., Addison-Wesley, Reading, Mass., 1980.

D. T. Greenwood, *Principles of Dynamics*, Prentice-Hall, Englewood Cliffs, N.J., 1965.

L. Meirovitch, *Methods of Analytical Dynamics*, McGraw-Hill, New York, 1970.

C. E. Smith, *Applied Mechanics: Dynamics*, John Wiley & Sons, New York, 1982.

HOMEWORK PROBLEMS

4.1 A gyropendulum consists of a flywheel that rotates at constant angular speed ω_1 relative to shaft BC. This shaft is pinned to the vertical shaft, which rotates at constant angular speed ω_2. The angle β measuring the inclination of shaft BC is an arbitrary function of time. Use the Eulerian angle formulas for angular velocity and angular acceleration, Eqs. (4.16) and (4.17), to derive expressions for the velocity and acceleration of point D, which coincides with the horizontal diameter at the instant of interest.

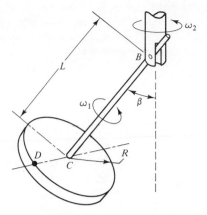

Problem 4.1

4.2 Consider a body whose orientation is described by Eulerian angles. Derive the transformation for a sequence beginning with precession $\psi = 20°$, followed by nutation $\theta = -60°$, then spin $\phi = 140°$. Is it possible to obtain the same transformation with a different sequence beginning with nutation θ', followed by spin ϕ', then precession ψ'? If so, determine the values of θ', ϕ', and ψ'.

4.3 Bar AB rotates at the constant rate of 30 rad/s, causing collar B to slide over curved bar CD. Determine the angular velocity and angular acceleration of CD in the position shown.

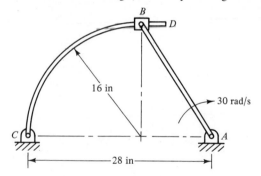

Problem 4.3

4.4 Collar C slides over bar AB. When the system is in the position shown, slider A is moving downward at 600 mm/s and its speed is decreasing at 15 m/s². Determine the corresponding angular velocity and angular acceleration of each bar.

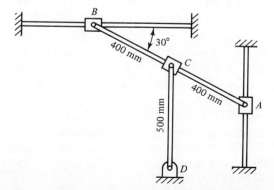

Problem 4.4

4.5 Pin B slides through groove CD in a plate that translates upward at speed v. The groove forms the parabolic curve $y = 1 - x^2/4$, where x and y have units of feet. In the position shown, bar AB is rotating clockwise at 40 rad/s, and that rate is decreasing at 160 rad/s^2. Determine the corresponding values of v and $\dot{v}$.

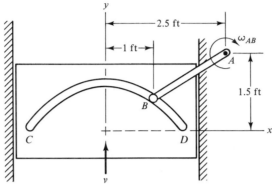

Problem 4.5

4.6 Collar A is connected to bar AB by a ball-and-socket joint, whereas the connection between collar B and bar AB is a forked pin. For the position shown, $v_B = 3$ m/s. Determine the velocity of slider A and the value of $\dot{\beta}$, where β is the angle between AB and the horizontal guide.

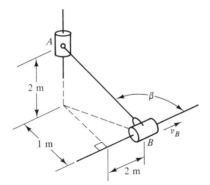

Problem 4.6

4.7 Determine the angular acceleration of bar AB in the linkage in Problem 4.6, for the case where $v_B = 5$ m/s and $\dot{v}_B = -20$ m/s^2 in the given position.

4.8 Collar A is pushed upward at $v_A = 30$ m/s, while the entire system precesses about the vertical axis at 2400 rev/min. Determine the velocity of the midpoint C of the bar in the position where $\beta = 53.13°$. The length of the bar is $L = 600$ mm.

4.9 Collar A is pushed upward at constant speed v_A, while the entire system precesses about the vertical axis at ω_1. Determine the angular velocity and angular acceleration of bar AB in the position where $\beta = 90°$.

4.10 Bead C slides relative to the curved guide bar AB, which rotates about the vertical axis at the constant rate Ω. The movement of the slider is actuated by arm DEF, which pushes the collar outward from the vertical axis at a constant rate u. Determine the velocity and acceleration of the slider in the position where $\theta = 30°$.

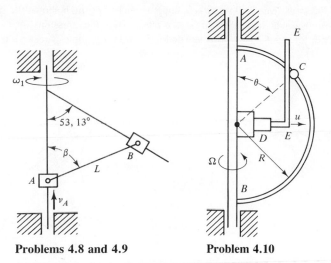

Problems 4.8 and 4.9   **Problem 4.10**

4.11 Bar AB rotates at the constant rate of 200 rev/min. Bar BC is connected to the other links by ball-and-socket joints. Determine the velocity and acceleration of joint C at the instant shown.

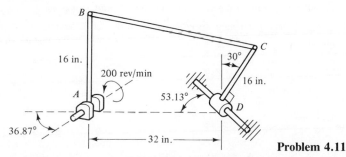

Problem 4.11

4.12 Bar BC is attached by a fork-and-clevis joint to cap B, which is free to rotate about the axis of bar AB. The connection between bar BC and collar C is a ball-and-socket joint. The guide bar for collar C is horizontal, as is the fixed horizontal shaft about which bar AB rotates; the rotation rate is $\omega_{AB} = 200$ rad/s, which is constant. Determine the angular velocity and angular acceleration of bar BC in the position shown.

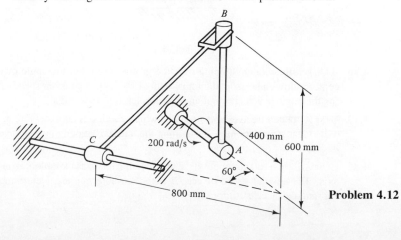

Problem 4.12

4.13 A disk rolls over the hill without slipping. In the position shown the center of the disk has a speed v, which is increasing at the rate $\dot{v}$. Derive expressions for the velocity and acceleration of point D, which is situated at an arbitrary angle β relative to the line of centers.

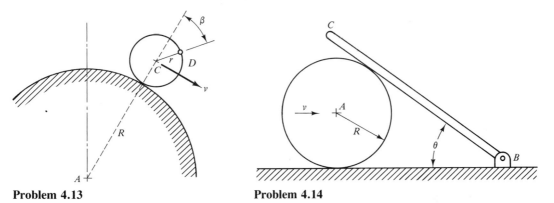

Problem 4.13   **Problem 4.14**

4.14 In the position shown cylinder A is moving to the right such that its center has a speed v. There is no slipping between the cylinder and bar BC, but there is slipping between the cylinder and the ground. Determine the angular velocity of bar BC, and the velocity of the cylinder at the point where it contacts the ground.

4.15 Collar C has a constant speed v to the right, and the rack is stationary. Determine the angular velocity and angular acceleration of gear A at the instant shown.

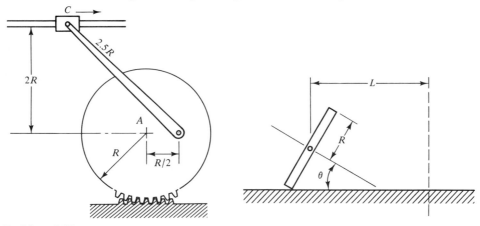

Problem 4.15   **Problem 4.16**

4.16 A disk rolls without slipping over the ground such that the angle of tilt θ is constant. The center follows a horizontal circular path of radius ρ at constant speed v. Derive an expression for the angular velocity and angular acceleration of the disk.

4.17 Gear A rotates freely about its shaft, which rotates at constant rate ω_1 about the horizontal axis. Gear B is stationary. Determine the angular velocity and angular acceleration of gear A.

4.18 Gear A is free to spin about its shaft, which rotates at variable rate ω_1 about the horizontal axis. The angular speed of gear B is the variable rate ω_2. Determine the angular velocity and angular acceleration of gear A.

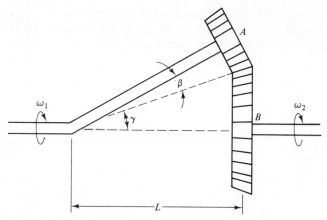

Problems 4.17 and 4.18

4.19 The sphere rolls without slipping over the interior wall of a hollow cylinder that rotates about its axis at ω_2. The angular speed of the vertical shaft driving the sphere is ω_1. Both rotation rates are constant. Determine the angular velocity and angular acceleration of the sphere.

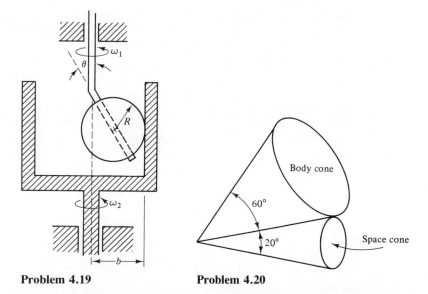

Problem 4.19   **Problem 4.20**

4.20 The body cone executes three revolutions about the stationary space cone in a 1-s interval. Determine the angular velocity and angular acceleration of the body cone.

4.21 The disk rolls without slipping over the horizontal XY plane. At the instant when $\theta = 36.87°$, the X and Y components of the velocity of point B on the horizontal diameter of the disk are 8 and -4 m/s, respectively. The X and Y components of the velocity of center A at this instant are 4 and 2 m/s. Determine the precession angle ψ between the horizontal diameter BAC and the X axis, and also evaluate the precession, nutation, and spin rates.

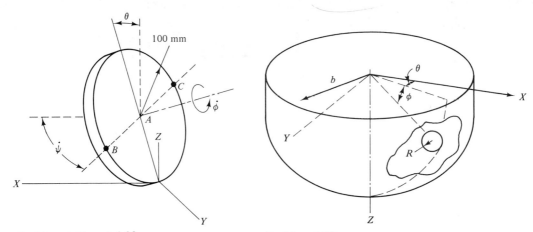

Problem 4.21 and 4.22   **Problem 4.23**

4.22 The disk is rolling without slipping. At the instant when the angle of inclination $\theta = 30°$, the disk is observed to be spinning at $\dot{\phi} = 15$ rad/s. At this instant, the speeds of points B and C on its horizontal diameter are 1 and 2 m/s, respectively. Determine the corresponding precession and nutation rates.

4.23 A sphere of radius R rolls without slipping over the interior of a hemispherical shell of radius b that rotates about the vertical axis at constant rate Ω. The polar and azimuth angles locating the center of the sphere are ϕ and θ, neither of which is constant. Derive expressions for the angular velocity and angular acceleration of the sphere.

4.24 Shaft BC is pinned to the T-bar, which rotates at the constant angular speed ω_1. Wheel C rotates freely relative to shaft BC. The platform, over which wheel C rolls, is raised at the constant speed u, causing angle β to decrease. The wheel does not slip relative to the platform in the direction transverse to the diagram, but slipping in the radial direction is observed to occur. Derive expressions for the angular velocity and the angular acceleration of the wheel.

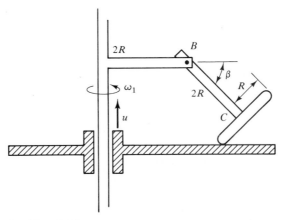

Problem 4.24

Chapter 5

Newtonian Kinetics of a Rigid Body

Chasle's theorem states that the general motion of a rigid body can be represented as a superposition of a translation following any point in the body and a pure rotation about that point. Kinematics studies are concerned only with the description of that motion. The developments in this chapter will disclose how the motion is related to the force system acting on the body. The resultant force may be regarded intuitively as the net tendency of the force system to push a body, so it may be expected to be related to the translational effect. Similarly, the resultant moment may be considered to be the net rotational effect. We shall confirm and quantify these expectations for general spatial motion in the following presentation, and then specialize the derived principles for the case of planar motion.

5.1 FUNDAMENTAL PRINCIPLES

Newton's laws govern the motion of a particle. A rigid body may be treated as a collection of particles whose motions are not independent. In the first part of this chapter, we shall derive the basic kinetics principles for rigid-body motion. The foundation for these derivations is Newton's second law, which describes inertial effects, and the third law, which describes the nature of the force system.

5.1.1 Basic Model

A rigid body is a collection of particles that are constrained to maintain a fixed relative position. For our purposes, it is initially useful to represent this collection as a continuum of small mass elements Δm_i. When deformation effects are to be modeled, the stresses are related to the strains through a constitutive law, such as the one for elasticity. In the modeling of a rigid body the constitutive law is replaced by the kinematical specification that all points in the body remain at a fixed relative distance. In this view, the stress resultants are constraint forces that maintain these fixed distances.

Figure 5.1 depicts three mass elements out of the full collection. Two types of force are depicted there. The internal force exerted on any element i by element j is denoted $\bar{f}_{ij}$. Such forces consist of stress resultants, which are associated with contact with neighboring elements, as well as gravitational and other body forces that are due to interactions with other elements in the body. The force $\bar{F}_i$ arises from the action of other objects on the ith element. Contributors to $\bar{F}_i$ may be body forces, applied loads, or constraint forces (that is, reactions) that result from the manner in which the body is supported.

A key aspect of the internal forces arises from Newton's third law (action and reaction). According to that law, each pair of forces $\bar{f}_{ij}$ and $\bar{f}_{ji}(i \neq j)$ are equal in magnitude, but oppositely directed, so

$$\bar{f}_{ij} + \bar{f}_{ji} = \bar{0} \tag{5.1}$$

In addition, the two forces forming each interaction pair share a common line of action, which is the line connecting elements i and j. When either force is moved along its line of action to the other particle for the purpose of computing the moment about an arbitrary point A, it becomes apparent that the two forces cancel each other in the moment sum. Thus

$$\bar{r}_{i/A} \times \bar{f}_{ij} + \bar{r}_{j/A} \times \bar{f}_{ji} = \bar{0} \tag{5.2}$$

As a result of Eqs. (5.1) and (5.2), the internal forces will not appear explicitly in the equations of motion. Nevertheless, they are important, because of their role in enforcing the rigidity of the body.

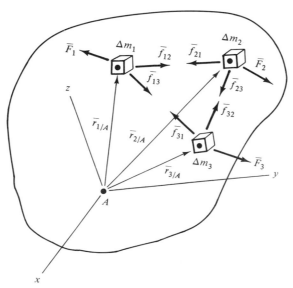

Figure 5.1   Infinitesimal mass elements.

5.1.2 Resultant Force and Point Motion

The resultant of all forces acting on any mass element must satisfy Newton's second law, which for element i is

$$\overline{F}_i + \sum_{\substack{j=1 \\ j \neq i}}^{N} \overline{f}_{ij} = \Delta m_i \overline{a}_i \tag{5.3}$$

where N is the number of elements into which the body is decomposed. Now add Eq. (5.3) for each element by summing over i.

$$\sum_{i=1}^{N} \overline{F}_i + \sum_{i=1}^{N} \sum_{\substack{j=1 \\ j \neq 1}}^{N} \overline{f}_{ij} = \sum_{i=1}^{N} \Delta m_i \overline{a}_i \tag{5.4}$$

Consider a specific set of values for each i, j pair in the double summation above, such as $i, j = 1, 2$. The corresponding contribution is $\overline{f}_{12} + \overline{f}_{21}$. By virtue of Eq. (5.1) these two terms cancel. The same result applies to every other pair, so the force sum may be rewritten as

$$\sum_{i=1}^{N} \overline{F}_i = \sum_{i=1}^{N} \Delta m_i \overline{a}_i \tag{5.5}$$

The left side is identifiable as the sum of the external forces exerted on the body by other objects. We shall now express the term on the right side in a more meaningful form.

First, we write the acceleration of each element as the second derivative of the position. Then the fact that each element has a constant mass makes it possible to sum prior to differentiation, with the result that

$$\sum_{i=1}^{N} \overline{F}_i = \frac{d^2}{dt^2} \sum_{i=1}^{N} \Delta m_i \overline{r}_{i/O} \tag{5.6}$$

The key observation about this form is that the right side is related to the position of the center of mass. Let the coordinates of each element relative to the center of mass be (X_i, Y_i, Z_i), and let the coordinates of the center of mass G be (X_G, Y_G, Z_G). The mass is the sum of the individual masses,

$$m = \sum_{i=1}^{N} \Delta m_i \tag{5.7}$$

In order to locate the center of mass G, the moment of the distributed gravitational force system is equated to the moment of the total weight acting at point G. That evaluation leads to the *first moments of mass* with respect to each coordinate, which are

$$mX_G = \sum_{i=1}^{N} \Delta m_i X_i \qquad mY_G = \sum_{i=1}^{N} \Delta m_i Y_i \qquad mZ_G = \sum_{i=1}^{N} \Delta m_i Z_i \tag{5.8}$$

When we multiply the above equations by $\overline{I}$, $\overline{J}$, and $\overline{K}$, respectively, and add them, we obtain the vector form of the first moment of mass

$$m\bar{r}_{G/O} = \sum_{i=1}^{N} \Delta m_i \bar{r}_{i/O} \tag{5.9}$$

This expression allows us to replace the summation on the right side of Eq. (5.6) with a single term involving the center of mass. The mass is constant, so the specified time derivatives in that equation lead to

$$\sum \bar{F} = m\bar{a}_G \tag{5.10}$$

where $\Sigma \bar{F}$ denotes the resultant (i.e., sum) of the external forces.

In terms of the kinematics of a rigid body, Eq. (5.10) specifies the motion of a point in the body. The choice for a point is arbitrary in Chasle's theorem, but it is always the center of mass when the resultant of the external forces is evaluated.

The similarity of Eq. (5.10) to Newton's second law, which only treats a particle, is important. It shows that modeling an object as a particle is equivalent to focusing attention on the motion of the center of mass of that object. In contrast, a rigid-body model is used to evaluate the rotational effects. The next task is to quantify the relation between the moment exerted by the force system and the corresponding rotational motion.

5.1.3 Resultant Moment and Rotation

The total moment exerted on a rigid body by a system of forces is obtained by summing the contribution of each element. A typical situation, in which moments about arbitrary point A will be evaluated, is depicted in Figure 5.2. In view of Newton's second law, the moment of the forces acting on element i about this point must equal the corresponding moment of the $\Delta m_i \bar{a}_i$ vector, that is,

$$(\bar{M}_A)_i = \bar{r}_{i/A} \times \left(\bar{F}_i + \sum_{\substack{j=1 \\ j \neq 1}}^{N} \bar{f}_{ij} \right) = \bar{r}_{i/A} \times \Delta m_i \bar{a}_i \tag{5.11}$$

Because the moment is being computed about point A, it is logical to employ point A as the reference point for the kinematical description. This introduces the first restriction. Point A must be a point in the body if the kinematical relationship between the acceleration of two points in a rigid body is to be used. When this condition applies, it follows that

$$\bar{a}_i = \bar{a}_A + \bar{\alpha} \times \bar{r}_{i/A} + \bar{\omega} \times (\bar{\omega} \times \bar{r}_{i/A}) \tag{5.12}$$

Substitution of this expression into the right side of Eq. (5.11) yields

$$(\bar{M}_A)_i = \bar{r}_{i/A} \times \left(\bar{F}_i + \sum_{\substack{j=1 \\ j \neq 1}}^{N} \bar{f}_{ij} \right)$$

$$= \bar{r}_{i/A} \times \Delta m_i [\bar{a}_A + \bar{\alpha} \times \bar{r}_{i/A} \times \bar{\omega} \times (\bar{\omega} \times \bar{r}_{i/A})] \tag{5.13}$$

The first manipulation of this expression results from recalling that because of the

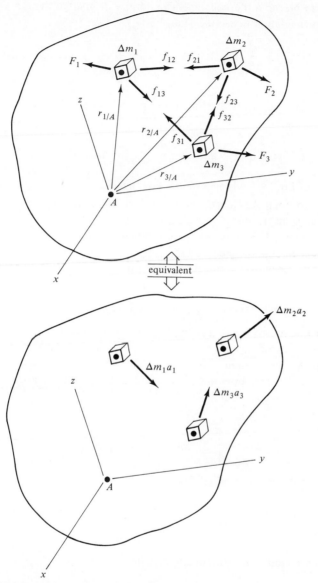

Figure 5.2 Forces acting on the mass elements in a rigid body.

rigidity of the body, $\bar{r}_{i/A}$ has constant components relative to a coordinate system that is attached to the body. This means that $\dot{\bar{r}}_{i/A} = \bar{\omega} \times \bar{r}_{i/A}$. It follows that

$$(\bar{M}_A)_i = \bar{r}_{i/A} \times \Delta m_i \bar{a}_A + \bar{r}_{i/A} \times \left[\Delta m_i \frac{d}{dt} (\bar{\omega} \times \bar{r}_{i/A}) \right] \qquad (5.14)$$

The time derivative may be performed after the cross product, if an appropriate term is inserted in order to preserve the equality. This term will involve the derivative of $\bar{r}_{i/A}$, for which the earlier relation is useful. Thus

$$(\overline{M}_A)_i = \bar{r}_{i/A} \times \Delta m_i \bar{a}_A + \frac{d}{dt}[\bar{r}_{i/A} \times \Delta m_i(\overline{\omega} \times \bar{r}_{i/A})] - (\overline{\omega} \times \bar{r}_{i/A}) \times \Delta m_i(\omega \times \bar{r}_{i/A}) \tag{5.15}$$

Because the latter term involves the cross product of parallel vectors, it vanishes, so that

$$(\overline{M}_A)_i = \bar{r}_{i/A} \times \Delta m_i \bar{a}_A + \frac{d}{dt}[\bar{r}_{i/A} \times \Delta m_i(\overline{\omega} \times \bar{r}_{i/A})] \tag{5.16}$$

The interpretation of Eq. (5.16) hinges on the fact that $\overline{\omega} \times \bar{r}_{i/A}$ is the (absolute) velocity of point i relative to point A. Hence, $\Delta m_i(\overline{\omega} \times \bar{r}_{i/A})$ is the momentum of the ith mass element relative to point A, and $\bar{r}_{i/A} \times \Delta m_i(\overline{\omega} \times \bar{r}_{i/A})$ is the moment about point A of the relative momentum. This *moment of momentum* is usually called the *angular momentum* about point A; it is denoted by the symbol $\overline{H}$ with a subscript corresponding to the reference point.† Because the quantity appearing here is associated with the element Δm_i, the definition is

$$(\overline{H}_A)_i = \bar{r}_{i/A} \times \Delta m_i(\overline{\omega} \times \bar{r}_{i/A}) \tag{5.17}$$

We now see that Eq. (5.16) states that the moment of the forces acting on element i is the sum of the moment of the $\Delta m_i \bar{a}_A$ vector, applied on the element, and the rate of change of the angular momentum.

In order to apply this principle in a useful manner, the relations for all mass elements must be combined. Adding Eq. (5.16) for each element i and factoring $\bar{a}_A$ out of the summation yields

$$\sum \overline{M}_A = \left[\sum_{i=1}^{N} \Delta m_i \bar{r}_{i/A} \right] \times \bar{a}_A + \frac{d}{dt}\overline{H}_A \tag{5.18}$$

where $\sum \overline{M}_A$ is the total moment of all forces,

$$\sum \overline{M}_A = \sum_{i=1}^{N} \bar{r}_{i/A} \times \overline{F}_i + \sum_{i=1}^{N} \sum_{\substack{j=1 \\ j\neq 1}}^{N} \bar{r}_{i/A} \times \bar{f}_{ij} \tag{5.19}$$

and $\overline{H}_A$ is the total angular momentum about point A.

$$\overline{H}_A = \sum_{i=1}^{N} \Delta m_i \bar{r}_{i/A} \times (\overline{\omega} \times \bar{r}_{i/A}) \tag{5.20}$$

Two simplifications may be introduced. First, consider a specific (i,j) pair for the double sum in Eq. (5.19). According to Eq. (5.2), the moment of $\bar{f}_{ij}$ cancels the amount of $\bar{f}_{ji}$, so the double sum vanishes. This means that only forces exerted from outside the

---

†We shall only develop principles governing angular momentum relative to a reference point on the body. Related principles governing angular momentum of a system of particles relative to a point that is fixed in the inertial reference frame can also be developed. However, they are less suitable for the study of rigid-body motion.

body contribute to the moment sum. Second, note that the summation in the right side of Eq. (5.18) is the first moment of mass about point A. In view of Eq. (5.9) this term locates the center of mass G relative to point A. These observations lead to

$$\sum \overline{M}_A = \sum_{i=1}^{N} \overline{r}_{i/A} \times \overline{F}_i = \overline{r}_{G/A} \times m\overline{a}_A + \frac{d}{dt}\overline{H}_A \qquad (5.21)$$

This is close to the desired principle governing the angular motion. However, it requires consideration of the (linear) acceleration of point A, thereby coupling linear and angular motion. For this reason, the selection of point A will be restricted beyond the requirement that it be a point in the body. We shall require that this point be selected such that $\overline{r}_{G/A} \times m\overline{a}_A = \overline{0}$. Then the moment equation reduces to

$$\sum \overline{M}_A = \dot{\overline{H}}_A \qquad \overline{H}_A = \sum_{i=1}^{N} \Delta m_i \overline{r}_{i/A} \times (\overline{\omega} \times \overline{r}_{i/A}) \qquad (5.22)$$

The condition $\overline{r}_{G/A} \times m\overline{a}_A = \overline{0}$, required for validity of Eqs. (5.22), is obtained when point A satisfies one of the following criteria:

1. Point A is the center of mass G: $\overline{r}_{G/A} = \overline{0}$. The center of mass is always an admissible point for summing moments.
2. Point A has no acceleration: $\overline{a}_A = \overline{0}$. This situation could arise if there is a point on the body that is constrained to follow a straight path at a constant speed, but such cases are comparatively rare. The more usual situation arises when a body is in pure rotation about point A.
3. Point A is accelerating directly toward or away from the center of mass. In this case, $\overline{a}_A$ is parallel to $\overline{r}_{G/A}$, so their cross product vanishes.

Very few types of motion fit the third criterion. One exception arises in planar motion when a disk or sphere rolls without slipping. Even then the body must be balanced; i.e., the center of mass and the geometric centroid must coincide. The acceleration of a disk rolling in a plane is depicted in Figure 5.3. The center C is accelerating, but the no-slip condition requires that the acceleration of the contact point A be normal to the

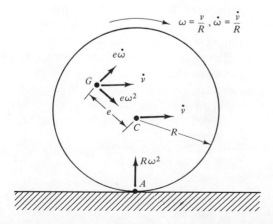

Figure 5.3 Eccentric disk.

surface. Because $\bar{a}_A$ is directed toward the center C, it would be permissible to formulate a moment sum about the contact point, provided that point C is the center of mass. However, if the wheel is unbalanced, so that the center of mass G is eccentric, then $\bar{a}_A$ will usually not be parallel to $\bar{r}_{G/A}$.

Because of its lack of generality, we will not consider the third criterion in selecting a point for summing moments. In contrast, there are strong reasons for selecting the fixed point for a body in pure rotation, in accord with the second criterion. In order for a point to be held stationary, reaction forces must be exerted at that point. This is exemplified by a ball-and-socket joint, which exerts an arbitrary reaction force having three components. These unknown reactions do not contribute to a moment sum about the stationary point. The ability to eliminate unknown reactions from the moment equation for a body in pure rotation makes it worthwhile to formulate moments about the stationary point. Note that this is the only case where the point for the moment sum is selected on the basis of eliminating reactions. In static systems, of course, all points are stationary, so any point is permissible for the moment sum.

Because the angular moment $\bar{H}_A$ is a function of $\bar{\omega}$, the moment equation (5.22) is the principle that we need to study the angular motion of a rigid body. Note its analogy to Eq. (5.10) for the motion of the center of mass. The total linear momentum of the system is $\bar{P} = m\bar{v}_G$, as may be seen by differentiating Eq. (5.9) with respect to time. Then the equation governing the motion of the center of mass may be rewritten as

$$\sum \bar{F} = \dot{\bar{P}} \qquad \bar{P} = m\bar{v}_G = \sum_{i=1}^{N} \Delta m_i \bar{v}_i \qquad (5.23)$$

In other words, the linear or angular effect of the external force system equals the rate of change of the corresponding type of momentum for the body.

The actual formulation of the moment equation requires expressions whereby the angular momentum of a rigid body may be formulated. Before we proceed to that evaluation, let us consider the impulse-momentum and work-energy principles for a rigid body.

5.1.4 Momentum Principles

The force and moment equations discussed thus far govern the linear and angular acceleration of a body. Momentum and energy principles, which represent standard integrals of these equations, may be used to relate the linear and angular velocity of the body at successive instants or locations. The evaluation of the associated impulse and work quantities in some situations requires knowledge of the body's motion, so these integral relations supplement, rather than replace, the basic acceleration equations.

Equations (5.22) and (5.23) are the time-derivative forms of the impulse-momentum relations. Definite integration of them between two instants t_1 and t_2 leads to

$$\bar{P}_2 = \bar{P}_1 + \int_{t_1}^{t_2} \sum \bar{F}\, dt$$

$$(\bar{H}_A)_2 = (\bar{H}_A)_1 + \int_{t_1}^{t_2} \sum \bar{M}_A\, dt \qquad (5.24)$$

Both of the above state that the final value of the respective momenta exceeds the initial value by the corresponding *impulse*, that is, the time integral of the resultant force or moment.

Both momentum principles are vector equations, so they each yield three scalar equations obtained from equating like components. If the impulses can be evaluated, the scalar equations fully define the corresponding change in the linear or angular velocities. The difficulty lies in that evaluation. The resultant force and moment acting on a body are seldom known in advance because the reactions are unknown. Furthermore, it is not sufficient to know the force or moment in terms of components relative to the body because the corresponding unit vectors are not constant. Carrying out the impulse integrals in this case requires knowledge of the orientation of the unit vectors, and of the components, as functions of time. Such information is not usually available, because it depends on the bodily motion being studied.

One situation where the momentum-impulse relations are useful is to treat *impulsive forces*, such as those generated by impacts and explosions. Impulsive forces are defined to impart very large accelerations to a body over a very short time interval. The velocity change (linear and angular) may be evaluated in this case by considering the time interval to be sufficiently short to neglect any change in position during the action of the impulsive forces. Further simplifications stem from the fact that only the impulse of the forces, as opposed to their true time dependence, needs to be known to evaluate the velocities. Also, the impulsive forces are usually large enough that the influence of nonimpulsive forces is negligible during the brief interval of the impulse.

5.1.5 Energy Principles

Energy principles begin from a more fundamental viewpoint than did the momentum principles. For mass element Δm_i in the rigid body, the work-energy equation is

$$\tfrac{1}{2}\Delta m_i(v_i^2)_2 = \tfrac{1}{2}\Delta m_i(v_i^2)_1 + \oint_1^2 \left(\bar{F}_i + \sum_{\substack{j=1 \\ j \neq i}}^{N} \bar{f}_{ij} \right) \cdot d\bar{r}_i \tag{5.25}$$

The line integral expresses the *work* done by the forces acting on mass element i in the sequence of infinitesimal displacements $d\bar{r}_i$ that the element undergoes as it moves from its initial position 1 to final position 2. Note that the process of evaluating a line integral merely refers to the fact that the integrand's dependence on position must be described. The quantity $\tfrac{1}{2}\Delta m_i v_i^2$ is, of course, the kinetic energy.

Since Eq. (5.25) applies to every mass element, it is valid for the whole collection of elements. The total kinetic energy of the system will be denoted as T and the work done by the forces acting on the system in a displacement from position 1 to position 2 will be denoted as $W_{1 \rightarrow 2}$. The result of adding Eq. (5.25) for each particle may then be written as

$$T_2 = T_1 + W_{1 \rightarrow 2} \tag{5.26}$$

where

$$T = \frac{1}{2} \sum_{}^{N} \Delta m_i v_i^2 = \frac{1}{2} \sum_{}^{N} \Delta m_i \bar{v}_i \cdot \bar{v}_i \tag{5.27}$$

$$W_{1 \to 2} = \oint_1^2 \sum_{i=1}^N \overline{F}_i \cdot d\overline{r}_i + \oint_1^2 \sum_{i=1}^N \sum_{\substack{j=1 \\ j \neq 1}}^N \overline{f}_{ij} \cdot d\overline{r}_i \tag{5.28}$$

The next step is to express the motion of the mass elements in terms of basic kinematical variables for rigid-body motion. For this, let B denote the point for translational motion in Chasle's theorem. (We do not require at this juncture that point B be one of the three permissible points for a moment sum.) Because $\overline{v}_B$ and $\overline{\omega}$ are independent of the mass element, they may be factored out of the sums. Equation (5.27) for the kinetic energy therefore becomes

$$T = \frac{1}{2} \sum_{i=1}^N \Delta m_i (\overline{v}_B + \overline{\omega} \times \overline{r}_{i/B}) \cdot (\overline{v}_B + \overline{\omega} \times \overline{r}_{i/B})$$

$$= \frac{1}{2} \left(\sum_{i=1}^N \Delta m_i \right) (\overline{v}_B \cdot \overline{v}_B) + \overline{v}_B \cdot \left(\overline{\omega} \times \sum_{i=1}^N \Delta m_i \overline{r}_{i/B} \right)$$

$$+ \frac{1}{2} \sum_{i=1}^N \Delta m_i (\overline{\omega} \times \overline{r}_{i/B}) \cdot (\overline{\omega} \times \overline{r}_{i/B}) \tag{5.29}$$

The first sum above is the total mass. The second sum is the first moment of mass relative to point B, so it locates the center of mass G relative to point B. The third sum may be written in a more recognizable form by using an identity for the scalar triple product,

$$\overline{a} \times \overline{b} \cdot \overline{c} = \overline{a} \cdot \overline{b} \times \overline{c}$$

Applying this to the sequence of vectors $\overline{\omega}$, $\overline{r}_{i/B}$, and $(\overline{\omega} \times \overline{r}_{i/B})$ yields

$$(\overline{\omega} \times \overline{r}_{i/B}) \cdot (\overline{\omega} \times \overline{r}_{i/B}) = \overline{\omega} \cdot [\overline{r}_{i/B} \times (\overline{\omega} \times \overline{r}_{i/B})]$$

The result is that the kinetic energy reduces to

$$T = \tfrac{1}{2} m \overline{v}_B \cdot \overline{v}_B + m \overline{v}_B \cdot \overline{\omega} \times \overline{r}_{G/B} + \frac{1}{2} \sum_{i=1}^N \Delta m_i \overline{\omega} \cdot [\overline{r}_{i/B} \times (\overline{\omega} \times \overline{r}_{i/B})]$$

$$= \tfrac{1}{2} m \overline{v}_B \cdot \overline{v}_B + m \overline{v}_B \cdot \overline{\omega} \times \overline{r}_{G/B} + \tfrac{1}{2} \overline{\omega} \cdot \overline{H}_B \tag{5.30}$$

where $\overline{H}_B$ is the angular momentum relative to point B, as defined in Eq. (5.22).

The evaluation of the second term in Eq. (5.30) may be avoided if point B is restricted to be either the center of mass of the body ($\overline{r}_{G/B} = \overline{0}$), or else the stationary point for a body in pure rotation ($\overline{v}_B = \overline{0}$). The latter condition also holds whenever point B is situated on the instantaneous axis of rotation for a body in general motion. However, such a point is not fixed relative to the body, so the inertia properties required to form $\overline{H}_B$ will not be constant. For that reason, we shall not use the instant center to formulate kinetic energy. Thus, the alternative expressions for kinetic energy are

$$T = \tfrac{1}{2} m \overline{v}_G \cdot \overline{v}_G + \tfrac{1}{2} \overline{\omega} \cdot \overline{H}_G : \text{all motions} \tag{5.31a}$$

$$T = \tfrac{1}{2} \overline{\omega} \cdot \overline{H}_C : \text{pure rotation about point } C \tag{5.31b}$$

The kinematical relation for velocity may also be applied to Eq. (5.28) for the work. For this, we note that $d\overline{r}_i$ is the infinitesimal displacement during an infinitesimal time interval, so

$$d\bar{r}_i = \bar{v}_i \, dt = (\bar{v}_B + \bar{\omega} \times \bar{r}_{i/B}) \, dt = d\bar{r}_B + \overline{d\theta} \times \bar{r}_{i/B} \qquad (5.32)$$

where $\overline{d\theta}$ is the corresponding infinitesimal rotation of the body. In the foregoing, point B will again be allowed to be arbitrary. Substituting Eq. (5.32) into Eq. (5.28) yields

$$W_{1\to2} = \oint_1^2 \sum_{i=1}^N \bar{F}_i \cdot d\bar{r}_B + \oint_1^2 \sum_{i=1}^N \bar{F}_i \cdot (\overline{d\theta} \times \bar{r}_{i/B})$$

$$+ \oint_1^2 \sum_{i=1}^N \sum_{\substack{j=1 \\ i\neq j}}^N \bar{f}_{ij} \cdot d\bar{r}_B + \oint_1^2 \sum_{i=1}^N \sum_{\substack{j=1 \\ i\neq j}}^N \bar{f}_{ij} \cdot (\overline{d\theta} \times \bar{r}_{i/B}) \qquad (5.33)$$

The vector identity for the rearrangement of a scalar triple product again produces substantial simplifications. Specifically,

$$\bar{f}_{ij} \cdot (\overline{d\theta} \times \bar{r}_{i/B}) = (\bar{r}_{i/B} \times \bar{f}_{ij}) \cdot \overline{d\theta}$$

Equation (5.33) then becomes

$$W_{1\to2} = \oint_1^2 \left[\sum_{i=1}^N \bar{F}_i \right] \cdot d\bar{r}_B + \oint_1^2 \left(\sum_{i=1}^N \bar{r}_{i/B} \times \bar{F}_i \right) \cdot \overline{d\theta}$$

$$+ \oint_1^2 \left(\sum_{i=1}^N \sum_{\substack{j=1 \\ j\neq 1}}^N \bar{f}_{ij} \right) \cdot d\bar{r}_B + \oint_1^2 \left(\sum_{i=1}^N \sum_{\substack{j=1 \\ j\neq 1}}^N \bar{r}_{i/B} \times \bar{f}_{ij} \right) \cdot \overline{d\theta} \qquad (5.34)$$

Note that the third and fourth bracketed terms describe the resultant force and moment exerted by the internal forces, both of which vanish. Similarly, the first and second terms describe the resultant force, $\Sigma \bar{F}$, and moment about point B, $\Sigma \overline{M}_B$, exerted by the external forces. Thus, the work may be computed according to

$$W_{1\to2} = \oint_1^2 \Sigma \bar{F} \cdot d\bar{r}_B + \oint_1^2 \Sigma \overline{M}_B \cdot \overline{d\theta} \qquad (5.35)$$

This expression could have been anticipated from Chasle's theorem. It shows that the total work is the sum of the work done by the resultant of the external forces in moving an arbitrary point B, and the work done by the moment of the external forces in the rotation about that point. Equation (5.35) provides an alternative to computing the work done by an external force as an integral along the path of its point of application, as originally expressed in Eq. (5.28).

Another alternative to direct evaluation of the work arises when a force is *conservative*. Consider a force that is exerted on a particle that moves over a closed path from position 1 to position 2, and then back to position 1, as in Figure 5.4. By definition, a conservative force does no work in the overall movement, so

$$W_{1\to2} + W_{2\to1} = 0 \qquad (5.36)$$

This must be true for all paths between the two positions. Hence, the work done in each phase of the movement can depend only on the initial and final positions. Furthermore,

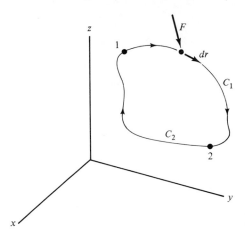

Figure 5.4   Work done by a force.

Eq. (5.36) must be valid for any pair of positions. These conditions can only be satisfied if the work done by a conservative force is determined by the change in the value of a function of position.

This function, which is called the *potential energy*, is defined such that

$$W_{1\to2} = V_1 - V_2 \tag{5.37}$$

In other words, the work done by a conservative force equals the amount by which its potential energy is depleted.

For further insight, suppose that position 2 is infinitesimally close to position 1, so that Eq. (5.37) may be written in differential form as

$$dW = -dV = -\frac{\partial V}{\partial X}\,dX - \frac{\partial V}{\partial Y}\,dY - \frac{\partial V}{\partial Z}\,dZ \tag{5.38a}$$

where the chain rule has been introduced because the potential energy is a function of position. This relation may be rewritten in the notation of vector calculus as

$$dW = -\nabla V \cdot d\bar{r}_{P/O} \tag{5.38b}$$

where ∇V represents the gradient of the potential energy,

$$\nabla V = \frac{\partial V}{\partial X}\bar{I} + \frac{\partial V}{\partial Y}\bar{J} + \frac{\partial V}{\partial Z}\bar{K} \tag{5.39}$$

However, the infinitesimal work done by a force is

$$dW = \bar{F} \cdot d\bar{r}_{P/O} = \bar{F} \cdot (dX\bar{I} + dY\bar{J} + dZ\bar{K}) \tag{5.40}$$

Comparison of Eqs. (5.38a) and (5.40) leads to

$$F_X = -\frac{\partial V}{\partial X} \qquad F_Y = -\frac{\partial V}{\partial Y} \qquad F_Z = -\frac{\partial V}{\partial Z} \tag{5.41a}$$

which is equivalent to the statement that a conservative force is the negative of the gradient of the potential energy function,

$$\overline{F}_{\text{conservative}} = -\nabla V \qquad (5.41b)$$

The potential energy function may be derived by integrating either of Eqs. (5.41). However, for the common forces, it is often easier to derive it by evaluating the work done by the force between two arbitrary positions. Comparison of the resulting expression with the general expression for the work done by a conservative force, Eq. (5.37), then permits identification of the potential energy function. The conservative forces that most commonly arise in mechanical systems are

Gravitational attraction near the earth's surface—the distance Z is measured vertically from a datum.

$$\overline{F}_{\text{grav}} = -mg\overline{K} \Leftrightarrow V_{\text{grav}} = mgZ \qquad (5.42)$$

Gravitational attraction between the earth and an orbiting body—the distance r is measured between the center of the earth and the body, and m is the mass of the body.

$$\overline{F}_{\text{grav}} = -\frac{GMm}{r^2}\overline{e}_r \Leftrightarrow V = -\frac{GMm}{r} \qquad (5.43)$$

where GM is the product of the universal gravitational constant and the mass of the earth.

$$GM = 5.990(10^{14}) \text{ m}^3/\text{s}^2 = 1.4083(10^{16}) \text{ ft}^3/\text{s}^2$$

Linear elastic spring—k is the stiffness of the spring, Δ is its elongation, L is the length of the spring, and L_O is the unstretched length.

$$\overline{F}_{\text{spr}} = k\Delta\overline{e}_s \Leftrightarrow V_{\text{spr}} = \tfrac{1}{2}k\Delta^2 \qquad \Delta = L = L_O \qquad (5.44)$$

where the unit vector $\overline{e}_s$ extends from the end where the force $\overline{F}_{\text{spr}}$ is applied to the other end.

The work-energy principle may be reformulated to account explicitly for conservative forces. When the work done by those forces is computed according to Eq. (5.37), the result is

$$T_2 + V_2 = T_1 + V_1 + W_{1\to2}^{(nc)} \qquad (5.45)$$

where $W_{1\to2}^{(nc)}$ represents the work done by all forces that are not included in the potential energy.

Equation (5.45) is a scalar equation relating the motion at two positions. It is adequate by itself only when there is one unknown, for example, a parameter describing a force or a speed. Additional equations relating motion at two positions might be available from conservation of momentum, particularly the angular momentum, or from one of the components of the equations of motion.

The ability to select point B in Eq. (5.35) arbitrarily can be a significant aid, since constrained points usually follow comparatively simple paths. It also leads to another relation between the motions at two positions. Suppose that point B is selected to be the center of mass G. The equation of motion for the center of mass, $\Sigma \bar{F} = m\bar{a}_G$, may be integrated to form an energy relation directly. A dot product with the displacement of the center of mass leads to

$$\oint_1^2 \Sigma \bar{F} \cdot d\bar{r}_G = \oint_1^2 m\bar{a}_G \cdot d\bar{r}_G = \oint_1^2 m\dot{\bar{v}}_G \cdot (\bar{v}_G \, dt)$$

$$= \oint_1^2 \frac{1}{2} m \frac{d}{dt} (\bar{v}_G \cdot \bar{v}_G) \, dt$$

The last integrand above is a perfect differential, so the relation reduces to

$$\tfrac{1}{2}m(\bar{v}_G \cdot \bar{v}_G)_2 = \tfrac{1}{2}m(\bar{v}_G \cdot \bar{v}_G)_1 + \oint_1^2 \Sigma \bar{F} \cdot d\bar{r}_G \qquad (5.46)$$

Now use Eqs. (5.31a) and (5.35) to formulate the basic work-energy principle, Eq. (5.26), and subtract Eq. (5.46) from that expression. The result is that

$$\tfrac{1}{2}(\bar{\omega} \cdot \bar{H}_G)_2 = \tfrac{1}{2}(\bar{\omega} \cdot \bar{H}_G)_1 + \oint_1^2 \Sigma \bar{M}_G \cdot d\theta \qquad (5.47)$$

The conclusion to be drawn from Eqs. (5.46) is that the increase in the *translational kinetic energy*, $\tfrac{1}{2}m\bar{v}_G \cdot \bar{v}_G$, which is associated with motion of the center of mass, results from the work of the resultant force in moving that point. Similarly, Eq. (5.47) shows that the increase in the *rotational kinetic energy*, $\tfrac{1}{2}\bar{\omega} \cdot \bar{H}_G$, associated with rotation about the center of mass results from the work done by the moment exerted by the force system about that point.

Another viewpoint is obtained when Eq. (5.35) is used to resolve the force system to a force and couple acting at the fixed point O for a body in pure rotation. Then substituting Eqs. (5.31b) and (5.35) into the work-energy principle yields

$$\tfrac{1}{2}(\bar{\omega} \cdot \bar{H}_O)_2 = \tfrac{1}{2}(\bar{\omega} \cdot \bar{H}_O)_1 + \oint_1^2 \Sigma \bar{M}_O \cdot d\theta \qquad (5.48)$$

In this case, the kinetic energy is entirely associated with rotation about the fixed point.

According to the foregoing discussion, there are two independent work-energy equations for a rigid body. They can consist of Eq. (5.26) for the overall motion, which would be formed by using Eq. (5.35) and either of Eqs. (5.31), in combination with the relation for the motion of the center of mass, Eq. (5.46). As an alternative, Eq. (5.46) may be employed in conjunction with Eq. (5.47) for the rotational kinetic energy about the center of mass. Another alternative, which is useful for pure rotational motion, is to form Eqs. (5.46) and (5.48). The latter two approaches treat the kinetic energy in terms of different rotational effects, which emphasizes that the concept of rotational kinetic energies is only meaningful if the point of reference is specified.

At this juncture we have developed basic principles governing the motion of a rigid body. The foundation for these principles is the basic laws of Newton. Our task now is to relate the quantities appearing in those principles to the inertial properties of the body and the kinematical properties of its motion. These relationships were derived by Euler, so the overall formulation is often referred to as the *Newton-Euler laws of rigid-body motion*. This distinction is important, because we shall derive alternative kinetics principles in Chapters 6 and 7.

5.2 ANGULAR MOMENTUM AND INERTIA PROPERTIES

The basic task here is to convert the summations over the mass elements to integral relationships that are amenable to quantitative evaluations. Our approach is consistent with the usual ways that finite sums extending over numerous small elements are treated in engineering-oriented calculus courses.

5.2.1 Moments and Products of Inertia

The rotational terms in each kinetics principle depend on the angular momentum. The expression for the angular momentum $\overline{H}_A$ given by Eq. (5.22) was

$$\overline{H}_A = \sum_{i=1} \overline{r}_{i/A} \times (\overline{\omega} \times \overline{r}_{i/A}) \, \Delta m_i$$

The summation form is acceptable for conceptual studies, but it is not suitable for computations. The first step in deriving a more useful form is to replace the small elements Δm_i by infinitesimal elements dm filling the region occupied by the body. The summation then becomes an integral over this domain.

In Figure 5.5, *the xyz coordinate system is fixed to the body with its origin at point A*. The unit vector $\overline{r}_{i/A}$ and angular velocity $\overline{\omega}$ may then be described in terms of coor-

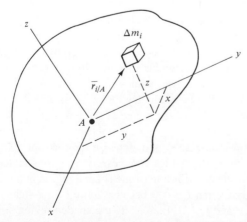

Figure 5.5 Position of a mass element.

dinates relative to xyz as

$$\bar{r}_{i/A} = x\bar{i} + y\bar{j} + z\bar{k}$$
$$\bar{\omega} = \omega_x\bar{i} + \omega_y\bar{j} + \omega_z\bar{k} \tag{5.49}$$

We substitute these expressions into the relation for the angular momentum, and convert the summation to an integral. This transforms the general relation to

$$\overline{H}_A = \iiint (x\bar{i} + y\bar{j} + z\bar{k}) \times [(\omega_x\bar{i} + \omega_y\bar{j} + \omega_z\bar{k}) \times (x\bar{i} + y\bar{j} + z\bar{k})]\, dm \tag{5.50}$$

The rotation rates are overall properties of the motion—they are independent of the position relative to the body. Hence, ω_x, ω_y, and ω_z may be factored out of the integral after the cross products have been evaluated. The result is that

$$\overline{H}_A = (I_{xx}\omega_x - I_{xy}\omega_y - I_{xz}\omega_z)\bar{i} + (I_{yy}\omega_y - I_{xy}\omega_x - I_{yz}\omega_z)\bar{j}$$
$$+ (I_{zz}\omega_z - I_{xz}\omega_x - I_{yz}\omega_y)\bar{k} \tag{5.51}$$

where

$$I_{xx} = \iiint (y^2 + z^2)\, dm \qquad I_{yy} = \iiint (x^2 + z^2)\, dm$$
$$I_{zz} = \iiint (x^2 + y^2)\, dm \tag{5.52a}$$

and

$$I_{xy} = \iiint xy\, dm \qquad I_{yz} = \iiint yz\, dm \qquad I_{xz} = \iiint xz\, dm \tag{5.52b}$$

The terms I_{pp} (repeated subscripts) are *moments of inertia* about the three coordinate axes, and the terms I_{pq} (nonrepeated subscripts) are *products of inertia*. The former are the inertia properties encountered in planar motion, but they are now defined for rotations about three axes. The similarity of any I_{pp} to the parameter for planar motion may be realized by looking down the p axis. Such a view for I_{zz} is shown in Figure 5.6. The distance $R = (x^2 + y^2)^{1/2}$ is the perpendicular distance from the z axis to the mass element dm. Thus, I_{zz} is the sum of the $R^2\, dm$ values for all elements.

The products of inertia describe the symmetry of the mass distribution relative to the coordinate planes. In Figure 5.7, the mirror image of the left region $x < 0$ is outlined in the right region $x > 0$. If the density is also symmetrical, every mass element at (x, y, z) within the outlined region in $x > 0$ is matched by a corresponding element at $(-x, y, z)$ to the left. Thus, the $xy\, dm$ values for these two regions cancel, as do the $xz\, dm$ values. All that remains is the contribution of the shaded region outside the outline. The situation in the figure suggests that $I_{xy} < 0$ because more of the shaded area seems to lie in the octants where $x > 0$ and $y < 0$. Increasing the shaded region would increase I_{xy} in magnitude. However, the actual sign of I_{xy} for the situation in Figure 5.7 cannot

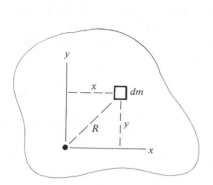

Figure 5.6 Contribution of a mass element to a moment of inertia.

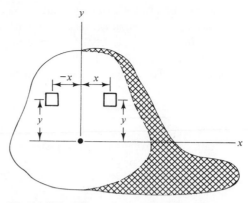

Figure 5.7 Effect of symmetry on a product of inertia.

be judged solely from the drawing because it depends on the variation of shape and density in the z direction.

A corollary of the foregoing is that if the yz plane is a plane of symmetry, then $I_{xy} = 0$ and $I_{xz} = 0$. The fact that the x axis is normal to the plane of symmetry leads to the generalization that

If two coordinate axes form a plane of symmetry for a body, then all products of inertia involving the coordinate normal to that plane are zero.

A further corollary is:

If at least two of the three coordinate planes are planes of symmetry for a body, then all products of inertia are zero.

Whenever the coordinate axes correspond to vanishing values of all products of inertia, they are said to be *principal axes*. We shall see that it is possible to identify principal axes for all bodies, not only symmetric ones.

The inertia properties of homogeneous bodies are tabulated for a variety of common shapes; see Appendix A. Properties for bodies that are not tabulated may be evaluated by treating them as composites of basic shapes or, as a last resort, by carrying out the integrals in Eqs. (5.52).

Once the inertia properties are known, the angular momentum may be evaluated according to Eq. (5.51). For this, the components of $\overline{\omega}$ would be found by the methods for spatial kinematics established in Chapters 3 and 4. The vector relation may alternatively be written in matrix form as

$$\{H_A\} = [I]\{\omega\} \tag{5.53}$$

where $\{H_A\}$ and $\{\omega\}$ consist of the components of $\overline{H}_A$ and $\overline{\omega}$, respectively, and $[I]$ is the *inertia matrix*,

$$[I] = \begin{bmatrix} I_{xx} & -I_{xy} & -I_{xz} \\ -I_{xy} & I_{yy} & -I_{yz} \\ -I_{xz} & -I_{yz} & I_{zz} \end{bmatrix} \tag{5.54}$$

This square array of moments and products of inertia, combined with the mass and the location of the center of mass, fully characterizes the inertia properties of a rigid body.

In some situations the orientation of the desired xyz coordinate system might not match the one appearing in a tabulation. It is necessary in that case to convert the inertia properties. The appropriate transformations will be developed in the next section.

EXAMPLE 5.1 _____

Derive the inertia matrix of the quarter sphere about the xyz axes; then use that result to obtain the inertia matrix for a quarter-spherical shell. Express each result in terms of the mass m of that body.

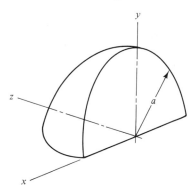

Example 5.1

Solution. Spherical cooordinates are ideal here. Any coordinate axis may be employed as the reference for the polar angle. We select the y axis, so that

$$x = r \sin \phi \cos \theta \qquad y = r \cos \phi \qquad z = r \sin \phi \sin \theta$$

The body occupies the domain $0 \leq r \leq a$, $0 \leq \phi \leq \pi/2$, $0 \leq \theta \leq \pi$, and a differential element of mass is

$$dm = \rho r^2 \sin \phi \, dr \, d\theta \, d\phi$$

Since this body is one-quarter of a sphere, the density ρ is

$$\rho = \frac{m}{V} = 3 \frac{m}{\pi} a^3$$

The coordinate axes are such that the body is symmetric with respect to the yz plane, so

$$I_{xy} = I_{xz} = 0$$

Also, $I_{zz} = I_{yy}$ by symmetry, so it is only necessary to compute I_{xx}, I_{yy}, and I_{yz}. The integral definitions, Eqs. (5.52), give

$$I_{xx} = \int_0^a \int_0^{\pi/2} \int_0^{\pi} (y^2 + z^2)(\rho r^2 \sin \phi) \, d\theta \, d\phi \, dr$$

$$= \rho \int_0^a r^4 \int_0^{\pi/2} \int_0^{\pi} (\cos^2 \phi \sin \phi + \sin^3 \phi \sin^2 \theta) \, d\theta \, d\phi \, dr$$

$$= \tfrac{2}{15} \pi \rho a^5$$

$$I_{yy} = \int_0^a \int_0^{\pi/2} \int_0^{\pi} (x^2 + z^2)(\rho r^2 \sin \phi) \, d\theta \, d\phi \, dr$$

$$= \rho \int_0^a r^4 \int_0^{\pi/2} \int_0^{\pi} \sin^3 \phi \, d\theta \, d\phi \, dr = \tfrac{2}{15}\pi \rho a^5$$

$$I_{yz} = \int_0^a \int_0^{\pi/2} \int_0^{\pi} yz(\rho r^2 \sin \phi) \, d\theta \, d\phi \, dr$$

$$= \rho \int_0^a r^4 \int_0^{\pi/2} \int_0^{\pi} \sin^2 \phi \cos \phi \sin \theta \, d\theta \, d\phi \, dr = \tfrac{2}{15}\rho a^5$$

In order to express these results in terms of the mass m, we substitute the earlier expression for ρ to find

$$[I] = \tfrac{2}{5}ma^2 \begin{bmatrix} 1 & 0 & 0 \\ 0 & 1 & -1/\pi \\ 0 & -1/\pi & 1 \end{bmatrix}$$

Results for a shell may be obtained by considering a composite body that is formed by removing a quarter sphere of radius $a - \Delta$ from the given body. The mass is then

$$m = \rho \frac{\pi}{3} [a^3 - (a - \Delta)^3] = \rho \frac{\pi}{3} (3a^2 \Delta - 3a\Delta^2 + \Delta^3)$$

For a thin shell $\Delta/a \ll 1$, so the above gives

$$\rho = \frac{m}{\pi a^2 \Delta}$$

The comparable differences for the inertia properties must be formed using the original results, which depended on ρ. Thus

$$I_{xx} = I_{yy} = I_{zz} = \tfrac{2}{15}\pi\rho[a^5 - (a - \Delta)^5] = \tfrac{2}{3}\pi\rho a^4 \Delta$$
$$I_{yz} = \tfrac{2}{15}\rho[a^5 - (a - \Delta)^5] = \tfrac{2}{3}\rho a^4 \Delta$$

Eliminating ρ in favor of the mass of the shell yields

$$[I] = \tfrac{2}{3}ma^2 \begin{bmatrix} 1 & 0 & 0 \\ 0 & 1 & -1/\pi \\ 0 & -1/\pi & 1 \end{bmatrix}$$

5.2.2 Transformation of Inertia Properties

Suppose that the moments and products of inertia of the body in Figure 5.8 are known relative to the xyz system. The origin O of xyz might not be acceptable for formulating the equation of rotational motion, whereas point O' is acceptable. Then it will be necessary to transfer the inertia properties to an $x'y'z'$ coordinate system having its origin at point O'. Furthermore, it also might be desirable for the $x'y'z'$ axes to be rotated relative to xyz. The general task involves translational and rotational transformations of the inertia properties.

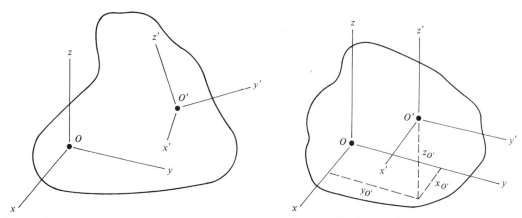

Figure 5.8   Alternative coordinate systems.   **Figure 5.9**   Translation transformation.

Figure 5.9 depicts a translational transformation. The inertia properties with respect to *xyz* are known, and the coordinates of the origin O' relative to *xyz* are denoted as $x_{O'}$, $y_{O'}$, $z_{O'}$. The coordinate transformation in this case is

$$x' = x - x_{O'} \qquad y' = y - y_{O'} \qquad z' = z - z_{O'} \tag{5.55}$$

Consider the definition of a moment of inertia, for example, $I_{x'x'}$. Substitution of Eqs. (5.55) into the first of Eqs. (5.52a) yields

$$\begin{aligned}
I_{x'x'} &= \iiint [(y')^2 + (z')^2]\, dm \\
&= \iiint [y^2 + z^2 - 2y_{O'}y - 2z_{O'}z + (y_{O'})^2 + (z_{O'})^2]\, dm \\
&= I_{xx} - 2y_{O'} \iiint y\, dm - 2z_{O'} \iiint z\, dm + m[(y_{O'})^2 + (z_{O'})^2] \tag{5.56}
\end{aligned}$$

The integrals remaining in Eq. (5.56) are first moments of mass with respect to the *y* and *z* coordinates. They locate the center of mass *G* relative to the origin of *xyz*. Thus

$$I_{x'x'} = I_{xx} - 2\, my_{O'}y_G - 2\, mz_{O'}z_G + m(y_{O'}^2 + z_{O'}^2) \tag{5.57}$$

This translation transformation may be simplified, if the origin *O* is restricted to being the center of mass. Then $x_G = y_G = z_G = 0$. Repetition of the derivation for the other terms leads to the *parallel-axis theorems for moments of inertia*,

$$\begin{aligned}
I_{x'x'} &= I_{xx} + m(y_{O'}^2 + z_{O'}^2) \\
I_{y'y'} &= I_{yy} + m(x_{O'}^2 + z_{O'}^2) \\
I_{z'z'} &= I_{zz} + m(x_{O'}^2 + y_{O'}^2)
\end{aligned} \tag{5.58}$$

Note that the sums of squares of the O' coordinates appearing in Eq. (5.58) are actually the square of the perpendicular distance between the parallel axes for the respective moments of inertia. Since the *x*, *y*, and *z* axes are required to intersect the center of mass, it is clear that the moments of inertia for centroidal axes are smaller than those about parallel noncentroidal axes.

The transformation of the products of inertia is obtained in a similar manner. The result is the *parallel-axis theorems for products of inertia*.

$$I_{x'y'} = I_{xy} + mx_{O'}y_{O'}$$
$$I_{y'z'} = I_{yz} + my_{O'}z_{O'} \qquad (5.59)$$
$$I_{x'z'} = I_{xz} + mx_{O'}z_{O'}$$

It is important to realize that although the signs of the coordinates of point O' are unimportant to the transformation of moments of inertia, they must be considered when transforming the products of inertia. It is useful in this regard to remember that

1. The origin of xyz is the center of mass of the body.
2. The values $(x_{O'}, y_{O'}, z_{O'})$ are the coordinates relative to xyz of the origin O' of the translated coordinate system $x'y'z'$.

The derivation of the rotation transformation follows a different approach. Let the transformation matrix from xyz to $x'y'z'$ in Figure 5.10 be $[R]$, where

$$\begin{Bmatrix} x' \\ y' \\ z' \end{Bmatrix} = [R] \begin{Bmatrix} x \\ y \\ z \end{Bmatrix} \qquad (5.60)$$

The rotational kinetic energy of a rigid body was shown in an earlier section to be

$$T_{\text{rot}} = \tfrac{1}{2}\overline{\omega} \cdot \overline{H}_A = \tfrac{1}{2}\{\omega\}^{\text{T}}\{H_A\} \qquad (5.61)$$

Using the matrix form of $\overline{H}_A$ in Eq. (5.53) leads to

$$T_{\text{rot}} = \tfrac{1}{2}\{\omega\}^{\text{T}}[I]\{\omega\} \qquad (5.62)$$

The key feature of the foregoing is that the kinetic energy is a scalar quantity. Therefore, it is independent of the orientation of the reference frame. If the angular velocity and the inertia properties are referred to the $x'y'z'$ axes, this invariance can be

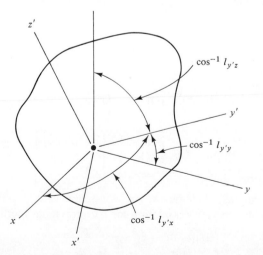

Figure 5.10 Rotation transformation.

satisfied only if

$$\{\omega'\}^{\mathrm{T}}[I']\{\omega'\} = \{\omega\}^{\mathrm{T}}[I]\{\omega\} \tag{5.63}$$

where $[I']$ denotes the inertia matrix associated with $x'y'z'$. The rotation transformation

$$\{\omega'\} = [R]\{\omega\} \tag{5.64}$$

may be solved for $\{\omega\}$ with the aid of the orthonormal property: $[R]^{-1} = [R]^{\mathrm{T}}$. Substitution of that expression into Eq. (5.63), followed by equating the inner matrices in the products yields

$$[I'] = [R][I][R]^{\mathrm{T}} \tag{5.65}$$

Any quantity transforming according to Eq. (5.65) is said to be a *tensor of the second rank*. In this viewpoint, vectors, whose components transform according to Eq. (5.64), are tensors of the first rank.

The transformation in Eq. (5.65) may be resolved into individual inertia values. The transformation matrix may be written in partition form as a sequence of rows as

$$[R] = \begin{bmatrix} l_{x'x} & l_{x'y} & l_{x'z} \\ l_{y'x} & l_{y'y} & l_{y'z} \\ l_{z'x} & l_{z'y} & l_{z'z} \end{bmatrix} = \begin{bmatrix} \{e_{x'}\}^{\mathrm{T}} \\ \{e_{y'}\}^{\mathrm{T}} \\ \{e_{z'}\}^{\mathrm{T}} \end{bmatrix} \tag{5.66}$$

where the column array $\{e_{p'}\}$, $p' = x'$, y', or z', consists of the direction cosines of axis p' relative to xyz or, equivalently, the components of the unit vector $\bar{e}_{p'}$ along the axes of xyz.

$$\{e_{p'}\} = \begin{Bmatrix} l_{p'x} \\ l_{p'y} \\ l_{p'z} \end{Bmatrix} \qquad p' = x', y', \text{ or } z' \tag{5.67}$$

Let us now employ the partitioned form of $[R]$ to evaluate the inertia transformation in Eq. (5.65). The columns forming $[R]^{\mathrm{T}}$ consist of the rows of $[R]$, so we compute the products in the following manner.

$$[I'] = \begin{bmatrix} \{e_{x'}\}^{\mathrm{T}} \\ \{e_{y'}\}^{\mathrm{T}} \\ \{e_{z'}\}^{\mathrm{T}} \end{bmatrix} [I][\{e_{x'}\} \quad \{e_{y'}\} \quad \{e_{z'}\}]$$

$$= \begin{bmatrix} \{e_{x'}\}^{\mathrm{T}} \\ \{e_{y'}\}^{\mathrm{T}} \\ \{e_{z'}\}^{\mathrm{T}} \end{bmatrix} [[I]\{e_{x'}\} \quad [I]\{e_{y'}\} \quad [I]\{e_{z'}\}]$$

$$= \begin{bmatrix} \{e_{x'}\}^{\mathrm{T}}[I]\{e_{x'}\} & \{e_{x'}\}^{\mathrm{T}}[I]\{e_{y'}\} & \{e_{x'}\}^{\mathrm{T}}[I]\{e_{z'}\} \\ \{e_{y'}\}^{\mathrm{T}}[I]\{e_{x'}\} & \{e_{y'}\}^{\mathrm{T}}[I]\{e_{y'}\} & \{e_{y'}\}^{\mathrm{T}}[I]\{e_{z'}\} \\ \{e_{z'}\}^{\mathrm{T}}[I]\{e_{x'}\} & \{e_{z'}\}^{\mathrm{T}}[I]\{e_{y'}\} & \{e_{z'}\}^{\mathrm{T}}[I]\{e_{z'}\} \end{bmatrix} \tag{5.68}$$

Each of the products appearing as an element above is a scalar value, so matching like elements on the left and right side yields a relation for the individual inertia properties.

$$\begin{aligned} I_{p'p'} &= \{e_{p'}\}^{\mathrm{T}}[I]\{e_{p'}\} \\ I_{p'q'} &= -\{e_{p'}\}^{\mathrm{T}}[I]\{e_{q'}\} = -\{e_{q'}\}^{\mathrm{T}}[I]\{e_{p'}\} \end{aligned} \tag{5.69}$$

EXAMPLE 5.2

The x axis lies in the plane of the upper face of the 100-lb homogeneous box, and the y axis is a main diagonal. Determine the inertia matrix of the box relative to xyz.

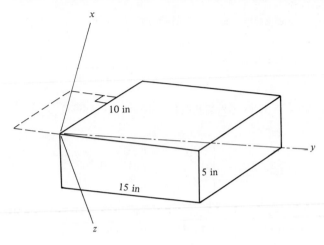

Example 5.2

Solution. Appendix A gives the inertia properties for centroidal axes of a rectangular parallelepiped. These inertia properties may be transferred to a coordinate system whose origin is corner A by means of the parallel axis theorems. For the $x'y'z'$ coordinate axes in the sketch, the coordinates of corner A relative to parallel centroidal axes are $(x_A, y_A, z_A) = (-5, -7.5, -2.5)$ in. Thus,

$$I_{x'x'} = \tfrac{1}{12}m(15^2 + 5^2) + m(7.5^2 + 2.5^2) = 83.33m$$
$$I_{y'y'} = \tfrac{1}{12}m(10^2 + 5^2) + m(5^2 + 2.5^2) = 41.67m$$
$$I_{z'z'} = \tfrac{1}{12}m(10^2 + 15^2) + m(5^2 + 7.5^2) = 108.33m$$
$$I_{x'y'} = 0 + m(-5.0)(-7.5) = 37.50m$$
$$I_{x'z'} = 0 + m(-5.0)(-2.5) = 12.50m$$
$$I_{y'z'} = 0 + m(-7.5)(-2.5) = 18.75m$$

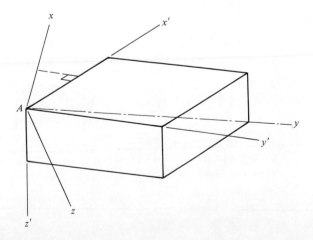

Coordinate systems.

where the mass is

$$m = \frac{100}{12(32.17)} = 0.2590 \text{ lb-s}^2/\text{in.}$$

In order to evaluate the transformation $[R]$ from $x'y'z'$ to xyz, we form the unit vector $\bar{j}$.

$$\bar{j} = \frac{10\bar{i}' + 15\bar{j}' + 5\bar{k}'}{(10^2 + 15^2 + 5^2)^{1/2}} = 0.5345\bar{i}' + 0.8018\bar{j}' + 0.2673\bar{k}'$$

Since the x axis lies in the $x'y'$ plane, the direction cosine $l_{xz'} = 0$. Thus, it follows that

$$[R] = \begin{bmatrix} l_{xx'} & l_{xy'} & 0 \\ 0.5345 & 0.8018 & 0.2673 \\ l_{zx'} & l_{zy'} & l_{zz'} \end{bmatrix}$$

In order for $[R]$ to be an orthogonal transformation, we set $[R][R]^{\mathrm{T}} = [U]$. The equations obtained from the elements of this product are

$$(l_{xx'})^2 + (l_{xy'})^2 = 1$$
$$0.5345 l_{xx'} + 0.8018 l_{xy'} = 0$$
$$l_{xx'} l_{zx'} + l_{xy'} l_{zy'} = 0$$
$$0.5345 l_{zx'} + 0.8018 l_{zy'} + 0.2673 l_{zz'} = 0$$
$$(l_{zx'})^2 + (l_{zy'})^2 + (l_{zz'})^2 = 1$$

The given diagram of the system indicates that $l_{xx'} > 0$ and $l_{zz'} > 0$, so the unknown direction cosines obtained from the above conditions are

$$l_{xx'} = 0.8321 \qquad l_{xy'} = -0.5547$$
$$l_{zx'} = -0.14825 \qquad l_{zy'} = -0.2224 \qquad l_{zz'} = 0.9636$$

The corresponding transformation is

$$[R] = \begin{bmatrix} 0.8321 & -0.5547 & 0 \\ 0.5345 & 0.8018 & 0.2673 \\ -0.1483 & -0.2224 & 0.9636 \end{bmatrix}$$

Since $[R]$ transforms $x'y'z'$ to xyz, the inertia transformation yields

$$[I] = [R][I'][R]^{T} = m[R] \begin{bmatrix} 83.33 & -37.50 & -12.50 \\ -37.50 & 41.67 & -18.75 \\ -12.50 & -18.75 & 108.33 \end{bmatrix} [R]^{\mathrm{T}}$$

$$= \begin{bmatrix} 105.14 & 4.63 & -1.29 \\ 4.63 & 14.58 & 3.47 \\ -1.29 & 3.47 & 113.62 \end{bmatrix} \text{lb-s}^2\text{-in.}$$

5.2.3 Inertia Ellipsoid

Comparison of Eqs. (5.62) and (5.69) leads to an interesting construction that shows the relationship between the moment of inertia about an axis and the kinetic energy. (In Chapter 8, we will find this construction to be useful to our understanding of free motion of a rigid body.) Let us represent the instantaneous angular velocity of a body as a rotation at rate ω about axis p', so that the components of the angular velocity are given in matrix form as

$$\{\omega\} = \omega\{e_{p'}\} \tag{5.70}$$

Substitution of this expression into Eq. (5.62), followed by application of Eq. (5.69), yields

$$2T_{\text{rot}} = \{\omega\}^{\text{T}}[I]\{\omega\} = I_{p'p'}\omega^2 \tag{5.71}$$

Suppose we impart a pure rotation about the origin of xyz, such that the angular velocity is parallel to a specified direction $\bar{e}_{p'}$. How should we adjust the rotation rate in order to obtain a specified value for the rotational kinetic energy? The above relation indicates that the appropriate angular speed is

$$\omega = \left(\frac{2T_{\text{rot}}}{I_{p'p'}}\right)^{1/2} \tag{5.72}$$

Equations (5.70) and (5.72) define the required angular velocity. Let us plot a point P in space that represents the tip of this angular velocity vector when the tail of the vector is placed at the origin. A typical point is shown in Figure 5.11. In order to standardize the definition, let us perform the construction for the special case of an angular velocity $\bar{\rho}$ that gives $T_{\text{rot}} = \frac{1}{2}$. We may represent this vector in either vector or matrix form.

$$\bar{\rho} = \frac{1}{\sqrt{I_{p'p'}}}\bar{e}_{p'} \qquad \{\rho\} = \frac{1}{\sqrt{I_{p'p'}}}\{e_{p'}\} \tag{5.73}$$

When we use the second relation to eliminate $\{e_{p'}\}$ from Eq. (5.70), and then substitute that expression for $\bar{\omega}$ into Eq. (5.71), we find that

$$\{\rho\}^{\text{T}}[I]\{\rho\} = 1 \tag{5.74}$$

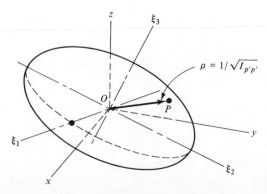

Figure 5.11 Inertia ellipsoid.

In view of the way in which we have defined point P, the (x, y, z) coordinates of the point are identical to the components of $\bar{\rho}$. Hence, expanding Eq. (5.74) leads to

$$I_{xx}x^2 + I_{yy}y^2 + I_{zz}z^2 - 2I_{xy}xy - 2I_{xz}xz - 2I_{yz}yz = 1 \qquad (5.75)$$

This is the equation for an ellipsoidal surface whose centroid coincides with the origin O of the xyz system of axes. This surface is called the *ellipsoid of inertia*. According to Eq. (5.73), the ellipsoid of inertia is the locus of points P for which the distance to the origin is the reciprocal of the square root of the moment of inertia about the axis intersecting point P and origin O. In other words the distance is inversely proportional to the radius of gyration about the axis. If we alter the rotation rate of a body about various axes having a common origin to match this inverse proportionality in accord with Eq. (5.72), we will find that the rotational kinetic energy is constant for all rotation axes.

The major, minor, and intermediate axes of the ellipsoid of inertia, along which the distance from the origin is an extreme value, do not necessarily coincide with the xyz coordinate system. Suppose that all products of inertia with respect to some coordinate system $\xi_1\xi_2\xi_3$ are zero, which means that they are *principal axes of inertia*. Let us denote the corresponding principal moments of inertia as I_1, I_2, and I_3. The equation for the inertia ellipsoid relative to principal axes is then given by Eq. (5.75) to be

$$I_1(\xi_1)^2 + I_2(\xi_2)^2 + I_3(\xi_3)^2 = 1 \qquad (5.76)$$

where (ξ_1, ξ_2, ξ_3) are the coordinates of point P on the inertia ellipsoid relative to the principal axes.

Figure 5.11 shows the inertia ellipsoid relative to principal and nonprincipal axes. Note that Eq. (5.76), which describes the ellipsoid in terms of principal axes, contains only sums of squares. This means that the *major, minor, and intermediate axes of the ellipsoid of inertia coincide with the principal axes*. We could use analytical geometry concepts to evaluate the orientation of the principal axes of inertia. In the next section, we shall instead develop a more direct method for that determination by returning to the fundamental inertia transformation in Eq. (5.65).

5.2.4 Principal Axes

Suppose we know the inertia properties relative to a set of axes xyz that are not principal ones, so that $[I]$ has nondiagonal elements. If we can find a transformation matrix $[R]$ for which $[I']$ is diagonal, then the corresponding $\xi_1\xi_2\xi_3$ axes will be principal. Consistent with the use of subscripted numerals to denote principal parameters, let $\{e_j\}$, $j = 1, 2,$ or 3, denote the (unknown) columns of direction cosines of ξ_1, ξ_2, and ξ_3, respectively, relative to xyz. According to Eq. (5.66), the partitioned form of the transformation matrix from the xyz coordinate system to the principal axes is

$$[R] = \begin{bmatrix} \{e_1\}^T \\ \{e_2\}^T \\ \{e_3\}^T \end{bmatrix} \qquad (5.77)$$

Premultiplication of Eq. (5.65) by $[R]^T$ leads to

$$[R]^T[I'] = [I][R]^T \qquad (5.78)$$

Let us now follow steps similar to those leading to Eq. (5.68) to express Eq. (5.78) explicitly in terms of the $\{e_j\}$. This yields

$$[\{e_1\} \ \{e_2\} \ \{e_3\}] \begin{bmatrix} I_1 & 0 & 0 \\ 0 & I_2 & 0 \\ 0 & 0 & I_3 \end{bmatrix} = [I][\{e_1\} \ \{e_2\} \ \{e_3\}]$$

$$[I_1\{e_1\} \ \ I_2\{e_2\} \ \ I_3\{e_3\}] = [[I]\{e_1\} \ \ [I]\{e_2\} \ \ [I]\{e_3\}] \tag{5.79}$$

In order for this equality to hold, corresponding columns must be identical, so that

$$I_i\{e_i\} = [I]\{e_i\} \qquad i = 1, 2, 3 \tag{5.80}$$

In other words, I_i are the eigenvalues λ, and $\{e_i\}$ are the eigenvectors $\{e\}$ of the matrix equation

$$[[I] - \lambda[U]]\{e\} = \{0\} \tag{5.81}$$

where $[U]$ is the unitary (i.e., identity) matrix.

The solution of matrix eigenvalue problems is a topic in linear algebra, as well as in linear vibration theory. We shall only highlight the concepts here. Equation (5.81) represents three simultaneous equations for the components of $\{e\}$, which are the direction cosines between a principal axis and the axes associated with inertia matrix $[I]$. If those equations are solvable, the only solution is the trivial one, $\{e\} = \{0\}$. The equations cannot be solved for a unique value of $\{e\}$ when the coefficient matrix $[[I] - \lambda[U]]$ cannot be inverted. Hence, nontrivial solutions for $\{e\}$ arise only if the value of λ is selected to satisfy the characteristic equation corresponding to vanishing of the determinant of the coefficients,

$$|[I] - \lambda[U]| = \{0\} \tag{5.82}$$

Evaluation of the determinant with λ as an algebraic parameter leads to a cubic equation for λ. The eigenvalues, which are the three roots of the characteristic equation, are the principal moments of inertia, $\lambda = I_1, I_2, I_3$.

Let us consider first the case where the principal moments of inertia are distinct values. Equating λ to one of these values in that case makes one of the three simultaneous equations represented by Eq. (5.81) a linear combination of the others. Because of this loss of an independent equation, any nonzero component of $\{e\}$ may be chosen arbitrarily. The other components may then be found in terms of the arbitrary one by solving the independent equations. The eigenvector associated with each principal moment of inertia I_i is the set of direction cosines $\{e_i\}$ locating that principal axis relative to the original xyz coordinate system.

Although the solution to the eigenvalue problem leaves an element of the eigenvector undetermined, we must remember that $\{e_i\}$ represents the components relative to xyz of a unit vector oriented parallel to the ith principal axis. The condition that such a vector has a unit magnitude is written in matrix form as

$$\{e_i\}^{\mathrm{T}}\{e_i\} = 1 \tag{5.83}$$

which provides the additional equation required to uniquely evaluate $\{e_i\}$.

The case where two of the principal moments of inertia are identical is very much like the above, except that one more equation is lost for each repetition of the principal value. Thus, the number of times a root is repeated is the same as the number of components of the eigenvector $\{e\}$ that are arbitrary. It follows that equating λ in Eq. (5.81) to a principal value and applying Eq. (5.83) is not sufficient to uniquely specify the $\{e_i\}$ in this case.

The method in which to proceed when this ambiguity arises becomes apparent when we consider a body of revolution. Such a body has identical principal moments of inertia for all axes that perpendicularly intersect the axis of symmetry at a common point. The axes that we wish to construct should be mutually orthogonal. This suggests that if a specific choice for the (several) arbitrary elements is made to solve for one eigenvector $\{e_i\}$ associated with the repeated principal value, then the other one, $\{e_{i+1}\}$, may be constructed by satisfying the orthogonality condition,

$$\{e_i\}^{\mathrm{T}}\{e_{i+1}\} = 0 \qquad (5.84)$$

This provides the extra equation required to define the otherwise arbitrary elements of the eigenvector.

We see from this development that the greater degree of arbitrariness associated with identical principal moments of inertia arises because the corresponding mutually perpendicular principal axes may be arbitrarily selected. The case where all three principal values are identical merely means that any set of axes are principal. There is then no need to solve an eigenvalue problem. This feature is exemplified by a homogeneous sphere or cube when the origin is placed at the centroid.

An important aspect of eigenvalue problems is the concept of orthogonality. The matrix $[R]$ formed from the eigenvectors $\{e_i\}$ is an orthogonal transformation in the usual sense, $[R]^{-1} = [R]^{\mathrm{T}}$. However, Eq. (5.69) shows that there is an extended orthogonality that also applies. Because the $\xi_1\xi_2\xi_3$ axes are principal, that relation for specific values of the inertia properties shows that

$$\{e_j\}^{\mathrm{T}}[I]\{e_k\} = I_j\delta_{jk} \qquad j, k = 1, 2, 3 \qquad (5.85)$$

where δ_{jk} is the Kronecker delta symbol. For comparison, the unit vectors of two orthogonal axes satisfy

$$\{e_j\}^{\mathrm{T}}\{e_k\} = \delta_{jk} \qquad j, k = 1, 2, 3 \qquad (5.86)$$

Hence, we say that the principal coordinate directions are orthogonal with respect to the inertia matrix, as well as with respect to the physical space.

The conversion to principal axes will simplify all equations involving the angular momentum $\overline{H}_A$. The benefits of such simplifications are countered by the need to solve an eigenvalue problem in order to locate the principal axes. Furthermore, the principal axes may not be convenient for the evaluation of $\overline{\omega}$, $\overline{\alpha}$, and $\Sigma\overline{M}_A$. For this reason, we shall make it a practice to formulate problems using coordinate axes having the most convenient orientation. Identification of principal axes on the basis of symmetry will be useful, but we usually will not solve the eigenvalue problem. However, we will find it useful to discuss several cases of motion in Chapter 8 in terms of principal axes and moments of inertia. If we wish to apply those results to an arbitrary body, we must be able to locate the principal axes.

EXAMPLE 5.3

The orthogonal tetrahedron shown has a mass of 60 kg. For axes having origin at the center of mass determine the principal moments of inertia and the rotation transformation locating the principal axes.

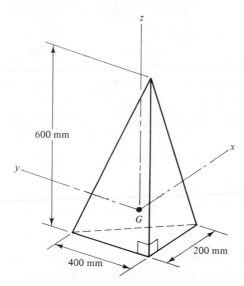

Example 5.3

Solution. Appendix A gives the inertia properties of an orthogonal tetrahedron with respect to centroidal axes parallel to the orthogonal edges. The values in the present case, where $m = 60$ kg, are

$$[I] = \begin{bmatrix} 1.170 & 0.060 & 0.090 \\ 0.060 & 0.900 & 0.180 \\ 0.090 & 0.180 & 0.450 \end{bmatrix} \text{kg-m}^2$$

The eigenvector equation for the principal axes is

$$[[I] - \lambda[U]]\{e\} = \{0\}$$

Expanding the characteristic equation $|[I] - \lambda[U]| = 0$ yields

$$\lambda^3 - 2.520\,\lambda^2 + 1.9404\,\lambda - 0.428976 = 0$$

The eigenvalues are the principal moments of inertia. In order of decreasing magnitude, they are

$$I_1 = 1.20592 \qquad I_2 = 0.93268 \qquad I_3 = 0.38140 \text{ kg-m}^2$$

For the first eigenvalue the eigenvector equation becomes

$$[[I] - \lambda_1[U]]\{e_1\} = \{0\}$$

$$\begin{bmatrix} -0.03592 & 0.060 & 0.090 \\ 0.060 & -0.30592 & 0.180 \\ 0.090 & 0.180 & -0.75592 \end{bmatrix} \begin{Bmatrix} e_{11} \\ e_{12} \\ e_{13} \end{Bmatrix} = \begin{Bmatrix} 0 \\ 0 \\ 0 \end{Bmatrix}$$

The solution of the first two equations is

$$e_{11} = 5.1880e_{13} \qquad e_{12} = 1.6059e_{13}$$

The condition that the values of e_{jk} are the components of unit vector $\bar{e}_j$ yields

$$e_{11}^2 + e_{12}^2 + e_{13}^2 = (5.1880^2 + 1.6059^2 + 1)e_{13}^2 = 1$$

Choosing the positive sign for e_{13} then gives

$$e_{13} = 0.18109 \qquad e_{11} = 0.93948 \qquad e_{12} = 0.29081$$

When the same procedure is carried out for $\lambda = \lambda_2$ and $\lambda = \lambda_3$, we obtain $\{e_2\}$ and $\{e_3\}$, respectively. Thus

$$[R] = \begin{bmatrix} \{e_1\}^T \\ \{e_2\}^T \\ \{e_3\}^T \end{bmatrix} = \begin{bmatrix} 0.9395 & 0.2908 & 0.1811 \\ -0.3322 & 0.9023 & 0.2746 \\ -0.0836 & -0.3181 & 0.9444 \end{bmatrix}$$

EXAMPLE 5.4

A homogeneous disk of mass m and radius R spins at rate ω_1 about its skew axis, which rotates about the horizontal at rate ω_2. Derive an expression for the kinetic energy of the disk.

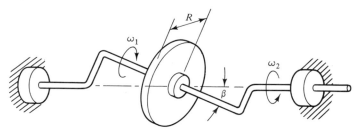

Example 5.4

Solution. In order to avoid transforming inertia properties, we align the z axis with the normal to the disk, and place the origin of xyz at the center of mass. Because of the axisymmetry of the disk, we may align the x and y axes in a manner that facilitates the description of $\bar{\omega}$; our choice is shown in the sketch. The inertia

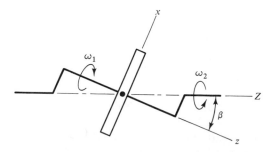

Coordinate systems.

properties in Appendix A give

$$I_{xx} = I_{yy} = \tfrac{1}{4}mR^2 \qquad I_{zz} = \tfrac{1}{2}mR^2 \qquad I_{xy} = I_{yz} = I_{xz} = 0$$

The angular velocity is

$$\overline{\omega} = \omega_2 \overline{K} - \omega_1 \overline{k} = \omega_2 \sin \beta \overline{i} + (\omega_2 \cos \beta - \omega_1)\overline{k}$$

The corresponding angular momentum relative to the center of mass C is

$$\overline{H}_C = I_{xx}\omega_x \overline{i} + I_{zz}\omega_z \overline{k} = mR^2[\tfrac{1}{4}\omega_2 \sin \beta \overline{i} + \tfrac{1}{2}(\omega_2 \cos \beta - \omega_1)\overline{k}]$$

Point C is fixed, so there is only rotational kinetic energy relative to the center of mass,

$$T = \tfrac{1}{2}\overline{\omega} \cdot \overline{H}_C = \tfrac{1}{2}mR^2[\tfrac{1}{4}\omega_2^2 \sin^2 \beta + \tfrac{1}{2}(\omega_2 \cos \beta - \omega_1)^2]$$

5.3 RATE OF CHANGE OF ANGULAR MOMENTUM

The angular momentum is a function of the inertia properties and the angular velocity, as expressed in Eq. (5.51). That form describes the components relative to the xyz reference frame, which is fixed to the body. Thus, the unit vectors of $\overline{H}_A$ change with time. The angular velocity components appearing in the components of $\overline{H}_A$ also change with time. However, the fact that xyz is fixed to the body is a substantial benefit, because it yields constant values for all inertia properties. Accordingly, the total derivative of $\overline{H}_A$ is related to the derivative relative to the moving reference frame by

$$\dot{\overline{H}}_A = \frac{\delta \overline{H}_A}{\delta t} + \overline{\omega} \times \overline{H}_A \tag{5.87}$$

where

$$\frac{\delta \overline{H}_A}{\delta t} = (I_{xx}\dot{\omega}_x - I_{xy}\dot{\omega}_y - I_{xz}\dot{\omega}_z)\overline{i}$$
$$+ (I_{yy}\dot{\omega}_y - I_{xy}\dot{\omega}_x - I_{yz}\dot{\omega}_z)\overline{j} + (I_{zz}\dot{\omega}_z - I_{xz}\dot{\omega}_x - I_{yz}\dot{\omega}_y)\overline{k} \tag{5.88}$$

Evaluation of Eq. (5.88) involves determination of the time derivatives of the components of angular velocity. One method for obtaining such terms is to describe the angular velocity in a general functional form that may be differentiated. However, a shortcut is available, because $\overline{\omega}$, which is the quantity to be differentiated, is also the angular velocity of the reference frame for the components. It follows that the derivatives of $\overline{\omega}$ relative to the fixed and moving reference frames are identical,

$$\frac{d\overline{\omega}}{dt} = \frac{\delta \overline{\omega}}{\delta t} + \overline{\omega} \times \overline{\omega} = \frac{\delta \overline{\omega}}{\delta t} \tag{5.89a}$$

By definition, the absolute derivative of $\overline{\omega}$ is the angular acceleration $\overline{\alpha}$. Hence, Eq.

(5.89a) may be written in component form as

$$\alpha_x = \dot{\omega}_x \qquad \alpha_y = \dot{\omega}_y \qquad \alpha_z = \dot{\omega}_z \qquad \text{(5.89b)}$$

The significance of these relations is that they enable us to evaluate the kinematical quantities affecting $\overline{H}_A$ by the method developed in Chapter 3. Recall that the procedure begins by writing a general equation for $\overline{\omega}$ that employs unit vectors associated with fixed and moving coordinate systems. A corresponding expression for $\overline{\alpha}$ is then found by differentiating each unit vector using the angular velocity of its reference frame. Finally, xyz components of $\overline{\omega}$ and $\overline{\alpha}$ are obtained by resolving the various unit vectors, either by inspection or through the use of rotation transformation matrices.

EXAMPLE 5.5 _____

Determine the rate of change of the angular momentum of the disk in Example 5.4 about its center of mass. Both ω_1 and ω_2 are constant.

Solution. Most of the parameters required to form $\dot{\overline{H}}_A$ were obtained in the solution to Example 5.4. It only remains to form $\overline{\alpha}$. Toward this end, we recall the general expression for $\overline{\omega}$.

$$\overline{\omega} = -\omega_1 \overline{k} + \omega_2 \overline{K}$$

Since both rotation rates are constant, we find that

$$\overline{\alpha} = -\omega_1 (\overline{\omega} \times \overline{k})$$

which when resolved into xyz components yields

$$\overline{\omega} = \omega_2 \sin \beta \overline{i} + (\omega_2 \cos \beta - \omega_1)\overline{k}$$
$$\overline{\alpha} = \omega_1 \omega_2 \sin \beta \overline{j}$$

The inertia properties relative to the principal axes xyz were found previously to be

$$I_{xx} = I_{yy} = \tfrac{1}{4} mR^2 \qquad I_{zz} = \tfrac{1}{2} mR^2$$

Thus, we find

$$\overline{H}_A = I_{xx}\omega_x \overline{i} + I_{zz}\omega_z \overline{k}$$
$$= mR^2 \left[\tfrac{1}{4}\omega_2 \sin \beta \overline{i} + \tfrac{1}{2}(\omega_2 \cos \beta - \omega_1)\overline{k} \right]$$

$$\frac{\delta \overline{H}_A}{\delta t} = I_{xx}\alpha_y \overline{j} = mR^2 \left(\tfrac{1}{4}\omega_1 \omega_2 \sin \beta \right) \overline{j}$$

Then

$$\dot{\overline{H}}_A = \frac{\delta \overline{H}_A}{\delta t} + \overline{\omega} \times \overline{H}_A$$
$$= mR^2 \left[\tfrac{1}{4}\omega_1 \omega_2 \sin \beta + (\tfrac{1}{4} - \tfrac{1}{2})(\omega_2 \sin \beta)(\omega_2 \cos \beta - \omega_1) \right] \overline{j}$$
$$= mR^2 \, \omega_2 \sin \beta \left(\tfrac{1}{2}\omega_1 - \tfrac{1}{4}\omega_2 \cos \beta \right) \overline{j}$$

EXAMPLE 5.6

The square plate is pinned at corner A to the vertical shaft, which rotates at the constant angular speed Ω. The angle θ is an abitrary function of time. Determine $\bar{H}_A$ for the plate as a function of θ.

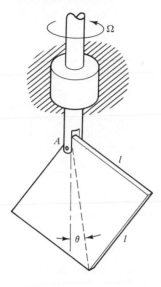

Example 5.6

Solution. In order to form $\bar{H}_A$, the origin of xyz must coincide with point A. Aligning x and y with the edges of the plate, in accord with the entry in Appendix A, yields the reference frame shown. The inertia properties obtained from the parallel-axis theorems are

$$I_{xx} = I_{yy} = \tfrac{1}{3}ml^2 \qquad I_{zz} = \tfrac{2}{3}ml^2 \qquad I_{xy} = \tfrac{1}{4}ml^2 \qquad I_{xz} = I_{yz} = 0$$

We next form $\bar{\omega}$ and $\bar{\alpha}$. The general expressions are

$$\bar{\omega} = \Omega\bar{K} + \dot{\theta}\bar{k} \qquad \bar{\alpha} = \ddot{\theta}\bar{k} + \dot{\theta}(\bar{\omega} \times \bar{k})$$

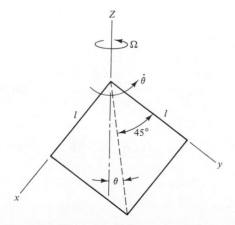

Coordinate system.

It is convenient to resolve these into *xyz* components using the substitution $\gamma = 45° + \theta$, which yields

$$\bar{\omega} = \Omega(-\sin\gamma\,\bar{i} - \cos\gamma\,\bar{j}) + \dot{\theta}\bar{k}$$
$$\bar{\alpha} = \Omega\dot{\theta}(-\cos\gamma\,\bar{i} + \sin\gamma\,\bar{j}) + \ddot{\theta}\bar{k}$$

from which it follows that

$$\bar{H}_A = (I_{xx}\omega_x - I_{xy}\omega_y)\bar{i} + (I_{yy}\omega_y - I_{xy}\omega_x)\bar{j} + I_{zz}\omega_z\bar{k}$$
$$= ml^2\,[\Omega\,(-\tfrac{1}{3}\sin\gamma + \tfrac{1}{4}\cos\gamma)\bar{i} + \Omega\,(-\tfrac{1}{3}\cos\gamma + \tfrac{1}{4}\sin\gamma)\bar{j} + \tfrac{2}{3}\dot{\theta}\bar{k}]$$

$$\frac{\delta\bar{H}_A}{\delta t} = (I_{xx}\alpha_x - I_{xy}\alpha_y)\bar{i} + (I_{yy}\alpha_y - I_{xy}\alpha_x)\bar{j} + I_{zz}\alpha_z\bar{k}$$
$$= ml^2[\Omega\dot{\theta}\,(-\tfrac{1}{3}\cos\gamma - \tfrac{1}{4}\sin\gamma)\bar{i} + \Omega\dot{\theta}\,(\tfrac{1}{3}\sin\gamma + \tfrac{1}{4}\cos\gamma)\bar{j} + \tfrac{2}{3}\ddot{\theta}\bar{k}]$$

In turn, these expressions yield

$$\dot{\bar{H}}_A = \frac{\delta H_A}{\delta t} + \bar{\omega} \times \bar{H}_A$$
$$= ml^2\,\{\Omega\dot{\theta}\,(-\tfrac{2}{3}\cos\gamma - \tfrac{1}{2}\sin\gamma)\bar{i} + \Omega\dot{\theta}\,(\tfrac{2}{3}\sin\gamma + \tfrac{1}{2}\cos\gamma)\bar{j}$$
$$+ [\tfrac{2}{3}\ddot{\theta} + \tfrac{1}{4}\Omega^2\,(\cos^2\gamma - \sin^2\gamma)]\bar{k}\}$$

Since $\gamma = 45° + \theta$, the trigonometric identities for the sine and cosine of the sum of two angles yield

$$\dot{\bar{H}}_A = \frac{\sqrt{2}}{12}\,ml^2\Omega\dot{\theta}[(-7\cos\theta + \sin\theta)\bar{i} + (7\cos\theta + \sin\theta)\bar{j}]$$
$$+ ml^2\,(\tfrac{2}{3}\ddot{\theta} - \tfrac{1}{4}\Omega^2\sin 2\theta)\bar{k}$$

5.4 EQUATIONS OF MOTION

The developments in the preceding section provide the foundation for synthesizing the relationship between the properties of the external force system and the motion of the body. Such relations are called *equations of motion*.

We employed Eqs. (5.89b) to evaluate $\delta\bar{H}_A/\delta t$ in Eq. (5.88), and then used that expression to form $\dot{\bar{H}}_A$ in Eq. (5.87). Equating $\dot{\bar{H}}_A$ to $\Sigma\bar{M}_A$ leads to the equation of rotational motion. In addition, the equation for the motion of the center of mass must be satisfied. For completeness, we summarize below the expressions needed to formulate the equations of translational and rotational acceleration.

$$\sum\bar{F} = m\bar{a}_G \tag{5.90}$$

$$\sum \overline{M}_A = \frac{\delta \overline{H}_A}{\delta t} + \overline{\omega} \times \overline{H}_A \tag{5.91}$$

$$\overline{H}_A = (I_{xx}\omega_x - I_{xy}\omega_y - I_{xz}\omega_z)\overline{i} + (I_{yy}\omega_y - I_{xy}\omega_x - I_{yz}\omega_z)\overline{j}$$
$$+ (I_{zz}\omega_z - I_{xz}\omega_x - I_{yz}\omega_y)\overline{k} \tag{5.92}$$

$$\frac{\delta \overline{H}_A}{\delta t} = (I_{xx}\alpha_x - I_{xy}\alpha_y - I_{xz}\alpha_z)\overline{i} + (I_{yy}\alpha_y - I_{xy}\alpha_x - I_{yz}\alpha_z)\overline{j}$$
$$+ (I_{zz}\alpha_z - I_{xz}\alpha_x - I_{yz}\alpha_y)\overline{k} \tag{5.93}$$

It must be emphasized that the moment equation should be formulated such that point A is selected to be either

1. The center of mass of the body, or
2. A fixed point in a body that is in a state of pure rotation.

Matrix notation offers a compact scheme for performing calculations. The angular momentum was written in this form in Eq. (5.53) in terms of the inertia matrix $[I]$, which was defined in Eq. (5.54). The corresponding forms of the equations of motion are

$$\begin{Bmatrix} \sum F_x \\ \sum F_y \\ \sum F_z \end{Bmatrix} = m \begin{Bmatrix} a_{Gx} \\ a_{Gy} \\ a_{Gz} \end{Bmatrix} \tag{5.94}$$

$$\begin{Bmatrix} \sum M_{Ax} \\ \sum M_{Ay} \\ \sum M_{Az} \end{Bmatrix} = [I]\begin{Bmatrix} \alpha_x \\ \alpha_y \\ \alpha_z \end{Bmatrix} + \begin{bmatrix} 0 & -\omega_z & \omega_y \\ \omega_z & 0 & -\omega_x \\ -\omega_y & \omega_x & 0 \end{bmatrix}[I]\begin{Bmatrix} \omega_x \\ \omega_y \\ \omega_z \end{Bmatrix} \tag{5.95}$$

In the foregoing, $\sum F_x$, $\sum F_y$, and $\sum F_z$ are the sum of the external forces acting on the body in the direction of the corresponding axes. Similarly, $\sum M_{Ax}$, $\sum M_{Ay}$, and $\sum M_{Az}$ represent the sum of the moments of the external forces about the corresponding axes whose origin is point A.

The special case where xyz are principal axes leads to explicit formulas for the equations of rotational motion, which are known as *Euler's equations*. They are

$$\sum M_{Ax} = I_{xx}\alpha_x - (I_{yy} - I_{zz})\omega_y\omega_z$$
$$\sum M_{Ay} = I_{yy}\alpha_y - (I_{zz} - I_{xx})\omega_x\omega_z \tag{5.96}$$
$$\sum M_{Az} = I_{zz}\alpha_z - (I_{xx} - I_{yy})\omega_x\omega_y$$

The repetitive pattern of Euler's equations can be used to remember the individual components by a mnemonic algorithm based on permutations of the alphabetical order. Euler's equations are particularly useful when it is only necessary to address the moment exerted about one axis.

An aspect of the moment equation of motion that puzzles a novice is the presence of a moment, even when the rotation rates are constant. This effect arises because the orientation of $\bar{\omega}$ is not constant, and also because $\bar{H}_A$ is usually not parallel to $\bar{\omega}$. Both effects cause the orientation of the angular momentum vector to change with time, so that $\dot{\bar{H}}_A \neq 0$. The moment equation merely requires that the force system apply a moment that balances the rate at which the angular momentum changes. (It is irrelevant to the discussion whether the moment is considered to sustain the angular motion, or the angular motion is considered to be produced by the moment.) The portion of $\dot{\bar{H}}_A$ that is present when the rotation rates are constant is often referred to as the *gyroscopic moment*.

Various questions may be investigated using the equations of motion. In the simplest case, the motion of a rigid body is fully specified. This permits complete evaluation of the right side of the translational and rotational equations. The forcing effects, which appear on the left side of the equations, originated from known loads, as well as reactions. The latter are particularly important to characterize. A *free body diagram*, in which the body is isolated from its surroundings, is essential to the correct description of the reactions.

As an aid in drawing a free body diagram, recall that reactions are the physical manifestations of kinematical constraints. Thus, if a support prevents a point in the body from moving in a certain direction, then at that point there must be a reaction force exerted on the body in that direction. Similarly, a kinematical constraint on rotation about an axis is imposed by a reaction couple exerted about that axis. In general, the reactions are not known in advance—they are unknown values that will appear in some or all of the equations of motion.

There are only six scalar equations of motion for each body contained in the system (three force sums and three moment sums). It is possible for the number of unknown reactions to exceed the number of available equations. Assuming that this condition does not result from erroneous omission of some characteristic of the supports, it can result from *redundant constraints*. That is the dynamic analog of the condition of *static indeterminacy*, whose resolution requires consideration of deformation effects.

In some situations, the qualitative features of the motion are known except for a value of an angle of orientation or a rotation rate. Such conditions lead to the same type of formulation as that in which the motion is fully specified, except that the list of unknowns will also contain the unspecified kinematical parameters.

The more difficult situation arises when the nature of the motion is not known in advance. The orientation of the body may then be described in terms of Eulerian angles (precession, nutation, and spin). The result will be differential equations for the Eulerian angles. Recall that $\bar{\alpha}$ depends on these angles and their first and second derivatives. Also, the product $\bar{\omega} \times \bar{H}_A$ enters into the evaluation of $\dot{\bar{H}}_A$. As a result, the equations of rotational motion will usually be coupled, nonlinear, second-order differential equations.

Analytical solutions of such equations are available in limited situations, but numerical techniques are often necessary. In any case, it is standard practice when the motion is unknown to eliminate the reactions from the equations of motion. The reactions enter into the equations of motion algebraically through the force and moment sums. Hence, their elimination involves, at the worst, a process of simultaneous solution of algebraic equations. (This, of course, assumes that a condition of redundant constraint does not exist.)

EXAMPLE 5.7

The cylinder, whose mass is M, is welded to the shaft such that its center is situated on the axis of rotation. The presence of a torque Γ causes the rotation rate Ω to vary. Derive expressions for Γ and the reactions at bearings A and B in terms of Ω and $\dot{\Omega}$.

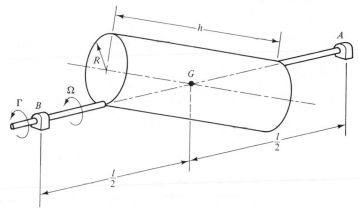

Example 5.7

Solution. The first step is to draw a free body diagram of the cylinder and the shaft. This diagram shows two transverse reactions at each bearing, and a thrust reaction at bearing A. We shall ignore the weight, because it is a static force. The corresponding static reactions divide equally between the bearings, and superpose on the dynamic reactions we shall evaluate.

It is reasonable to assume that the mass of the shaft is negligible in comparison with the mass of the cylinder. The xyz axes we select match those in Appendix A, so the inertia properties for the centroidal principal axes are

$$I_{xx} = I_{yy} = m\left(\tfrac{1}{4}R^2 + \tfrac{1}{12}h^2\right) \qquad I_{zz} = \tfrac{1}{2}mR^2 \qquad I_{xy} = I_{xz} = I_{yz} = 0$$

The center of mass is on the axis of rotation so $\bar{a}_G = \bar{0}$. The angular velocity and angular acceleration are

$$\bar{\omega} = -\Omega\bar{K} \qquad \bar{\alpha} = -\dot{\Omega}\bar{K}$$

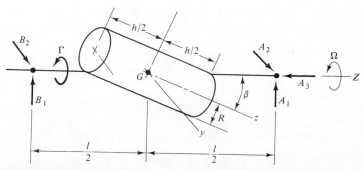

Free body diagram.

which when resolved into xyz components become

$$\overline{\omega} = -\Omega(\sin \beta \,\overline{i} + \cos \beta \,\overline{k}) \qquad \overline{\alpha} = -\dot{\Omega}(\sin \beta \,\overline{i} + \cos \beta \,\overline{k})$$

Since xyz are principal axes, we may employ Euler's equations. For this, the applied couple $\overline{\Gamma}$ must be resolved into components. Referring to the free body diagram for the lever arms of the forces yields

$$\sum M_{Gx} = -\Gamma \sin \beta + (B_2 - A_2)\frac{l}{2}\cos \beta = I_{xx}\alpha_x$$

$$= m\,(\tfrac{1}{4}R^2 + \tfrac{1}{12}h^2)(-\dot{\Omega}\sin \beta)$$

$$\sum M_{Gy} = (A_1 - B_1)\frac{l}{2} = -(I_{zz} - I_{xx})\omega_x\omega_z$$

$$= -m(\tfrac{1}{4}R^2 - \tfrac{1}{12}h^2)\,\Omega^2 \sin \beta \cos \beta$$

$$\sum M_{Gz} = -\Gamma \cos \beta + (A_2 - B_2)\frac{l}{2}\sin \beta = I_{zz}\alpha_z$$

$$= \tfrac{1}{2}mR^2(-\dot{\Omega}\cos \beta)$$

The force sums may be formed in terms of components relative to any set of axes, because $\overline{a}_G = \overline{0}$. Using the two transverse directions leads to

$$A_3 = 0 \qquad A_1 + B_1 = 0 \qquad A_2 + B_2 = 0$$

When we consider Ω to be a known function of time, we have three force equations and three moment equations for the six unknowns A_1, A_2, B_1, B_2, A_3, and Γ. The solutions are

$$A_1 = -B_1 = \frac{1}{6}m\frac{\Omega^2}{l}(h^2 - 3R^2)\sin 2\beta$$

$$A_2 = -B_2 = \frac{1}{6}m\frac{\dot{\Omega}}{l}(h^2 - 3R^2)\sin 2\beta$$

$$\Gamma = \tfrac{1}{12}m\dot{\Omega}[6R^2 + (h^2 - 3R^2)\sin^2 \beta]$$

Each of these results may be readily explained from the viewpoint of changes in the angular momentum $\overline{H}_G$. Let us suppose for this discussion that $h^2 > 3R^2$, so that $I_{xx} > I_{zz}$. Then at any instant, such as the one depicted in the sketch, the angle from the negative z axis to $\overline{H}_G$ is larger than β. In this sketch $(\Delta\overline{H}_G)_1$ and $(\Delta\overline{H}_G)_2$ represent increments in $\overline{H}_G$ that would be observed over a small time interval. The effect of the rotation about the horizontal axis is to change the direc-

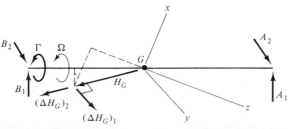

Effect of rotation on the angular momentum.

tion of $\overline{H}_G$, such that $(\Delta\overline{H}_G)_1$ is parallel to the y direction. The force system must exert a corresponding net moment about the y axis, which can only be produced by the reactions A_1 and B_1. The increase in Ω due to its time dependence results in an increase in the magnitude of $\overline{H}_G$, which corresponds in the sketch to $(\Delta\overline{H}_G)_2$, parallel to $\overline{H}_G$. The component of this change parallel to the horizontal axis of rotation must be matched by the torque Γ, which is the only portion of the force system that exerts a moment about the rotation axis. In addition, $(\Delta\overline{H}_G)_2$ also has a component transverse to the rotation axis in the zZ plane. The corresponding moment is obtained from the forces A_2 and B_2. Because the center of mass is on the rotation axis, A_1 and B_1 must form a couple, as must A_2 and B_2. The sense of each couple suggested by the increments in $\overline{H}_G$ appearing in the sketch is consistent with the results of our analysis.

A convenient additional check on the solutions is whether the results are physically sensible when $\beta = 0°$ and $\beta = 90°$. Since the fixed axis of rotation Z coincides with a principal axis in either case, $\overline{H}_A$ is then always parallel to the Z axis. Correspondingly, we expect that the reactions should vanish, and that $M = I_{ZZ}\dot{\Omega}$. Both expectations are borne out by the above expressions.

EXAMPLE 5.8 _____

A thin homogeneous disk of mass m rolls without slipping on a horizontal plane such that the center A has a constant speed v as it follows a circular path of radius ρ. The angle of inclination of the axis relative to the vertical is a constant value θ. Derive an expression relating v to the other parameters.

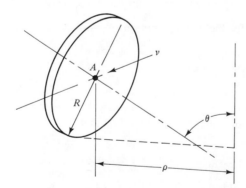

Example 5.8

Solution. In addition to the gravitational force, there are reactions at the contact between the disk and the ground. We have depicted in the free body diagram the frictional forces lying in the horizontal plane as a component F_n toward the vertical axis about which the disk rotates, and a component F_t opposite the velocity of point A. We select a body-fixed coordinate system xyz such that the x and y axes lie in the plane of the disk. At the instant of interest the y axis is aligned horizontal, in the sense of the velocity of point A. The rotation of the disk consists of a precession $\dot{\psi}$ about the vertical axis and a spin $\dot{\phi}$ about the z axis; the speed of the

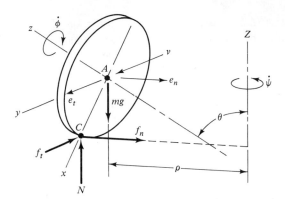

Free body diagram.

center is related to the precession rate by

$$\dot{\psi} = \frac{v}{\rho}$$

Superposing the rotations about the two axes leads to the following for the angular velocity of the disk:

$$\overline{\omega} = \dot{\psi}\overline{K} + \dot{\phi}\overline{k}$$

The corresponding angular acceleration is

$$\overline{\alpha} = \dot{\phi}\overline{\omega} \times \overline{k}$$

At the instant we have depicted in the free body diagram we have

$$\overline{K} = \cos\theta\overline{k} - \sin\theta\overline{i}$$

which leads to expressions for the angular motion of the disk.

$$\overline{\omega} = -\dot{\psi}\sin\theta\overline{i} + (\dot{\phi} + \dot{\psi}\cos\theta)\overline{k}$$
$$\overline{\alpha} = \dot{\psi}\dot{\phi}\sin\theta\overline{j}$$

The rotation rates are related to the speed v by the no-slip condition at the contact point C, which requires that $\overline{v}_C = \overline{0}$. We may describe the velocity of the center A by relating it to point C and by using the fact that it is undergoing circular motion. Thus,

$$\overline{v}_A = \overline{\omega} \times \overline{r}_{A/C} = v\overline{j} = \rho\dot{\psi}\overline{j}$$

Substituting $\overline{r}_{A/C} = -R\overline{i}$ yields

$$\rho\dot{\psi} = -R(\dot{\phi} + \dot{\psi}\cos\theta) \Rightarrow \dot{\phi} = -\left(\frac{\rho}{R} + \cos\theta\right)\dot{\psi}$$

We substitute this relation into the angular motion expressions, with the result that

$$\overline{\omega} = \dot{\psi}(-\sin\theta\overline{i} - \frac{\rho}{R}\overline{k})$$

$$\overline{\alpha} = -\dot{\psi}^2\left(\frac{\rho}{R} + \cos\theta\right)\sin\theta\overline{j}$$

The *xyz* axes are principal, with $I_{xx} = I_{yy} = \frac{1}{4}mR^2$, $I_{zz} = \frac{1}{2}mR^2$. Combining the inertia properties with the angular rotation components leads to

$$\overline{H}_A = mR^2\dot{\psi}\left(-\frac{1}{4}\sin\theta\bar{i} - \frac{1}{2}\frac{\rho}{R}\bar{k}\right)$$

$$\frac{\delta\overline{H}_A}{\delta t} = -\frac{1}{4}mR^2\dot{\psi}^2\left(\frac{\rho}{R} + \cos\theta\right)\sin\theta\bar{j}$$

The moment equation of motion is formed by substituting the foregoing expressions into Eq. (5.91) and matching the result to the moments exerted by the actual force system. The latter quantities must be described as moments about the *x*, *y*, and *z* axes, in order to match them to the components of $\overline{H}_A$. The result of this step is

$$\sum\overline{M}_A = (F_nR\sin\theta - NR\cos\theta)\bar{j} - F_tR\bar{k}$$

$$= \frac{\delta\overline{H}_A}{\delta t} + \overline{\omega} \times \overline{H}_A$$

$$= -\frac{1}{4}mR^2\dot{\psi}^2\left(\frac{\rho}{R} + \cos\theta\right)\sin\theta\bar{j} + mR^2\dot{\psi}^2\left(-\frac{1}{4}\frac{\rho}{R}\sin\theta\right)\bar{j}$$

Since $\dot{\psi} = v/\rho$, the corresponding component equations are

$$F_nR\sin\theta - NR\cos\theta = -\frac{1}{4}mv^2\left(\frac{R}{\rho}\right)^2\left(2\frac{\rho}{R} + \cos\theta\right)\sin\theta$$

$$F_t = 0$$

We must eliminate the dependence on the reaction forces if we are to obtain the desired relationship for the speed *v*. Additional equations of motion are those for the resultant force on the body. The center of mass follows a horizontal circular path of radius ρ at constant speed *v*, so it undergoes a centripetal acceleration. Rather than resolving $\overline{a}_A$ into components relative to the *xyz* reference frame, a representation in terms of the path variables for that point yields equations that have a more convenient form. Thus, we have

$$\overline{a}_A = \frac{v^2}{\rho}\overline{e}_n$$

The corresponding force equations are

$$\sum F_n = F_n = m\frac{v^2}{\rho} \qquad \sum F_t = F_t = 0 \qquad \sum F_b = N - mg = 0$$

Substituting $F_n = mv^2/\rho$ and $N = mg$ into the moment equations yields

$$v^2 = \frac{4gR\rho^2\cot\theta}{6\rho + R\cos\theta}$$

There is a simple explanation for this steady motion. The gravitational force and normal reaction form a couple because the disk is tilted. The frictional reaction required to impart the centripetal acceleration to the center of mass also exerts a moment. The net moment must be matched by a change in the angular momentum.

The latter effect is achieved by the precession, which alters the true direction of the angular momentum, even though its components relative to the x and z axes are constant.

EXAMPLE 5.9

A servomotor maintains the spin rate $\dot{\phi}$ at which the disk rotates relative to the pivoted shaft AB at a constant value. The precession rate $\dot{\psi}$ about the vertical axis is also held constant by a torque $M(t)$. Derive the differential equation governing the nutation angle θ. Also derive an expression for M.

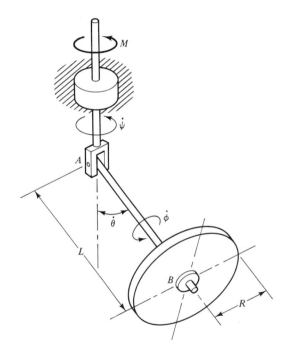

Example 5.9

Solution. A free body diagram of the disk and shaft assembly must account for the reactions. Toward this end, we also draw a free body diagram of the vertical shaft. The pin exerts an arbitrary force $\bar{A}$, which we decompose into components parallel and transverse to shaft AB. The couple $\bar{\Gamma}$ exerted by the pin has no component about the axis of the pin (assuming there is no friction). The vertical shaft carries equal, but opposite, reactions at the pin, as well as transverse forces and couples at its bearing. We assume that both shafts are massless. Then equilibrium of the vertical shaft requires that

$$\sum (M_{\text{vert shaft}})_Z = M - \Gamma_x \sin \theta + \Gamma_z \cos \theta = 0$$

The *xyz* axes selected for this resolution of the forces is parallel to the axes appearing in Appendix A for a disk. However, the origin is placed at the fixed

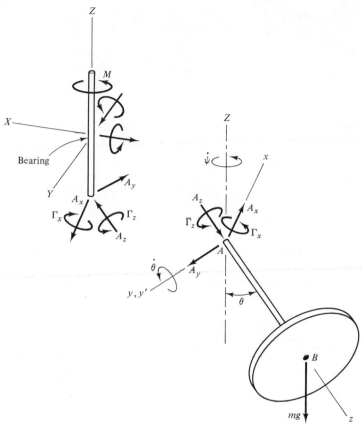

Free body diagrams.

point A in order to eliminate the reaction at the pin. Also, because the disk is axisymmetric, we may define the y axis to be instantaneously normal to the plane of the two shafts, without loss of generality. The parallel-axis theorems provide

$$I_{xx} = I_{yy} = m\left(\tfrac{1}{4}R^2 + L^2\right) \qquad I_{zz} = \tfrac{1}{2}mR^2 \qquad I_{xy} = I_{yz} = I_{xz} = 0$$

We could employ the Eulerian angles in Chapter 4 to form $\overline{\omega}$ and $\overline{\alpha}$, but it is just as easy to rederive the results here using the y' axis as the line of nodes for the nutation. Thus, for constant $\dot{\psi}$ and $\dot{\phi}$, we have

$$\overline{\omega} = \dot{\psi}\overline{K} + \dot{\theta}\overline{j}' + \dot{\phi}\overline{k} \qquad \overline{\omega}' = \dot{\psi}\overline{K}$$
$$\overline{\alpha} = \ddot{\theta}\overline{j}' + \dot{\theta}(\overline{\omega}' \times \overline{j}') + \dot{\phi}(\overline{\omega} \times \overline{k})$$

At this instant $\overline{K} = \sin\theta\,\overline{i} - \cos\theta\,\overline{k}$ and $\overline{j}' = \overline{j}$, so

$$\overline{\omega} = \dot{\psi}\sin\theta\,\overline{i} + \dot{\theta}\,\overline{j} + (-\dot{\psi}\cos\theta + \dot{\phi})\overline{k} \qquad \overline{\omega}' = \dot{\psi}(\sin\theta\,\overline{i} - \cos\theta\,\overline{k})$$
$$\overline{\alpha} = (\dot{\psi}\dot{\theta}\cos\theta + \dot{\theta}\dot{\phi})\overline{i} + (\ddot{\theta} - \dot{\psi}\dot{\phi}\sin\theta)\overline{j} + \dot{\psi}\dot{\theta}\sin\theta\,\overline{k}$$

Euler's equations are applicable because xyz are principal axes. Using the free body diagram of the disk and shaft AB to form the moment sums leads to

$$\sum M_{Ax} = \Gamma_x = I_{xx}\alpha_x - (I_{yy} - I_{zz})\omega_y\omega_z$$
$$= m\,[2L^2\dot\psi\dot\theta\cos\theta + \tfrac{1}{2}R^2\dot\theta\dot\phi]$$
$$\sum M_{Ay} = -mgL\sin\theta = I_{yy}\alpha_y - (I_{zz} - I_{xx})\omega_x\omega_z$$
$$= m\,[(L^2 + \tfrac{1}{4}R^2)\ddot\theta - (L^2 - \tfrac{1}{4}R^2)\,\dot\psi^2\sin\theta\cos\theta - \tfrac{1}{2}R^2\dot\psi\dot\theta\sin\theta]$$
$$\sum M_{Az} = \Gamma_z = I_{zz}\alpha_z - (I_{xx} - I_{yy})\omega_x\omega_y = \tfrac{1}{2}mR^2\dot\psi\dot\theta\sin\theta$$

There is no need to form $\Sigma\overline{F} = m\overline{a}_G$ because those equations would merely lead to relations for the reaction forces A_x, A_y, and A_z. The equation for ΣM_{Ay} yields the differential equation of motion.

$$(L^2 + \tfrac{1}{4}R^2)\ddot\theta - (L^2 - \tfrac{1}{4}R^2)\,\dot\psi^2\sin\theta\cos\theta + (gL - \tfrac{1}{2}R^2\dot\psi\dot\phi)\sin\theta = 0$$

The equations for $\Sigma\,M_{Ax}$ and $\Sigma\,M_{Az}$ describe the couple reactions Γ_x and Γ_z. Substituting those expressions into the moment equilibrium equation for the vertical shaft yields the couple M required to sustain the motion.

$$M = \Gamma_x\sin\theta - \Gamma_z\cos\theta$$
$$= 2m\,(L^2 - \tfrac{1}{4}R^2)\,\dot\psi\dot\theta\sin\theta\cos\theta + \tfrac{1}{2}mR^2\dot\theta\dot\phi\sin\theta$$

It is interesting to note that couples must be applied about both shafts in order to sustain the precession and spin rates. In particular, if a servomotor was not used to maintain a constant spin rate, that is, if Γ_z was identically zero, the spin angle would be an unknown variable.

5.5 PLANAR MOTION

The principles governing spatial motion provide an interesting perspective for the kinetics of planar motion. Let x and y represent convenient directions in the plane, so that $\overline{\omega}$ is parallel to the z axis. Then

$$\overline{a}_G = a_{Gx}\overline{i} + a_{Gy}\overline{j} \qquad \overline{\omega} = \omega\overline{k} \qquad \overline{\alpha} = \dot\omega\overline{k} \qquad (5.97)$$

The corresponding angular momentum for a coordinate system whose origin is at an allowable point is

$$\overline{H}_A = -I_{xz}\omega\overline{i} - I_{yz}\omega\overline{j} + I_{zz}\omega\overline{k} \qquad (5.98)$$

This shows that, if z is not a principal axis, then the angular momentum is not parallel to the angular velocity. The consequence of this feature is that couples in the plane of motion (in the vectorial sense) are required to sustain the planar motion. The equations of motion corresponding to Eqs. (5.97) and (5.98) are

$$\sum F_x = ma_{Gx} \qquad \sum F_y = ma_{Gy} \qquad \sum F_z = 0$$
$$\sum M_{Ax} = -I_{xz}\dot\omega + I_{yz}\omega^2$$
$$\sum M_{Ay} = -I_{yz}\dot\omega - I_{xz}\omega^2 \qquad (5.99)$$
$$\sum M_{Az} = I_{zz}\dot\omega$$

The three force equations and the equation governing the moment about the z axis are the same as those developed in elementary courses. The moments about the x and y axes are the gyroscopic moments. They are the result of unsymmetrical distributions of mass relative to the xy plane, corresponding to nonzero values of I_{xz} and I_{yz}. The dependence of these moments on the square of the rate of rotation means that very large couples must be exerted by the external force system in order to sustain a planar motion at large values of ω. This has serious implications for balancing rotating machinery, because the reactions must be provided by the supports, e.g., bearings. This condition, which is known as *dynamic imbalance*, can occur even though the center of mass is on a fixed axis of rotation ($\bar{a}_G = \bar{0}$), as it would be after the rotating system has been statically balanced.

EXAMPLE 5.10

The bar of mass m is placed horizontally on the semicylinder, such that contact is below the centroid G. Assuming that the bar does not slip, derive the differential equation of motion governing the angle θ.

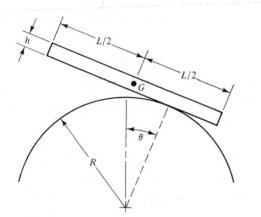

Example 5.10   Free body diagram.

Solution. The free body diagram of the bar must show the bar at an arbitrary angle of elevation θ. The friction and normal forces act at the contact point C. Owing to the absence of slippage, point C is at a distance $R\theta$ from the center of mass G. The moment equation must be formulated relative to point G because the body is in general motion.

We attach xyz to the bar in order to perform a kinematical analysis. For the assumed sense of the rotation, the angular velocity is

$$\bar{\omega} = -\dot{\theta}\bar{k} \qquad \bar{\alpha} = -\ddot{\theta}\bar{k}$$

The velocity of point C is zero, so the velocity of the center of mass is

$$\bar{v}_G = \bar{\omega} \times \bar{r}_{G/C} = (-\dot{\theta}\bar{k}) \times \left(-R\theta\bar{i} + \frac{h}{2}\bar{j}\right) = \frac{h}{2}\dot{\theta}\bar{i} + R\theta\dot{\theta}\bar{j}$$

Since this expression is generally valid, it may be differentiated. Thus, the acceleration of the center of mass is

$$\bar{a}_G = \frac{\delta \bar{v}_G}{\delta t} + \bar{\omega} \times \bar{v}_G$$

$$= \frac{h}{2}\ddot{\theta}\bar{i} + R(\theta\ddot{\theta} + \dot{\theta}^2)\bar{j} + (-\dot{\theta}\bar{k}) \times \left(\frac{h}{2}\dot{\theta}\bar{i} + R\theta\dot{\theta}\bar{j}\right)$$

$$= \left(R\theta\dot{\theta}^2 + \frac{h}{2}\ddot{\theta}\right)\bar{i} + \left[R\theta\ddot{\theta} + \left(R - \frac{h}{2}\right)\dot{\theta}^2\right]\bar{j}$$

For the purpose of evaluating the moment of inertia, we assume that the cross section of the bar is rectangular. Then

$$I_{zz} = \tfrac{1}{12}m(L^2 + h^2)$$

The equations of the motion are

$$\sum M_{Gz} = N(R\theta) - f\left(\frac{h}{2}\right) = \tfrac{1}{12}m(L^2 + h^2)(-\ddot{\theta})$$

$$\sum F_x = mg \sin \theta - f = m\left(R\theta\dot{\theta}^2 + \frac{h}{2}\ddot{\theta}\right)$$

$$\sum F_y = N - mg \cos \theta = m\left[R\theta\ddot{\theta} + \left(R - \frac{h}{2}\right)\dot{\theta}^2\right]$$

In order to obtain the desired differential equation, we eliminate the reactions. The force equations give

$$f = m\left(g \sin \theta - R\theta\dot{\theta}^2 - \frac{h}{2}\ddot{\theta}\right)$$

$$N = m\left[g \cos \theta + R\theta\ddot{\theta} + \left(R - \frac{h}{2}\right)\dot{\theta}^2\right]$$

which when substituted into the moment equations yields

$$[\tfrac{1}{12}(L^2 + 4h^2) + R^2\theta^2]\ddot{\theta} + R^2\theta\dot{\theta}^2 + gR\theta \cos \theta - g\frac{h}{2}\sin \theta = 0$$

When θ is small, we may linearize the differential equation by introducing the approximations $\cos \theta = 1$, $\sin \theta \approx \theta$, and dropping any terms that have quadratic or higher powers of θ. The equation for small rotations is found in this manner to be

$$\tfrac{1}{12}(L^2 + 4h^2)\ddot{\theta} + g\left(R - \frac{h}{2}\right)\theta = 0$$

When $h < 2R$, the response obtained from this equation is sinusoidal, corresponding to oscillations about a stable static equilibrium position. In contrast, when $h > 2R$, the solution of the linearized equation of motion is exponential, corresponding to continuous movement away from an unstable static equilibrium position. The transition from stability to instability has a simple explanation. In the case where the bar is slender, $h < 2R$, the center of mass rises as θ increases. Thus, $\theta = 0$ is the position of minimum potential energy. In contrast, when $h > 2R$, the center of mass descends with movement away from the equilibrium position, which means that $\theta = 0$ corresponds to maximum potential energy.

Note that the foregoing stability transition is independent of the value of the length L. The magnitude of L affects the moment of inertia of the bar, which is the coefficient of the angular acceleration term in the equation of motion. Thus, when the equilibrium position is stable, the value of L will affect the frequency of the stable oscillation.

EXAMPLE 5.11

The 4-kg bar lies in the vertical plane. The masses of collars A and B are negligible, and the guide bars are smooth. Collar A is pushed to the right at a constant speed of v by the horizontal force F. Determine the value of F as a function of the angle of elevation θ.

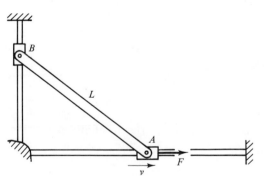

Example 5.11   Free body diagram.

Solution. It is convenient to define the body-fixed reference frame such that the x axis is horizontal at the instant under consideration. The kinematical relationship between the motion of collar A, the center of mass G, and the rotation of the bar is readily obtained by differentiating their positions relative to the fixed reference frame XYZ. For the constrained point A, we have

$$\bar{r}_{A/O} = L \cos \theta \bar{I} \qquad \bar{v}_A = -L \dot{\theta} \sin \theta \bar{I}$$

We also know that $\bar{v}_A = v\bar{I}$. Matching the two descriptions leads to

$$\dot{\theta} = -\frac{v}{L \sin \theta}$$

which, when differentiated again with v held constant, yields

$$\ddot{\theta} = \frac{v\dot{\theta}\cos\theta}{L\sin^2\theta} = -\frac{v^2\cos\theta}{L^2\sin^3\theta}$$

Then, for point G, we find

$$\bar{r}_{G/A} = \frac{L}{2}(\cos\theta\bar{I} + \sin\theta\bar{J})$$

$$\bar{v}_G = \dot{\bar{r}}_{G/A} = \frac{L}{2}\dot{\theta}(-\sin\theta\bar{I} + \cos\theta\bar{J}) = \tfrac{1}{2}v(\bar{I} - \cot\theta\bar{J})$$

$$\bar{a}_G = \dot{\bar{v}}_G = \frac{1}{2}v\frac{\dot{\theta}}{\sin^2\theta}\bar{J} = -\frac{1}{2}\frac{v^2}{L\sin^3\theta}\bar{J}$$

The foregoing gives $\bar{a}_G$ in terms of horizontal and vertical components. The angular acceleration is

$$\bar{\alpha} = -\ddot{\theta}\bar{k}$$

The corresponding equations of motion are

$$\sum M_{Gz} = N_A\left(\frac{L}{2}\cos\theta\right) + (F - N_B)\left(\frac{L}{2}\sin\theta\right) = \tfrac{1}{12}m\,L^2(-\ddot{\theta})$$

$$= \frac{mv^2\cos\theta}{12\sin^3\theta}$$

$$\sum F_x = F + N_B = 0$$

$$\sum F_y = N_A - mg = m\left(-\frac{v^2}{2L\sin^3\theta}\right)$$

We solve the force equations for the reactions, and then use those expressions to eliminate the reactions from the moment equation. The resulting expression for F is

$$F = \frac{mv^2\cos\theta}{3\,L\sin^4\theta} = \tfrac{1}{2}mg\cot\theta$$

5.6 APPLICATION OF IMPULSE-MOMENTUM AND WORK-ENERGY PRINCIPLES

The difficulties entailed in using the impulse-momentum equations were addressed earlier. Basically, the problem is associated with the rotation of the coordinate axes, which makes the unit vectors, as well as the components, of the impulse quantities functions of time.

In some situations, the component of the resultant force in a specific fixed direction, which will be denoted $\bar{e}_F$, might be known as a function of time. Similarly, the resultant moment about a fixed axis, denoted $\bar{e}_M$, might be known. The corresponding type of impulse in these directions may therefore be evaluated by application of the appropriate impulse-momentum principle.

$$m(\bar{v}_G)_2 \cdot \bar{e}_F = m(\bar{v}_G)_1 \cdot \bar{e}_F + \int_{t_1}^{t_2} \sum \bar{F} \cdot \bar{e}_F \, dt$$

$$(\bar{H}_A)_2 \cdot \bar{e}_M = (\bar{H}_A)_1 \cdot \bar{e}_M + \int_{t_1}^{t_2} \sum \bar{M}_A \cdot \bar{e}_M \, dt$$

(5.100)

where point A is one of the allowable points for formulating the equations of rotational motion. Equations (5.100) are often employed as conservation principles, corresponding to the case where the force or moment sum is zero in a specific direction.

As is true for momentum principles, the work-energy principles in Eqs. (5.45) to (5.48) have inherent limitations. Most profound of these is the necessity to evaluate the work. According to Eq. (5.35), the external force system can be resolved into a force-couple system acting at an arbitrary point. Obviously, the motion of such a point must be known in order to evaluate the path integrals. It is equally important to know how the resultant varies as the position of the selected point changes, and how the moment depends on the angle of orientation. If the forces are explicit functions of time or velocity, such dependencies can only be expressed after the motion has been evaluated. Even when it might seem that the work could be evaluated, the line integrals may be prohibitively complicated.

Of course, if a force is conservative, the work it does may be described in terms of its potential energy. However, it is necessary to recognize that the potential energy was derived by following the displacement of the actual point at which the force is applied. It might be necessary to rederive the energy function if the force is transferred to a different point.

In closing, it must be emphasized that the impulse-momentum and work-energy principles offer possible approaches for avoiding the solution of differential equations of motion. The conditions for usefully employing those principles are quite restrictive. Although it might be more difficult to solve differential equations, the derivation of the equations of motion is often the only valid approach.

EXAMPLE 5.12 _____

The coin is rolling without slipping, but the angle θ at which the plane of the coin is inclined is not constant. Evaluate the kinetic energy of the disk in terms of θ, the precession rate $\dot{\psi}$, and spin rate $\dot{\phi}$. Also, prove that the work done by the friction and normal forces is zero.

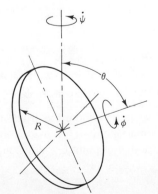

Example 5.12

Solution.   Rotational symmetry allows us to select body-fixed *xyz* axes whose instantaneous orientation is such that the *y* axis coincides with the line of nodes, which is the *y'* axis in the sketch.

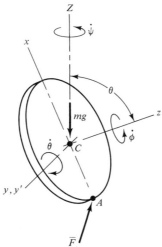

Free body diagram.

The inertia properties are

$$I_{xx} = I_{yy} = \tfrac{1}{4}mR^2 \qquad I_{zz} = \tfrac{1}{2}mR^2 \qquad I_{xy} = I_{yz} = I_{xz} = 0$$

The angular velocity is

$$\overline{\omega} = \dot{\psi}\overline{K} - \dot{\theta}\overline{j}' + \dot{\phi}\overline{k} = \dot{\psi}\sin\theta\,\overline{i} - \dot{\theta}\overline{j} + (\dot{\psi}\cos\theta + \dot{\phi})\overline{k}$$

We shall form the kinetic energy using the center of mass as the reference point. Since there is no slippage at point *A*, the velocity of the center of mass is

$$\overline{v}_C = \overline{\omega} \times \overline{r}_{C/A} = R(\dot{\psi}\cos\theta + \dot{\phi})\overline{j} + R\dot{\theta}\overline{k}$$

The angular momentum about the center of mass is

$$\overline{H}_C = I_{xx}\omega_x\overline{i} + I_{yy}\omega_y\overline{j} + I_{zz}\omega_z\overline{k}$$
$$= \tfrac{1}{4}mR^2[\dot{\psi}\sin\theta\,\overline{i} - \dot{\theta}\overline{j} + 2(\dot{\psi}\cos\theta + \dot{\phi})\overline{k}]$$

Then the kinetic energy is

$$T = \tfrac{1}{2}m\overline{v}_C \cdot \overline{v}_C + \tfrac{1}{2}\overline{\omega} \cdot \overline{H}_C$$
$$= \tfrac{1}{2}mR^2[(\dot{\psi}\cos\theta + \dot{\phi})^2 + \dot{\theta}^2]$$
$$+ \tfrac{1}{8}mR^2[\dot{\psi}^2\sin^2\theta + \dot{\theta}^2 + 2(\dot{\psi}\cos\theta + \dot{\phi})^2]$$
$$= \tfrac{1}{8}mR^2[\dot{\psi}^2\sin^2\theta + 5\dot{\theta}^2 + 6(\dot{\psi}\cos\theta + \dot{\phi})^2]$$

For the evaluation of the work, we let $\overline{F}$ denote the frictional and normal components of the reaction. Since the contact point *A* follows a complicated path, we replace these reactions by a force $\overline{F}$ acting at point *C* and a couple $\overline{M}$ equal to the moment exerted by $\overline{F}$ about point *C*. Therefore,

$$\overline{M} = \overline{r}_{A/C} \times \overline{F}$$

The infinitesimal work done by the reactions is

$$dW = \overline{F} \cdot d\overline{r}_C + \overline{M} \cdot d\overline{\theta}$$

where

$$\overline{d\theta} = \overline{\omega}\, dt \qquad d\overline{r}_C = \overline{v}_C\, dt = (\overline{\omega} \times \overline{r}_{C/A})\, dt = \overline{d\theta} \times \overline{r}_{C/A}$$

Thus

$$dW = \overline{F} \cdot (\overline{d\theta} \times \overline{r}_{C/A}) + (\overline{r}_{A/C} \times \overline{F}) \cdot \overline{d\theta}$$

Since $\overline{r}_{A/C} = -\overline{r}_{C/A}$, rearranging the second scalar triple product leads to

$$dW = \overline{F} \cdot (\overline{d\theta} \times \overline{r}_{C/A}) - \overline{F} \cdot (\overline{d\theta} \times \overline{r}_{C/A}) = 0$$

This proves that the friction and normal forces acting on a rolling disk never do work if there is no slippage. Indeed, the same result may be proved for any situation in which rigid bodies roll without slipping.

EXAMPLE 5.13 ————————————————————————————————

A 10-kg square plate suspended by a ball-and-socket joint is at rest when it is struck by a hammer. The impulsive force $\overline{F}$ generated by the hammer is normal to the surface of the plate, and its average value during the 4-ms interval that it acts is 5000 N. Determine the angular velocity of the plate at the instant following the impact, and the average reaction at the support.

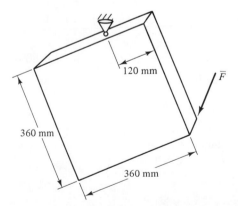

120 mm

$\overline{F}$

360 mm

360 mm

Example 5.13

Solution. The force F is much larger than the weight of the plate, so the latter is omitted from the free body diagram. In contrast, the reaction exerted by the ball-and-socket joint is impulsive, because it must be as large as necessary to prevent movement of point A. We place the origin of xyz at point A in order to eliminate the angular impulse of this reaction. The coordinates of point A relative to parallel centroidal axes are $(-0.18, 0.06, 0)$ m, so the inertia properties are

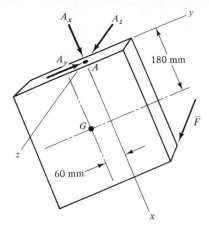

Free body diagram.

$$I_{xx} = \tfrac{1}{12}(10)(0.36^2) + 10(0.06^2) = 0.144 \text{ kg-m}^2$$

$$I_{yy} = \tfrac{1}{12}(10)(0.36^2) + 10(0.18^2) = 0.432$$

$$I_{zz} = I_{xx} + I_{yy} = 0.576$$

$$I_{xy} = 0 + 10(-0.18)(0.06) = -0.108 \qquad I_{xz} = I_{yz} = 0$$

The angular velocity initially is zero. Let $\bar{\omega}_2 = \omega_x\bar{i} + \omega_y\bar{j} + \omega_z\bar{k}$ denote the angular velocity at the termination of the impulsive action. Then the corresponding velocity of the center of mass is

$$(\bar{v}_G)_2 = \bar{\omega}_2 \times \bar{r}_{G/A} = 0.06\omega_z\bar{i} + 0.18\omega_z\bar{j} - (0.06\omega_x + 0.18\omega_y)\bar{k}$$

The final angular momentum about pivot A is

$$(\bar{H}_A)_2 = (0.144\omega_x + 0.108\omega_y)\bar{i} + (0.432\omega_y + 0.108\omega_x)\bar{j} + 0.576\omega_z\bar{k}$$

Applying the angular impulse-momentum principle to the 4-ms interval of the force leads to

$$(\bar{H}_A)_2 = \sum \bar{M}_A\Delta t = [(0.36\bar{i} + 0.12\bar{j}) \times 5000\bar{k}](0.004)$$
$$= 2.4\bar{i} - 7.2\bar{j} \text{ N-s}$$

The result of matching like components is

$$0.144\omega_x + 0.108\omega_y = 2.4$$
$$0.432\omega_y + 0.108\omega_x = -7.2$$
$$0.576\omega_y = 0$$

from which we obtain

$$\bar{\omega}_2 = 35.90\bar{i} - 25.64\bar{j} \text{ rad/s}$$

The next step is to form the linear impulse-momentum principle in order to determine the reaction. Using the earlier expression for $(\bar{v}_G)_2$ leads to

$$m(\bar{v}_G)_2 = \sum \bar{F}\,\Delta t$$
$$10[0.06\omega_z\bar{i} + 0.18\omega_z\bar{j} - (0.06\omega_x + 0.18\omega_y)\bar{k}] = [A_x\bar{i} + A_y\bar{j} + (A_z + F)\bar{k}]\,\Delta t$$

After substitution of the result for the angular velocity $\bar{\omega}_2$, the components of this equation yield

$$A_x = A_y = 0$$
$$A_z = -F - 10(0.06\omega_x + 0.18\omega_y)/\Delta t = 1153 \text{ N}$$

These are average values over the 4-ms interval. The maximum values exceed these.

It might surprise you that the reaction is in the same sense as the impulsive force $\bar{F}$. This result indicates that if the ball-and-socket joint were not present, the plate would rotate about its mass center because of the moment of the force, such that point A moves in the negative z direction. It is possible to locate a curve on the plate representing the locus of points at which the force can be applied without generating a dynamic reaction at the joint. Any such point is sometimes referred to as a center of percussion.

EXAMPLE 5.14 ———————————————————————————————

The sphere, whose mass is m, is fastened to the end of bar AB. The connection at end A is an ideal pin, which allows θ to vary. The vertical shaft rotates freely, and its mass, as well as that of bar AB, are negligible. Initially, θ is held constant at 90° by a string, and the initial rotation rate about the vertical axis is $\Omega_1 = 5$ rad/s. Determine the minimum value of θ in the motion following breakage of the string. Then determine the values of $\dot{\theta}$ and Ω_2 at the instant when θ is 10° greater than its minimum value.

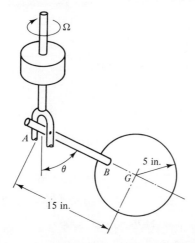

Example 5.14

Solution. A free body diagram of the entire system, including the vertical shaft, shows that there is no moment exerted about the fixed axis of rotation. The force-couple reaction at the bearing, $\bar{F}$ and $\bar{\Gamma}$, does no work, and gravity is conservative. Therefore, both angular momentum about the Z axis and mechanical energy are conserved throughout the motion. We shall develop an expression for $\bar{H}_A$ under

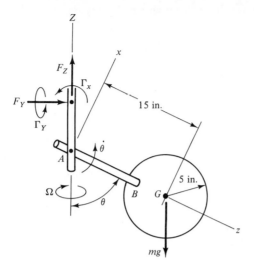

Free body diagram.

arbitrary conditions in order to address all aspects of the problem. The inertia properties for the principal body-fixed axes in the sketch are

$$I_{xx} = I_{yy} = \tfrac{2}{5}m(5^2) + m(15^2) = 235m \qquad I_{zz} = \tfrac{2}{5}m(5^2) = 10m$$

The angular velocity of the sphere is

$$\bar{\omega} = -\Omega\bar{K} - \dot{\theta}\bar{j} = -\Omega \sin\theta\,\bar{i} + \dot{\theta}\bar{j} + \Omega \cos\theta\,\bar{k}$$

The corresponding angular momentum is

$$\bar{H}_A = -235\,m\Omega \sin\theta\,\bar{i} + 235m\dot{\theta}\bar{j} + 10m\Omega \cos\theta\,\bar{k}$$

The component of $\bar{H}_A$ parallel to the Z axis is

$$\bar{H}_A \cdot \bar{K} = -235m\Omega \sin^2\theta - 10m\Omega \cos^2\theta$$

and the kinetic energy is

$$T = \tfrac{1}{2}\bar{\omega} \cdot \bar{H}_A = \tfrac{1}{2}m(235\Omega^2 \sin^2\theta + 235\dot{\theta}^2 + 10\Omega^2 \cos^2\theta)$$

If the datum for the gravitational potential energy is placed at the elevation of pin A, then

$$V = -m(32.17)(12)(15 \cos\theta) \text{ in.-lb}$$

The initial condition is that $\Omega_1 = 5$ rad/s, $\theta_1 = 90°$, and $\dot{\theta}_1 = 0$. Thus,

$$(\bar{H}_A \cdot \bar{K})_1 = 1175m \text{ lb-in.-s}$$
$$T_1 = 2937.5m \text{ in.-lb} \qquad V_1 = 0$$

The first question is the minimum value of θ, which corresponds to $\dot{\theta} = 0$. Conservation of angular momentum requires that

$$(\bar{H}_A \cdot \bar{K})_2 = (\bar{H}_A \cdot \bar{K})_1$$

$$\Omega(235 \sin^2\theta + 10 \cos^2\theta) = 1175$$

while energy is conserved if

$$T_2 + V_2 = T_1 + V_1$$
$$\Omega^2(117.5 \sin^2 \theta + 5 \cos^2 \theta) - 5790.6 \cos \theta = 2937.5$$

Eliminating Ω between the angular momentum and energy equations yields

$$117.5 \sin^2 \theta + 5 \cos^2 \theta$$
$$= (2937.5 + 5790.6 \cos \theta) \left(\frac{235 \sin^2 \theta + 10 \cos^2 \theta}{1175} \right)^2$$

This equation may be converted to a fifth-order polynomial in $\cos \theta$. The root giving the largest value of $\cos \theta$, which corresponds to the minimum angle, is

$$\theta_{min} = 0.6446 \text{ rad} = 36.933°$$

For the second part, we must find Ω_2 and $\dot{\theta}$ when $\theta_2 = \theta_{min} + 10°$. The earlier equation for conservation of angular momentum is generally valid. In this case it gives

$$\Omega_2 = \frac{1175}{235 \sin^2 \theta_2 + 10 \cos^2 \theta} = 9.033 \text{ rad/s}$$

We now determine the corresponding value of $\dot{\theta}$ from the energy conservation equation, for which we use the general expression for kinetic energy. The result is

$$(117.5\Omega_2^2 \sin^2 \theta_2 + 117.5\dot{\theta}^2 + 5\Omega_2^2 \cos^2 \theta_2) - 5790.6 \cos \theta_2 = 2937.5$$

which leads to

$$\dot{\theta}_2 = 3.673 \text{ rad/s}$$

5.7 SYSTEMS OF RIGID BODIES

The representation of a collection of particles as a rigid body offers two primary advantages: an enormous reduction in the number of kinematical variables and a commensurate reduction in the reaction forces that appear in the equations of motion. Both gains result from the recognition that the particles forming a rigid body are mutually constrained. In the same manner, we also occasionally find it useful to consider interacting rigid bodies as a unified system. Consider the pair of bodies in Figure 5.12 which are loaded by a set of external forces, as well as by reaction forces associated with their interaction.

Without loss of generality, we may resolve the external forces applied to body j into an equivalent force-couple system, $\Sigma \overline{F}_j$ and $\Sigma \overline{M}_j$, respectively, acting at the center of mass G_j. Similarly, the interaction forces exerted on body j by body k may be represented as a force-couple system, $\Sigma \overline{F}_{jk}$ and $\Sigma \overline{M}_{jk}$, acting at point G_j. A comparable representation also applies to the forces acting on body k.

Note that the individual internal force and moment resultants are equivalent to the

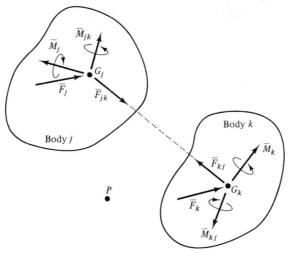

Figure 5.12   Forces acting on interacting bodies.

system of distributed forces exerted by the particles of one body on the particles of the other body. Hence, according to Newton's third law, the internal forces $\Sigma \overline{F}_{kj}$ and $\Sigma \overline{F}_{jk}$ cancel in the force sum. For the same reason, the net moment of the interaction forces and moments about an arbitrary point P also cancel. Let $\Sigma \overline{F}$ denote the resultant of all external forces exerted on the system by bodies not included in the system. Similarly, let $\Sigma \overline{M}_P$ denote the total moment about point P of all external forces.

Each body in Figure 5.12 must obey its own equation of motion. Hence, the forces acting on body j are equivalent to a force-couple system consisting of the linear inertia vector, $m_j \overline{a}_{Gj}$, and the rotational inertia vector $\dot{\overline{H}}_{Gj}$. Similarly, the forces acting on body k are equivalent to $m_k \overline{a}_{Gk}$ and $\dot{\overline{H}}_{Gk}$. Both equivalencies are illustrated in Figure 5.13. As

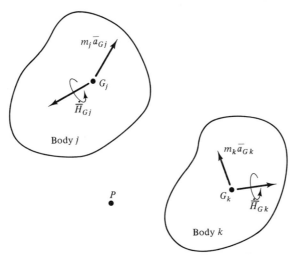

Figure 5.13   Inertial equivalent force-couple systems for interacting rigid bodies.

we did for the actual forces acting on the system of rigid bodies, we may combine the equivalent inertial force-couple systems into a single resultant. Note that the moment sum must account for the moment of the inertial forces $m_j(a_G)_j$ about point P. Thus, we find that the forces exerted on the system must satisfy

$$\sum \bar{F} = \sum_j [m_j(\bar{a}_{Gj})]$$

$$\sum \bar{M}_P = \sum_j \dot{\bar{H}}_{Gj} + \sum_j [\bar{r}_{Gj/P} \times m_j(\bar{a}_{Gj})] \tag{5.101}$$

The interpretation of this principle is that

> The forces and couples exerted on a system of rigid bodies resolve into a force-couple system acting at an arbitrary point P. The force in the equivalent system is the sum of the linear inertia vectors, $m_j(\bar{a}_{Gj})$, associated with the motion of the center of mass of each body. The couple in the equivalent system is the sum of the rotational inertia vector for each body, $\bar{H}_{Gj}$, and the moment of the linear inertia vectors about point P.

The usefulness of Eqs. (5.101) becomes apparent when we recall the techniques for static equilibrium. The main consideration in selecting the point for a moment sum in the static case is the ability to prevent forces (usually reactions) from appearing in the moment equilibrium equations. Since the point P in Eqs. (5.101) was arbitrarily selected, we have developed here a comparable ability for dynamic systems.

Despite the apparent simplicity of this principle, it should not be relied on too heavily. One pitfall lies in the need to include the linear inertia vectors in the formation of the moment equations of motion. A comparable operation arose in the formulation of the equations of motion for a single rigid body. We chose there to restrict the point for the moment sum. That decision was made on the basis that the formulation associated with an arbitrary selection of the point would be more prone to error, particularly in situations where the motion is to be determined. Another shortcoming is that Eqs. (5.101) provide only six scalar equations for the entire system, while the individual equations of motion provide six equations of motion for each body.

EXAMPLE 5.15

The gimbal supporting the flywheel is suspended from pivot A by two cables of equal length. The flywheel, which may be modeled as a thin disk of mass m_1, spins at the constant rate $\dot{\phi}$. The mass of the gimbal is m_2. It is observed that under the appropriate conditions the system will precess about the vertical axis through pivot A at a constant rate Ω, such that the angle of inclination θ is constant and the axis of the flywheel is horizontal. Derive expressions for the required precession and spin rates as function of the angle θ and the other system parameters.

Solution. In order to avoid considering the bearing forces exerted between the gimbal and the flywheel, we draw a free body diagram of the assembly. Let body 1 be the flywheel and body 2 be the gimbal . The $x_2y_2z_2$ reference frame is fixed

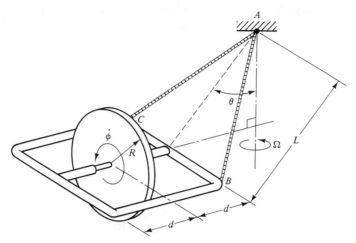

Example 5.15

to the gimbal, while $x_1y_1z_1$ is fixed to the flywheel. Letting the instantaneous orientation of $x_1y_1z_1$ coincide with $x_2y_2z_2$ expedites the overall description of the kinematic and kinetic properties.

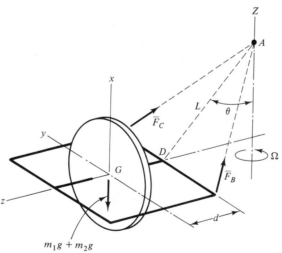

Free body diagram.

The center of mass for each body is point G, whose path is circular. Thus

$$\bar{a}_{G1} = \bar{a}_{G2} = -(L \sin \theta + d)\Omega^2 \bar{k}$$

The angular motion of the flywheel is

$$\bar{\omega}_1 = \Omega \bar{K} + \dot{\phi}\bar{k} = \Omega \bar{i} + \dot{\phi}\bar{k}$$
$$\bar{\alpha}_1 = \dot{\phi}(\bar{\omega} \times \bar{k}) = -\Omega\dot{\phi}\bar{j}$$

The corresponding expression for $\dot{\bar{H}}_G$ may be obtained from Euler's equations because xyz are principal axes. Setting $I_{xx} = I_{yy} = \frac{1}{2}I_{zz} = \frac{1}{4}m_1R^2$ yields

$$(\dot{\bar{H}}_G)_1 = [I_{yy}(\alpha_1)_y - (I_{zz} - I_{xx})(\omega_1)_x(\omega_1)_z]\bar{j} = -\frac{1}{2}m_1R^2\Omega\dot{\phi}\bar{j}$$

The corresponding analysis for the gimbal shows that it only precesses about the Z axis. By symmetry, the x axis, which is parallel to Z, is a principal axis for the gimbal, and the precession rate is constant. Thus, $(\dot{\bar{H}}_G)_2 = \bar{0}$.

We may now form the equations of motion for the system. Point A is convenient for the moment sum, because both cable forces intersect there. The moment of the actual force system about point A must match the sum of the moments about point A of the $m\bar{a}_G$ vector for each body, and of the couple $\dot{\bar{H}}_G$ for each body. Thus,

$$\sum \bar{M}_A = -(m_1g + m_2g)(L\sin\theta + d)\bar{j}$$

$$= (\dot{\bar{H}}_G)_1 + (\dot{\bar{H}}_G)_2 + \bar{r}_{G/A} \times [m_1(\bar{a}_G)_1 + m_2(\bar{a}_G)_2]$$

$$= [-\tfrac{1}{2}m_1R^2\Omega\dot{\phi} - (m_1 + m_2)(L\sin\theta + d)\Omega^2(L\cos\theta)]\bar{j}$$

This yields only one algebraic equation. Additional equations may be obtained from the resultant force. As an aid in forming those equations, we note that the system is symmetric with respect to the zZ plane, and neither center of mass has an acceleration transverse to that plane. Therefore, the tensile forces $\bar{F}_B$ and $\bar{F}_C$ have equal magnitude. We may replace them by an equivalent force $\bar{F}$ directed from point D to point A. The force sums are

$$\sum F_x = F\cos\theta - (m_1g + m_2g) = 0$$

$$\sum F_z = -F\sin\theta = m_1(a_G)_{1z} + m_2(a_G)_{2z}$$

$$= -(m_1 + m_2)(L\sin\theta + d)\Omega^2$$

Eliminating F from the force equations leads to a solution for Ω.

$$\Omega = \left(\frac{g\tan\theta}{L\sin\theta + d}\right)^{1/2}$$

We substitute this expression for Ω into the moment equation, and solve for $\dot{\phi}$, which results in

$$\dot{\phi} = \left(1 + \frac{m_2}{m_1}\right)\frac{2d}{R}\left[\frac{g}{R}\left(\frac{L\sin\theta + d}{R\tan\theta}\right)\right]^{1/2}$$

EXAMPLE 5.16 ——————————————————————————————

An orbiting space capsule is spinning about its axis of symmetry at 2 rad/s, and its velocity is 12 km/s parallel to the axis. The mass of the capsule is 2000 kg, and its radii of gyration for centroidal axes are 1.5 m about the axis of symmetry and 2.5 m transverse to that axis. A 1-kg meteorite traveling at 30 km/s impacts the capsule as shown, and then is embedded in the capsule's wall. Determine the velocity of the center of mass and the angular velocity immediately after the collision.

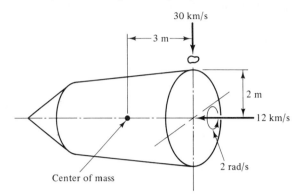

Example 5.16

Solution. The only force that is significant during the impact is the impulsive interaction force $\bar{F}$, which is shown in the free body diagrams of the space capsule and of the meteorite. The reference frame *xyz* is fixed to the capsule. Furthermore, owing to the short duration of the impact, we model the process by considering the positions to not change. Thus, we may also employ *xyz* as space-fixed axes for the analysis.

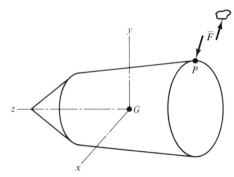

Free body diagrams.

Since F is an interaction force, we may avoid its appearance in the formulation by considering the capsule and the meteorite as a system. The external force and moment resultants vanish for this system, so both linear and angular momentum are conserved. The appropriate conservation principle may be obtained by multiplying Eqs. (5.101) by dt, then integrating over the interval Δt for the impact. The center of mass G is a convenient reference point for the angular momentum, because the moment of the linear momentum of the capsule about that point is identically zero. Also, we may accurately treat the meteorite as a particle, which means that its angular momentum about its own center of mass is also zero. The conservation principles for the system therefore reduce to

$$m_c(\bar{v}_G)_2 + m_m(\bar{v}_P)_2 = m_c(\bar{v}_G)_1 + m_m(\bar{v}_P)_1$$
$$(\bar{H}_{Gc})_2 + \bar{r}_{P/G} \times m_m(\bar{v}_P)_2 = (\bar{H}_{Gc})_1 + \bar{r}_{P/G} \times m_m(\bar{v}_P)_1$$

In these expressions, the subscript c refers to parameters for the space capsule, while the subscript m is for the meteorite. Note that we were able to treat the position $\bar{r}_{P/G}$ as a constant for the integration because of our idealization that changes in position are insignificant during the impact.

It is given that the space capsule is axisymmetric, so xyz are principal axes. The inertia properties are found from the given inertia properties to be

$$I_{xx} = I_{yy} = m_c(2.5^2) \qquad I_{zz} = m_c(1.5^2) \text{ kg-m}^2$$

Let the angular velocity of the capsule after the impacet be $\bar{\omega} = \omega_x \bar{i} + \omega_y \bar{j} + \omega_2 \bar{k}$. The angular momentum terms are then found to be

$$(\bar{H}_{Gc})_2 = m_c(6.25\ \omega_x \bar{i} + 6.25\omega_y \bar{j} + 2.25\omega_z \bar{k})$$
$$(\bar{H}_{Gc})_1 = m_c(2.25)(-2)\bar{k} \text{ kg-m}^2$$

We have not yet used the fact that the meteorite is embedded after the impact, which means that it has the same velocity as point P on the space capsule. Therefore,

$$(\bar{v}_P)_2 = (\bar{v}_G)_2 + \bar{\omega}_2 \times \bar{r}_{P/G}$$

Next, we substitute this expression and the expressions for $\bar{H}_{Gc}$ into the conservation principles. Dividing each equation by $m_c + m_m$ yields

$$(\bar{v}_G)_2 + \sigma\bar{\omega}_2 \times \bar{r}_{P/G} = -30{,}000\sigma\bar{j} + 12{,}000(1 - \sigma)\bar{k}$$
$$(1 - \sigma)(6.25\omega_x \bar{i} + 6.25\omega_y \bar{j} + 2.25\omega_z \bar{k}) + \sigma\bar{r}_{P/G} \times (\bar{v}_G)_2$$
$$+ \sigma\bar{r}_{P/G} \times (\bar{\omega}_2 \times \bar{r}_{P/G}) = -4.50(1 - \sigma)\bar{k} + \bar{r}_{P/G} \times (-30{,}000\sigma\bar{j})$$

where σ is the ratio of the mass of the meteorite to the total mass,

$$\sigma = \frac{m_m}{m_c + m_m}$$

We could represent $\bar{v}_G$ in terms of an unknown set of components, as we did for $\bar{\omega}$. Correspondingly, we would obtain six simultaneous equations by matching like components in each conservation equation. However, it is possible to simplify the equations to be solved. We solve the linear momentum equation for $(\bar{v}_G)_2$, and substitute the result into the angular momentum equation. After like terms are collected, this operation gives

$$(\bar{v}_G)_2 = -\sigma\bar{\omega}_2 \times \bar{r}_{P/G} - 30{,}000\sigma\bar{j} + 12{,}000(1 - \sigma)\bar{k}$$
$$6.25\omega_x \bar{i} + 6.25\omega_y \bar{j} + 2.25\omega_z \bar{k} + \sigma\bar{r}_{P/G} \times (\bar{\omega}_2 \times \bar{r}_{P/G})$$
$$= -4.50\ \bar{k} + \sigma\bar{r}_{P/G} \times (-30{,}000\bar{j} - 12{,}000\bar{k})$$

Upon substitution of the component form of $\bar{\omega}$, as well as $\bar{r}_{P/G} = 2\bar{j} - 3\bar{k}$ meters, we obtain a vector equation whose only unknowns are ω_x, ω_y, and ω_z. Matching like components then yields

$$(6.25 + 13\sigma)\omega_x = -114{,}000\ \sigma$$
$$(6.25 + 9\sigma)\omega_y + 6\sigma\omega_z = 0$$
$$(2.25 + 4\sigma)\omega_z + 6\sigma\omega_y = -4.5$$

Since $\sigma = \frac{1}{2001}$ in the present case, these equations yield

$$\bar{\omega}_2 = -9.106\bar{i} + 0.00096\bar{j} - 1.9982\bar{k} \text{ rad/s}$$

We obtain the corresponding velocity of the center of mass by substitution, which yields

$$(\bar{v}_G)_2 = 0.002\bar{i} - 14.979\bar{j} + 11{,}994.0\bar{k} \text{ m/s}$$

These results have a ready explanation. The velocity in the z direction is decreased by the impact because the mass of the capsule increased, while the y velocity component is attributable to the linear momentum transfer from the initial motion of the meteorite. The small change in the x component of velocity may be traced to the fact that the center of mass of the system shifts as a result of the capture of the meteorite. For the angular velocity, we note that the predominant effect is to induce a rotation about the x axis. The small changes in the rotation about the y and z axes are a consequence of the alteration of the mass distribution due to the embedded meteorite, which results in xyz no longer being principal axes.

REFERENCES

R. N. Arnold and M. Maunder, *Gyrodynamics and Its Engineering Applications*, Academic Press, New York, 1961.

E. A. Desloge, *Classical Mechanics*, vol. 1, John Wiley & Sons, New York, 1982.

J. H. Ginsberg and J. Genin, *Dynamics*, 2d ed., John Wiley & Sons, New York, 1984.

D. T. Greenwood, *Principles of Dynamics*, Prentice-Hall, Englewood Cliffs, N.J., 1965.

J. B. Marion, *Classical Dynamics of Particles and Systems*, Academic Press, New York, 1960.

E. J. Routh, *Dynamics of a System of Rigid Bodies*, Part I, *Elementary Part*, 7th ed., Macmillan, New York, 1905.

E. J. Routh, *Dynamics of a System of Rigid Bodies*, Part II, *Advanced Part*, 6th ed., Macmillan, New York, 1905.

J. L. Synge and B. A. Griffith, *Principles of Mechanics*, 3d ed., McGraw-Hill, New York, 1959.

HOMEWORK PROBLEMS

5.1 Derive the centroidal location and centroidal inertia properties of a homogeneous semicone, as tabulated in Appendix A. Use those results to derive comparable expressions for a semiconical shell.

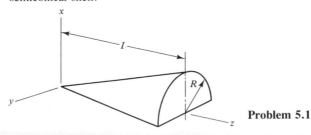

Problem 5.1

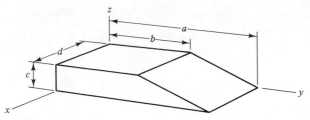

Problem 5.2

5.2 Determine I_{xx} and I_{xy} for the truncated rectangular parallelipiped relative to the xyz coordinate axes shown in the sketch.

5.3 The length of the homogeneous cylinder is twice its radius. Consider an xyz coordinate system whose origin is located on the perimeter of one end, whose xy plane contains the axis of the cylinder, and whose x axis intersects the centroid. Determine the inertia properties of the cylinder relative to xyz.

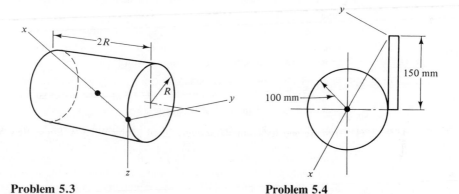

Problem 5.3   **Problem 5.4**

5.4 The 6-kg bar is welded to the 10-kg disk. Determine the inertia properties of this system relative to the xyz coordinate system shown in the sketch.

5.5 The x axis lies in the plane of the 10-kg plate, and the y axis is elevated at 36.87° above the diagonal. Determine the inertia matrix of the plate relative to xyz.

5.6 Consider the plate in Problem 5.5. Determine the principal moments of inertia relative to corner O.

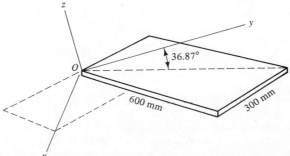

Problems 5.5 and 5.6

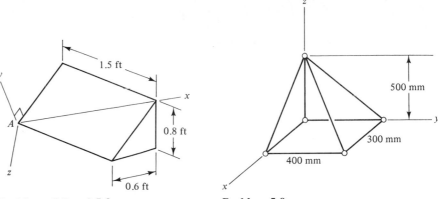

Problems 5.7 and 5.8   Problem 5.9

5.7 The x axis coincides with the inclined face of the 100-lb homogeneous prism, and the y axis is normal to that face. Determine the inertia matrix of the prism relative to this coordinate system.

5.8 Consider the prism in Problem 5.7. For an origin coincident with corner A, determine the principal moments of inertia. Also determine the rotation transformation for the principal axes relative to coordinate axes aligned with the orthogonal edges.

5.9 A rigid body consists of five small spheres of mass m mounted at the corners of a lightweight wire frame in the shape of a pyramid. Determine the principal moments of inertia and the rotation matrix of the principal axes relative to the given xyz coordinate system.

5.10 The 24-kg block is welded to a shaft that rotates about bearings A and B at a constant rate ω. The shaft is collinear with the diagonal to a face of the block, as shown. Determine the inertia properties of the block relative to the xyz coordinate system, whose origin is situated at the midpoint of the diagonal. The x axis is aligned with the shaft, and the z axis is normal to the face of the block. Then use these properties to evaluate the dynamic reactions.

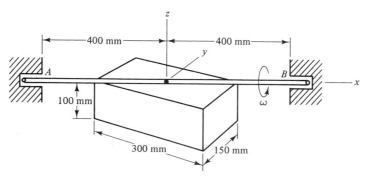

Problem 5.10

5.11 The gyroscopic turn indicator consists of a 1-kg flywheel whose principal radii of gyration are $k_x = 50$ mm, $k_y = k_z = 40$ mm. The center of mass of the flywheel coincides with the intersection of axes AB and CD. The flywheel spins relative to the gimbal at the constant rate $\omega_1 = 10,000$ rev/min. A couple $\overline{M}$ acts about shaft CD, which supports the gimbal, in order to hold constant the angle β between the gimbal and the horizontal. Determine $\overline{M}$ when the rotation rate about the vertical axis is $\Omega = 0.8$ rad/s.

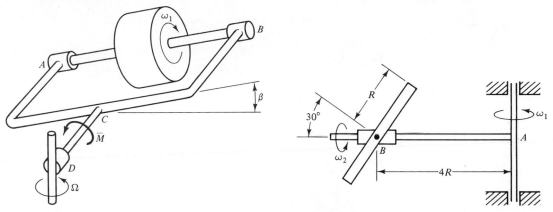

Problem 5.11   **Problem 5.12**

5.12 The disk of mass m freely spins about the horizontal shaft AB at angular rate ω_2, and the precession rate is held constant at ω_1. The angle between the centerline of the disk and the shaft is $30°$, so that the illustrated position occurs only once for each rotation of the disk relative to shaft AB. Derive an expression for the couple reaction at joint A at the instant depicted in the sketch. Then use that answer to explain physically whether or not a moment about the vertical shaft is required to hold ω_1 constant at this instant.

5.13 A 50-kg rectangular plate is mounted diagonally on a shaft whose mass is negligible. The system was initially at rest when a constant torque of 5 kN-m was applied to the shaft. Determine the reactions at bearings A and B 4 s after the application of the torque.

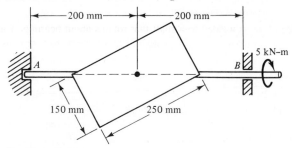

Problem 5.13

5.14 The device shown is a wobble plate, in which the rotation rate ω_1 of the disk relative to the shaft, and the precession rate ω_2 of the shaft are constant. The mass of shaft AB is negligible. For the case where $\omega_2 = 3\omega_1$, determine **(a)** the angle between shaft AB and the angular velocity $\bar{\omega}$, and between shaft AB and the angular momentum $\bar{H}_G$ of the disk, **(b)** the gyroscopic moment $\dot{\bar{H}}_G$, **(c)** the reactions at bearings A and B.

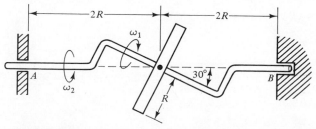

Problem 5.14

5.15 The disk spins at rate $\dot{\phi}$ relative to its shaft, which is pinned to the vertical shaft. The

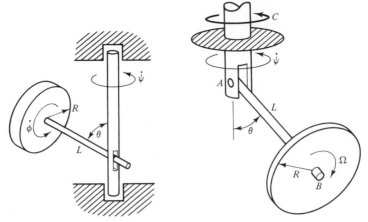

Problem 5.15   **Problem 5.16**

system rotates freely about the vertical axis, and the mass of both shafts is negligible. The nutation angle θ is initially held constant by a cable, in which condition the precession rate is $\dot{\psi}$. Derive an expression for $\ddot{\theta}$ at the instant after the cable is cut.

5.16 A flywheel, whose mass is m, is mounted on shaft AB, which is made to spin at the constant rate Ω by a servomotor. A torque $C(t)$ is applied to the vertical shaft, and pivot A has ideal properties. Derive differential equations of motion for the precession angle ψ and the nutation angle θ.

5.17 The slender bar is mounted on a gimbal that rotates about the horizontal axis at constant rate Ω due to torque Γ. Derive the differential equation governing the angle β between the bar and the horizontal axis, and also derive an expression for Γ.

5.18 Bar BC is pivoted from the end of the T-bar, which rotates about the vertical axis at the constant speed Ω. Derive the differential equation of motion for the angle of elevation θ.

5.19 The system in Problem 5.12 rotates freely about the vertical axis at $\omega_1 = \dot{\psi}$, while the spin rate $\omega_2 = \dot{\phi}$ is held constant by a servomotor. The spin angle $\phi = 0$ in the illustrated position. Derive the differential equation of motion of ψ.

5.20 A thin disk of mass m rolls over the ground without slipping as it rotates freely relative to bent shaft AB. The precession rate of the bent shaft about the vertical axis is the constant value Ω. Determine the magnitude of the normal force $\overline{N}$ exerted between the disk and the ground.

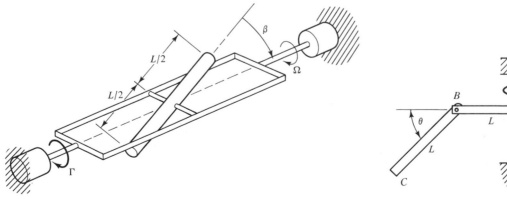

Problem 5.17   **Problem 5.18**

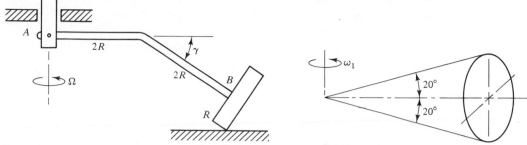

Problem 5.20   Problem 5.21

5.21 The 2-lb cone, whose apex angle is 40°, rolls without slipping over the horizontal surface. The rolling motion of the cone is such that it rotates about a fixed vertical axis intersecting its apex at constant angular rate ω_1. Determine the maximum value of ω_1 for which the cone will not tip over its rim in this motion. Also, determine the minimum coefficient of static friction corresponding to that value of ω_1.

5.22 The sphere, whose mass is m, spins freely relative to shaft AB, whose mass is negligible. The system precesses about the vertical axis at the constant rate ω_1, and joint A is an ideal pin. Consider the possibility that the sphere rolls over the horizontal surface without slipping. Determine whether there is a range of values of ω_1 for which such a motion can occur.

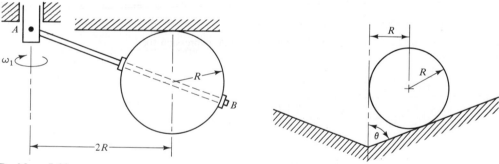

Problem 5.22   Problem 5.23

5.23 As shown in the cross-sectional view, a sphere of radius R rolls without slipping over the interior surface of a cone, such that its center is always at distance R from the vertical axis of the conical surface. Derive expressions for the speed v of the center of the sphere, and the components of the contact force exerted between the sphere and the conical surface.

5.24 The bar of mass m is falling toward the horizontal surface. Friction is negligible. Derive

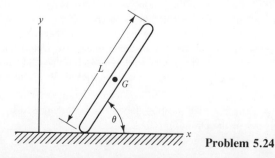

Problem 5.24

differential equations of motion for the position coordinates (x_G, y_G) of the center of mass of the bar, and for the angle of inclination θ. Also obtain an expression for the contact force exerted by the ground on the bar in terms of x_G, y_G, θ, and their derivatives.

5.25 The 4-kg bar lies in the vertical plane, and the masses of collars A and B are negligible. Collar A has a constant velocity of 10 m/s to the right. Determine the required value of the force $\overline{F}$ acting on collar A for the position shown.

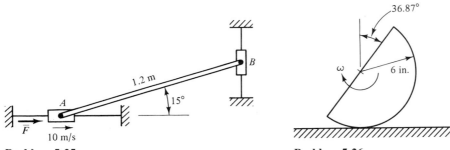

Problem 5.25   **Problem 5.26**

5.26 The 20-lb semicylinder has an angular speed $\omega = 10$ rad/s in the position shown. The coefficient of static friction between the ground and the semicylinder is μ. Determine the minimum value of μ for which slipping between the semicylinder and the ground will not occur in this position. What is the corresponding angular acceleration $\dot{\omega}$ of the semicylinder?

5.27 The bar is pinned at its left end to a ring. The system is initially at rest in the position shown, when a horizontal force $\overline{F}$, whose magnitude is $\frac{3}{7}mg$, is applied to the right end of the bar. The ring does not slip over the ground in the ensuing motion, while friction between the bar and the ground is negligible. Determine the acceleration of the center of the ring at the instant the force is applied. The mass of the ring is negligible.

5.28 Solve Problem 5.27 for the case where the ring and the bar have equal masses.

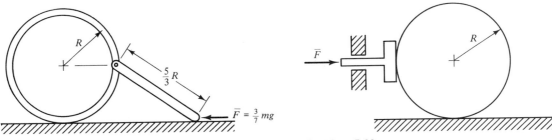

Problems 5.27 and 5.28   **Problem 5.29**

5.29 Horizontal force $\overline{F}$ is applied to the piston, whose mass is small compared with that of the circular cylinder of mass m. The coefficients of friction are μ and ν for static and kinetic friction, respectively. **(a)** For the case where there is no slipping relative to the ground, determine the translational acceleration of the cylinder. **(b)** Determine the largest value of $\overline{F}$ for which the motion in part (a) is possible. **(c)** Determine the translational and angular acceleration of the disk when the magnitude of $\overline{F}$ exceeds the value in part (b).

5.30 Bar AB has a mass of 40 kg, and the stiffness of the spring is 20 kN/m. The bar is released from rest at $\theta = 30°$, at which position the spring is compressed by 100 mm. Determine the angular velocity of the bar at $\theta = 0$.

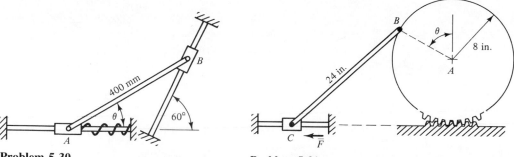

Problem 5.30

Problem 5.31

5.31 Gear A, which weighs 20 lb and has a radius of gyration of 6 in., rolls over the horizontal rack owing to a constant horizontal force $\bar{F}$ acting on collar C. Connecting bar BC weighs 15 lb, and the weight of the collar is negligible. The system was at rest at $\theta = 0$. Determine the magnitude of $\bar{F}$ for which the velocity of the center of gear A when $\theta = 90°$ is 20 ft/s.

5.32 Bar AB is pinned to the vertical shaft, which rotates freely. When the bar is inclined at $\theta = 10°$ from the vertical, the rotation rate about the vertical axis is $\Omega = 10$ rad/s, and $\dot{\theta} = 4$ rad/s at that instant. Determine the maximum value of θ in the subsequent motion.

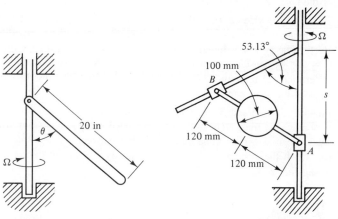

Problem 5.32

Problems 5.33 and 5.34

5.33 Collars A and B are interconnected by a bar on which a 100-kg sphere of 100-mm diameter is mounted. The mass of the collars and the bar is negligible. The system rotates freely about the vertical axis. Initially, $s = 240$ mm, and the collars are not moving relative to the guide bars. What minimum value of the initial angular speed Ω about the vertical axis, if any, is required for collar A to reach $s = 180$ mm?

5.34 Initially, the system in Problem 5.33 is at $s = 180$ mm, at which position the rotation rate about the vertical axis is 30 rev/min. Determine whether collar A attains the position $= 300$ mm in the subsequent motion. If so, what is the angular velocity of the sphere at that position?

5.35 A semicircular plate is falling at speed v with its plane oriented horizontally. It strikes the ledge at corner A, and the impact is perfectly elastic. The interval of the collision is Δt. Derive expressions for the velocity of the center of mass and the angular velocity at the

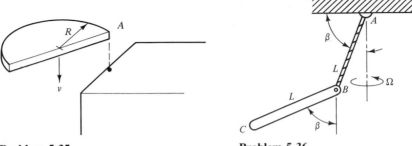

Problem 5.35   **Problem 5.36**

instant following the collision. Also, derive an expression for the collision force exerted between the plate and the ledge.

5.36 A slender bar of mass m, which is suspended by a cable from pivot A, executes a steady precession about the vertical axis at angular speed Ω as it maintains the orientation shown. **(a)** Determine Ω and the angle of inclination β. **(b)** An impulsive force $\overline{F}$ at end B, parallel to the initial velocity of that end, acts over a short time interval Δt. Determine the magnitude of $\overline{F}$ for which the angular velocity of the bar at the conclusion of the impulsive action is horizontal. What are the corresponding velocity of the center of mass G and angular velocity of the bar?

5.37 The flyball device consists of two 500-g spheres connected to the vertical shaft by a parallelogram linkage. This shaft, which passes through collar C supporting the linkage, rotates freely. The system is initially rotating steadily at 900 rev/min about the vertical axis, with $\theta = 75°$. A constant upward force $\overline{F}$ is applied to the vertical shaft, causing θ to decrease. Determine the minimum magnitude of $\overline{F}$ for which the system will reach $\theta = 15°$. The mass of the shaft and the bars in the linkage is negligible.

5.38 Identical disks A and B are separated by distance L on a massless, rigid shaft, about which they may spin freely. The system is suspended at the $L/3$ position from a pivot on a vertical shaft, as shown. Determine the relationship between the spin rates ω_B and ω_A of the disks for which the system will precess at a steady rate about the vertical axis, with shaft AB always horizontal.

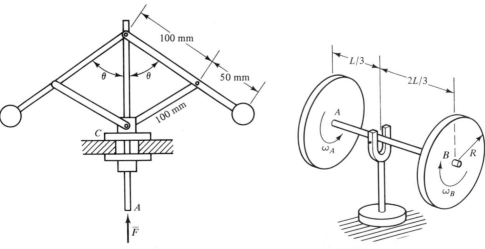

Problem 5.37   **Problem 5.38**

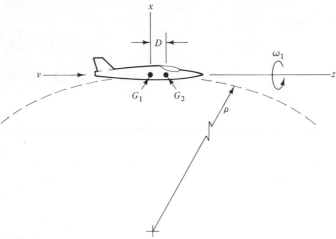

Problem 5.39

5.39 A single-engine turbojet airplane has its minimum speed v at the top of a vertical circle of radius ρ. At this instant, the airplane is executing a roll, clockwise as viewed by the pilot, at angular speed ω_1. The engine turns at angular speed Ω, counterclockwise from the pilot's viewpoint. The rotating parts of the engine have mass m_2, and moments of inertia J about the rotation axis and J' transverse to the rotation axis. The mass of the airplane, excluding the rotating parts of the engine, is m_1, and the corresponding moments of inertia about centroidal xyz axes are I_x, I_y, and I_z. The spin axis of the engine is collinear with the z axis of the airplane, and the centers of mass G_1 and G_2, associated, respectively, with m_1 and m_2, both lie on this axis. Derive expressions for the aerodynamic force and moment about the center of mass of the airplane required to execute this maneuver.

5.40 The tires of an automobile have mass m and radius of gyration κ about their respective axles. The wheelbase of the automobile is L and the track is w. The center of mass G of the automobile is on the centerline of the automobile at distance σ behind the front axle and height h above the ground. Consider the situation in which the automobile executes a right turn of radius ρ (measured to point G) at constant speed v. Determine the change in the normal reaction exerted between each tire and the ground resulting from the rotatory inertia of the wheels.

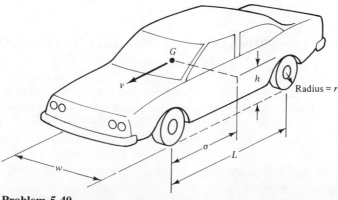

Problem 5.40

Chapter 6

Introduction to Analytical Mechanics

The constraints imposed on the motion of a system enter the Newtonian formulation of the equations of motion in two ways. The kinematical relations must account for the restrictions imposed on the motion, while the kinetics principles must account for the reaction force (or moment) associated with each constraint. When the system consists of more than one body, the need to account individually for the constraints associated with each connection substantially enhances the level of difficulty.

The Lagrangian formulation we develop in this chapter takes a different view of systems. One important feature is that the principles are based on an overview of the system and its mechanical energy (kinetic and potential). In contrast, the Newtonian equations of motion are time derivatives of momentum principles. Another, and perhaps the most important, difference is that in the Lagrangian viewpoint, the reactions exerted by supports will usually not appear in the formulation. This is a consequence of the fact that the reactions and the geometrical description of the system are two manifestations of the same physical feature. It is from the kinematical perspective that we begin our study.

6.1 GENERALIZED COORDINATES AND DEGREES OF FREEDOM

Suppose the reference location of a system is given. (Such a location might be the starting position or the static equilibrium position.) We must select a set of geometrical parameters whose value uniquely defines a new position of the system relative to the initial position. For example, it should be possible to draw a diagram of the system in its current position by knowing only the fixed dimensions and the position parameters. Geometrical quantities that meet this specification are called *generalized coordinates*. The minimum number of generalized coordinates required to specify the position of the system are the *number of degrees of freedom* of that system.

A simple system consisting of a single rigid bar in planar motion is adequate to develop these concepts. According to Chasle's theorem, the general motion of a rigid body is a superposition of a translation following any specified point and a rotation about that point. In Figure 6.1, the movement of end A is described by its position coordinates, x_A and y_A, and the rotation is described by angle θ measured from the horizontal. We always consider fixed parameters, such as L, to be known, so they are system properties. Thus, the generalized coordinates that have been selected here are (x_A, y_A, θ). Since it is necessary to know the value of all three parameters to draw Figure 6.1, this bar has three degrees of freedom.

Generalized coordinates do not form a unique set—other parameters may be equally suitable for describing the motion. In Figure 6.2 the generalized coordinates (x_B, y_B, θ) describe the motion in terms of the position of end B and the rotation. In general, it must be possible to express one set of generalized coordinates in terms of another, for each must be capable of describing the position of all points in the system. The transformation from the set in Figure 6.1 to that in Figure 6.2 is

$$x_B = x_A + L \cos \theta \qquad y_B = y_A + L \sin \theta \qquad \theta = \theta$$

Another choice, shown in Figure 6.3, leads to a difficulty. The three generalized coordinates depicted there are (x_A, y_A, x_B). As shown in the figure, the difficulty is that for given values of (x_A, y_A, x_B), the bar can have one of two orientations. Specifically, since the length L is a fixed parameter, the vertical position of end B is given by $y_B = y_A \pm [L^2 - (x_B - x_A)^2]^{1/2}$. Recall that the generalized coordinates must *uniquely* specify the location. This means that (x_A, y_A, x_B) can serve as generalized coordinates only if the case where $y_B > y_A$ (positive sign in the above relation for y_B) is to be considered, or alternatively, only $y_B < y_A$ (negative sign). In most cases where the orientation of a body is significant, it is best to actually select an angle as a generalized coordinate.

The situations in Figures 6.1 to 6.3 correspond to cases where the number of generalized coordinates equals the number of degrees of freedom. The generalized coordinates in such cases are *unconstrained*. This means that their values may be set independently, without influencing each other. (Indeed, the generalized coordinates in this case are sometimes called *independent coordinates*.) When the number of generalized coordinates exceeds the number of degrees of freedom, the generalized coordinates are

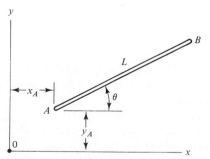

Figure 6.1 Generalized coordinates for a bar in arbitrary planar motion.

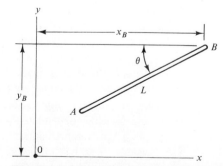

Figure 6.2 Generalized coordinates for a bar in arbitrary planar motion.

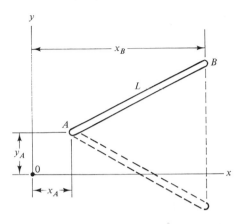

Figure 6.3 Ambiguous generalized coordinates.

constrained, because they must satisfy additional conditions other than those arising from kinetics principles.

A set of constrained generalized coordinates for the bar in Figures 6.1 to 6.3 could be the position coordinates of each end and the angle of orientation, $(x_A, y_A, x_B, y_B, \theta)$. There are two independent relations that may be written between these five variables, for example,

$$x_B - x_A = L \cos \theta \qquad y_B - y_A = L \sin \theta$$

These relations are consistent with the fact that the system has three degrees of freedom, since the existence of two relations between five variables means that only three variables may be selected independently. The relations between constrained generalized coordinates are called *constraint equations*. The number of degrees of freedom equals the number of generalized coordinates minus the number of constraint equations. The question of constrained and unconstrained generalized coordinates will be examined in greater detail in the next section.

Other than the restriction to planar motion, the bar that has been discussed thus far is free to move in space. Any constraint imposed on its motion by supporting it in some manner alters the number of degrees of freedom and, therefore, the selection of unconstrained generalized coordinates. Figure 6.4 shows a common way in which a bar might be supported. The pin at end A prevents movement of that end in both the x and y directions. This reduces the number of degrees of freedom to one, since the position of the bar is now completely specified by the value of θ.

Another way in which we could regard the bar in Figure 6.4 is to say that the set of generalized coordinates (x_A, y_A, θ) for the bar are now constrained to satisfy $x_A = 0$ and $y_A = 0$. The latter are two constraint equations, which confirms that two of the three generalized coordinates selected to represent this one-degree-of-freedom system cannot be assigned independent values.

A different manner of supporting the bar also leads to a one-degree-of-freedom system. The bar in Figure 6.5 is constrained by the collars at its ends. A suitable unconstrained generalized coordinate is the angle θ. If this parameter is known, then the (x, y) coordinates of the ends may be evaluated with the aid of the law of sines. (Recall

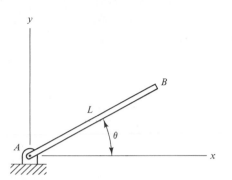

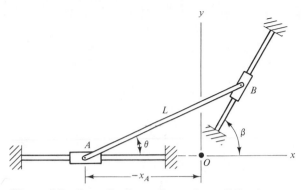

Figure 6.4 Generalized coordinate for a pinned bar.

Figure 6.5 Generalized coordinates for a sliding bar.

that the constant geometrical parameters, such as L and β in this system, are always considered to be known.)

An aspect of Figure 6.5 that should be noted is the selection of the origin of xyz. Because it is desirable that xyz be fixed, it is useful to place its origin at a point in the system that is stationary. This practice will avoid the possibility of inadvertently writing position coordinates relative to a moving coordinate system when it is coordinates relative to a fixed system that we need.

We could, of course, select constrained generalized coordinates for the system in Figure 6.5. For example, the set (x_A, y_A, θ) used earlier are related by

$$x_A = -L \frac{\sin(\beta - \theta)}{\sin \beta} \qquad y_A = 0$$

These constraint equations are more complicated than were those for the previous system. We shall see in later sections that there are situations where it might be necessary or desirable to use constrained generalized coordinates. Nevertheless, a set of unconstrained generalized coordinates, if available, will almost certainly lead to the most straightforward formulation of the equations of motion.

The discussion thus far has only dealt with a planar system consisting of a single body. A useful set of generalized coordinates for a body in spatial motion consists of the xyz coordinates of the center of mass, and the Eulerian angles defined relative to convenient sets of axes. In other words, a body in unconstrained spatial motion has six degrees of freedom. This number is decreased by the number of constraints that are imposed.

A common system consisting of a multitude of bodies is a mechanical linkage. A typical one is depicted in Figure 6.6. If nothing is specified regarding the motion, this linkage has two degrees of freedom. (One way of recognizing this number is to ask what motion parameters are required to draw a picture of the system.) For example, in order to specify the position of bar AB, it is necessary to know s_A and θ. This will define the location of end B. Then the orientation of bar BC may be established by seeking the intersection of an arc of length L_2, centered at end B, with the inclined guide bar. Note that two, or possibly no, intersections might occur. This means that (s_A, θ) are not a

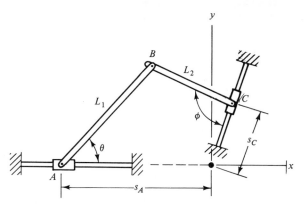

Figure 6.6   Generalized coordinates for a linkage.

suitable set of generalized coordinates unless their range of values is restricted and it is stipulated which of the intersections (upper or lower) is of interest.

   A better set of generalized coordinates for this system would have been the distances s_A and s_C locating both ends of the linkage. This would eliminate the question of which intersection should be considered. Nevertheless, the set (s_A, s_C) is still limited in its range of values, since the largest possible distance between ends A and C is $L_1 + L_2$. Indeed, other combinations of variables, such as (θ, ϕ), might be preferable.

   We remarked that the system in Figure 6.6 has two degrees of freedom, provided that nothing is specified about the motion. This is not necessarily the case. For example, collar A in Figure 6.6 might be required to move in a given manner along its guide, that is, $s_A = s_A(t)$ is given. This is a constraint on the motion of the system, so that the system then has only one degree of freedom.

6.2 CONSTRAINTS—HOLONOMIC AND NONHOLONOMIC

The discussion in the previous section was purposefully qualitative in order to focus on the important concept of generalized coordinates. It is useful now to change the approach. Suppose we select a set of M constrained generalized coordinates to represent a system having N degrees of freedom. Because the coordinates are a constrained set, $M > N$, there must be $M - N$ constraint equations. If the constrained equations are like those arising in the previous section, each may be written in the functional form

$$f_i(q_1, q_2, \ldots, q_M, t) = 0 \tag{6.1}$$

where the subscript i denotes which of the $M - N$ constraints are under consideration. A relation such as Eq. (6.1) is sometimes referred to as a *configuration constraint*. This term stems from the fact that any limitation imposed on the generalized coordinates restricts the overall position that the system can attain. In the most general situation the value of time must be specified, since a motion imposed on one of the physical supports can move the system, even though the generalized coordinates do not vary.

   We may equivalently replace a configuration constraint by a *velocity constraint*, which is a restriction on the velocity that a system may have when it is in a specified position. This viewpoint is obtained when Eq. (6.1) is differentiated with respect to time. The chain rule for differentiation must be employed because the generalized coordinates are (unknown) functions of time. The time derivative of constraint equation (6.1) is therefore

$$\dot{f}_i = \sum_{j=1}^{M} \left[\frac{\partial}{\partial q_j} f_i(q_1, q_2, \ldots, q_M, t) \right] \dot{q}_j + \left[\frac{\partial}{\partial t} f_i(q_1, q_2, \ldots, q_M, t) \right] = 0 \quad (6.2)$$

When the values of the generalized coordinates and time are specified, Eq. (6.2) represents one relation among the M rates $\dot{q}_j$, which are called *generalized velocities*. Equations (6.1) and (6.2) are fully equivalent in the restriction they impose, provided that the initial position is specified.

A simple example of this dual viewpoint is to consider using as constrained generalized coordinates x and y for a point following the circular path in Figure 6.7. Clearly, one form of the constraint equation is the equation for a circle

$$x^2 + y^2 = R^2$$

Differentiation of this relation yields

$$\dot{y} = -\frac{x}{y} \dot{x}$$

which is merely a statement that the velocity of a point following the circle must be parallel to the tangent to the circle at the instantaneous location of the point. Note that the radius R does not appear in the velocity constraint form. However, R must be known in order to select the x and y values when the motion begins.

Equation (6.2) does not represent the most general type of kinematical constraint that can be imposed on the motion of a system. Some types of motion restrictions cannot be treated in any general manner. Instead, they require an ad hoc approach based on a case-by-case study. One such situation arises in treating inequality relationships, such as the limitation that the brakes of a car reaching the top of a hill are functional only if the wheels remain in contact with the road.

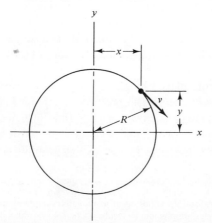

Figure 6.7 Constraint condition for circular motion.

Fortunately, a large class of problems involving mechanical systems may be treated as arbitrary velocity constraints. The derivatives in Eq. (6.2) may be considered to be the coefficients of a linear set of simultaneous equations for the generalized velocities. A more general form would replace the derivatives by arbitrary coefficients that depend only on the generalized coordinates and time. Such constraint equations have the form†

$$\sum_{j=1}^{M} a_{ij}(q_1, q_2, \ldots, q_M, t)\dot{q}_j + b_i(q_1, q_2, \ldots, q_M, t) = 0 \qquad (6.3)$$

The restrictions imposed by Eqs. (6.2) and (6.3) are equivalent if corresponding coefficients of each generalized velocity, and of the velocity-independent term, are identical to within a multiplicative factor. This factor may be a function g_i of the generalized coordinates and time, $g_i = g_i(q_1, q_2, \ldots, q_M, t)$. Hence, we may conclude that a velocity constraint is derivable from a configuration constraint if, and only if,

$$a_{ij}g_i = \frac{\partial f_i}{\partial q_j} \qquad b_i g_i = \frac{\partial f_i}{\partial t} \qquad (6.4)$$

The constraint equation(s) relating the generalized velocities are said to be *holonomic* (which may be taken to mean "integrable") if they satisfy Eq. (6.4). If Eq. (6.4) is not satisfied, the constraint is *nonholonomic*.

This terminology refers to the *Pfaffian* form of a constraint equation, which is the differential form obtained by multiplying Eq. (6.3) by dt, specifically,

$$\sum_{j=1}^{M} a_{ij}(q_1, q_2, \ldots, q_M, t) \, dq_j + b_i(q_1, q_2, \ldots, q_M, t) \, dt = 0 \qquad (6.5)$$

When Eq. (6.4) is true, multiplying Eq. (6.5) by the function g_i converts the Pfaffian form to a perfect differential of the function f_i. In other words:

If a velocity constraint is holonomic, there exists an integrating factor g_i for which the Pfaffian form of the constraint equation becomes a perfect differential.

In that case the constraint may be integrated to obtain the configuration constraint imposed on the generalized coordinates. When Eq. (6.4) is not valid, the kinematical relation between the generalized coordinates can only be established after those parameters have been solved as functions of time, in other words, after the equations of motion have been solved.

A different viewpoint for the role played by constraints may be gained by introducing the concept of the *configuration space*. The position of a point in real space is associated with the (x, y, z) coordinates it occupies at each time instant. In the same manner the position of any system may be associated with an M-dimensional Euclidean space, each of whose axes represents the value of one of the generalized coordinates

---

†Constraint conditions that match Eq. (6.3) are said to be linear velocity constraints, because they depend linearly on the generalized velocities. In addition to the inequality constraints mentioned earlier, some systems feature constraints in which the generalized velocities occur nonlinearly. Another generalization is to allow accelerations to be constrained. Both cases are possible in feedback control systems.

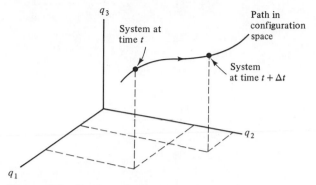

Figure 6.8 Configuration space.

$(q_1, q_2, \ldots, q_M)$. The path in the configuration space is the locus of points formed as the motion evolves in time. Figure 6.8 depicts the path in configuration space for a three-degree-of-freedom system; the picture may be conceptually extended to systems for which $M > 3$.

Consider a holonomic constraint in the form $f(q_1, q_2, \ldots, q_M, t) = 0$. At any instant t, this restricts the position in the configuration space to be somewhere along a surface, shown shaded in Figure 6.9. At any point along the path in the configuration space, a multitude of displacements are possible. The configuration constraint requires that the next point also be situated on the constraint surface. A displacement in which the new point in the configuration space is on the constraint surface is said to satisfy the constraint condition. It represents a *kinematically admissible movement* of the system.

The corresponding Pfaffian form of the velocity constraint, Eq. (6.5), is merely a statement that the infinitesimal displacement of the system must be along the plane in the configuration space tangent to the constraint surface. Any other type of displacement would move the point in the configuration space off the constraint surface.

When the constraint is nonholonomic, it is not possible to identify a constraint surface. Nevertheless, the effect of the Pfaffian constraint equation is to restrict infinitesimal displacements of the system to lie on a common tangent plane that is dictated by

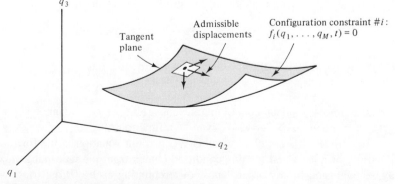

Figure 6.9 Configuration constraint.

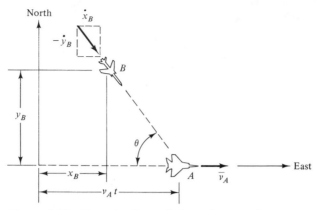

Figure 6.10   Example of a nonholonomic constraint.

the current state of motion. Such a plane may be considered to be a local manifestation of a constraint surface.

There are two types of holonomic constraints. If the constraint is time-independent, the holonomic constraint is said to be *scleronomic*; otherwise it is *rheonomic*. It follows that the integrated form of a scleronomic constraint is $f_i(q_1, q_2, \ldots, q_M) = 0$. Another classification of velocity constraints pertains to the presence of the coefficient b_i. If $b_i = 0$, the equation is a *catastatic* constraint, whereas $b_i \neq 0$ means that the generalized coordinates are related by an *acatastatic* constraint equation. It is evident that a scleronomic constraint is catastatic, while a rheonomic constraint is acatastatic. However, it is possible in the case of a nonholonomic constraint for the coefficients a_{ij} to be time-dependent, even though the relation is catastatic.

An interesting example of a nonholonomic constraint, which does not usually arise in a course in mechanics, is a pursuit problem. Consider Figure 6.10, which depicts an airplane A that flies eastward at a constant velocity $\bar{v}_A$. Airplane B has a laser mounted parallel to its axis. This airplane must always keep its laser aimed at airplane A. This restriction constrains the path that airplane B follows, but it is only possible to express the restriction on the generalized coordinates (x_B, y_B) as a velocity constraint.

The constraint equation associated with the situation in Figure 6.10 may be formulated by noting that the angle θ which defines the orientation of the velocity vector also may be expressed geometrically in terms of the position coordinates. This yields

$$\tan \theta = -\frac{\dot{y}_B}{\dot{x}_B} = \frac{y_B}{v_A t - x_B}$$

which may be written in the standard form of Eq. (6.3) as

$$y_B \dot{x}_B + (v_A t - x_B)\dot{y}_B = 0$$

This is a time-dependent nonholonomic constraint, since it cannot be integrated with respect to time unless x_B or y_B is given as a function of t.

Sometimes it is not clear whether a constraint equation in the form of Eq. (6.3) is holonomic or not. A test for this condition comes from the fact that a mixed derivative may be evaluated in any order. Thus, differentiating each of Eqs. (6.4) with respect to an arbitrarily selected generalized coordinate q_k leads to the conclusion that if constraint equation number i is holonomic, then

$$\frac{\partial}{\partial q_k}(g_i a_{ij}) = \frac{\partial}{\partial q_j}(g_i a_{ik})$$

$$\frac{\partial}{\partial q_k}(g_i b_i) = \frac{\partial}{\partial t}(g_i a_{ik}) \qquad j, k = 1, 2, \ldots, M \qquad j \neq k$$

(6.6)

In practice, it usually is easier to try to determine whether there is an integrating factor g_i that converts the Pfaffian form in Eq. (6.5) to a perfect differential, rather than to determine if Eqs. (6.6) are valid.

It is important to know that a constraint is holonomic because such constraints may be used to eliminate excess generalized coordinates. Suppose that H is the number of holonomic constraints, so that $M - N - H$ is the number of nonholonomic constraints. If the holonomic constraints were originally stated as velocity constraints, they may be integrated to obtain configurational constraints in the form of Eq. (6.1). Since these represent H equations, they may be solved to express H generalized coordinates in terms of the remaining $M - H$ generalized coordinates.

A *holonomic system* is one in which $H = M - N$; that is, the number of generalized coordinates in excess of the number of degrees of freedom equals the number of holonomic constraints. It is always possible to describe a holonomic system by a set of unconstrained generalized coordinates that satisfy all kinematical conditions arising from the physical manner in which the system is supported. In contrast, a *nonholonomic system* must always be described by a set of constrained generalized coordinates. It is necessary in that case to supplement the equations of motion with explicit statements of the kinematical constraint equations. Nonholonomic constraints commonly arise in systems having parts that roll.

EXAMPLE 6.1 _____

An insect walks along the surface of a spherical balloon whose radius is a specified function $r(t)$. Describe in terms of both rectangular Cartesian and cylindrical coordinates the constraint imposed on the motion of the insect by the condition that it remain on the surface. Express the result for each set of coordinates as a configuration constraint and as a velocity constraint.

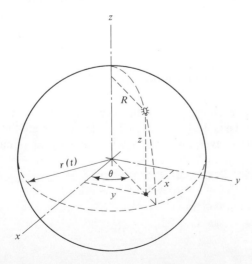

Position coordinates.

Solution. We may obtain the configuration constraint corresponding to a set of Cartesian coordinates by placing the origin at the center of the sphere, as shown in the sketch. The equation of a sphere describes the distance from a point on the sphere to the center.

$$f(x, y, z, t) = x^2 + y^2 + z^2 - r(t)^2 = 0$$

The corresponding velocity constraint results from a differentiation of the configuration constraint. This yields

$$x\dot{x} + y\dot{y} + z\dot{z} - r(t)\frac{d}{dt}r(t) = 0$$

For cylindrical coordinates, we note that the radius of the sphere may be expressed in terms of the transverse distance R and axial distance z by the Pythagorean theorem. The configurational constraint therefore is

$$R^2 + z^2 - r(t)^2 = 0$$

Differentiation of the configurational constraint leads to the velocity constraint that

$$R\dot{R} + z\dot{z} - r(t)\frac{d}{dt}r(t) = 0$$

EXAMPLE 6.2

Two bars, pinned at joint B, move in the horizontal plane subject only to the restriction that the velocity of end C must be directed toward end A. Determine the corresponding velocity constraint. Is this constraint holonomic?

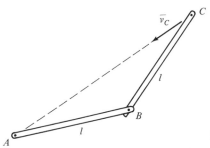

Example 6.2

Solution. The position of each bar is uniquely specified by the coordinates x_B, y_B of the pin connection and the angles of rotation θ_1 and θ_2; these are the generalized coordinates we select. The given condition on the velocity of point C may be written in vector form as

$$\bar{v}_C = v_C \bar{e}_{A/C}$$

which leads to the constraint condition

$$\bar{v}_C \times \bar{r}_{A/C} = \bar{0}$$

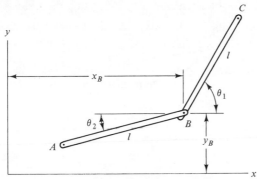

Generalized coordinates.

We must express this condition in terms of the generalized coordinates. Because points B and C are common to the same body, we have

$$\bar{v}_C = \bar{v}_B + (\dot{\theta}_1 \bar{k} \times \bar{r}_{C/B}) = (\dot{x}_B - l\,\dot{\theta}_1 \sin \theta_1)\bar{i} + (\dot{y}_B + l\,\dot{\theta}_1 \cos \theta_1)\bar{j}$$

Also, the position vector is

$$\bar{r}_{A/C} = -l(\cos \theta_1 + \cos \theta_2)\bar{i} - l(\sin \theta_1 + \sin \theta_2)\bar{j}$$

Hence, the constraint equation leads to

$$(\bar{v}_C \times \bar{r}_{A/C}) \cdot \bar{k} = l(\dot{y}_B + l\dot{\theta}_1 \cos \theta_1)(\cos \theta_1 + \cos \theta_2)$$
$$- l(\dot{x}_B - l\dot{\theta}_1 \sin \theta_1)(\sin \theta_1 + \sin \theta_2) = 0$$

$$\dot{y}_B(\cos \theta_1 + \cos \theta_2) - \dot{x}_B (\sin \theta_1 + \sin \theta_2)$$
$$+ l\dot{\theta}_1(\cos \theta_1 \cos \theta_2 + \sin \theta_1 \sin \theta_2 + 1) = 0$$

The relation has the standard form of a velocity constraint,

$$a_{11}\dot{x}_B + a_{12}\dot{y}_B + a_{13}\dot{\theta}_1 + a_{14}\dot{\theta}_2 + b_1 = 0$$

where

$$a_{11} = -\sin \theta_1 - \sin \theta_2 \qquad a_{12} = \cos \theta_1 + \cos \theta_2$$
$$a_{13} = l[\cos (\theta_1 - \theta_2) + 1] \qquad a_{14} = b_1 = 0$$

In order to test whether the constraint is holonomic, we shall assume that it is, and determine if there are any contradictions. Applying Eqs. (6.4) in the present case yields

$$\frac{\partial f_1}{\partial x_b} = g_1 a_{11} \qquad \frac{\partial f_1}{\partial y_B} = g_1 a_{12}$$

$$\frac{\partial f_1}{\partial \theta_1} = g_1 a_{13} \qquad \frac{\partial f_1}{\partial \theta_2} = g_1 a_{14} = 0$$

Integrating the first with respect to x_B yields

$$f_1 = -(\sin \theta_1 + \sin \theta_2) \int g_1 \, dx_B + h_1(y_B, \theta_1, \theta_2)$$

where h_1 is an arbitrary function. In contrast, integrating the last assumed equality leads to

$$f_1 = h_4(x_B, y_B, \theta_1)$$

which states that f_1 must be independent of θ_2. Clearly, this is incompatible with the first form of f_1. We therefore conclude that f_1 does not exist, which means that the constraint is nonholonomic.

6.3 VIRTUAL DISPLACEMENTS

The concept of a virtual movement of a system plays a central role in the kinetics principles in analytical mechanics. In a sense, the word "virtual" may be thought to mean "fictitious." It represents an alteration in the position of the system that would result if the values of the generalized coordinates were changed. This position change is based on time being constant. Thus, the old and new positions are alternative configurations of the system at that time instant. The specific definition is

In a virtual movement the generalized coordinates of the system are considered to be incremented by an infinitesimal amount from the values they have at an arbitrary instant, with time held constant.

If this change were actually introduced into the system, the points in the system would move by an infinitesimal amount. The *virtual displacement* of points resulting from changes in the generalized coordinates are not actual movements. However, we shall soon see that they are related to the difference between possible displacements that may be imparted to a system owing to different force systems. The evaluation of these virtual displacements is an important aspect of the analytical formulation.

6.3.1 Analytical Method

In order to characterize a virtual displacement, consider an arbitrary point A in a system. The position vector $\bar{r}_{A/O}$ relative to a fixed origin depends on the generalized coordinates and time, so

$$\bar{r}_{A/O} = \bar{r}_A(q_1, q_2, \ldots, q_M, t) \tag{6.7}$$

The infinitesimal virtual change in any quantity is denoted by the symbol δ in order to emphasize that the change is not the result of an actual motion of the system. Thus, the generalized coordinate values are incremented in the virtual movement by δq_1, δq_2, etc. The analytical method for virtual displacements evaluates the change in position resulting from these increments by differentiating an algebraic expression for position, such as Eq. (6.7).

It is important to recognize that time is held fixed at arbitrary t in a virtual movement. This is significant to two aspects of Eq. (6.7). First, since t is arbitrary, the generalized coordinates have arbitrary values, which means that the generalized coordinates must be treated as algebraic, rather than numerical, parameters. Also, since time is held constant, the explicit dependence of $\bar{r}_{A/O}$ on t should not be considered.

The change in the position of point A in a virtual movement is the virtual displacement $\delta\bar{r}_A$. Partial differentiation of Eq. (6.7) shows that

$$\delta\bar{r}_A = \sum_{j=1}^{M} \frac{\partial\bar{r}_{A/O}}{\partial q_j} \delta q_j \qquad (6.8a)$$

which may also be written in component form as

$$\delta\bar{r}_A = \sum_{j=1}^{M} \left(\frac{\partial x_A}{\partial q_j}\bar{i} + \frac{\partial y_A}{\partial q_j}\bar{j} + \frac{\partial z_A}{\partial q_j}\bar{k} \right) \delta q_j \qquad (6.8b)$$

Before we apply these relations, we shall develop them by a different method, whose viewpoint provides insight into why virtual displacements are important. Suppose we consider two kinematically admissible displacements of point A at an arbitrary instant t, each of which could be produced by applying a different set of forces to the system. Let superscript 1 or 2 denote variables associated with each set. The chain rule for differentiation indicates that the differential displacement in each case is

$$d\bar{r}_A^{(k)} = \sum_{j=1}^{M} \frac{\partial\bar{r}_{A/O}}{\partial q_j} dq_j^{(k)} + \frac{\partial\bar{r}_{A/O}}{\partial t} dt \qquad k = 1, 2 \qquad (6.9)$$

The difference between the two possible displacements is

$$d\bar{r}_A^{(2)} - d\bar{r}_A^{(1)} = \sum_{j=1}^{M} \frac{\partial\bar{r}_{A/O}}{\partial q_j} (dq_j^{(2)} - dq_j^{(1)}) \qquad (6.10)$$

If we define the difference between the differential increments in the generalized coordinates as the virtual increment, that is,

$$\delta q_j = dq_j^{(2)} - dq_j^{(1)} \qquad (6.11)$$

then we recover Eq. (6.8a). In other words,

> When we form a virtual displacement, we are studying the differences in the movement of a system that possibly could result from the action of different sets of forces.

Equations (6.8) are the essence of the analytical approach to evaluating the virtual displacement of a point. They require that the position coordinates of a point be expressed as algebraic functions of the generalized coordinates. (Such dependencies may usually be obtained from the laws of geometry.) Also, in order to assure that the expression for $\bar{r}_{A/O}$ is actually an absolute position, the origin O should be selected to be an actual fixed point in the system, as was mentioned earlier.

It will be necessary in many cases to express the change in the angles of orientation of various bodies in the system. This will usually involve using the law of sines and/or cosines to relate such angles to the generalized coordinates. We may then evaluate the virtual rotation by a differentiation process, such as that followed in Eqs. (6.8). Two cases arise, depending on whether the angle of orientation is expressed in explicit or implicit form. Let β be the angle of orientation. In the explicit case, $\beta = f(q_1, q_2, \ldots, q_M, t)$ has been determined. Then

$$\delta\beta = \sum_{j=1}^{M} \frac{\partial f}{\partial q_j} \delta q_j \tag{6.12}$$

The case where β is known implicitly has the functional form $g(\beta, q_1, q_2, \ldots, q_M, t) = 0$. The virtual change in the function g resulting from incrementing all variables except time is

$$\delta g = \frac{\partial g}{\partial \beta} \delta\beta + \sum_{j=1}^{M} \frac{\partial g}{\partial q_j} \delta q_j = 0$$

which yields

$$\delta\beta = -\frac{1}{\partial g/\partial \beta} \sum_{j=1}^{M} \frac{\partial g}{\partial q_j} \delta q_j \tag{6.13}$$

EXAMPLE 6.3 _____

The horizontal distance x between pin A and roller B is selected as the generalized coordinate for the parallelogram linkage. Describe the virtual displacement of pin F and the virtual rotation of bar EF resulting from a virtual increment δx.

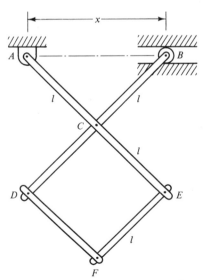

Example 6.3

Solution. Virtual displacements are found in the analytical method by differentiating expressions for the position parameters. We place the origin at pin A because that is the only fixed point in the system. The horizontal and vertical distances

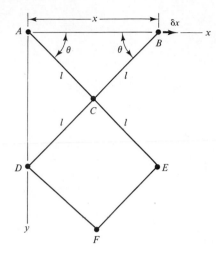

Coordinate system.

between two joints on a diagonal bar are $x/2$ and $(l^2 - x^2/4)^{1/2}$, respectively, so

$$\theta = \cos^{-1}\left(\frac{x/2}{l}\right) \qquad \bar{r}_{F/A} = \frac{x}{2}\bar{i} + 3\left(l^2 - \frac{x^2}{4}\right)^{1/2}\bar{j}$$

Then

$$\delta\theta = \frac{d}{dx}\left[\cos^{-1}\left(\frac{x}{2l}\right)\right]\delta x = -\frac{\delta x}{(4l^2 - x^2)^{1/2}}$$

$$\delta\bar{r}_F = \frac{d}{dx}(\bar{r}_{F/A})\,\delta x = \left[\frac{1}{2}\bar{i} - \frac{3x}{2(4l^2 - x^2)^{1/2}}\bar{j}\right]\delta x$$

6.3.2 Kinematical Method

The analytical method for evaluating virtual displacements relies on the laws of geometry for the derivation of differentiable expressions for position. This is the primary source of difficulty encountered in the application of the method. Simple systems, such as those whose parts form isosceles triangles or right angles, are relatively easy to describe geometrically. However, increasing the complexity of the geometry can lead to substantial complication.

Consider the common task of evaluating the velocity relationships for a linkage such as the one in Figure 6.11. We could, in principle, write general expressions for the position of the joints, and then differentiate those expressions with respect to time in order to determine the velocity. However, to do so would ignore other kinematical techniques that expedite the analysis. The kinematical method for virtual displacement employs such techniques.

The essence of the approach is to recognize the analogous relationship between virtual and real displacements. Suppose that the position coordinates of some point A in an arbitrary system were known as a function of the generalized coordinates and time.

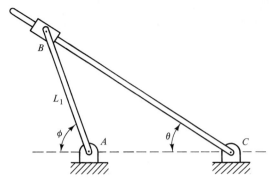

Figure 6.11 Typical linkage.

Then the velocity of this point would be

$$\bar{v}_A = \frac{d\bar{r}_A}{dt} = \sum_{j=1}^{M} \frac{\partial \bar{r}_A}{\partial q_j} \dot{q}_j + \frac{\partial \bar{r}_A}{\partial t} \tag{6.14}$$

The actual displacement of this point in an infinitesimal time interval dt would be

$$d\bar{r}_A = \bar{v}_A \, dt = \sum_{j=1}^{M} \frac{\partial \bar{r}_{A/O}}{\partial q_j} \, dq_j + \frac{\partial \bar{r}_{A/O}}{\partial t} \, dt \tag{6.15}$$

This expression is reminiscent of Eq. (6.8a) for virtual displacements. One difference is that the infinitesimal increments in the generalized coordinates were denoted as δq_j for virtual increments, whereas they are now dq_j for actual increments. Another difference is that, by definition, time is held constant in a virtual change. Hence, Eq. (6.8a) does not contain a time-derivative term. However, the time-derivative term also is not present in the case of a time-independent system. (Recall that such a system corresponds to a situation where all kinematical conditions imposed on the system are independent of time.)

This similarity between $\delta \bar{r}_A$ and $d\bar{r}_A$ can be exploited. Suppose we perform a conventional velocity analysis of $\bar{v}_A$ at an arbitrary position and time. If the system is time-dependent, we retain in that expression only those terms that are proportional to the generalized velocities. The result, which we denote as $\bar{v}_A^c$, represents the velocity resulting from changes in the generalized coordinates with the physical constraints imposed on the system held constant. Hence, the "velocity" will be

$$\bar{v}_A^c = \sum_{j=1}^{M} \frac{\partial \bar{r}_{A/O}}{\partial q_j} \dot{q}_j \tag{6.16}$$

This expression for $\bar{v}_A^c$ is identical to Eq. (6.8a), aside from the presence of rates of change instead of virtual increments. Thus,

An expression for a virtual displacement may be obtained directly from a velocity relation by replacing generalized velocities with virtual displacements. The only restriction on this approach is that, when a system is time-dependent, all terms in the velocity that do not contain a generalized velocity should be dropped.

As an example, suppose the generalized coordinate for a particle as it moves along a specified path is the arclength s. Then the path variable formula for velocity leads to

$$\delta \bar{r}_A = \delta s \bar{e}_t \qquad (6.17)$$

In the same way, the velocity formulas in cylindrical and spherical coordinates may be adapted to virtual displacement, as follows:

Cylindrical coordinates, Figure 6.12a:

$$\delta \bar{r}_A = \delta R \bar{e}_R + R\, \delta\theta \bar{e}_\theta + \delta z \bar{k} \qquad (6.18)$$

Spherical coordinates, Figure 6.12b:

$$\delta \bar{r}_A = \delta r \bar{e}_r + r\, \delta\phi \bar{e}_\phi + r \sin \phi\, \delta\theta \bar{e}_\theta \qquad (6.19)$$

Similarly, we may relate the virtual displacements of two points in a rigid body by modifying their velocity relationship. Let $\overline{\delta\theta}$ represent the infinitesimal rotation (a vector quantity according to the right-hand rule) in a virtual movement. In the velocity equation for rigid-body motion, we replace velocities of points by virtual displacements, and angular velocity by virtual rotation. The result is that

$$\delta \bar{r}_B = \delta \bar{r}_A + \overline{\delta\theta} \times \bar{r}_{B/A} \qquad (6.20)$$

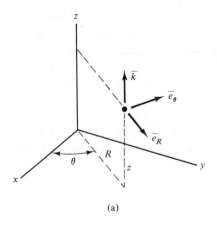

(a)

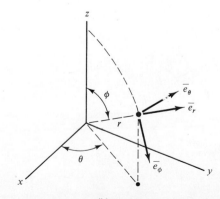

(b)

Figure 6.12 Curvilinear coordinate systems.

The only subtlety involved in using this concept is the question of eliminating the effect of time dependence of the constraints. The velocity analog to each of Eqs. (6.17) to (6.20) must have been obtained for the time-independent version of the constraint condition. For example, if the transverse distance R in cylindrical coordinates is constrained to be a specified function of time, then the velocity equation for a time-independent constraint would hold R constant at an arbitrary value, very much like a partial derivative. The result would be

$$\bar{v}_A^c = R\dot{\theta}\bar{e}_\theta + \dot{z}\bar{k} \Rightarrow \delta\bar{r}_A = R\,\delta\theta\bar{e}_\theta + \delta z\bar{k}$$

The primary utility of Eq. (6.20) is that it leads to techniques for virtual displacements that parallel those for velocity analysis. Notable among these is the method of instant centers, which is particularly useful for planar situations involving linkages and rolling without slipping. As an illustration of this concept, consider the linkage in Figure 6.13. This system was considered earlier in the discussion of generalized coordinates, where s_A and s_C were suggested as generalized coordinates. Suppose it is desired to evaluate the virtual rotations $\delta\theta$ and $\delta\phi$ resulting from a virtual displacement δs_A. However, let us now consider the case where the movement of end C is a time-dependent constraint, such that s_C is a specified function of time. Therefore, that end is held fixed in a virtual displacement, and $\delta s_C = 0$.

In order to exploit the analogy between virtual displacements and velocities, we perform an instant-center analysis on the basis that end C does not move from its current position. The corresponding instant center for the virtual movement is at point D for bar AB and at point C for bar BC. (Note that the virtual rotation depicted in the figure corresponds to an increase in the generalized coordinate s_A. We make it a standard practice to depict positive increments of the generalized coordinates in order to avoid sign errors.)

The laws of trigonometry yield the distances l_A and l_B in terms of s_A and s_C, and the system parameters β, L_1, and L_2. Then the desired virtual displacements may be constructed by considering the virtual movement of bar AB to be a rotation about point D. This yields

$$\delta s_A = -l_A\,\delta\theta \qquad |\delta\bar{r}_B| = l_B\,\delta\theta = L_2\,\delta\phi$$

These relations may be solved for the virtual rotations.

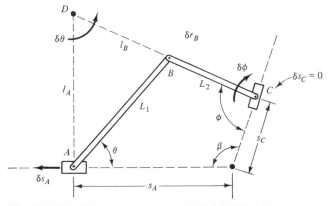

Figure 6.13   Instant center analysis of virtual displacements.

$$\delta\theta = -\frac{1}{l_A}\delta s_A \qquad \delta\phi = -\frac{l_B}{l_A L_2}\delta s_A$$

If we were to use these expressions in the context of an overall formulation of the equations of motion, we would need to employ the expression for l_A and l_B in terms of s_A and s_C. The derivation of such expressions is not a trivial task, but at least the kinematical method avoids the need to differentiate the corresponding position vectors.

A significant aid to the kinematical method is the principle of superposition, which permits the effects of changing each generalized coordinate to be considered individually. Indeed, the basic relation, Eq. (6.8a), represents a superposition of virtual displacements, since it may be rewritten as

$$\delta\bar{r}_A = \sum_{j=1}^{M}(\delta\bar{r}_A)_j \qquad (\delta\bar{r}_A)_j = \frac{\partial\bar{r}_{A/O}}{\partial q_j}\delta q_j \qquad (6.21)$$

The jth contribution, $(\delta\bar{r}_A)_j$, is the virtual displacement obtained when only the corresponding q_j is incremented. This converts the kinematical analysis to investigations of a sequence of one-degree-of-freedom systems associated with each generalized coordinate. The overall virtual displacements would then be obtained by a *vectorial* superposition of the individual contributions.

The linkage in Figure 6.13 serves to illustrate the superposition principle. Suppose that the collars at both ends of the linkage were free to slide over their guide bars. The total virtual movement of the system is the superposition of the effects of incrementing s_A and s_C. The first increment was treated above. A comparable analysis could be performed to increment s_C with s_A held fixed. The overall virtual displacements and rotations would then be the vector sum of the individual effects.

EXAMPLE 6.4 _____

The crankshaft AB is given a virtual rotation $\delta\phi$ when it is at an arbitrary orientation ϕ. Determine the corresponding virtual displacement of the piston.

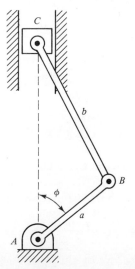

Example 6.4

Solution. This linkage has one degree of freedom because the other angles may be evaluated when ϕ is known. Specifically, the law of sines gives

$$\sin \theta = \frac{a}{b} \sin \phi$$

We shall evaluate the virtual displacement by analogy with the method of instant centers. The sketch shows the construction of the virtual displacements and rotations corresponding to an increase in the generalized coordinate ϕ. (Considering $\delta\phi$ to be positive avoids uncertainties in sign.)

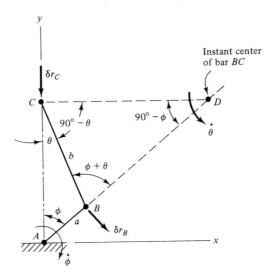

Instant-center diagram.

First, we locate the instant center D of bar BC by drawing perpendiculars to the virtual displacements of joint B, which must rotate about point A, and of piston C, which is constrained to translate vertically. The virtual displacement of the pins is then

$$\delta r_B = |\bar{r}_{B/A}| \, \delta\phi = |\bar{r}_{B/D}| \, \delta\theta \qquad \delta \bar{r}_C = |\bar{r}_{C/D}| \, \delta\theta \, (-\bar{j})$$

This leads to

$$\delta\theta = \frac{|\bar{r}_{B/A}|}{|\bar{r}_{B/D}|} \, \delta\phi \qquad \delta \bar{r}_C = - \frac{|\bar{r}_{C/D}| \, |\bar{r}_{B/A}|}{|\bar{r}_{B/D}|} \bar{j} \, \delta\phi$$

Our last task is to express the geometrical parameters in terms of the generalized coordinate ϕ. The law of sines applied to triangle BCD yields

$$\frac{|\bar{r}_{C/D}|}{|\bar{r}_{B/D}|} = \frac{\sin (\phi + \theta)}{\cos \theta} = \sin \phi + \cos \phi \tan \theta$$

We related θ to ϕ at the beginning of the solution. That expression yields

$$\tan \theta = \frac{b \sin \theta}{b(1 - \sin^2 \theta)^{1/2}} = \frac{a \sin \phi}{(b^2 - a^2 \sin^2 \phi)^{1/2}}$$

We use this to eliminate θ from the distance ratio, which leads to

$$\frac{|\bar{r}_{C/D}|}{|\bar{r}_{B/D}|} = \sin \phi + \frac{a \sin \phi \cos \phi}{(b^2 - a^2 \sin^2 \phi)^{1/2}}$$

$$\delta \bar{r}_C = -a \sin \phi \left[1 + \frac{a \cos \phi}{(b^2 - a^2 \sin^2 \phi)^{1/2}} \right] \bar{j} \, \delta \phi$$

EXAMPLE 6.5

In the position shown gear B is falling as it rolls over gear A, which is rolling over rack C. Generalized coordinates are the horizontal distance x to gear A and the angle of elevation θ for the line connecting the centers. Determine the virtual displacement of the center of each gear and the virtual rotation of each gear resulting from virtual increments in the generalized coordinates.

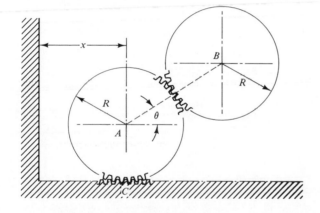

Example 6.5

Solution. The analogy with the relative-velocity equation is useful for evaluating virtual displacements of rolling bodies. We shall use the vector equations to satisfy the constraint that the virtual displacements of contacting points must match because the gear teeth prevent slippage. We therefore write

$$\delta \bar{r}_C = \bar{0} \qquad \delta \bar{r}_A = \delta x \bar{i} = \delta \theta_A \bar{k} \times \bar{r}_{A/C} \qquad \delta \bar{r}_B = \delta \bar{r}_A + \delta \theta \bar{k} \times \bar{r}_{B/A}$$
$$\delta \bar{r}_D = \delta \bar{r}_A + \delta \theta_A \bar{k} \times \bar{r}_{D/A} = \delta \bar{r}_B + \delta \theta_B \bar{k} \times \bar{r}_{D/B}$$

Note that the angles of rotation of the gears, θ_A and θ_B, are different variables from θ.

We describe the position vectors in terms of θ and x.

$$\bar{r}_{A/C} = R \bar{j} \qquad \bar{r}_{D/A} = -\bar{r}_{D/B} = \tfrac{1}{2} \bar{r}_{B/A} = R \cos \theta \bar{i} + R \sin \theta \bar{j}$$

Then matching the two expressions for $\delta \bar{r}_A$ leads to

$$\delta \theta_A = -\frac{\delta x}{R}$$

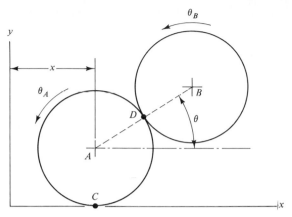

Kinematical parameters.

The above position vectors show the virtual displacement of the center of gear B to be

$$\delta \bar{r}_B = [\delta x - (2R \sin \theta) \, \delta\theta] \bar{i} + (2R \cos \theta) \, \delta\theta \bar{j}$$

which when substituted into the two descriptions of the virtual displacement of the contact point D yields

$$\delta \bar{r}_D = [\delta x - (R \sin \theta) \, \delta\theta_A] \bar{i} + (R \cos \theta) \, \delta\theta_A \bar{j}$$
$$= [\delta x - (2R \sin \theta) \, \delta\theta] \bar{i} + (2R \cos \theta) \, \delta\theta \bar{j} + (R \sin \theta \bar{i} - R \cos \theta \bar{j}) \, \delta\theta_B$$

Matching either set of components in the above leads to

$$-\delta\theta_A = -2 \, \delta\theta + \delta\theta_B$$

$$\delta\theta_B = 2 \, \delta\theta + \frac{1}{R} \, \delta x$$

6.4 GENERALIZED FORCES

The selection of a set of generalized coordinates and the evaluation of the virtual displacements in terms of those quantities are primary aspects of the Lagrangian approach to the derivation of equations of motion for a system. It is necessary to recognize what parameters are appropriate to the kinematical description. By doing so, we create the model on which the rest of the analysis will be based. The kinematics phase of the formulation is essentially complete when the physical velocities and virtual displacements have been related to the generalized coordinates. The kinetics principles, which we shall derive in the following sections, tend to be much more straightforward. The first task is to represent the effect of the forces exerted on a system.

6.4.1 Virtual Work

Consider the particle in Figure 6.14 that is subjected to a variety of forces. When this particle is given a virtual displacement $\delta \bar{r}$, the forces acting on the particle do *virtual*

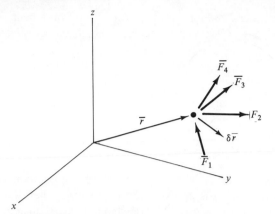

Figure 6.14 Virtual work.

work, which will be denoted δW. Because the virtual displacement is infinitesimal, the virtual work is also infinitesimal. Also, since the change is virtual, time is held constant at an arbitrary value. This means that the force is constant throughout the virtual displacement.

The virtual work of the force system in Figure 6.14 may be evaluated by taking the dot product of $\delta \bar{r}$ with each individual force. Alternatively, the resultant of the force system, $\Sigma \bar{F}$, may be formed first. Thus,

$$\delta W = \sum_i (\bar{F}_i \cdot \delta \bar{r}) = \left(\sum \bar{F} \right) \cdot \delta \bar{r}$$

Equation (6.8a) for virtual displacement then yields

$$\delta W = \sum \bar{F} \cdot \sum_{j=1}^{M} \frac{\partial \bar{r}}{\partial q_j} \delta q_j \tag{6.22}$$

The resultant force may be brought inside the sum over the generalized coordinates, with the result that

$$\delta W = \sum_{j=1}^{M} \left(\sum \bar{F} \cdot \frac{\partial \bar{r}}{\partial q_j} \right) \delta q_j \tag{6.23}$$

Equation (6.23) shows that the virtual work is a sum of force terms multiplying the virtual change in each generalized coordinate. The definition of a *generalized force*, which is denoted as Q_i, is that it is the coefficient of the corresponding increment δq_i in the expression for virtual work. Thus

$$\delta W = \sum_{i=1}^{M} Q_i \, \delta q_i \tag{6.24}$$

A comparison of Eqs. (6.23) and (6.24) shows that for a single particle

$$Q_i = \sum \bar{F} \cdot \frac{\partial \bar{r}}{\partial q_i} \tag{6.25}$$

The reason for calling Q_i a generalized force may be recognized from Eq. (6.25). Suppose that Cartesian coordinates (x, y, z) are selected as the generalized coordinates. Then

$$\frac{\partial \bar{r}}{\partial x} = \bar{i} \qquad \frac{\partial \bar{r}}{\partial y} = \bar{j} \qquad \frac{\partial \bar{r}}{\partial z} = \bar{k}$$

so that

$$Q_1 = \sum \bar{F} \cdot \bar{i} = \sum F_x \qquad Q_2 = \sum \bar{F} \cdot \bar{j} = \sum F_y$$
$$Q_3 = \sum \bar{F} \cdot \bar{k} = \sum F_z$$

In other words the generalized forces in this case are the force components in the respective directions. Equation (6.24) is an extension to the case where the generalized coordinates are any type of geometrical variable. For example, if q_j is an angle of rotation, Q_j will be a moment. [This is recognizable from the fact that δW in Eq. (6.24) must have units of work.]

Generalized forces may be regarded as the analog in the configuration space of the Cartesian components of an actual force. In general, we denote a vector in the configuration space by a circumflex (^). Let $\hat{u}_i$ be the unit vector in the configuration space that is aligned with the q_i axis. Time is constant in a virtual change, so the virtual displacement of a system is defined in the configuration space by

$$\delta \hat{r} = \delta q_1 \hat{u}_1 + \delta q_2 \hat{u}_2 + \cdots + \delta q_M \hat{u}_M \qquad (6.26)$$

The generalized forces appear in the configuration space as components parallel to the respective axes of a force vector $\hat{Q}$ defined as

$$\hat{Q} = Q_1 \hat{u}_1 + Q_2 \hat{u}_2 + \cdots + Q_M \hat{u}_M \qquad (6.27)$$

The definition of generalized force, Eq. (6.24), leads to the recognition that the virtual work is the dot product of the force and virtual displacement vectors. Specifically,

$$\delta W = \hat{Q} \cdot \delta \hat{r} \qquad (6.28)$$

Constraint forces (which are synonymous with reactions) play a special role. Consider the case where a particle is constrained to move along a specified curve. As shown in Figure 6.15, this situation leaves only a single degree of freedom, for which the arclength s is a convenient generalized coordinate. The resultant force may be resolved into tangent, normal, and binormal components. The latter two are constraint forces because they prevent the particle from moving perpendicular to the path. In a virtual movement that increases s by δs, the particle moves in the tangential direction by that amount, so $\delta \bar{r} = \delta s \bar{e}_t$. The corresponding virtual work is $\delta W = F_t \delta s$, since the virtual displacement is perpendicular to the normal and binormal force components. This result that the constraint forces do no virtual work is not a chance occurrence.

Our earlier interpretation of a constraint force is that it is the force required to enforce a restriction on the movement of the system. Suppose we were to impose a virtual movement on the system. If the movement does not violate the constraint condition, the corresponding reaction will not resist the motion. This means that:

A constraint force does no virtual work when a virtual displacement satisfies the constraint condition it imposes.

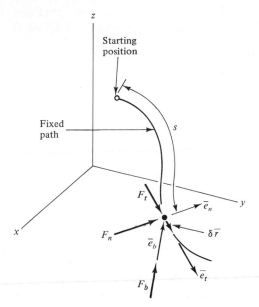

Figure 6.15 Virtual work in movement along a constrained path.

This statement may considered to be the definition of a constraint force. An important corollary comes from the observation that unconstrained generalized coordinates satisfy all constraint equations. Thus,

The virtual work done by the reactions is always zero in a holonomic system that is described by unconstrained generalized coordinates.

It is possible to make general statements regarding how reactions enter into the generalized forces when the generalized coordinates do not identically satisfy all constraint conditions. Let $\hat{R}$ be the vector in the configuration space representing the generalized forces attributable to one of the constraints that is not satisfied. The components R_j are the contributions of the reaction associated with this constraint to the generalized forces Q_j. According to the definition of a constraint force given earlier, $\delta W = 0$ if $\delta \hat{r}$ satisfies the constraint condition. It follows that $\hat{R}$ is perpendicular in the configuration space to the virtual displacement of the system, $\delta \hat{r}$. This orthogonality property enables us to relate the generalized constraint forces R_j to the coefficients in the velocity constraint equation.

Since time is held constant in a virtual displacement, the Pfaffian constraint equation (6.5) may be employed with $dt = 0$ and all differential increments replaced by virtual increments. Thus, Eq. (6.5) requires that the components of $\delta \hat{r}$ in the configuration space satisfy

$$\sum_{j=1}^{M} a_{ij}(q_1, q_2, \ldots, q_M, t)\, \delta q_j = 0 \qquad (6.29)$$

where the first subscript, i, serves to denote which constraint is being considered. However, $\delta W = 0$, so it also must be that

$$\sum_{j=1}^{M} R_j\, \delta q_j = 0 \qquad (6.30)$$

Equations (6.29) and (6.30) both restrict the value of one of the δq_i, for example, δq_1, in terms of the other $M - 1$ values. Matching the value of δq_1 obtained from each relation yields

$$\sum_{j=2}^{M} \frac{R_j}{R_1} \delta q_j = \sum_{j=2}^{M} \frac{a_{ij}}{a_{i1}} \delta q_j \qquad (6.31)$$

The δq_j appearing in this relation are arbitrary, so the equality can only be satisfied if like coefficients match. This leads to

$$\frac{R_j}{R_1} = \frac{a_{ij}}{a_{i1}}$$

which is equivalent to

$$R_j = \lambda_i a_{ij} \qquad j = 1, 2, \ldots, M \qquad (6.32)$$

The coefficient λ_i is the *Lagrangian multiplier* for the ith constraint.[†] We shall see later how to determine these quantities, and thereby evaluate the reactions through Eq. (6.32).

The significance of these theorems becomes apparent when we consider a Newtonian formulation of the equations of motion. A free body diagram of each body in a system will generally display a multitude of reactions. The force and moment equations of motion for each body yield a set that relates the acceleration parameters to the reactions. A large part of the solution process must be devoted to solving for, or eliminating, the reactions.

The physical principles developed in the next section circumvent these difficulties, because the applied forces are described in terms of generalized forces. Since reactions do not work in a compatible virtual displacement, they usually will not occur in the equations of motion.

Let us examine a few situations that illustrate when constraint forces may be avoided in the formulation. First, consider Figure 6.16a, where two small spheres, which may be modeled as particles, are constrained to move in the xy plane. The spheres are connected by a massless rigid bar. Since the bar is rigid, unconstrained generalized coordinates are (x_A, y_A, ϕ). The free body diagram for each particle (Figure 6.16b) shows that there is an axial force $\bar{F}$ exerted by the rigid bar on each particle. The position vectors for the particles are

$$\bar{r}_A = x_A \bar{i} + y_A \bar{j} \qquad \bar{r}_B = \bar{r}_A + L \bar{e}_R$$

Since L is constant, the virtual displacement $\delta \bar{r}_B$ differs from $\delta \bar{r}_A$ solely as a result of the change in ϕ. Thus,

$$\delta \bar{r}_A = \delta x_A \bar{i} + \delta y_A \bar{j} \qquad \delta \bar{r}_B = \delta \bar{r}_A + L \, \delta \phi \bar{e}_\phi$$

The virtual work done by the axial force is therefore

$$\delta W = (-F \bar{e}_R) \cdot \delta \bar{r}_A + (F \bar{e}_R) \cdot \delta \bar{r}_B = FL \, \delta \phi \bar{e}_R \cdot \bar{e}_\phi = 0$$

[†]In a strict sense, the derivation actually is only valid for the case of a single constraint force. When more than one constraint force must be satisfied, only a subset of N virtual increments δq_j are arbitrary. However, the choice of which δq_j should form that subset is also arbitrary. The only way in which Eq. (6.29) and (6.30) can remain valid under such conditions is if Eq. (6.32) holds.

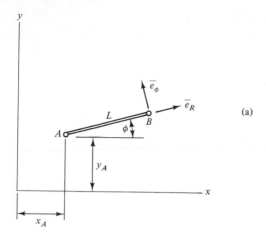

(a)

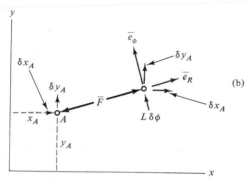

(b)

Figure 6.16 Virtual work for connected particles.

The axial force does no work in this situation because it is the constraint force required to keep L constant. If we wish to violate this constraint, we must employ a set of constrained generalized coordinates. Suppose that (x_A, y_A, R, ϕ) are used as generalized coordinates. Then the virtual displacements would satisfy

$$\delta \bar{r}_B = \delta \bar{r}_A + \delta R \bar{e}_R + L\,\delta\phi\,\bar{e}_\phi$$

so that

$$\delta W = F\,\delta R$$

If $\delta R \neq 0$ in violation of the constraint imposed by $\overline{F}$, then $\overline{F}$ does virtual work.

The case where the spheres in Figure 6.16a are connected by a spring offers an instructive contrast. Instead of being an unknown reaction that restricts motion, a spring force is known in terms of position. There would be no constraints on the motion of the spheres in this case, so the system would have four degrees of freedom. The set (x_A, y_A, R, ϕ) would then be unconstrained generalized coordinates. When the spring force is written as $F = k\Delta = k(R - R_0)$, the virtual work done by the spring force is $\delta W = F\,\delta R = k(R - R_0)\,\delta R$.

This system provides an important analogy for the general task of modeling. Let the spheres represent two adjacent particles in a body. Correspondingly, the force $\overline{F}$

represents the stress resultant exerted between them. When a body is considered to be rigid, the internal stress resultants are equivalent to reactions that maintain the particles at fixed relative distances. These forces do no virtual work. In a deformable body model, the internal stress resultants are equivalent to spring forces. An analysis of such a system requires consideration of the internal stress distribution.

Connections between bodies are an important element in most dynamic systems. In the absence of friction, that is, if the connection is ideal, the reactions associated with the connections will do no virtual work. In Figure 6.17, two bars are connected by a pin. Let $B1$ and $B2$ denote the connection point on the respective bars. The corresponding virtual work is

$$\delta W = (B_x \bar{i} + B_y \bar{j}) \cdot \delta \bar{r}_{B1} + (-B_x \bar{i} - B_y \bar{j}) \cdot \delta \bar{r}_{B2}$$

The pin connection constrains the points to displace by the same amount. If that constraint is not violated, then $\delta \bar{r}_{B2} = \delta \bar{r}_{B1}$, which leads to $\delta W = 0$. Thus, if the system is described by a set of unconstrained generalized coordinates, there is no need to explicitly consider the pin reactions.

Another connection that is commonly encountered is a sliding collar, such as the one in Figure 6.18. Since the collar can only move inward or outward relative to bar CD, the virtual displacements of point $B1$ on bar CD, and of point $B2$ on the collar satisfy

$$\delta \bar{r}_{B2} = \delta \bar{r}_{B1} + \delta R \bar{e}_R$$

If the connection is ideal, the only force that is developed at the collar is the normal reaction $\overline{N}$. The virtual work is therefore

$$\delta W = (-N \bar{e}_\phi) \cdot \delta \bar{r}_{B1} + N \bar{e}_\phi \cdot \delta \bar{r}_{B2} = N \bar{e}_\phi \cdot (\delta R \bar{e}_R) = 0$$

The reaction $\overline{N}$ is the constraint force that enforces the kinematical requirement that the collar and bar CD execute the same movement in the direction normal to bar CD. This condition was satisfied by the virtual displacement, so the reaction $\overline{N}$ did no virtual work.

Suppose that friction is present. Then a friction force $\bar{f}$ parallel to bar CD would act in opposition to the sliding motion. The situation appearing in Figure 6.19 is based on the collar sliding outward, $\dot{R} > 0$. The virtual work in this case would be

$$\delta W = (-N \bar{e}_\phi + f \bar{e}_R) \cdot \delta \bar{r}_{B1} + (N \bar{e}_\phi - f \bar{e}_R) \cdot \delta \bar{r}_{B2} = -f \, \delta R$$

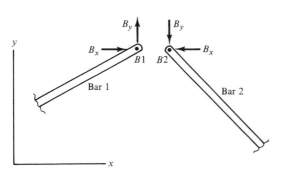

Figure 6.17   Reactions for a pin constraint.

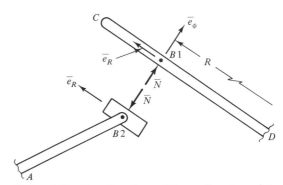

Figure 6.18   Reactions for a sliding-collar constraint.

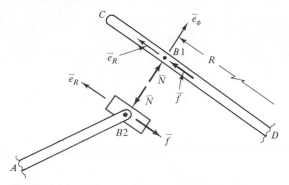

Figure 6.19 Effect of friction in a sliding collar.

Hence, $\bar{f}$ does work in a virtual displacement that is consistent with the constraint imposed by the collar. Note that friction does not represent a constraint force, since it does not prevent sliding motion. (The exception is static Coulomb friction. However, in that case the connection acts like a pin, because the collar does not move relative to the bar.)

Many other types of constraint forces could be considered at this juncture, for example, the reaction forces associated with rolling motion. The normal force is a constraint force that prevents interpenetration of the contacting surfaces. In the case of no slipping, the tangential force, developed by friction or gear teeth, makes the points of contact on the two bodies move by the same amount. Hence, it too is a constraint force. It follows that the forces exerted between rolling bodies will do no virtual work if there is no slippage. Conversely, the tangential force will do virtual work when there is slippage, because it is then not constraining the motion.

These examples emphasize the fact that reactions do no work in a compatible virtual displacement. They also illustrate that some internal forces exerted between parts of a system do virtual work (friction and elastic forces in the examples). Thus, before using the theorem regarding virtual work done by reactions, it is important to characterize the various forces. Free body diagrams are valuable aids for this task.

EXAMPLE 6.6

A force $\bar{F}$ having constant magnitude is applied to end C of the linkage such that it always is perpendicular to link BC. Consider the alternative choice for the generalized coordinate as either the angle ϕ or the distance x. Determine the corresponding generalized force.

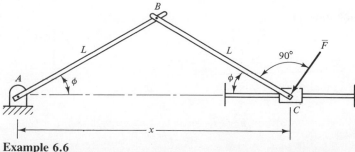

Example 6.6

Solution. The virtual work done by the external force at end C is $\overline{F} \cdot \delta \overline{r}_C$. Our selection of the generalized coordinate affects the description of both $\overline{F}$ and $\delta \overline{r}_C$, although the latter is directed along the x axis in any case. When ϕ is the generalized coordinate, the analytical method for virtual displacement gives

$$\overline{F} = -F \sin \phi \overline{i} - F \cos \phi \overline{j} \qquad \overline{r}_{C/A} = 2 L \cos \phi \overline{i}$$

$$\delta \overline{r}_C = \frac{d \overline{r}_{C/A}}{d\phi} \delta \phi = -(2 L \sin \phi) \delta \phi \overline{i}$$

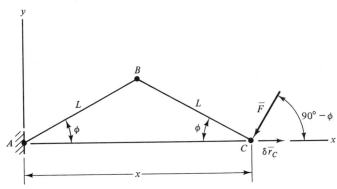

Kinematical parameters.

The corresponding virtual work is

$$\delta W = (-F \sin \phi \overline{i} - F \cos \phi \overline{j}) \cdot (-2 L \sin \phi \, \delta \phi \overline{i}) = 2 FL \sin^2 \phi \, \delta \phi$$

Matching this to the standard form $\delta W = Q_\phi \delta \phi$ then yields

$$Q_\phi = 2 FL \sin^2 \phi$$

When x is used as the generalized coordinate, the virtual displacement of point C is

$$\delta \overline{r}_C = \delta x \overline{i}$$

The angle ϕ is related to this generalized coordinate by

$$\cos \phi = \frac{x}{2 L} \qquad \sin \phi = \frac{1}{L} \left(L^2 - \frac{x^2}{4} \right)^{1/2}$$

so the force is

$$\overline{F} = -\frac{F}{L} \left[\left(L^2 - \frac{x^2}{4} \right)^{1/2} \overline{i} + \frac{x}{2} \overline{j} \right]$$

The virtual work in this case is

$$\delta W = \overline{F} \cdot \delta \overline{r}_C = -\frac{F}{L} \left(L^2 - \frac{x^2}{4} \right)^{1/2} \delta x = Q_x \delta x$$

Matching the two expressions for δW yields

$$Q_x = -\frac{F}{L}\left(L^2 - \frac{x^2}{4}\right)^{1/2}$$

EXAMPLE 6.7

A horizontal force $F(t)$ is applied to the end of the compound pendulum whose pivot is given a specified horizontal displacement $x(t)$. Generalized coordinates are the absolute angle of rotation θ_1 for the upper bar and the relative angle θ_2 for the lower arm. Determine the corresponding generalized forces.

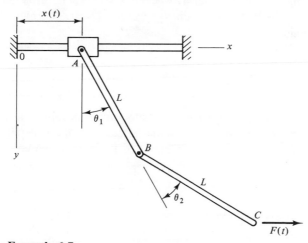

Example 6.7

Solution. The virtual work is

$$\delta W = \bar{F} \cdot \delta\bar{r}_C$$

We shall employ the analytical method to determine the virtual displacement. The position of point C with respect to the fixed support is

$$\bar{r}_{C/O} = [x + L \sin \theta_1 + L \sin (\theta_1 + \theta_2)]\bar{i} + [L \cos \theta_1 + L \cos (\theta_1 + \theta_2)]\bar{j}$$

It is important to the evaluation of the virtual displacement that time is held constant. Hence, in this evaluation collar A is held stationary, $\delta x = 0$. The chain rule for differentiation then yields the displacement resulting from virtual increments in θ_1 and θ_2, according to

$$\delta\bar{r}_C = \frac{\partial \bar{r}_{C/O}}{\partial \theta_1} \delta\theta_1 + \frac{\partial \bar{r}_{C/O}}{\partial \theta_2} \delta\theta_2$$

$$= L\{[\cos \theta_1 + \cos (\theta_1 + \theta_2)]\bar{i} - [\sin \theta_1 + \sin (\theta_1 + \theta_2)]\bar{j}\} \delta\theta_1$$
$$+ L[\cos (\theta_1 + \theta_2)\bar{i} - \sin (\theta_1 + \theta_2)\bar{j}] \delta\theta_2$$

Because the force $\overline{F}$ is horizontal, $\overline{F} = F\overline{i}$, only the horizontal component of $\delta\overline{r}_C$ affects the virtual work,

$$\delta W = FL[\cos\theta_1 + \cos(\theta_1 + \theta_2)]\,\delta\theta_1 + FL\cos(\theta_1 + \theta_2)\,\delta\theta_2$$

We may identify the generalized forces by matching the coefficients of $\delta\theta_1$ and $\delta\theta_2$ above to the definition,

$$\delta W = Q_1\,\delta\theta_1 + Q_2\,\delta\theta_2$$

This yields

$$Q_1 = FL[\cos\theta_1 + \cos(\theta_1 + \theta_2)] \qquad Q_2 = FL\cos(\theta_1 + \theta_2)$$

These expressions have a simple interpretation. $L\cos\theta_1 + L\cos(\theta_1 + \theta_2)$ is the lever arm of $\overline{F}$ about collar A, whereas $L\cos(\theta_1 + \theta_2)$ is the lever arm of the force about pin B. Correspondingly, Q_1 and Q_2 are the moments about the respective points.

EXAMPLE 6.8 ——

A cable is tied to pin B on pinion gear A. A tensile force $\overline{F}$ is applied to the free end of the cable such that the cable remains horizontal. Determine the generalized force corresponding to the choice of the rotation of gear A as the generalized coordinate.

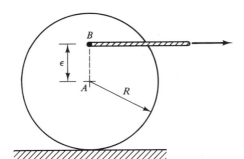

Example 6.8

Solution. An important aspect of the evaluation of any generalized forces is that the analysis should be performed when a system is at an arbitrary position. Thus, we draw a sketch with the radial line to the pin rotated by an angle θ away from the vertical.

The kinematical method is appropriate to the description of rolling motion. The analogy with a velocity analysis, combined with the observation that contact point C does not slip when its constraint condition is satisfied, leads to

$$\delta\overline{r}_C = \overline{0} \qquad \delta\overline{r}_A = \delta x_A\overline{i} = \delta\theta(-\overline{k}) \times \overline{r}_{A/C}$$
$$\delta\overline{r}_B = \delta\overline{r}_A + \delta\theta(-\overline{k}) \times \overline{r}_{B/A} \qquad \overline{r}_{B/A} = \varepsilon(\sin\theta\,\overline{i} + \cos\theta\,\overline{j})$$

Substitution for the position vectors yields

$$\delta x_A = R\,\delta\theta \qquad \delta\overline{r}_B = [(R + \varepsilon\cos\theta)\overline{i} - \varepsilon\sin\theta\,\overline{j}]\,\delta\theta$$

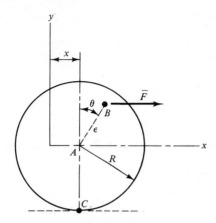

Kinematical parameters.

The cable is maintained at a horizontal orientation, so the corresponding virtual work is

$$\delta W = F\bar{i} \cdot \delta\bar{r}_B = F(R + \varepsilon \cos \theta)\, \delta\theta$$

Matching this expression to the standard form $\delta W = Q_1\, \delta\theta$ leads to

$$Q_1 = F(R + \varepsilon \cos \theta)$$

This term is recognizable as the moment of $\bar{F}$ about contact point C.

6.4.2 Conservative Forces

The ability to avoid the virtual work done by constraint forces in a compatible virtual displacement substantially simplifies the evaluation of the generalized forces. Another saving stems from the potential energy function associated with a conservative force. By definition, a conservative force does no work when the point at which it is applied follows a closed path, such as curve C_1 from point 1 to point 2 and then curve C_2 back to point 1 in Figure 6.20. The consequence of this property is that the work done by a conservative force is expressible as the difference of the potential energy at two positions

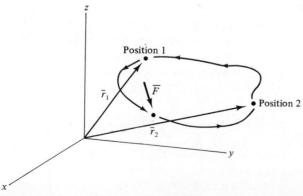

Figure 6.20 Work along a closed path.

$$W_{1 \to 2} = V_1 - V_2 \tag{6.33}$$

The potential energy V depends only on the position of the system, so it is an explicit function of the generalized coordinates. In a system whose constraints are time-dependent, the position is also an explicit function of time, so that $V = V(q_1, q_2, \ldots, q_m, t)$.

A virtual movement is obtained by incrementing each of the generalized coordinates. The corresponding virtual work is found from Eq. (6.33) to be

$$\delta W = V(q_1, q_2, \ldots, q_M, t) - V(q_1 + \delta q_1, q_2 + \delta q_2, \ldots, t)$$

$$\delta W = -\delta V = -\sum_{j=1}^{M} \frac{\partial V}{\partial q_j} \delta q_j \tag{6.34}$$

The definition of the generalized forces is that

$$\delta W = \sum_{j=1}^{M} Q_j \, \delta q_j$$

Both expressions must yield the same value for δW regardless of the increments given to each generalized coordinate. This condition can only be fulfilled if the corresponding coefficients of each δq_j match, so that

$$Q_j^{(\text{conservative})} = -\frac{\partial V}{\partial q_j} \tag{6.35}$$

Of course, not all forces are conservative. The virtual work in a general situation may be apportioned between conservative and nonconservative effects. The corresponding expression for the generalized forces may be written as

$$Q_j^{(\text{total})} = -\frac{\partial V}{\partial q_j} + Q_j \tag{6.36}$$

Here, and in all future developments, the symbol Q_j without a subscript denotes generalized forces associated with forces that are not described by the potential energy. This provides a degree of flexibility. It is not necessary to formulate the potential energy of a conservative force. If the nature of a force is uncertain, or if it is straightforward to evaluate the virtual work of a force that is known to be conservative, then that force may be considered to be nonconservative. Henceforth, the generalized forces Q_j will be considered to include all forces, conservative and nonconservative, whose effect is derived from an analysis of the virtual work. Obviously, it would not be correct to include the conservative force both in the potential energy and in the generalized force Q_j.

EXAMPLE 6.9 _____

The parallelogram linkage, which is situated in the vertical plane, is steadied by identical springs that are fastened across the diagonals. The stiffness of the springs is k and the springs are unstretched when the angle of elevation $\beta = 90°$. The mass per unit length of each bar is σ. Determine the potential energy of this system as a function of β.

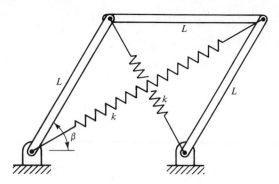

Example 6.9

Solution. The free body diagram shows the conservative forces exerted by the springs and gravity. We designate the elongation of the springs (from their unstretched length) as Δ_1 and Δ_2. The mass of each bar is σL, and the individual weight forces act at the center of mass of the respective bars. The elevation of the fixed pins A and D provides a useful datum for the potential energy of gravity. The total potential energy is the sum of the individual effects, so

$$V = \tfrac{1}{2}k\Delta_1^2 + \tfrac{1}{2}k\Delta_2^2 + 2(\sigma Lg)\left(\frac{L}{2}\sin\beta\right) + (\sigma Lg)(L\sin\beta)$$

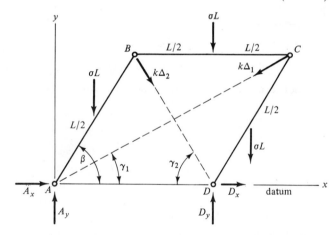

Kinematical parameters.

It still remains to express the elongations in terms of β. For this, we note that because $ABCD$ is an equilateral parallelogram, the diagonals intersect perpendicularly and bisect the interior angles. Thus, $\gamma_1 = \beta/2$, $\gamma_2 = 90° - \beta/2$. We form the elongation at each spring by subtracting the unstretched length L_0 from the respective diagonal length, with the result that

$$\Delta_1 = 2L\sin\gamma_2 - L_0 = 2L\cos\left(\frac{\beta}{2}\right) - L_0$$

$$\Delta_2 = 2L\sin\gamma_1 - L_0 = 2L\sin\left(\frac{\beta}{2}\right) - L_0$$

The value of L_0 was not given explicitly. Instead we were told that $\Delta_1 = \Delta_2 = 0$

when $\beta = 90°$, from which we find that

$$L_0 = 2 L \sin 45° = \sqrt{2}\, L$$

When we substitute the expressions for Δ_1 and Δ_2 into V, we obtain

$$V = \tfrac{1}{2} k\, L^2 \left[2 \cos \left(\frac{\beta}{2} \right) - \sqrt{2} \right]^2 + \tfrac{1}{2} kL^2 \left[2 \sin \left(\frac{\beta}{2} \right) - \sqrt{2} \right]^2 + 2\sigma\, L^2 g \sin \beta$$

$$= \tfrac{1}{2} kL^2 \left[8 - 4\sqrt{2} \left(\cos \frac{\beta}{2} + \sin \frac{\beta}{2} \right) \right] + 2\sigma L^2 g \sin \beta$$

6.5 HAMILTON'S PRINCIPLE

We will soon see that the equations of motion for a system of rigid bodies may be obtained by evaluating a standard set of relations known as Lagrange's equations. It is possible to derive the latter directly from Newton's laws. Instead, we shall first obtain Hamilton's principle. This principle has much wider applicability, in that it is also valid for deformable continuous media. Also, Hamilton's principle may be extended to relativistic systems, provided that the mechanical energy is suitably redefined.

The derivation begins by considering a single particle. Newton's second law was reformulated by d'Alembert[†] as

$$\overline{F} - m\overline{a} = \overline{0} \tag{6.37}$$

The essence of this transposition is that it converts a dynamic system into a static one, with $-m\overline{a}$ considered to represent an inertial force. The virtual work done by this force system in a virtual displacement of the particle is

$$(\overline{F} - m\overline{a}) \cdot \delta\overline{r} = \delta W - (m\overline{a} \cdot \delta\overline{r}) = 0 \tag{6.38}$$

where δW is the virtual work done by the force acting on the particle.

The inertial force term may be expressed in a more convenient form by recalling that δ is a differential operator. It represents the differential increase due to a virtual increment of the generalized coordinates in which time is held constant. In contrast, a time derivative evaluates the change that occurs in an actual time increment. Let us consider the effect of applying both operations to $\overline{r}$. We have

$$\frac{d}{dt} (\delta\overline{r}) = \frac{d}{dt} \sum_{i=1}^{M} \frac{\partial \overline{r}}{\partial q_i} \delta q_i$$

$$= \sum_{j=1}^{M} \frac{\partial}{\partial q_j} \left(\sum_{i=1}^{M} \frac{\partial \overline{r}}{\partial q_i} \delta q_i \right) \dot{q}_j + \frac{\partial}{\partial t} \left(\sum_{i=1}^{M} \frac{\partial \overline{r}}{\partial q_i} \delta q_i \right)$$

$$= \sum_{i=1}^{M} \frac{\partial}{\partial q_i} \left(\sum_{j=1}^{M} \frac{\partial \overline{r}}{\partial q_j} \dot{q}_j + \frac{\partial \overline{r}}{\partial t} \right) \delta q_i = \delta \left(\frac{d\overline{r}}{dt} \right) = \delta\overline{v}$$

†As explained by Rosenberg, this actually is an oversimplification of d'Alembert's principle. Essentially, d'Alembert categorized forces as to whether they are the given forces inducing the motion or constraint forces. He grouped only the given forces with the $-m\overline{a}$ terms, and then employed the principle of virtual work for static systems. Since constraint forces do no virtual work, this procedure enabled him to formulate equations of motion in which only the given forces and inertial effects appear.

In other words, it does not matter whether the virtual increment or the time differentiation is done first. With the aid of the foregoing, Eq. (6.38) may be rewritten as

$$\delta W - m \frac{d\bar{v}}{dt} \cdot \delta \bar{r} = \delta W - m \frac{d}{dt} (\bar{v} \cdot \delta \bar{r}) + m\bar{v} \cdot \delta \bar{v}$$

$$= \delta W - m \frac{d}{dt} (\bar{v} \cdot \delta \bar{r}) + \delta T = 0 \tag{6.39}$$

where

$$T = \tfrac{1}{2} m\bar{v} \cdot \bar{v} \tag{6.40}$$

is the kinetic energy of the particle.

We may treat a system of particles by modifying Eq. (6.39). Let i denote the particle number in the system. Then, addition of Eq. (6.39) for each particle leads to redefinition of T as the kinetic energy of all particles in the system, and δW as the virtual work done by all forces. The latter may be decomposed into the work done by conservative and nonconservative forces. The virtual work done by the conservative forces is the negative of the virtual change in the potential energy, according to Eq. (6.34). Henceforth, δW will denote the portion of the virtual work done by nonconservative forces. Addition of Eq. (6.39) for each particle in the system therefore yields

$$\delta T - \delta V + \delta W - \sum_i m_i \frac{d}{dt} (\bar{v}_i \cdot \delta \bar{r}_i) = 0 \tag{6.41}$$

The last step is to integrate over the time interval $t_1 \leq t \leq t_2$, where t_1 and t_2 are arbitrary. The result is

$$\int_{t_1}^{t_2} (\delta T - \delta V + \delta W) \, dt - \sum_i m_i \bar{v}_i \cdot \delta \bar{r}_i \Big|_{t=t_1}^{t=t_2} = 0 \tag{6.42}$$

This relation is best understood by considering the motion of the system in configuration space. In Figure 6.21, points P_1 and P_2 represent the values of the generalized coordinates at the initial and final instants, t_1 and t_2, respectively. The solid curve C represents the

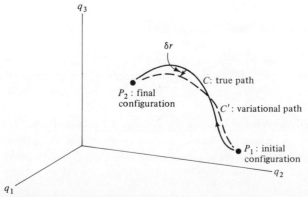

Figure 6.21 Variational path.

actual evolution of the generalized coordinates as time elapses. (Since the equations of motion have not yet been derived, curve C is not yet known.) Curve C' represents a kinematically admissible motion that would be obtained if the set of external loads were infinitesimally different from their given values. The terminology is to say that curve C' is the *variational path*, because it is obtained from an infinitesimal variation of the generalized coordinates away from their values along the true path C.

The virtual displacement imparted to each particle at each instant is arbitrary, with two exceptions. Any change in the initial condition described by point P_1 would produce a different state of motion. In other words, if the initial conditions are specified, then $\delta \bar{r}_i = \bar{0}$ for each particle at time t_1. Also, the virtual displacement must be such that the alternative curve C' leads to the actual final position represented by point P_2. In order to arrive at the true final position P_2, the virtual displacement must be such that $\delta \bar{r}_i = \bar{0}$ when $t = t_2$.

Because $\delta \bar{r}_i = \bar{0}$ for each particle at the initial and final positions, Eq. (6.42) reduces to

$$\int_{t_1}^{t_2} (\delta T - \delta V + \delta W) \, dt = 0 \qquad (6.43)$$

This is *Hamilton's principle*, after Sir W. R. Hamilton. It states that among all possible variational paths, the true path is the one along which the virtual increment $\delta T - \delta V + \delta W$ has a zero mean value. Note that δT and δV represent virtual increments in the values of the corresponding quantities at an arbitrary position, whereas there is no "work" quantity W from which the virtual work of a nonconservative force may be derived. Suppose we consider the restricted group of systems for which $\delta W = 0$. Clearly, this can only be the case if a system is conservative, but we must impose the further restriction that the system is holonomic. Otherwise, the constraint forces will do work when the generalized coordinates are given arbitrary virtual increments. Thus, in the case of a conservative holonomic system, we obtain a more enlightening view of Hamilton's principle. Specifically, among all variational paths connecting the initial and final position, the true one is the one for which the *action integral*

$$I = \int_{t_1}^{t_2} (T - V) \, dt \qquad (6.44)$$

is *stationary*, which means that it has an extreme value (maximum, minimum, or inflection point).

It is logical at this juncture to question the significance of these results, since Hamilton's principle seems to represent only one relation. For example, the work-energy principle $\Delta T + \Delta V = W_{1 \to 2}^{(nc)}$ is not adequate by itself to solve problems involving several generalized coordinates. The difference is that Eq. (6.43) leads to several relations, because the virtual movement is arbitrary except at the initial and final instants. An infinite number of variational curves C' can be constructed, and Hamilton's principle must be satisfied for each. For example, we can construct different variational paths by holding all generalized coordinates except one constant, and imparting any virtual change to the remaining generalized coordinate.

In the next section, we will derive Lagrange's equations for the generalized co-ordinates in a system having a finite number of degrees of freedom. Systems containing deformable bodies may be conceptualized as having an infinite number of degrees of freedom. The equations of motion for such systems, which are partial differential equations in space and time for the displacements, may be obtained by applying the calculus of variations in conjunction with Hamilton's principle. Also, for static systems, the kinetic energy vanishes and the system is independent of time. Hamilton's principle then reduces to

$$\delta V = \delta W \tag{6.45}$$

which is the *principle of virtual work and stationary potential energy*. It is particularly useful for the analysis of statically indeterminate structures and machines.

6.6 LAGRANGE'S EQUATIONS

Hamilton's principle may be specialized to a system having a discrete number of degrees of freedom by accounting for the functional form of the energies and virtual work. The dependence of the kinetic energy may be recognized by considering an arbitrary particle, numbered k. The position $\bar{r}_k$ in a general situation is a function of the generalized co-ordinates and time, $\bar{r}_k = \bar{r}_k(q_1, q_2, \ldots, q_M, t)$. The corresponding velocity expression is

$$\bar{v}_k = \frac{d\bar{r}_k}{dt} = \sum_{j=1}^{M} \frac{\partial \bar{r}_k}{\partial q_j} \dot{q}_j + \frac{\partial \bar{r}_k}{\partial t} \tag{6.46}$$

In later developments, it will be important to identify whether the time-derivative term, representing the effect of the explicit time dependence of the kinematical constraints imposed on time, actually occurs. The requirement that the system be time-independent is a sufficient condition for the absence of this term. The necessary condition merely requires that all kinematical constraints imposed on the motion be catastatic, which is the definition of a *catastatic system*.

The kinetic energy of this particle may now be expressed in terms of the generalized coordinates as

$$
\begin{aligned}
T &= \tfrac{1}{2} m_k \sum_{j=1}^{M} \left(\frac{\partial \bar{r}_k}{\partial q_j} \dot{q}_j + \frac{\partial \bar{r}_k}{\partial t} \right) \cdot \sum_{j=1}^{M} \left(\frac{\partial \bar{r}_k}{\partial q_j} \dot{q}_j + \frac{\partial \bar{r}_k}{\partial t} \right) \\
&= \tfrac{1}{2} m_k \sum_{i=1}^{M} \sum_{j=1}^{M} \left(\frac{\partial \bar{r}_k}{\partial q_i} \cdot \frac{\partial \bar{r}_k}{\partial q_j} \right) \dot{q}_i \dot{q}_j + m_k \sum_{j=1}^{M} \left(\frac{\partial \bar{r}_k}{\partial q_j} \cdot \frac{\partial \bar{r}_k}{\partial t} \right) \dot{q}_j + \tfrac{1}{2} m_k \left(\frac{\partial \bar{r}_k}{\partial t} \cdot \frac{\partial \bar{r}_k}{\partial t} \right)
\end{aligned}
\tag{6.47}
$$

The total kinetic energy of the system is the sum of that for each particle, although such a sum for rigid bodies involves an integration over the differential mass elements. In any case, the summation over the particles removes the dependence on the particle number k.

We may conclude that the kinetic energy of a system consists of three groups of terms: T_2 is quadratic in the generalized velocities, T_1 is linear in the generalized velocities, and T_0 is independent of the generalized velocities. The general form is

$$T = T_2 + T_1 + T_0 \tag{6.48a}$$

where

$$T_2 = \frac{1}{2} \sum_{i=1}^{M} \sum_{j=1}^{M} M_{ij}\, \dot{q}_i \dot{q}_j \qquad T_1 = \sum_{j=1}^{M} N_j \dot{q}_j \tag{6.48b}$$

A useful property is that the coefficient of the quadratic terms, M_{ij}, obeys the symmetry property,

$$M_{ij} = M_{ji} \qquad i, j = 1, 2, \ldots, M \tag{6.49}$$

because the sequence in which the product $\dot{q}_i \dot{q}_j$ is formed is unimportant. Note that the coefficients M_{ij} and N_j, as well as T_0, might depend on the generalized coordinates and time, because that is the dependence of the partial derivatives appearing within each bracket in Eq. (6.47). However, in the special case of a catastatic system, the velocity in Eq. (6.46) does not contain the time-derivative term. In that case T_0 and all N_j are identically zero, and the M_{ij} depend only on the generalized coordinates.

We will employ the specific quadratic form, Eqs. (6.48), in our later studies. The above observations regarding the properties of the coefficients will be needed there. It is adequate for our current purposes to note that the kinetic energy may be an explicit function of the generalized coordinates q_j, generalized velocities $\dot{q}_j$, and time t. Hence, the virtual change in the kinetic energy resulting from virtual changes in the q_j and $\dot{q}_j$ is given by

$$\delta T = \sum_{j=1}^{M} \left(\frac{\partial T}{\partial q_j} \delta q_j + \frac{\partial T}{\partial \dot{q}_j} \delta \dot{q}_j \right) \tag{6.50}$$

We seek to express δT in terms of virtual increments in the generalized coordinates only. We obtain such a form by using the interchangeability of a virtual increment and a time derivative, so that $\delta \dot{q}_i = d(\delta q_i)/dt$. Then manipulating the derivative of a product leads to

$$
\begin{aligned}
\delta T &= \sum_{j=1}^{M} \left(\frac{\partial T}{\partial q_j} \delta q_j + \frac{\partial T}{\partial \dot{q}_j} \frac{d}{dt} \delta q_j \right) \\
&= \sum_{j=1}^{M} \left[\frac{\partial T}{\partial q_j} - \frac{d}{dt} \left(\frac{\partial T}{\partial \dot{q}_j} \right) \right] \delta q_j + \sum_{j=1}^{M} \frac{d}{dt} \left(\frac{\partial T}{\partial \dot{q}_j} \delta q_j \right)
\end{aligned}
\tag{6.51}
$$

The virtual change in the potential energy and the virtual work done by nonconservative forces were related previously to the generalized coordinates, specifically,

$$\delta V = \sum_{j=1}^{M} \frac{\partial V}{\partial q_j} \delta q_j \qquad \delta W = \sum_{j=1}^{M} Q_j\, \delta q_j$$

When these expressions are substituted into Hamilton's principle, the coefficients of δq_j may be collected, with the result that

$$\int_{t_1}^{t_2} \sum_{j=1}^{M} \left[\frac{\partial T}{\partial q_j} - \frac{d}{dt} \left(\frac{\partial T}{\partial \dot{q}_j} \right) - \frac{\partial V}{\partial q_j} + Q_j \right] \delta q_j\, dt + \int_{t_1}^{t_2} \sum_{j=1}^{M} \frac{d}{dt} \left(\frac{\partial T}{\partial \dot{q}_j} \delta q_j \right) dt = 0 \tag{6.52}$$

The integrand in the second term is a perfect differential; its integration yields a sum of terms $(\partial T/\partial q_j)\ \delta q_j$ evaluated at $t = t_1$ and $t = t_2$. According to the derivation of Hamilton's principle, the variational path must be such that the virtual displacement at the initial and final instants is zero, so $\delta q_i = 0$ at $t = t_1$ and $t = t_2$. Consequently, the second integral vanishes, which reduces Eq. (6.52) to

$$\int_{t_1}^{t_2} \sum_{j=1}^{M} \left[\frac{\partial T}{\partial q_j} - \frac{d}{dt}\left(\frac{\partial T}{\partial \dot{q}_j}\right) - \frac{\partial V}{\partial q_j} + Q_j \right] \delta q_j\ dt = 0 \qquad (6.53)$$

Recall that the virtual increments assigned to each generalized coordinate in an unconstrained set are arbitrary. Even if the set is constrained, the increments assigned to any unconstrained subset is arbitrary. In terms of the configuration space in Figure 6.21, the δq_j have different time dependencies along each of the variational paths neighboring the true path. The only way Hamilton's principle can be satisfied under these conditions is if the bracketed term in Eq. (6.53) is zero for each generalized coordinate. This term forms *Lagrange's equations of motion*.

$$\frac{d}{dt}\left(\frac{\partial T}{\partial \dot{q}_j}\right) - \frac{\partial T}{\partial q_j} + \frac{\partial v}{\partial q_j} = Q_j \qquad j = 1, 2, \ldots, M \qquad (6.54)$$

Note that Lagrange's equations yield M equations (usually differential type) for the M generalized coordinates, since the partial derivatives must be evaluated for each generalized coordinate. Suppose that a system is holonomic, so that a set of unconstrained generalized coordinates can be selected. In that case, the number of generalized coordinates matches the number of degrees of freedom, $M = N$. Also, a virtual movement of the system satisfies all kinematical constraint conditions, so none of the (unknown) reactions appear in the generalized forces. In this situation, Lagrange's equations fully define the motion of the system.

In contrast, Lagrange's equations must be supplemented by other relations when the generalized coordinates form a constrained set. Constrained generalized coordinates must be employed when a system is nonholonomic. They may also be desirable for holonomic systems. Consideration of these matters will be deferred until the next chapter.

An alternative form of Lagrange's equations, preferred by some practitioners, features the *Lagrangian function L*, which is defined to be

$$L = T - V \qquad (6.55)$$

Since the potential energy depends explicitly only on the current position, it is independent of the generalized velocities. Thus

$$\frac{\partial L}{\partial \dot{q}_j} = \frac{\partial T}{\partial \dot{q}_j} \qquad \frac{\partial L}{\partial q_j} = \frac{\partial T}{\partial q_j} - \frac{\partial V}{\partial q_j}$$

from which Lagrange's equations (6.54) convert to

$$\frac{d}{dt}\left(\frac{\partial L}{\partial \dot{q}_j}\right) - \frac{\partial L}{\partial q_j} = Q_j \qquad j = 1, 2, \ldots, M \qquad (6.56)$$

Although Eq. (6.56) has one fewer term than Eq. (6.54), both forms require equivalent

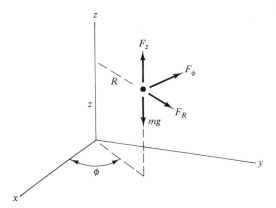

Figure 6.22 Force system in cylindrical coordinates.

mathematical evaluations. The primary reason for introducing the Lagrangian function is its utility for further development of principles.

The actual evaluation of either form of Lagrange's equations for a specific system is straightforward, provided that one is cognizant of the difference between partial and total derivatives. For the partial derivatives the generalized coordinates q_j and generalized velocities $\dot{q}_j$ are treated like independent variables. In the derivative with respect to time, all quantities that are time-dependent must be differentiated.

A simple example that demonstrates the equivalence of Lagrange's equations and Newton's second law is a particle in spatial motion under the influence of gravity and a force $\overline{F}$ that is described in terms of its cylindrical components (F_R, F_ϕ, F_z). Let the cylindrical coordinates (R, ϕ, z) in Figure 6.22 be the generalized coordinates. Since $\overline{v} = \dot{R}\overline{e}_R + R\dot{\phi}\overline{e}_\phi + \dot{z}\overline{k}$, the kinetic energy is

$$T = \tfrac{1}{2}m(\dot{R}^2 + R^2\dot{\phi}^2 + \dot{z}^2)$$

The partial derivatives are

$$\frac{\partial T}{\partial \dot{R}} = m\dot{R} \qquad \frac{\partial T}{\partial \dot{\phi}} = mR^2\dot{\phi} \qquad \frac{\partial T}{\partial \dot{z}} = m\dot{z}$$

$$\frac{\partial T}{\partial R} = mR\dot{\phi}^2 \qquad \frac{\partial T}{\partial \phi} = 0 \qquad \frac{\partial T}{\partial z} = 0$$

The time derivatives of the first group of terms are

$$\frac{d}{dt}\left(\frac{\partial T}{\partial \dot{R}}\right) = m\ddot{R} \qquad \frac{d}{dt}\left(\frac{\partial T}{\partial \dot{\phi}}\right) = 2\,mR\dot{R}\dot{\phi} + mR^2\ddot{\phi} \qquad \frac{d}{dt}\left(\frac{\partial T}{\partial \dot{z}}\right) = m\ddot{z}$$

Let the datum of the conservative gravitational force be the xy plane, so

$$V = mgz$$

The virtual work done by the force $\overline{F}$ is

$$\delta W = (F_R\overline{e}_R + F_\phi\overline{e}_\phi + F_z\overline{k}) \cdot (\delta R\,\overline{e}_R + R\,\delta\phi\,\overline{e}_\phi + \delta z\,\overline{k})$$
$$= F_R\,\delta R + RF_\phi\,\delta\phi + F_z\,\delta z$$

The generalized forces are the coefficients of the virtual increments of the generalized coordinates in δW. Thus,

$$Q_1 = F_R \qquad Q_2 = RF_\phi \qquad Q_3 = F_z$$

The corresponding set of Lagrange's equations is

1: $$m\ddot{R} - mR\dot{\phi}^2 = F_R$$
2: $$2mR\dot{R}\dot{\phi} + mR^2\ddot{\phi} = RF_\phi$$
3: $$m\ddot{z} + mg = F_z$$

Each of these is merely Newton's second law in terms of polar coordinates, with the exception of equation 2 for ϕ, which has an additional factor R. That form of the equation results from the fact that $Q_2 = RF_R$ represents the moment of the external force system about the z axis. Correspondingly, the left side of the second of Lagrange's equations is the derivative of the angular momentum, $mR^2\dot{\phi}$, about that axis.

The steps we followed in this simple example parallel those for all systems for which unconstrained generalized coordinates have been selected. The bulk of the analysis usually lies in the kinematical analysis of the virtual displacements and kinetic energy. Then, after the potential energy and generalized forces have been determined, Lagrange's equations directly yield the equations of motion.

EXAMPLE 6.10 _____

Determine the equations of motion for the homogeneous sphere of radius r that rolls without slipping along the interior of the semicylinder. The sphere is constrained to remain in the vertical plane shown.

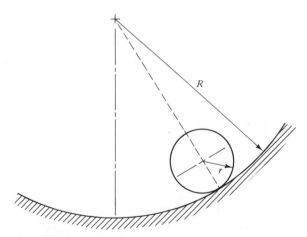

Example 6.10

Solution. A useful generalized coordinate for this system is the angle θ to the center of the sphere. It is specified that the sphere rolls without slipping. This constrains the absolute angle through which the sphere rotates. We shall impart to the sphere a virtual movement that satisfies this constraint. The normal force $\overline{N}$ and friction force $\overline{f}$ impose the constraint. Hence, they do no virtual work and will

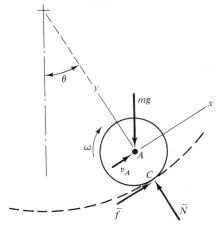

Free body diagram.

not appear in the formulation. The only other force acting on the sphere is gravity. We place the datum for the potential energy of gravity at the center of the semi-cylinder, because that is a fixed point of geometrical significance. Hence, we have

$$\delta W = 0 \qquad V = mg[-(R - r) \cos \theta]$$

where the negative sign in V arises because the center of mass is below the datum.

The kinetic energy is

$$T = \tfrac{1}{2}mv_A^2 + \tfrac{1}{2}I_{zz}\omega^2$$

where $I_{zz} = \tfrac{2}{5}mr^2$ for a sphere. We must express v_A in terms of the generalized coordinate. Because the center of the wheel follows a circular path of radius $R - r$, we write $v_A = (R - r)\dot{\theta}$. We obtain a comparable expression for ω by noting that the point of contact is the instant center. Hence, $\omega = v_A/r = (R - r)\dot{\theta}/r$. The resulting kinetic energy expression is

$$T = \frac{1}{2}\left[m(R - r)\dot{\theta}^2 + \left(\frac{2}{5}mr^2\right)\left(\frac{R - r}{r}\dot{\theta}\right)^2 \right] = \tfrac{7}{10}m(R - r)^2\dot{\theta}^2$$

Since the kinetic energy is independent of θ, and the generalized force vanishes because $\delta W = 0$, Lagrange's equation reduces to

$$\frac{d}{dt}\left(\frac{\partial T}{\partial \dot{\theta}}\right) + \frac{\partial V}{\partial \theta} = 0$$

$$\tfrac{7}{5}m(R - r)\ddot{\theta} + mg(R - r) \sin \theta = 0$$

$$\tfrac{7}{5}\ddot{\theta} + \frac{g}{R - r} \sin \theta = 0$$

The last form is recognizable as being comparable with the equation of motion for a simple pendulum.

EXAMPLE 6.11

The table rotates in a horizontal plane about bearing A due to a torque $\Gamma(t)$. The mass of the table is M and its radius of gyration about its center is κ. The slider, whose mass is m, moves within groove BC under the restraint of a pair of springs which are unstretched in the position shown. Derive the equations of motion for this system.

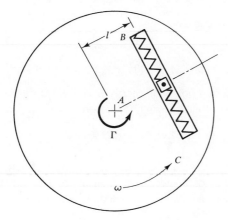

Example 6.11

Solution. Our selection of generalized coordinates for this system originates from the observation that the table is in pure rotation, while the slider executes a rectilinear motion relative to the table. Correspondingly, we select the angle of rotation, $q_1 = \theta$, and the displacement of the slider relative to the unstretched position of the springs, $q_2 = s$. The only nonconservative force that does work when the generalized coordinates are given virtual increments is the torque load $\Gamma(t)$. Hence,

$$\delta W = \Gamma\,\delta\theta \Rightarrow Q_1 = \Gamma(t) \qquad Q_2 = 0$$

The kinetic energy is the sum of the values for the table and for the slider.

$$T = \tfrac{1}{2}(I_{zz})_{\text{table}}\,\dot\theta^2 + \tfrac{1}{2}mv_B^2$$

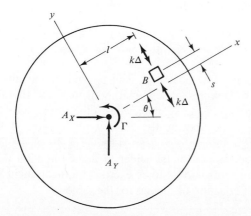

Free body diagram.

where $(I_{zz})_{\text{table}} = M\kappa^2$. We shall relate $\bar{v}_B$ to the generalized coordinates by using the relative-motion equation based on a moving reference frame xyz attached to the table. Then

$$\bar{v}_B = \bar{v}_A + (\bar{v}_B)_{\text{rel}} + \bar{\omega} \times \bar{r}_{B/A} = \bar{0} + \dot{s}\bar{j} + \dot{\theta}\bar{k} \times (l\bar{i} + s\bar{j})$$
$$= -s\dot{\theta}\bar{i} + (\dot{s} + l\dot{\theta})\bar{j}$$

The kinetic energy therefore becomes

$$T = \tfrac{1}{2}M\kappa^2\dot{\theta}^2 + \tfrac{1}{2}m[(s\dot{\theta})^2 + (\dot{s} + l\dot{\theta})^2]$$
$$= \tfrac{1}{2}(M\kappa^2 + ms^2 + ml^2)\dot{\theta}^2 + \tfrac{1}{2}m\dot{s}^2 + ml\dot{s}\dot{\theta}$$

It is a simple matter to derive the potential energy because the displacement s is, by definition, the deformation of each spring. Therefore,

$$V = 2(\tfrac{1}{2}k\Delta^2) = ks^2$$

We must evaluate Lagrange's equations corresponding to θ and s. We shall evaluate the various derivatives of T individually, in order to emphasize the difference between the partial derivatives and total derivatives. Thus

$$\frac{d}{dt}\left(\frac{\partial T}{\partial \dot{\theta}}\right) = \frac{d}{dt}[(M\kappa^2 + ms^2 + ml^2)\dot{\theta} + ml\dot{s}]$$

$$= (M\kappa^2 + ms^2 + ml^2)\ddot{\theta} + 2ms\dot{s}\dot{\theta} + ml\ddot{s}$$

$$\frac{\partial T}{\partial \theta} = \frac{\partial V}{\partial \theta} = 0$$

$$\frac{d}{dt}\left(\frac{\partial T}{\partial \dot{s}}\right) = \frac{d}{dt}(m\dot{s} + ml\dot{\theta}) = m\ddot{s} + ml\ddot{\theta}$$

$$\frac{\partial T}{\partial s} = ms\dot{\theta}^2 \qquad \frac{\partial V}{\partial s} = 2ks$$

The corresponding Lagrange's equations are

$$(M\kappa^2 + ml^2 + ms^2)\ddot{\theta} + 2ms\dot{s}\dot{\theta} + ml\ddot{s} = \Gamma$$
$$m\ddot{s} + ml\ddot{\theta} - ms\dot{\theta}^2 + 2ks = 0$$

Some of the terms in these equations may be understood intuitively. For example, $M\kappa^2 + m(l^2 + s^2)$ represents the total moment of inertia of the system about pivot A. Understanding of the significance of other terms may be obtained by resorting to Newton's laws. Specifically, when we isolate the slider from the table, we see that there is a normal force exerted by the walls of the groove. This force is related to the acceleration of the slider in the x direction through $\bar{F} = m\bar{a}$. In turn, the reaction to this normal force exerts a moment about point A that must be considered in the equation of motion for the table.

These observations demonstrate that although Lagrange's equations lead to the appropriate set of equations of motion, they often will not provide the physical insight that we obtain from Newtonian methods.

EXAMPLE 6.12 ——————————————————————————

A small sphere of mass m is suspended from the top of a hollow pole through which the cable passes. The length of the cable is a specified function $l(t)$ because its free end is pulled inward by the tensile force $\overline{F}$. The sphere is given an initial velocity that causes it to rotate about the pole, as well as to swing outward from the pole. Determine the equations of motion for the sphere.

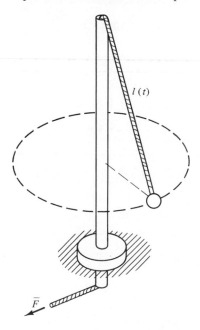

$l(t)$

$\overline{F}$

Example 6.12

Solution. The position of the sphere relative to the pivot may be conveniently defined in terms of spherical coordinates, which are depicted in the free body diagram. However, the radial distance l is a specified function of time, so it is not an unconstrained generalized coordinate. We therefore have $q_1 = \phi$, $q_2 = \theta$.

The only nonconservative force acting on the sphere is the tension $\overline{F}$. This force is the constraint force that imposes the restriction that $l(t)$ is specified. Therefore, $\overline{F}$ does no virtual work in a virtual movement that is consistent with the constraint, and

$$\delta W = 0 \Rightarrow Q_1 = Q_2 = 0$$

This same result may also be obtained in another way. We know that l is a specified function of time, and time is constant in a virtual movement. Therefore, the dis-

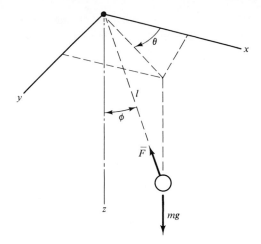

Free body diagram.

placement of the sphere resulting from virtual increments in the generalized co-ordinates ϕ and θ is

$$\delta \bar{r}_s = l \, \delta \phi \, \bar{e}_\phi + l \sin \phi \, \delta \theta \, \bar{e}_\theta$$

Since $\overline{F}$ acts in the radial direction, $\overline{F} = -F\bar{e}_r$, it follows that

$$\delta W = \overline{F} \cdot \delta \bar{r}_s = 0$$

The fact that the length l is not constant affects the kinetic energy. The velocity in terms of spherical coordinates is

$$\bar{v}_s = \dot{l}\bar{e}_r + l\dot{\phi}\bar{e}_\phi + l\dot{\theta} \sin \phi \, \bar{e}_\theta$$

The corresponding kinetic energy is

$$T = \tfrac{1}{2}m\bar{v}_s \cdot \bar{v}_s = \tfrac{1}{2}m(\dot{l}^2 + l^2\dot{\phi}^2 + l^2\dot{\theta}^2 \sin^2 \phi)$$

The elevation of the pivot O serves as a convenient datum for gravitational potential energy,

$$V = mg(-l \cos \theta)$$

The derivatives for Lagrange's equations are

$$\frac{d}{dt}\left(\frac{\partial T}{\partial \dot{\phi}}\right) = \frac{d}{dt}(ml^2\dot{\phi}) = m(l^2\ddot{\phi} + 2l\dot{l}\dot{\phi})$$

$$\frac{\partial T}{\partial \phi} = ml^2\dot{\theta}^2 \sin \phi \cos \phi \qquad \frac{\partial V}{\partial \phi} = mgl \sin \phi$$

$$\frac{d}{dt}\left(\frac{\partial T}{\partial \dot{\theta}}\right) = \frac{d}{dt}(ml^2\dot{\theta} \sin^2 \phi)$$

$$= m(l^2\ddot{\theta} \sin^2 \phi + 2l\dot{l}\dot{\theta} \sin^2 \phi + 2l^2\dot{\theta}\dot{\phi} \sin \phi \cos \phi)$$

$$\frac{\partial T}{\partial \theta} = \frac{\partial V}{\partial \theta} = 0$$

Note that in these derivatives, l is held constant in the partial differentiations, whereas the dependence of l must be recognized in the total differentiation. The corresponding Lagrange's equations are

$$l^2\ddot{\phi} + 2l\dot{l}\dot{\phi} - l^2\dot{\theta}^2 \sin\phi\cos\phi + gl\sin\phi = 0$$
$$l^2\ddot{\theta}\sin^2\phi + 2l\dot{l}\dot{\theta}\sin^2\phi + 2l^2\dot{\theta}\dot{\phi}\sin\phi\cos\phi = 0$$

We may verify that these equations are correct by recalling the relations for acceleration in terms of spherical coordinates. The first of the above equations is merely $\Sigma F_\phi = ma_\phi$, multiplied by a factor l. Similarly, the second equation is $\Sigma F_\theta = ma_\theta$, multiplied by a factor $l\sin\phi$.

It is possible to remove one equation of motion by a procedure that anticipates the treatment of ignorable generalized coordinates in Section 7.2.3. Since T and V do not explicitly depend on θ, and the generalized force $Q_\theta = 0$, the corresponding Lagrange equation may be integrated in time, with the result that

$$\frac{\partial T}{\partial \dot{\theta}} = ml^2\dot{\theta}\sin^2\phi = m\beta$$

where β is a constant. This relation states that angular momentum about the vertical axis is conserved. The values of the generalized coordinates and generalized velocities at the initial instant may be deduced from initial conditions, which must be specified if the response is to be uniquely defined. This, in turn, allows us to evaluate β. We therefore may solve the above expression for the value of $\dot{\theta}$ at any instant.

$$\dot{\theta} = \frac{\beta}{l^2\sin^2\phi}$$

Substitution of $\dot{\theta}$ into the Lagrange equation for ϕ yields

$$l\ddot{\phi} + 2\dot{l}\dot{\phi} - \frac{\beta^2\cos\phi}{l^3\sin^3\phi} + g\sin\phi = 0$$

It is less difficult to solve this equation than the two Lagrange equations we originally obtained.

EXAMPLE 6.13

The linkage, which lies in the vertical plane, is loaded by a force $\overline{F}(t)$ that is always parallel to bar BC. The torsional spring, whose stiffness is k, is undeformed when $\theta = 60°$. Determine the equations of motion for the system.

Solution. The angle θ is suitable as the generalized coordinate. The reaction forces at pin A and collar C, which are shown in the free body diagram, do no work in a virtual movement that increments θ. The weights of each bar, which we assume to be mg for each, are conservative. Therefore, the virtual work is

$$\delta W = \overline{F} \cdot \delta\overline{r}_B$$

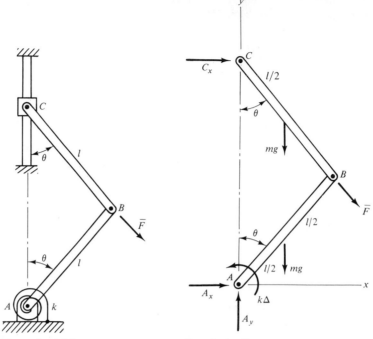

Example 6.13   Free body diagram.

Because the linkage forms an isosceles triangle, we shall develop the virtual displacement and velocity relations by the analytical method. For $\delta \bar{r}_B$, we have

$$\bar{r}_{B/A} = l \sin \theta \bar{i} + l \cos \theta \bar{j}$$

$$\delta \bar{r}_B = \frac{\delta \bar{r}_{B/A}}{\partial \theta} \delta \theta = (l \cos \theta \bar{i} - l \sin \theta \bar{j})\delta \theta$$

The corresponding virtual work is

$$\delta W = (F \sin \theta \bar{i} - F \cos \theta \bar{j}) \cdot (l \cos \theta \bar{i} - l \sin \theta \bar{j})\delta \theta$$
$$= 2\,Fl \sin \theta \cos \theta\,\delta \theta = Fl \sin 2\theta\,\delta \theta$$

so the generalized force is

$$Q_1 = Fl \sin 2\theta$$

Bar AB is in pure rotation about pin A, and bar BC is in general motion. The kinetic energy is therefore

$$T = \tfrac{1}{2}(I_A)_{AB}\omega_{AB}^2 + \tfrac{1}{2}m(v_G)_{BC}^2 + \tfrac{1}{2}(I_G)_{BC}\omega_{BC}^2$$

where $(I_A)_{AB}$ and $(I_G)_{BC}$ are the moments of inertia of bar AB about end A, and of bar BC about its center of mass, respectively, both about axes perpendicular to the plane of motion. Thus,

$$(I_A)_{AB} = \tfrac{1}{3}ml^2 \qquad (I_G)_{BC} = \tfrac{1}{12}ml^2$$

The angle θ defines the orientation of each bar, $\omega_{AB} = \omega_{BC} = \dot{\theta}$. For the velocity of the center of mass of bar BC, we write

$$(\bar{v}_G)_{BC} = \frac{d}{dt}\bar{r}_{G/A} = \frac{d}{dt}\left(\frac{l}{2}\sin\theta\bar{i} + \frac{3l}{2}\cos\theta\bar{j}\right)$$

$$= \frac{l}{2}\dot{\theta}(\cos\theta\bar{i} - 3\sin\theta\bar{j})$$

Thus the kinetic energy is

$$T = \tfrac{1}{2}(\tfrac{1}{3}ml^2)\,\dot{\theta}^2 + \tfrac{1}{2}m\left(\frac{l}{2}\dot{\theta}\right)^2(\cos^2\theta + 9\sin^2\theta) + \tfrac{1}{2}(\tfrac{1}{12}ml^2)\,\dot{\theta}^2$$

$$= \tfrac{1}{2}mL^2\,\dot{\theta}^2\,(\tfrac{2}{3} + \sin^2\theta)$$

The last step is to evaluate the potential energy. For this we observe that the torsional spring is undeformed when $\theta = 60° = \pi/3$. (We assume that the spring constant is expressed as moment units per radian.) Thus, the rotational deformation is $\Delta = \theta - \pi/3$. We place the datum for gravity at the elevation of the fixed pin, so that

$$V = \tfrac{1}{2}k\Delta^2 + mg\left(\frac{l}{2}\cos\theta\right) + mg\left(\frac{3l}{2}\cos\theta\right)$$

$$= \tfrac{1}{2}k\left(\theta - \frac{\pi}{3}\right)^2 + 2mgl\cos\theta$$

The derivatives of T for Lagrange's equations are

$$\frac{d}{dt}\left(\frac{\partial T}{\partial\dot{\theta}}\right) = \frac{d}{dt}\,[ml^2\dot{\theta}\,(\tfrac{2}{3} + 2\sin^2\theta)]$$

$$= ml^2\,[\ddot{\theta}\,(\tfrac{2}{3} + 2\sin^2\theta) + 4\dot{\theta}^2\sin\theta\cos\theta]$$

$$\frac{\partial T}{\partial\theta} = 2ml^2\dot{\theta}^2\sin\theta\cos\theta$$

The corresponding equation of motion is

$$ml^2\,[\ddot{\theta}\,(\tfrac{2}{3} + 2\sin^2\theta) + \dot{\theta}^2\sin 2\theta] + k\left(\theta - \frac{\pi}{3}\right) - 2\,mgl\sin\theta = Fl\sin 2\theta$$

EXAMPLE 6.14 _____

The disk spins about shaft AB at the constant rate ω_1, whereas the vertical shaft, to which shaft AB is pinned, rotates freely. The masses are m_1 for the disk and m_2 for shaft AB. Derive the equations of motion.

Solution. Since ω_1 is specified, the position of the system at any instant may be described by the precession angle ψ and the nutation angle θ. Hence, we choose

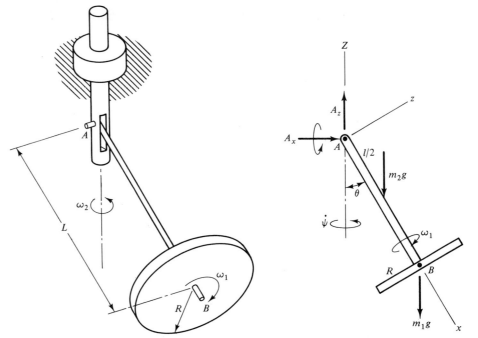

Example 6.14   Free body diagram.

these parameters as the generalized coordinates for this two-degree-of-freedom system.

We have isolated the system in a free body diagram, which is shown in side view. If we neglect the inertia of the vertical shaft, there is no couple affecting the precession. Consequently, no virtual work is done in a virtual movement that results from increments in ψ and θ, $\delta W = 0$, which leads to $Q_1 = Q_2 = 0$.

Since pin A has a fixed position relative to the disk, we may formulate the kinetic energy of each body relative to that point.

$$T = \tfrac{1}{2}(\overline{H}_A \cdot \overline{\omega})_{\text{disk}} + \tfrac{1}{2}(\overline{H}_A \cdot \overline{\omega})_{\text{shaft}}$$

We shall use the *xyz* coordinate system shown in the free body diagram to formulate the kinetic energy at the disk. (It is imperative to recognize that this coordinate system is acceptable only because the disk is axisymmetric. If the spinning body had arbitrary properties, we would need to develop expressions based on the *z* axis being at an arbitrary spin angle relative to the vertical plane.) Because *xyz* are principal axes for both bodies, the kinetic energy reduces to

$$T = \tfrac{1}{2}(I_{xx}\omega_x^2 + I_{yy}\omega_y^2 + I_{zz}\omega_z^2)_{\text{disk}} + \tfrac{1}{2}(I_{xx}\omega_x^2 + I_{yy}\omega_y^2 + I_{zz}\omega_z^2)_{\text{shaft}}$$

The angular velocity of the disk is the vector sum of the precession rate $\dot{\psi}$, the nutation rate $\dot{\theta}$, and the spin rate ω_1, all of which must be resolved into *xyz* components.

$$\begin{aligned}
\overline{\omega}_{\text{disk}} &= \dot{\psi}\overline{K} + \dot{\theta}(-\overline{j}) + \omega_1(-\overline{i})\\
&= -(\dot{\psi}\cos\theta + \omega_1)\overline{i} - \dot{\theta}\overline{j} + \dot{\psi}\sin\theta\,\overline{k}
\end{aligned}$$

The shaft is not spinning, so

$$\overline{\omega}_{\text{shaft}} = -\dot{\psi} \cos \theta \overline{i} - \dot{\theta} \overline{j} + \dot{\psi} \sin \theta \overline{k}$$

We obtain the respective moments of inertia from the tabulated properties and the parallel axis theorems. The result is

$$T = \tfrac{1}{2}(\tfrac{1}{2}m_1 R^2)(\dot{\psi} \cos \theta + \omega_1)^2 + \tfrac{1}{2}(\tfrac{1}{4}m_1 R^2 + m_1 L^2)[\dot{\theta}^2 + (\dot{\psi} \sin \theta)^2]$$

$$+ \tfrac{1}{2}(\tfrac{1}{3}m_2 L^2)[\dot{\theta}^2 + (\dot{\psi} \sin \theta)^2]$$

where we have considered $I_{xx} = 0$ for the slender shaft.

When we place the datum for gravitational potential energy at the elevation of pin A, we find that

$$V = m_1 g(-L \cos \theta) + m_2 g \left(-\frac{L}{2} \cos \theta \right) = -(m_1 + \tfrac{1}{2}m_2) gL \cos \theta$$

For brevity, let us define the following moment-of-inertia parameters:

$$I_1 = \tfrac{1}{2}m_1 R^2 \qquad I_2 = \tfrac{1}{4}m_1 R^2 + (m_1 + \tfrac{1}{3}m_2) L^2$$

Then, we have

$$\frac{d}{dt}\left(\frac{\partial T}{\partial \dot{\psi}} \right) = \frac{d}{dt} [I_1 (\dot{\psi} \cos \theta + \omega_1) \cos \theta + I_2 \dot{\psi} \sin^2 \theta]$$

$$= I_1(\ddot{\psi} \cos^2 \theta - 2\dot{\psi}\dot{\theta} \cos \theta \sin \theta - \omega_1 \dot{\theta} \sin \theta)$$

$$+ I_2 \ddot{\psi} \sin^2 \theta + 2I_2 \dot{\psi}\dot{\theta} \sin \theta \cos \theta$$

$$\frac{\partial T}{\partial \psi} = \frac{\partial V}{\partial \psi} = 0 \qquad \frac{d}{dt}\left(\frac{\partial T}{\partial \dot{\theta}} \right) = \frac{d}{dt}[I_2 \dot{\theta}] = I_2 \ddot{\theta}$$

$$\frac{\partial T}{\partial \theta} = I_1(\dot{\psi} \cos \theta + \omega_1)(-\dot{\psi} \sin \theta) + I_2(\dot{\psi}^2 \sin \theta \cos \theta)$$

$$\frac{\partial V}{\partial \theta} = (m_1 + \tfrac{1}{2}m_2) gL \sin \theta$$

The corresponding Lagrange equations are

$$(I_1 \cos^2 \theta + I_2 \sin^2 \theta)\ddot{\psi} - 2(I_1 - I_2) \sin \theta \cos \theta - I_1\omega_1\dot{\theta} = 0$$

$$I_2 \ddot{\theta} + (I_1 - I_2)\dot{\psi}^2 \sin \theta \cos \theta + I_1\omega_1\dot{\psi} \sin \theta + (m_1 + \tfrac{1}{2}m_2) gl \sin \theta = 0$$

In order to recognize the physical significance of these equations of motion, suppose that a nonconservative force were present. The generalized forces in that case would be the moments of the force about the precession (i.e., vertical) axis and about the nutation (i.e., y) axis, corresponding to the virtual work done when ψ and θ, respectively, are incremented. We therefore conclude that the equations of motion are the Z and y components of $\Sigma \overline{M}_A = \dot{\overline{H}}_A$. Only the latter is identical to what we would have obtained from Euler's equations for a rigid body.

REFERENCES

E. A. Desloge, *Classical Mechanics*, vol. 2, John Wiley & Sons, New York, 1982.

A. F. D'Souza and V. K. Garg, *Advanced Dynamics*, Prentice-Hall, Englewood Cliffs, N.J. 1984.

H. Goldstein, *Classical Mechanics*, 2d ed., Addison-Wesley, Reading, Mass., 1980.

D. T. Greenwood, *Principles of Dynamics*, Prentice-Hall, Inc., Englewood Cliffs, N.J., 1965.

J. B. Marion, *Classical Dynamics of Particles and Systems*, Academic Press, New York, 1960.

L. Meirovitch, *Methods of Analytical Dynamics*, McGraw-Hill, New York, 1970.

L. A. Pars, *A Treatise on Analytical Dynamics*, William Heinemann, Ltd., London, 1965.

R. M. Rosenberg, *Analytical Dynamics of Discrete Systems,* Plenum Press, New York, 1977.

K. R. Symon, *Mechanics*, 3d ed., Addison-Wesley, Reading, Mass., 1971.

HOMEWORK PROBLEMS

6.1 The bar is made to slide along the horizontal plane such that the velocity of end B is always directed at a constant angle γ relative to the bar. Describe the corresponding velocity constraint, and determine whether it is holonomic.

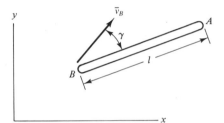

Problem 6.1

6.2 The slider descends along a curved guide in the shape of a parabola, $y = \beta x^2$, as the guide translates to the right at the constant speed u. Describe the constraint equations on the (absolute) Cartesian coordinates of the slider as a velocity constraint and show that the constraint is holonomic. Derive the corresponding configurational constraint by integration of the velocity constraint, and also by geometrical analysis of the position.

6.3 The bar remains in contact with the semicylinder of radius R as the collar slides over the vertical guide. Determine the velocity constraint condition between the distance Y locating the collar and the angle θ. Prove that this constraint condition is holonomic. Show that the same configuration constraint could have been obtained from a geometrical analysis.

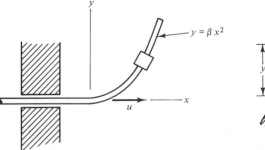 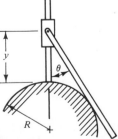

Problem 6.2   **Problem 6.3**

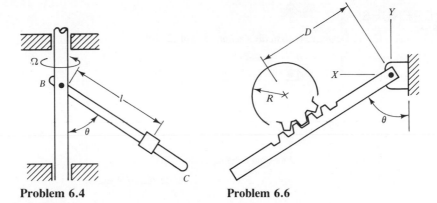

Problem 6.4   **Problem 6.6**

6.4 The system is forced to precess about the vertical axis at the constant rate Ω, but the nutation angle θ for bar BC and the distance l locating the collar are unknown. Perform a velocity analysis to derive the constraint conditions imposed on the cylindrical coordinates of the collar. Are they holonomic? If so, determine the corresponding configuration constraints.

6.5 The Cartesian coordinate (x, y, z) of a particle relative to a fixed reference frame are related by $(z/x - 2\alpha y)\dot{x} - \alpha x \dot{y} + \dot{z} = \dot{\alpha} xy + \dot{\beta}$ where α and β are specified functions of time. Prove that this constraint is holonomic, and derive the corresponding configuration constraint.

6.6 The gear rolls without slipping over the rack, which pivots about pin A. Let the angle of rotation θ of the rack and the distance D from the pivot be two generalized coordinates. Describe the velocity constraints relating the (X, Y) coordinates of the center of the gear and the angular velocity of the gear to θ and D. Are these constraints holonomic?

6.7 The slider in Problem 6.2 has mass m. Determine its equations of motion. Friction has negligible effect.

6.8 The collar of mass m slides over the smooth horizontal guide under the restraint of a spring whose stiffness is k. The unstretched length of the spring is $0.8L$. Determine the equations of motion for the system.

6.9 A couple load $\Gamma(t)$ is applied to rod AB, whose mass is m. Collar C is pinned to collar D. The mass of each collar is $m/4$. The system lies in the vertical plane, and the spring is unstretched in the position where $\theta = 20°$. Determine the equations of motion for the system.

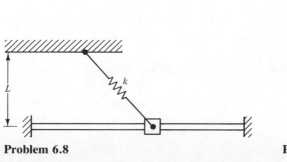

Problem 6.8

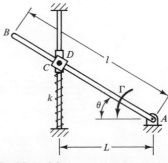

Problem 6.9

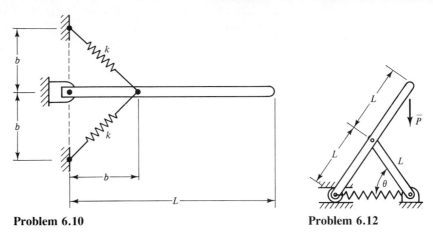

Problem 6.10   **Problem 6.12**

6.10 The bar is supported by two springs whose stiffness is k. The springs are unstretched when the bar is horizontal. Determine the equations of motion.

6.11 Determine the equations of motion of the compound pendulum in Example 6.7.

6.12 The linkage is braced by a spring of stiffness k in order to support the vertical force $\overline{P}$. The system lies in the vertical plane, and the spring is unstretched when $\theta = 45°$. Derive the equations of motion.

6.13 The force $\overline{P}(t)$ acting on link BC is always perpendicular to that link. The linear spring is unstretched in the position where $\beta = 53.13°$, and the identical bars each have mass m. If the spring can sustain both compressive and tensile forces, derive the equations of motion. The system is situated in the vertical plane.

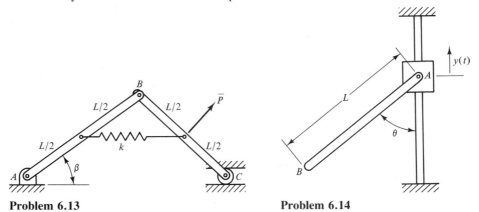

Problem 6.13   **Problem 6.14**

6.14 The collar supporting bar AB is given a specified displacement $y(t)$. The collar and the bar have equal mass m. Derive the equation of motion for the angle of rotation θ.

6.15 The collar, whose mass is m_1, supports a homogeneous bar whose mass is m_2. The springs

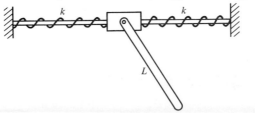

Problem 6.15

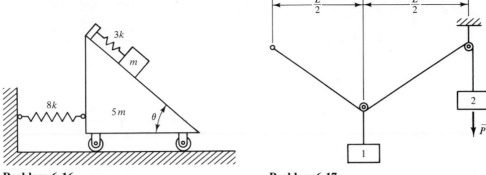

Problem 6.16   **Problem 6.17**

restraining the collar each have stiffness k. Determine the equations of motion for this system.

6.16 The stiffness of the horizontal spring is $8k$, whereas the spring holding the small block has stiffness $3k$, where k is a basic unit of stiffness. The masses are $5m$ and m for the cart and the block, respectively. Determine the equations of motion for the system.

6.17 A downward force $\overline{P}(t)$ is applied to block 2. The masses are $m_1 = 2m$, $m_2 = m$. Derive the equations of motion for the system.

6.18 A force $P(t)$ acting on the piston causes crankshaft AB to rotate. The mass per unit length of each bar is σ, and the mass of piston C is σL. The system lies in the horizontal plane. Derive the equations governing the generalized coordinate θ.

6.19 The cylinder is unbalanced such that its center of mass is situated at an eccentricity ε from the geometric center. Determine the equation of motion for arbitrarily large movements.

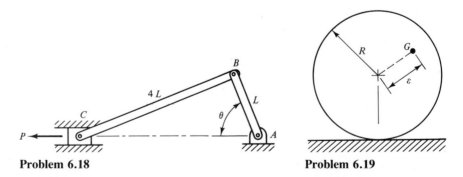

Problem 6.18   **Problem 6.19**

6.20 The pulleys roll over the rack without slipping. The masses and radii of gyration about the center of mass are m_i and σ_i ($i = 1$ for the left pulley and $i = 2$ for the right). The spring, whose stiffness is k, is capable of sustaining both compressive and tensile forces. Determine the equations of motion.

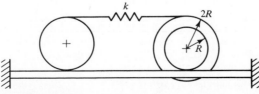

Problem 6.20

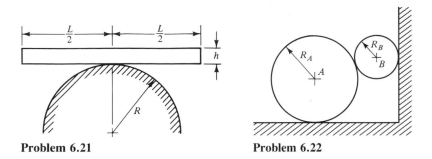

Problem 6.21   **Problem 6.22**

6.21 The bar is initially positioned horizontally at the top of the stationary semicylinder, as shown. It is then given a small push. There is no slipping between the bar and the semicylinder. Determine the equations of motion.

6.22 The cylinders roll in the vertical plane such that there is no slipping between the cylinders or between cylinder A and the ground. The vertical surface is smooth. The mass of each cylinder is m. Derive the equations of motion for the system.

6.23 Force $\overline{P}$ acts normal to bar AB, whose mass is m. This causes the disk, whose mass is $2m$, to move to the left. The disk does not slip relative to the bar, and friction between the disk and the ground is negligible. Derive the equations of motion for the system.

6.24 The collar of mass m slides over the circular guide bar that rotates about its pivot at the constant angular speed ω. The force applied to the free end of the cable after it passes through pivot A is known to be $F(t)$. Derive the equation of motion for this system, which lies in the vertical plane.

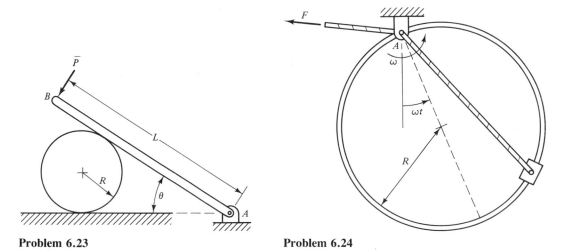

Problem 6.23   **Problem 6.24**

6.25 A circular disk of mass m is suspended in the horizontal plane by three cables of equal length L. The cables are vertical when the system is at its equilibrium position. Derive the equation of motion for the angle θ by which the disk rotates about its axis. Assume that all cables remain taut, and that $L \gg R\theta$.

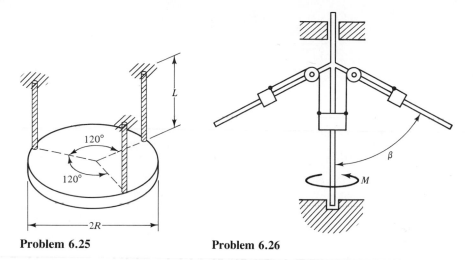

Problem 6.25   **Problem 6.26**

6.26 Each of the collars has mass m. The bar assembly on which they ride has negligible mass and rotates about the vertical axis because of a torsional load $M(t)$. Derive the equations of motion for the system.

6.27 The orientation of the homogeneous cylinder relative to the gimbal is described by the angle β. The torque Γ is such that the rotation rate Ω of the gimbal about the horizontal axis is constant. Derive the equation of motion for β.

6.28 Servomotors make the flywheel spin at a constant rate ω_2, and also impose a precession rate ω_1 that is a function of time. The center of mass of the flywheel is situated on the precession axis, and the centroidal moments of inertia are I_1 about the spin axis and I_2 transverse to that axis. Derive the equations of motion for the system.

6.29 Solve Problem 6.28 for the case where a known torque $Q(t)$ acts about the vertical axis. (The precession rate is not known in this case.)

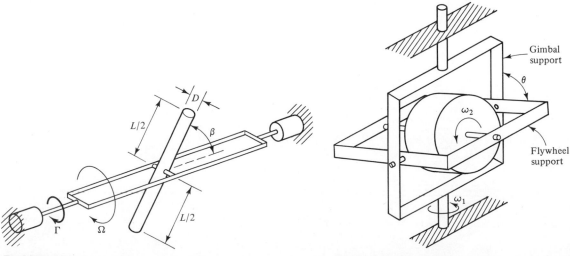

Problem 6.27   **Problems 6.28 and 6.29**

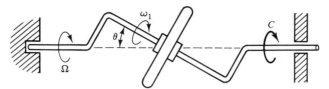

Problem 6.30

6.30 A servomotor forces the disk to spin at angular speed ω_1 which is a known function of time. The couple $C(t)$ induces rotation at rate Ω of the system about the horizontal shaft. Derive the differential equation for Ω.

6.31 The sphere spins at constant angular speed Ω relative to its shaft, which is connected to the vertical post by a ball-and-socket joint. Derive the equations of motion for the precession ψ and nutation θ of the system. Then establish the conditions for which the nutation angle is constant.

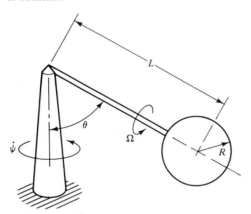

Problem 6.31

6.32 A force $\overline{F}$ is applied to the vertical control rod in the flyball speed governor. The system is made to precess at a constant rate Ω about the vertical axis by a torque $\overline{M}$. Determine

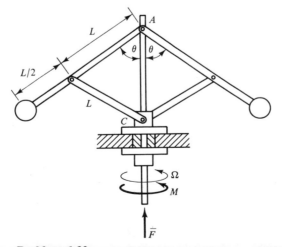

Problem 6.32

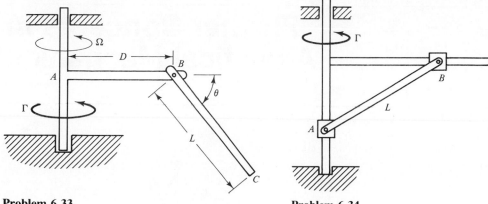

Problem 6.33   **Problem 6.34**

the equation of motion governing θ. The mass of each sphere is m, and the links have negligible mass.

6.33 Bar BC is pivoted from the end of the T-bar, which rotates about the vertical axis at the constant speed Ω. Derive the differential equation of motion for the angle of elevation θ.

6.34 A known couple $\Gamma(t)$ induces rotation of the system about the vertical axis. Collars A and B, each of whose mass is m, are interconnected by a rigid bar whose mass is $4m$. The moment of inertia of the T-bar about the vertical axis is I. Derive the equations of motion for this system.

6.35 The homogeneous bar, having mass m, is pinned to a collar that permits precessional rotation ψ about the vertical guide, as well as nutational rotation θ. The collar is attached to a spring whose extensional stiffness is k_e and whose torsional stiffness for precessional rotation is k_t. Derive the equations of motion for this system.

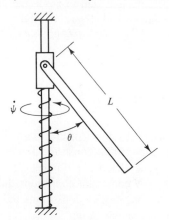

Problem 6.35

Chapter 7

Further Concepts in Analytical Mechanics

The basic principles in the preceding chapter provide a sufficient foundation to treat the majority of modeling tasks that arise in engineering practice. Our goal in this chapter is to expand these capabilities. The first priority is to be able to apply Lagrange's equations in situations where constrained generalized coordinates have been selected. We will find that such a description might be desirable, even if the system is holonomic, especially if friction is present in the system.

We will also develop alternative, and sometimes simpler, forms of the equations of motion. Those developments are partially intended to assist the phase of a dynamics study in which the equations of motion are solved. However, they also will enhance our understanding of the basic concepts of analytical mechanics, and their relationship to the principles of Newtonian mechanics.

7.1 CONSTRAINED GENERALIZED COORDINATES

A nonholonomic system must be described by a set of constrained generalized coordinates. The key feature of such formulations is that constraint forces appear in the equations of motion. When the number of generalized coordinates, M, exceeds the number of degrees of freedom, N, there are $M - N$ constraint conditions that must be explicitly satisfied. Such conditions may be written as velocity constraints having the form of Eq. (6.3), even if the constraint is holonomic. Specifically,

$$\sum_{k=1}^{M} a_{ik}\dot{q}_k + b_i = 0 \qquad i = 1, 2, \ldots, M - N \tag{7.1}$$

Each constraint condition that must be explicitly stated would be violated if the generalized coordinates were chosen arbitrarily. The constraint force associated with each is an unknown reaction that does virtual work, and therefore occurs as an unknown in the generalized force. Lagrange's equations may be employed in this case by defining $Q_j^{(a)}$ to be the contribution of the given applied forces to the jth generalized force, and R_j to be the contribution due to all reactions. Then

$$\frac{d}{dt}\left(\frac{\partial T}{\partial \dot{q}_j}\right) - \frac{\partial T}{\partial q_j} + \frac{\partial V}{\partial q_j} = Q_j^{(a)} + R_j \qquad j = 1, 2, \ldots, M \qquad (7.2)$$

The reaction force corresponding to each constraint condition that must be stated appears in some, or all, of the terms R_j. Thus, $M - N$ reactions and M unknown generalized coordinates appear in Lagrange's equations. The combination of the $M - N$ constraint equations (7.1) and the M Lagrange's equations (7.2) yields the required number of equations of motion.

We might not wish to formulate the generalized force terms R_j, particularly if we do not wish to determine the constraint forces. In such cases, it is possible to account for constraint forces indirectly through Lagrangian multipliers. These factors were introduced in Eq. (6.32), where constraint forces were related to their corresponding constraint conditions. The contribution to R_j from the ith constraint was found there to be $\lambda_i a_{ij}$, where λ_i is the Lagrangian multiplier for the constraint. The combined contribution resulting from each of the $M - N$ constraints that are not satisfied is the sum of the individual contributions, so that

$$R_j = \sum_{i=1}^{M-N} \lambda_i a_{ij} \qquad (7.3)$$

It follows immediately from the foregoing that Eq. (7.2) becomes

$$\frac{d}{dt}\left(\frac{\partial T}{\partial \dot{q}_j}\right) - \frac{\partial T}{\partial q_j} + \frac{\partial V}{\partial q_j} = \sum_{i=1}^{M-N} \lambda_i a_{ij} + Q_j^{(a)} \qquad j = 1, 2, \ldots, M \qquad (7.4)$$

The unknown quantities appearing in this form of Lagrange's equations are the M generalized coordinates and the $M - N$ Lagrangian multipliers. The M Lagrange's equations are supplemented by $M - N$ constraint equations, Eqs. (7.1), with the result that the number of system equations again balances the number of unknowns.

We generally employ the constraint force form, Eq. (7.2), rather than the Lagrangian multiplier formulation, Eq. (7.4), whenever we wish to study the reactions. In either approach, the constraint equations are auxiliary conditions that must be satisfied in addition to Lagrange's equations. Although such equations may always be written as velocity constraints in the form of Eq. (7.1), there is an advantage in describing a holonomic constraint configurationally as $f_i(q_1, q_2, \ldots, q_M, t) = 0$. This relation may be solved for one of the generalized coordinates. Substitution of that result into the equations of motion and the other constraint equations will remove the selected generalized coordinate from the formulation. The result will be a reduction in the number of system equations to be solved.

There are several reasons why we might choose to formulate the equations of motion of a holonomic system in terms of constrained generalized coordinates. Most common is the situation where it is necessary to evaluate a reaction. If the desired reaction is to appear in the equations of motion, a set of generalized coordinates that do not satisfy the corresponding constraint equation must be employed. Simultaneous solution of Lagrange's equations and the constraint equation would yield the reactions and the response at the same stage of the solution process. The alternative approach, in which unconstrained generalized coordinates are employed, would require a separate analysis using the Newton-Euler equations of motion in Chapter 5 after the response has been evaluated.

For example, consider the pendulum in Figure 7.1, whose length l is a specified functions of time, $l = L(t)$, due to the application of the tensile force $\overline{F}$. One approach is to treat the pendulum as a rheonomic, one-degree-of-freedom system that is described by the angle ϕ. In that approach $\overline{F}$ does no virtual work, because l does not change in a virtual displacement, which holds time constant. Lagrange's equation yields a single differential equation for ϕ. If it is necessary to determine $\overline{F}$, such information can be obtained from the radial component of Newton's second law.

An alternative approach to this problem employs ϕ and l as constrained generalized coordinates that must satisfy the condition $l = L(t)$. Since l increases by δL in a virtual movement, the force $\overline{F}$ does virtual work. Hence, it contributes to the generalized forces. The system equations in this formulation are the two Lagrange's equations and the constraint equation; the corresponding unknowns are the two generalized forces and the magnitude of the axial force. Note that we would formulate the equations of motion using Eq. (7.2), which employs the constraint forces, because we have a specific interest in the reaction force $\overline{F}$.

The need to evaluate a reaction force is often a discretionary matter that depends on the application. However, in situations involving Coulomb sliding friction, the evaluation of the normal force is an intrinsic part of the solution process, because the magnitude of the tangential (that is, friction) force depends on the normal force. Since the sliding friction force does not prevent motion, it acts like an applied force that does virtual work. Hence, the magnitude of the normal force will always occur in some of the generalized forces, even though the force itself is a reaction.

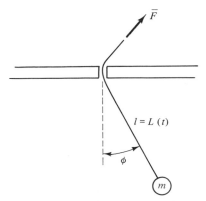

Figure 7.1 Rheonomic constraint vs. constrained generalized coordinates.

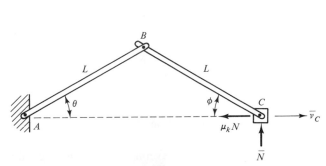

Figure 7.2 Usage of constrained generalized coordinates to account for friction.

A system illustrating this aspect is the linkage in Figure 7.2, which because of symmetry must satisfy the configurational constraint equation $\phi = \theta$. However, if the equations of motion for this one-degree-of-freedom system were to be formulated using either angle as the unconstrained generalized coordinate, we would find that the friction force $\mu_k N$ does work. (Strictly speaking, the friction force should be written as $-\mu_k N \bar{v}_C / |\bar{v}_C|$, because friction resists movement.) In a formulation using unconstrained generalized coordinates, the only system equation is the one derived from Lagrange's equations. There is no direct way in which to obtain an equation featuring the normal force.

An analysis using constrained generalized coordinates can treat this system effectively. Let θ and ϕ be the generalized coordinates. A virtual movement corresponding to arbitrary increments of both generalized coordinates produces a displacement of collar C perpendicular, as well as parallel, to the horizontal guide bar. The virtual work done by $\bar{N}$ in this displacement leads to terms R_1 and R_2 representing the contributions of $\bar{N}$ to the generalized forces. The contributions of the friction force $\mu_k |\bar{N}|$ would appear in the applied generalized forces, $Q_1^{(a)}$ and $Q_2^{(a)}$. The system of equations governing the two generalized coordinates and $|\bar{N}|$ will consist of two Lagrange's equations and the constraint equation; see Problem 7.2.

EXAMPLE 7.1 _____

A torque Γ applied to the vertical shaft of the T-bar causes the rotation rate Ω about the vertical axis to increase in proportion to the angle θ by which bar AB swings outward, that is, $\Omega = c\theta$. The mass of bar BC is m_1 and the moment of inertia of the T-bar about its axis of rotation is I_2. Determine the equations of motion for the system, and for the torque Γ.

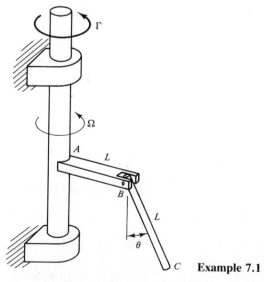

Example 7.1

Solution. The location of the bar is fully specified by the precession angle and the nutation angle. Since $\Omega = \dot{\psi}$, the given constraint on the motion is $\dot{\psi} = c\theta$, which is nonholonomic. In addition, we wish to obtain an equation for Γ, which imposes the constraint. Hence, we employ both angles as generalized coordinates, $q_1 = \psi$ and $q_2 = \theta$, even though the system has only one degree of freedom.

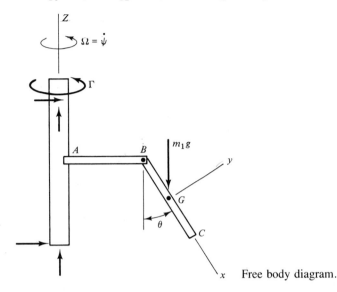

Free body diagram.

The kinetic energy of the T-bar is $\frac{1}{2}I_2\dot{\psi}^2$, to which we must add the kinetic energy of bar BC. Since this bar is in general motion, we have

$$T_{BC} = \tfrac{1}{2}\,m_1(\bar{v}_G \cdot \bar{v}_G) + \tfrac{1}{2}\,\bar{\omega}_{BC} \cdot \bar{H}_G$$

where the velocity parameters are

$$\bar{\omega}_{BC} = \dot{\psi}\bar{K} + \dot{\theta}\bar{k} = -\dot{\psi}\cos\theta\,\bar{i} + \dot{\psi}\sin\theta\,\bar{j} + \dot{\theta}\bar{k}$$

$$\bar{v}_G = \bar{v}_B + \bar{\omega}_{BC} \times \bar{r}_{G/B} = \dot{\psi}L(-\bar{k}) + \bar{\omega}_{BC} \times \left(\frac{L}{2}\bar{i}\right)$$

$$= \frac{L}{2}\dot{\theta}\bar{j} - L\dot{\psi}\left(1 + \frac{1}{2}\sin\theta\right)\bar{k}$$

Considering bar BC to be slender leads to

$$I_{yy} = I_{zz} = \tfrac{1}{12}\,m_1L^2 \qquad I_{xx} = I_{xy} = I_{yz} = I_{xz} = 0$$

so its angular momentum is

$$\bar{H}_G = I_{yy}\omega_y\bar{j} + I_{zz}\omega_z\bar{k} = \tfrac{1}{12}\,m_1L^2(\dot{\psi}\sin\theta\,\bar{j} + \dot{\theta}\bar{k})$$

The corresponding kinetic energy of the system is

$$T = \tfrac{1}{2}\,m_1\left[\frac{L^2}{4}\dot{\theta}^2 + L^2\dot{\psi}^2\,(1 + \tfrac{1}{2}\sin\theta)^2\right]$$

$$+ \tfrac{1}{2}\,(\tfrac{1}{12}\,m_1L^2)\,(\dot{\psi}^2\sin^2\theta + \dot{\theta}^2) + I_2\dot{\psi}^2$$

$$= \tfrac{1}{2}\,(m_1L^2 + I_2 + m_1L^2\sin\theta + \tfrac{1}{3}\,m_1L^2\sin^2\theta)\,\dot{\psi}^2 + \tfrac{1}{6}\,m_1L^2\dot{\theta}^2$$

We select the elevation of pin B as the datum for gravitational potential energy, so

$$V = -m_1 g \frac{L}{2} \cos \theta$$

In order to evaluate the torque Γ, we explicitly account for reactions in the virtual work, rather than using Lagrangian multipliers. Arbitrary increments $\delta\psi$ and $\delta\theta$ only violate the constraint on ψ imposed by Γ. Therefore, Γ is the only reaction that does work,

$$\delta W = \Gamma \, \delta\psi = Q_1 \, \delta\psi + Q_2 \, \delta\theta \Rightarrow Q_1 = \Gamma \qquad Q_2 = 0$$

Since T and V are independent of ψ, the first Lagrange equation is

$$\frac{d}{dt}\left(\frac{\partial T}{\partial \dot\psi}\right) = Q_1$$

$$(m_1 L^2 + I_2 + m_1 L^2 \sin\theta + \tfrac{1}{3} m_1 L^2 \sin^2\theta)\,\ddot\psi$$
$$+ m_1 L^2 (1 + \tfrac{2}{3}\cos\theta)(\sin\theta)\,\dot\psi\dot\theta = \Gamma \quad (1)$$

For the second Lagrange equation, we have

$$\frac{d}{dt}\left(\frac{\partial T}{\partial \dot\theta}\right) - \frac{\partial T}{\partial \theta} + \frac{\partial V}{\partial \theta} = 0$$

$$\tfrac{1}{3}\ddot\theta - (\tfrac{1}{2} + \tfrac{1}{3}\cos\theta)(\sin\theta)\,\dot\psi^2 + \frac{g}{2L}\sin\theta = 0 \quad (2)$$

These two Lagrange's equations, in combination with the constraint equation, $\dot\psi = c\theta$, govern the three unknowns, ψ, θ, and Γ. Substituting the constraint equation into Eq. (2) yields an ordinary differential equation for θ. After the response $\theta(t)$ has been obtained, the value of $\Gamma(t)$ may be found by substituting the response and the constraint equation into Eq. (1).

EXAMPLE 7.2 _____

The coefficient of sliding friction between the inclined wall and the bar is μ, but friction between collar A and the horizontal guide bar is negligible. The spring, whose stiffness is k, is unstretched when $\phi = 0$, and the mass of the bar is m. Determine the equations of motion of the system.

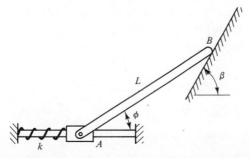

Example 7.2

Solution.   The position of the bar is fully specified by the angle ϕ, so this is a holonomic system with one degree of freedom. However, a virtual movement of the bar in which only ϕ is incremented will not violate the constraint that end B cannot move transversely to the incline. Correspondingly, the normal reaction N_B will not appear in an equation of motion, while the friction force, whose magnitude is μN_B, will appear in the (single) Lagrange equation. We therefore select two constrained generalized coordinates, for which $q_1 = \phi$ and the absolute position $q_2 = X_A$ are suitable.

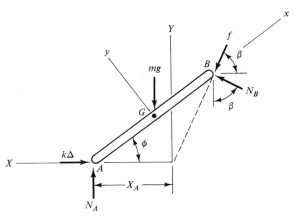

Free body diagram.

In order to derive the constraint equation, we first express the velocity of end B in terms of ϕ and X_A. Thus,

$$\bar{v}_B = \bar{v}_A + \bar{\omega}_{AB} \times \bar{r}_{B/A} = \dot{X}_A \bar{I} + (-\dot{\phi}\bar{K}) \times (-L \cos \phi \bar{I} + L \sin \phi \bar{J})$$
$$= (\dot{X}_A + L\dot{\phi} \sin \phi)\bar{I} + L\dot{\phi} \cos \phi \bar{J}$$

The normal to the incline is

$$\bar{n} = \sin \beta \bar{I} + \cos \beta \bar{J}$$

so the requirement that $\bar{v}_B$ is parallel to the incline is satisfied if

$$\bar{v}_B \cdot \bar{n} = L\dot{\phi} (\sin \phi \sin \beta + \cos \phi \cos \beta) + \dot{X}_A \sin \beta = 0$$
$$L\dot{\phi} \cos (\beta - \phi) + \dot{X}_A \sin \beta = 0 \tag{1}$$

Note that the expression for $\bar{v}_B$ is independent of time. Hence, the virtual displacement of end B may be described by forming $\bar{v}_B dt$, and then replacing differentials by virtual increments. This yields

$$\delta \bar{r}_B = (\delta X_A + L \,\delta\phi \sin \phi)\bar{I} + L \,\delta\phi \cos \phi \bar{J}$$

Since the spring and gravity forces are conservative, and the constraint at end A is not violated, the virtual work is

$$\delta W = [(f \cos \beta + N_B \sin \beta)\bar{I} + (-f \sin \beta + N_B \cos \beta)\bar{J}] \cdot \delta \bar{r}_B$$

Collecting coefficients of $\delta\phi$ and δX_A in the dot product yields the following generalized forces.

$$Q_1 = (\cos \beta \sin \phi - \sin \beta \cos \phi)fL + (\sin \beta \sin \phi + \cos \beta \cos \phi)N_B L$$
$$= -fL \sin (\beta - \phi) + N_B L \cos (\beta - \phi)$$
$$Q_2 = f \cos \beta + N_B \sin \beta$$

We eliminate f from the above by noting that $\bar{f} = \mu N_B$ in the direction opposite the velocity of end B. We describe this velocity by noting that, when the velocity constraint is satisfied,

$$\dot{X}_A = -L\dot{\phi} \frac{\cos (\beta - \phi)}{\sin \beta}$$

The expression for $\bar{v}_B$ in this case becomes

$$\bar{v}_B = L\dot{\phi} \left[\sin \phi - \frac{\cos (\beta - \phi)}{\sin \beta} \right]\bar{I} + L\dot{\phi} \cos \phi \bar{J}$$
$$= L\dot{\phi} \cos \phi (- \cot \beta \bar{I} + \bar{J})$$

This confirms our intuition that end B moves up and to the right if $\dot{\phi} > 0$. Such movement corresponds to $\bar{f}$ being down and to the left, as was assumed in the free body diagram. Thus, we set

$$f = \frac{\mu N_B \dot{\phi}}{|\dot{\phi}|}$$

We must express the kinetic energy for arbitrary ϕ and X_A. The velocity of the center of mass is

$$\bar{v}_G = \bar{v}_A + \bar{\omega}_{AB} + \bar{r}_{G/A} = \left(\dot{X}_A + \frac{L}{2} \dot{\phi} \sin \phi \right)\bar{I} + \frac{L}{2} \dot{\phi} \cos \phi \bar{J}$$

from which we obtain

$$T = \tfrac{1}{2} m\bar{v}_G \cdot \bar{v}_G + \tfrac{1}{2} I_{zz}\dot{\phi}^2 = \tfrac{1}{2} m (\tfrac{1}{3} L^2\dot{\phi}^2 + L\dot{\phi}\dot{X}_A \sin \phi + \dot{X}_A^2)$$

The spring and gravity contribute to the potential energy. We let the elevation of end A be the gravitational datum. The elongation of the spring is

$$\Delta = X_A|_{\phi=0} - X_A = L - X_A$$

so that

$$V = \tfrac{1}{2} k(L - X_A)^2 + mg \frac{L}{2} \sin \phi$$

We now form Lagrange's equations, using the earlier expressions for Q_1, Q_2, and f. For $q_1 = \phi$, we obtain

$$\tfrac{1}{3} mL^2\ddot{\phi} + \tfrac{1}{2} mL\ddot{X}_A \sin \phi + mg \frac{L}{2} \cos \phi$$

$$= \left[\cos (\beta - \phi) - \mu \frac{\dot{\phi}}{|\dot{\phi}|} \sin (\beta - \phi) \right] LN_B \quad (2)$$

while the equation for $q_2 = X_A$ is

$$m\ddot{X}_A + \tfrac{1}{2} mL\ddot{\phi} \sin \phi + \tfrac{1}{2} mL\dot{\phi}^2 \cos \phi - k(L - X_A)$$

$$= \left(\sin \beta + \mu \frac{\dot{\phi}}{|\dot{\phi}|} \cos \beta \right) N_B \quad (3)$$

There are three unknowns: ϕ, X_A, and N_B. The third equation is the constraint condition, Eq. (1). Since the system is holonomic, that equation may be integrated. Toward that end we write it in Pfaffian form:

$$L \, d\phi \cos (\beta - \phi) + dX_A \sin \beta = 0$$

Each term is a perfect differential, so the corresponding configuration constraint is

$$-L \sin (\beta - \phi) + X_A \sin \beta = C$$

The value of the constant of integration C must be such that $X_A = L$ when $\phi = 0$, which yields $C = 0$, and

$$X_A = L \frac{\sin (\beta - \phi)}{\sin \beta} \quad (4)$$

which is identical to the expression for X_A given by the law of sines.

If we wish, we may obtain a single differential equation of motion. Substituting Eq. (4) into the Lagrange's equations (2) and (3) would remove X_A. Then forming the ratio of Eqs. (2) and (3) would eliminate N_B. The resulting differential equation, which we shall not detail, is second-order and highly nonlinear.

EXAMPLE 7.3

A thin disk wobbles as it rolls without slipping along the ground. Consequently, the plane of the disk is inclined at an unsteady angle θ. Derive the differential equations of motion for the system. Then specialize the result to the case where θ is constant and the center A follows a circular path. The radius of gyration of the disk about its axis of symmetry is κ.

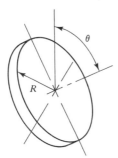

Example 7.3

Solution. The position of any rigid body may always be described in terms of three position coordinates for any point, such as the center of mass, and three

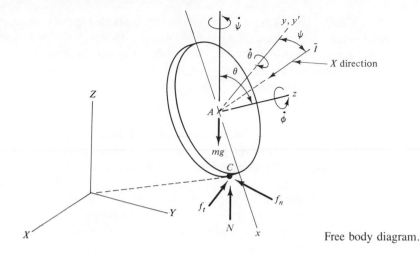

Free body diagram.

Eulerian angles. The circular shape of the disk and the absence of slipping constrain some of these variables, so it is not apparent at the outset which of the variables are independent. In order to identify the appropriate choice, we recall from Section 4.4 the kinematical analysis of a disk that wobbles as it rolls without slipping.

Let (X, Y, Z) be the Cartesian coordinates of the center of mass. The Eulerian angles (ψ, θ, ϕ) are the precession, nutation, and spin angles, respectively; they are defined by letting the vertical direction define the precession axis. Note that the y' axis is always horizontal (it is the line of nodes), whereas the y axis is a body-fixed axis. The angle between the y' axis and the negative X axis is ψ. The angular velocity of the disk is

$$\bar{\omega} = \dot{\psi}\bar{K} + \dot{\theta}\bar{j}' + \dot{\phi}\bar{k} = -\dot{\psi}\sin\theta\,\bar{i} + \dot{\theta}\bar{j} + (\dot{\psi}\cos\theta + \dot{\phi})\bar{k}$$

The velocity of the center obtained from the no-slip condition is

$$\bar{v}_A = \bar{\omega} \times \bar{r}_{A/C} = -R(\dot{\psi}\cos\theta + \dot{\phi})\bar{j} + R\dot{\theta}\bar{k}$$

The velocity of the center may also be described in terms of the Cartesian coordinates,

$$\bar{v}_A = \dot{X}\bar{I} + \dot{Y}\bar{J} + \dot{Z}\bar{K}$$

We match these two descriptions by resolving the unit vectors of one set of axes onto the other set of axes,

$$\bar{i} = -\sin\psi\cos\theta\,\bar{I} + \cos\psi\cos\theta\,\bar{J} - \sin\theta\,\bar{K}$$
$$\bar{j} = -\cos\psi\,\bar{I} - \sin\psi\,\bar{J}$$
$$\bar{k} = -\sin\psi\sin\theta\,\bar{I} + \cos\psi\sin\theta\,\bar{J} + \cos\theta\,\bar{K}$$

When these expressions are substituted into the first equation for $\bar{v}_A$, and compared with the second, we find that

$$\dot{X} = R(\dot{\psi}\cos\theta + \dot{\phi})\cos\psi - R\dot{\theta}\sin\psi\sin\theta$$
$$\dot{Y} = R(\dot{\psi}\cos\theta + \dot{\phi})\sin\psi + R\dot{\theta}\cos\psi\sin\theta$$
$$\dot{Z} = R\dot{\theta}\cos\theta$$

These relations are three velocity constraints that the six position variables must satisfy, so the disk only has three degrees of freedom. The constraints on $\dot{X}$ and $\dot{Y}$ are nonholonomic, but the one governing $\dot{Z}$ may be integrated. Multiplying each rate variable by dt shows that both sides are perfect differentials. Setting $Z = 0$ when $\theta = 0$ leads to

$$Z = R \sin \theta$$

This position constraint permits us to eliminate Z from the formulation. Hence, we shall employ five generalized coordinates in the sequence: X, Y, ψ, θ, ϕ.

The generalized coordinates are a constrained set that must satisfy the velocity constraints on $\dot{X}$ and $\dot{Y}$. We are not specifically interested in the reactions at the ground, which enforce these constraints. Therefore, we employ the Lagrangian multiplier formulation. In order to identify the coefficients that correspond to each multiplier, we adapt the standard form of a velocity constraint to the present system,

$$\sum_{k=1}^{5} a_{jk}\dot{q}_k + b_j = 0 \qquad j = 1, 2 \tag{1,2}$$

Comparing this form with the actual constraint equations shows that

$$
\begin{aligned}
&a_{11} = 1 \qquad a_{12} = 0 \qquad a_{13} = -R \cos \psi \cos \theta \\
&a_{14} = R \sin \psi \sin \theta \qquad a_{15} = -R \cos \psi \qquad b_1 = 0 \\
&a_{21} = 0 \qquad a_{22} = 1 \qquad a_{23} = -R \sin \psi \cos \theta \\
&a_{24} = -R \cos \psi \sin \theta \qquad a_{25} = -R \sin \psi \qquad b_2 = 0
\end{aligned}
$$

We now proceed to formulate the mechanical energies. Adding the translational kinetic energy associated with the center of mass to the rotational kinetic energy yields

$$
\begin{aligned}
T &= \tfrac{1}{2} m(\dot{X}^2 + \dot{Y}^2 + \dot{Z}^2) + \tfrac{1}{2} I'(\omega_x^2 + \omega_y^2) + \tfrac{1}{2} I\omega_z^2 \\
&= \tfrac{1}{2} m [\dot{X}^2 + \dot{Y}^2 + R^2\dot{\theta}^2 \cos^2 \theta + \tfrac{1}{2} \kappa^2\dot{\psi}^2 \sin^2 \theta + \tfrac{1}{2} \kappa^2\dot{\theta}^2 \\
&\qquad\qquad\qquad\qquad\qquad\qquad\qquad + \kappa^2(\dot{\psi} \cos \theta + \dot{\phi})^2]
\end{aligned}
$$

$$V = mgZ = mgR \sin \theta$$

Note that we have used the fact that the moment of inertia of a planar body about an axis in its plane is one-half the value about an intersecting normal axis to express $I_{xx} = I_{yy} = I'$. Also, note that we have not used the velocity constraint to remove the dependence on $\dot{X}$ and $\dot{Y}$ because Lagrange's equations must describe the effect of variations in each of the generalized coordinates.

The only nonconservative forces acting on the disk are the reactions at the ground, whose effect we shall describe by Lagrangian multipliers. Hence, we have $\delta W = 0$, which leads to $Q_j = 0$ for each generalized coordinate.

Applying the constrained Lagrange's equations (7.4) to the present system leads to a set of five differential equations

$$m\ddot{X} = \lambda_1 \qquad m\ddot{Y} = \lambda_2 \tag{3,4}$$

$$m\kappa^2[\ddot{\psi} (\tfrac{1}{2} \sin^2 \theta + \cos^2 \theta) - \dot{\psi}\dot{\theta} \sin \theta \cos \theta + \ddot{\phi} \cos \theta - \dot{\theta}\dot{\phi} \sin \theta]$$
$$= -\lambda_1 R \cos \psi \cos \theta - \lambda_2 R \sin \psi \cos \theta \tag{5}$$

$$mR^2\,\ddot{\theta}\cos^2\theta - mR^2\dot{\theta}^2\sin\theta\cos\theta + \tfrac{1}{2}\,m\kappa^2\ddot{\theta} + \tfrac{1}{2}\,m\kappa^2\dot{\psi}^2\sin\theta\cos\theta$$

$$+ m\kappa^2\dot{\psi}\dot{\phi}\sin\theta + mgR\cos\theta = \lambda_1 R\sin\psi\sin\theta - \lambda_2 R\cos\psi\sin\theta \quad (6)$$

$$m\kappa^2(\ddot{\psi}\cos\theta + \ddot{\phi} - \dot{\psi}\dot{\theta}\sin\theta) = -\lambda_1 R\cos\psi - \lambda_2 R\sin\psi \quad (7)$$

There are seven unknowns in this formulation: the five generalized coordinates and the two Lagrangian multipliers. These variables must satisfy the constraint equations (1) and (2), and the Lagrange equations (3) to (7).

A simplification that is obvious is to use equations of motion (3) and (4), which govern X and Y, to eliminate the Lagrangian multipliers. The resulting system of equations, including the constraints, is

$$\kappa^2\,[\tfrac{1}{2}\,\ddot{\psi}(1 + \cos^2\theta) + \ddot{\phi}\cos\theta - \dot{\psi}\dot{\theta}\sin\theta\cos\theta - \dot{\theta}\dot{\psi}\sin\theta]$$

$$+ R\ddot{X}\cos\psi\cos\theta + R\ddot{Y}\sin\psi\cos\theta = 0 \quad (1')$$

$$(\tfrac{1}{2}\,\kappa^2 + R^2\cos^2\theta)\ddot{\theta} + (\tfrac{1}{2}\,\kappa^2\dot{\psi}^2 - R^2\dot{\theta}^2)\sin\theta\cos\theta + \kappa^2\dot{\psi}\dot{\phi}\sin\theta$$

$$+ gR\cos\theta - R\ddot{X}\sin\psi\sin\theta + R\ddot{Y}\cos\psi\sin\theta = 0 \quad (2')$$

$$\kappa^2(\ddot{\psi}\cos\theta + \ddot{\phi} - \dot{\psi}\dot{\theta}\sin\theta) + R\ddot{X}\cos\psi + R\ddot{Y}\sin\psi = 0 \quad (3')$$

$$\dot{X} - R(\dot{\psi}\cos\theta + \dot{\phi})\cos\psi + R\dot{\theta}\sin\psi\sin\theta = 0 \quad (4')$$

$$\dot{Y} - R(\dot{\psi}\cos\theta + \dot{\phi})\sin\psi - R\dot{\theta}\cos\psi\sin\theta = 0 \quad (5')$$

We could simplify these equations further by using constraint equations (4') and (5') to remove the dependence on X and Y, but doing so merely complicates the other equations. We therefore proceed to consider the case where the nutation angle θ is constant.

When $\dot{\theta} = 0$, the center A of the disk follows a circular path. Let ρ be the radius of curvature of that path, and let the center of the path be situated on the Z axis. For the XYZ coordinate system we defined, the position of the center A is

$$X = -\rho\sin\psi \qquad Y = \rho\cos\psi$$

Both constraint equations, (4') and (5'), are satisfied by these expressions, provided that

$$\dot{\phi} = -\left(\frac{\rho}{R} + \cos\theta\right)\dot{\psi}$$

Next, we substitute these expressions into Lagrange's equations. Equations (1') and (3') are satisfied only if $\ddot{\psi} = 0$, which corresponds to a constant speed $\rho\dot{\psi}$ for the center A. The second Lagrange equation yields an expression relating $\dot{\psi}$ to the other parameters, specifically,

$$\dot{\psi}^2\left[\tfrac{1}{2}\,\kappa^2\sin\theta\cos\theta + \frac{\rho}{R}\,(R^2 + \kappa^2)\sin\theta\right] = gR\cos\theta$$

When the disk is homogeneous, the radius of gyration is $\kappa = R/\sqrt{2}$, which leads to

$$\dot{\psi}^2 = \frac{4g\cot\theta}{R\cos\theta + 6\rho}$$

This solution is in complete agreement with the result derived in Example

5.8 by using the Newton-Euler formulation. The earlier methods provide greater physical insight. However, it would have been much more difficult to use them to derive the equations of motion for the general case treated here, in which the nutation angle is not constant.

7.2 FIRST-ORDER-SYSTEM EQUATIONS

When the motion of a system is known, the equations of motion may be solved algebraically for the forces, such as applied loads, required to sustain that motion. A more interesting situation arises when the generalized coordinates are unknown, in which case equations of motion are differential equations. Many numerical methods and associated computer subroutines have been formulated for solving sets of first-order differential equations. We shall develop here two methods for representing the differential equations of motion in such a form.

Both methods rely on the fact that direct application of Lagrange's equations leads to equations of motion in which the highest-order derivatives are $\ddot{q}_i$, and further, that such derivatives occur linearly. In order to demonstrate this, we recall the general expression for kinetic energy, Eqs. (6.48), which displayed the most general form of the dependence on the generalized velocities.

$$T = T_2 + T_1 + T_0 \tag{7.5}$$

where T_2, T_1, and T_0 are, respectively, terms that are quadratic, linear, and independent of the generalized velocities. The specific forms identified in Eqs. (6.48) were

$$T_2 = \frac{1}{2} \sum_{i=1}^{M} \sum_{j=1}^{M} M_{ij} \dot{q}_i \dot{q}_j$$

$$T_1 = \sum_{j=1}^{M} N_j \dot{q}_j \tag{7.6}$$

The coefficients of the quadratic and linear terms, M_{ij} and N_j, respectively, as well as the zero-degree term T_0, may all be functions of the generalized coordinates and time, but not of the generalized velocities.

Let us consider the result of substituting this form of the kinetic energy into the first term of Lagrange's equations, $d(\partial T / \partial \dot{q}_n)/dt$, where n is arbitrarily selected. We begin with the quadratic terms, T_2. Whenever either of the summation indices in Eq. (7.6) matches the selected value of n, the corresponding derivative is nonzero. The derivative with respect to the generalized velocity $\dot{q}_n$ is therefore

$$\frac{\partial T_2}{\partial \dot{q}_n} = \frac{1}{2} \sum_{i=1}^{M} \sum_{j=1}^{M} M_{ij} \frac{\partial}{\partial \dot{q}_n} (\dot{q}_i \dot{q}_j)$$

$$= \frac{1}{2} \sum_{i=1}^{M} \sum_{j=1}^{M} M_{ij} \left(\frac{\partial \dot{q}_i}{\partial \dot{q}_n} \dot{q}_j + \dot{q}_i \frac{\partial \dot{q}_j}{\partial \dot{q}_n} \right)$$

$$= \frac{1}{2} \sum_{j=1}^{M} M_{nj} \dot{q}_j + \frac{1}{2} \sum_{i=1}^{M} M_{in} \dot{q}_i = \sum_{j=1}^{M} M_{nj} \dot{q}_j \tag{7.7a}$$

where the last step is a consequence of the symmetry of the coefficients M_{nj}. The corresponding terms obtained from T_1 and T_0 are

$$\frac{\partial T_1}{\partial \dot{q}_n} = N_n \qquad \frac{\partial T_0}{\partial \dot{q}_n} = 0 \qquad (7.7b)$$

Differentiation of the sum of Eqs. (7.7) with respect to time yields

$$\frac{d}{dt}\left(\frac{\partial T}{\partial \dot{q}_n}\right) = \sum_{j=1}^{M}\left(M_{nj}\ddot{q}_j + \frac{\partial M_{nj}}{\partial t}\dot{q}_j\right) + \sum_{i=1}^{M}\sum_{j=1}^{M}\frac{\partial M_{nj}}{\partial q_i}\dot{q}_i\dot{q}_j + \sum_{j=1}^{M}\frac{\partial N_n}{\partial q_j}\dot{q}_j + \frac{\partial N_n}{\partial t}$$

$$(7.8)$$

A similar analysis of $\partial T/\partial q_n$ based on Eqs. (7.5) and (7.6) leads to

$$\frac{\partial T}{\partial q_n} = \frac{1}{2}\sum_{i=1}^{M}\sum_{j=1}^{M}\frac{\partial M_{ij}}{\partial q_n}\dot{q}_i\dot{q}_j + \sum_{j=1}^{M}\frac{\partial N_j}{\partial q_n}\dot{q}_j + \frac{\partial T_0}{\partial q_n} \qquad (7.9)$$

The Lagrange's equations corresponding to the foregoing expressions are

$$\sum_{j=1}^{M}\left[M_{nj}\ddot{q}_j + \left(\frac{\partial M_{nj}}{\partial t} + \frac{\partial N_n}{\partial q_j} - \frac{\partial N_j}{\partial q_n}\right)\dot{q}_j\right] + \sum_{i=1}^{M}\sum_{j=1}^{M}\left(\frac{\partial M_{nj}}{\partial q_i} - \frac{1}{2}\frac{\partial M_{ij}}{\partial q_n}\right)\dot{q}_i\dot{q}_j + \frac{\partial N_n}{\partial t}$$

$$-\frac{\partial T_0}{\partial q_n} + \frac{\partial V}{\partial q_n} = Q_n \qquad n = 1, 2, \ldots, M \quad (7.10)$$

We observe from Eq. (7.10) that the equations of motion are a set of second-order differential equations for the unknown q_n, whose general form is

$$\sum_{j=1}^{M}M_{nj}\ddot{q}_j = F_n \qquad n = 1, 2, \ldots, M \qquad (7.11a)$$

where the functions F_n consist of all terms that do not contain second derivatives of the generalized coordinates. This expression may be written equivalently in matrix form as

$$[M]\{\ddot{q}\} = \{F\} \qquad (7.11b)$$

where the elements of the column array $\{\ddot{q}\}$ are the second derivatives of the sequence of generalized coordinates. It is significant to the developments that follow that the elements of the array $[M]$ are the respective coefficients M_{ij}. These elements only depend on the generalized coordinates and time. In contrast, the elements of $\{F\}$ might be functions of the generalized velocities, as well as the generalized coordinates and time.

If the generalized coordinates are a constrained set, the constraint equations may be written as velocity constraints; see Eq. (7.1). Since such relations are already first-order differential equations, they may be readily incorporated into the formulation at the appropriate juncture.

7.2.1 State-Space Formulation

We do not usually consider a derivative of a generalized coordinate to be a distinct variable, separate from the generalized coordinate itself. However, doing so leads to a simple transformation that converts Eq. (7.11b) into a system of first-order differential equations.

We define a set of $2M$ variables x_i, such that the first group of M variables are the generalized coordinates, while the second group are the generalized velocities. This may be describe in matrix form as upper and lower partitions of a column, according to

$$\{x\} = \begin{Bmatrix} \{q\} \\ \{\dot{q}\} \end{Bmatrix} \tag{7.12}$$

Note that second derivatives, $\{\ddot{q}\}$, are not considered to be new variables because their values are specified by the matrix equation of motion, Eq. (7.11b).

The derivative of $\{q\}$ obviously is $\{\dot{q}\}$. Since $\{q\}$ is the upper partition of $\{x\}$ and $\{\dot{q}\}$ is the lower partition, this identity for the derivative may be written in partitioned form. Recall that when the partitioning of a matrix equation is consistent, a product may be formed by treating the partitions as though they were individual elements. Thus, we have

$$\begin{bmatrix} [U] & [0] \end{bmatrix} \frac{d}{dt} \begin{Bmatrix} \{q\} \\ \{\dot{q}\} \end{Bmatrix} = \begin{bmatrix} [0] & [U] \end{bmatrix} \begin{Bmatrix} \{q\} \\ \{\dot{q}\} \end{Bmatrix}$$

where $[U]$ is the identity matrix. In view of the definition of $\{x\}$, the foregoing is equivalent to

$$\begin{bmatrix} [U] & [0] \end{bmatrix} \frac{d}{dt} \{x\} = \begin{bmatrix} [0] & [U] \end{bmatrix} \{x\} \tag{7.13}$$

The partitioned form of $\{x\}$ converts the equation of motion, Eq. (7.11b), to

$$\begin{bmatrix} [0] & [M] \end{bmatrix} \frac{d}{dt} \{x\} = \{F\} \tag{7.14}$$

Equations (7.13) and (7.14) may be combined as

$$\begin{bmatrix} [U] & [0] \\ [0] & [M] \end{bmatrix} \frac{d}{dt} \{x\} = \begin{Bmatrix} \begin{bmatrix} [0] & [U] \end{bmatrix} \{x\} \\ \{F\} \end{Bmatrix} \tag{7.15}$$

The foregoing represents a set of $2M$ first-order differential equations for the $2M$ elements of $\{x\}$. Note that the dependencies of $[M]$ and $\{F\}$ on the generalized coordinates and velocities must be expressed in terms of the appropriate elements of $\{x\}$, consistent with the overall change of variables. When the generalized coordinates are unconstrained, this is the full set of equations governing the motion of the system.

The set of variables x_i define the state of a system. Their instantaneous values define $\{\dot{x}\}$ according to Eq. (7.15). These values, in turn, define the value of $\{x\}$ at the next instant. The x_i variables form the *state space* for a system, just as the generalized coordinates q_i form the configuration space.

In the case of constrained generalized coordinates, the forcing array $\{F\}$ contains $M - N$ unknown reactions or Lagrangian multipliers, which are associated with the additional motion constraints. When expressed in velocity form, the constraint equations constitute a set of $M - N$ additional differential equations to be satisfied. In the current matrix notation these additional equations are

$$[a]\{\dot{q}\} = -\{b\}$$

or equivalently

$$[[a] \quad [0]] \frac{d}{dt} \{x\} = -\{b\} \tag{7.16}$$

The differential equations of constraint must be included in the scheme by which the equations of motion are solved. Let us focus our attention on the Lagrangian multiplier formulation, although the same treatment is applicable to the formulation using the reaction forces explicitly. The Lagrangian multipliers occur linearly in Eqs. (7.4). Thus, the procedure we follow is to solve algebraically (or numerically) $M - N$ Lagrange equations for the λ_j multipliers. That solution then allows us to eliminate the multipliers from the remaining N Lagrange equations. The resulting state-space equations of motion for a nonholonomic system still consist of $2M$ differential equations. The derivative identities represented by Eq. (7.13) provide M equations. Another N equations are the subset of the Lagrange's equations (7.14) from which the Lagrangian multipliers have been removed. The constraint equations represented by Eq. (7.16) provide the remaining $M - N$ equations.

EXAMPLE 7.4 _____

Write the state-space form of the equations of motion for the rolling disk in Example 7.3.

Solution. The equations of motion for a rolling, wobbling disk were numbered 1′ to 5′ in the solution to Example 7.3. The unknowns appearing in those equations are the three Eulerian angles, ψ, ϕ, and θ, and the two Cartesian coordinates, X and Y. We define a column array $\{q\}$ that contains the generalized coordinates as

$$\{q\} = \{\psi \quad \theta \quad \phi \quad X \quad Y\}^{\mathrm{T}}$$

The corresponding state-space vector $\{x\}$ is formed from $\{q\}$ and its first derivative.

$$\{x\} = \begin{Bmatrix} \{q\} \\ \{\dot{q}\} \end{Bmatrix}$$

The derivative identity provides

$$[[U] \quad [0]] \frac{d}{dt} \{x\} = [[0] \quad [U]]\{x\}$$

The equations of motion (1′) to (3′) in the previous solution are second-order in the generalized coordinates. Because we wish to avoid the appearance of such derivatives, we write that set of equations as

$$[M] \frac{d}{dt} \{q\} = \{F\}$$

which is equivalent to

$$[[0] \quad [M]] \frac{d}{dt} \{x\} = \{F\}$$

The coefficient array $[M]$ has three rows and five columns, and $\{F\}$ is a three-element column array. In listing these coefficients, we replace generalized coordinates by the corresponding state-space variable x_1 to x_5, while generalized velocities are replaced by the appropriate choice from x_6 to x_{10}. The first row comes from Eq. (1′) in Example 7.3:

$$M_{11} = \tfrac{1}{2}\,\kappa^2(1 + \cos^2 x_2) \qquad M_{12} = 0 \qquad M_{13} = \cos x_2$$

$$M_{14} = R \cos x_1 \cos x_2 \qquad M_{15} = R \sin x_1 \cos x_2$$

$$F_1 = \kappa^2(x_6 x_7 \sin x_2 \cos x_2 + x_7 x_8 \sin x_2)$$

while Eq. (2′) in Example 7.3 yields the second row,

$$M_{21} = 0 \qquad M_{22} = \tfrac{1}{2}\,\kappa^2 + R^2 \cos^2 x_2 \qquad M_{23} = 0$$

$$M_{24} = -R \sin x_1 \sin x_2 \qquad M_{25} = R \cos x_1 \sin x_2$$

$$F_2 = -(\tfrac{1}{2}\,\kappa^2 x_6^2 - R^2 x_7^2) \sin x_2 \cos x_2 - \kappa^2 x_6 x_8 \sin x_2 - gR \cos x_2$$

Similarly, the third row corresponds to Eq. (3′).

$$M_{31} = \kappa^2 \cos x_2 \qquad M_{32} = 0 \qquad M_{33} = \kappa^2$$

$$M_{34} = R \cos x_1 \qquad M_{35} = R \sin x_1$$

$$F_3 = \kappa^2 x_6 x_7 \sin x_2$$

The complete set of equations of motion includes the constraint equations (4′) and (5′) in the previous solution. Those equations already are in first-order form, so we consider all generalized velocities occurring there to be derivatives of the corresponding generalized coordinate. In other words, we write the constraint equations as

$$[a]\,\frac{d}{dt}\{q\} = -\{b\} = \{0\}$$

which when converted to state-space variables becomes

$$[[a] \quad [0]]\,\frac{d}{dt}\{x\} = \{0\}$$

where $[a]$ has two rows and five columns. As before, we express the elements of $[a]$ in terms of the state-space variables. The expressions are

$$a_{11} = -R \cos x_1 \cos x_2 \qquad a_{12} = R \sin x_1 \sin x_2$$
$$a_{13} = -R \cos x_1 \qquad a_{14} = 1 \qquad a_{15} = 0$$
$$a_{21} = -R \sin x_1 \cos x_2 \qquad a_{22} = -R \cos x_1 \sin x_2$$
$$a_{23} = -R \sin x_1 \qquad a_{24} = 0 \qquad a_{15} = 1$$

Finally, we form the full set of state-space equations. We let each of the matrix equations we have listed form a partitioned row, according to

$$\begin{bmatrix} [U] & [0] \\ [0] & [M] \\ [a] & [0] \end{bmatrix} \frac{d}{dt}\{x\} = \begin{Bmatrix} [[0] \quad [U]]\{x\} \\ \{f\} \\ \{0\} \end{Bmatrix}$$

Note that the coefficient array multiplying $\{\dot{x}\}$ has 10 rows and 10 columns, and that its elements depend on the current values of the elements of $\{x\}$.

7.2.2 Hamilton's Canonical Equations

Hamilton developed a standard (i.e., canonical) set of first-order equations of motion that are quite different from the state-space formulation in the preceding section. The derivation alters the appearance of Lagrange's equations of motion by relating momentum and kinetic energy. The motivation for this formulation may be recognized by referring back to Eq. (6.47), which described the kinetic energy of a single particle, numbered k, in terms of the generalized coordinates.

$$T_k = \frac{1}{2} m_k \sum_{i=1}^{M} \sum_{j=1}^{M} \left(\frac{\partial \bar{r}_k}{\partial q_i} \cdot \frac{\partial \bar{r}_k}{\partial q_j} \right) \dot{q}_i \dot{q}_j + m_k \sum_{j=1}^{M} \left(\frac{\partial \bar{r}_k}{\partial q_j} \cdot \frac{\partial \bar{r}_k}{\partial t} \right) \dot{q}_j + \frac{1}{2} m_k \left(\frac{\partial \bar{r}_k}{\partial t} \cdot \frac{\partial \bar{r}_k}{\partial t} \right)$$

This expression has the same form as Eqs. (7.5) and (7.6), except that the three bracketed terms above were replaced there by M_{ij}, N_j, and T_0, respectively. As a result of this similarity, the process whereby $\partial T/\partial \dot{q}_n$ was evaluated in Eqs. (7.7) may be applied directly to the above. The derivative is found to be

$$\frac{\partial T_k}{\partial \dot{q}_n} = \sum_{i=1}^{M} m_k \frac{\partial \bar{r}_k}{\partial q_i} \cdot \frac{\partial \bar{r}_k}{\partial q_n} \dot{q}_i + m_k \frac{\partial \bar{r}_k}{\partial t} \cdot \frac{\partial \bar{r}_k}{\partial q_n} \tag{7.17}$$

A simple rearrangement of terms leads to the observation that

$$\frac{\partial T_k}{\partial \dot{q}_n} = m_k \left(\sum_{i=1}^{M} \frac{\partial \bar{r}_k}{\partial q_i} \dot{q}_i + \frac{\partial \bar{r}_k}{\partial t} \right) \cdot \frac{\partial \bar{r}_k}{\partial q_n} = m_k \bar{v}_k \cdot \frac{\partial \bar{r}_k}{\partial q_n} \tag{7.18}$$

The term $m_k \bar{v}_k$ is the momentum of this particle. The partial derivative $\partial \bar{r}_k/\partial q_n$ serves to generalize the derivative to fit the type of parameter associated with q_n, for example, linear or angular motion.

For this reason, the derivative $\partial T/\partial \dot{q}_n$ for any system is called the *generalized momentum* p_n corresponding to q_n. Since the potential energy is independent of the generalized velocities, the generalized momentum may also be defined in terms of the Lagrangian function $L = T - V$. Thus

$$p_n = \frac{\partial T}{\partial \dot{q}_n} = \frac{\partial L}{\partial \dot{q}_n} \tag{7.19}$$

We now use the generalized momenta to form the *Hamiltonian function H* according to

$$H = \sum_{i=1}^{M} p_i \dot{q}_i - L \tag{7.20}$$

In this formulation the state variables are considered to be the generalized coordinates and momenta, so the generalized velocities must be removed from all relationships. Such a change of variables may be achieved in a general situation by a sequence of operations:

1. Form the Lagrangian $L = T - V$ in the usual manner, as a function of the generalized coordinates q_i, generalized velocities $\dot{q}_i$, and time t.
2. Derive expressions for the generalized momenta p_i as functions of the q_i, $\dot{q}_i$, and t according to Eq. (7.19).
3. Solve the equations found in the preceding step for the $\dot{q}_i$ in terms of the q_i, p_i, and t.
4. Substitute the expressions for the $\dot{q}_i$ into Eq. (7.20), thereby obtaining the functional form, $H = H(q_1, \ldots, q_M, p_1, \ldots, p_M, t)$.

Let us consider the time derivative of the Hamiltonian function. Since H depends on the generalized coordinates and momenta, we have

$$\dot{H} = \sum_{i=1}^{M} \left(\frac{\partial H}{\partial p_i} \dot{p}_i + \frac{\partial H}{\partial q_i} \dot{q}_i \right) + \frac{\partial H}{\partial t} \tag{7.21}$$

We can also use the definition of H, Eq. (7.20), to form the derivative. This yields

$$\dot{H} = \sum_{i=1}^{M} \left(\dot{p}_i \dot{q}_i + p_i \ddot{q}_i - \frac{\partial L}{\partial q_i} \dot{q}_i - \frac{\partial L}{\partial \dot{q}_i} \ddot{q}_i \right) - \frac{\partial L}{\partial t}$$

$$= \sum_{i=1}^{M} \left(\dot{p}_i \dot{q}_i - \frac{\partial L}{\partial q_i} \dot{q}_i \right) - \frac{\partial L}{\partial t} \tag{7.22}$$

where the simplified form results from substitution of Eq. (7.19). The two descriptions of $\dot{H}$ must match for any set of values of generalized velocities and momenta. Thus, a comparison of like terms in Eqs. (7.21) and (7.22) reveals that

$$\frac{\partial H}{\partial p_i} = \dot{q}_i \qquad \frac{\partial H}{\partial q_i} = -\frac{\partial L}{\partial q_i} \qquad \frac{\partial H}{\partial t} = -\frac{\partial L}{\partial t} \tag{7.23}$$

These identities and the definition of p_i make it possible to express Lagrange's equations in terms of H, rather than L. This yields

$$\dot{p}_i + \frac{\partial H}{\partial q_i} = Q_i \qquad i = 1, 2, \ldots, M \tag{7.24}$$

The combination of Eqs. (7.23) and (7.24) forms a set of first-order differential equations, which are called *Hamilton's canonical equations*.

$$\dot{q}_i = \frac{\partial H}{\partial p_i} \qquad i = 1, 2, \ldots, M$$

$$\dot{p}_i = -\frac{\partial H}{\partial q_i} + Q_i \qquad i = 1, 2, \ldots, M \tag{7.25}$$

These constitute $2M$ coupled, first-order differential equations for the M values of q_i and the M values of p_i. If the generalized coordinates are constrained, these equations must be supplemented by the constraint equations. The constraint forces would then appear in the generalized forces, or alternatively, the Lagrangian multiplier terms may be added to the generalized force array.

Once Hamilton's canonical equations are formulated, they are somewhat easier than the state-space form to implement for numerical solution. Equations (7.25) give the derivative of the state variable as an explicit function at each instant. In contrast, the state-space form, Eq. (7.15), derived in the previous section requires inversion of the square array on the left side in order to obtain the corresponding result. The avoidance of matrix inversion is a substantial benefit of Hamilton's canonical form, particularly when this coefficient array is not constant. This gain is balanced by the fact that the evaluation of the Hamiltonian function requires the intricate change of variables described above, in order to remove the generalized velocities in favor of the generalized momenta.

EXAMPLE 7.5

The double pendulum consists of identical bars of mass m connected at the ideal pin B. Derive the Hamiltonian equations of motion for the system.

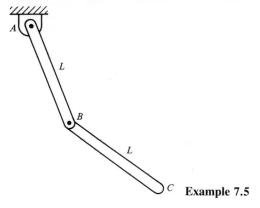

Example 7.5

Solution. The angles $q_1 = \theta_1$ and $q_2 = \theta_2$, which describe the orientation of each bar relative to the vertical, are convenient generalized coordinates for this holonomic, two-degree-of-freedom system. The first step in forming H is to express L in terms of the generalized coordinates and velocities. Bar AB is in pure rotation about end A, but bar BC is in general motion. We shall obtain the velocity of the center of mass G of bar BC by differentiating its position. Thus,

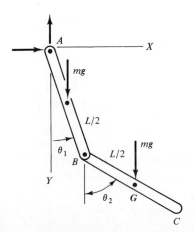

C Free body diagram.

$$\bar{r}_{G/A} = (L \sin \theta_1 + \tfrac{1}{2} L \sin \theta_2) \, \bar{I} + (L \cos \theta_1 + \tfrac{1}{2} L \cos \theta_2) \, \bar{J}$$

$$\bar{v}_G = L \, (\dot{\theta}_1 \cos \theta_1 + \tfrac{1}{2} \dot{\theta}_2 \cos \theta_2) \, \bar{I} - L(\dot{\theta}_1 \sin \theta_1 + \tfrac{1}{2} \dot{\theta}_2 \sin \theta_2) \, \bar{J}$$

The kinetic energy is the sum of the rotational energy of bar AB relative to end A, and of the translational and rotational energy of bar BC relative to point G. Thus,

$$T = \tfrac{1}{2} (\tfrac{1}{3} mL^2) \, \dot{\theta}_1^2 + \tfrac{1}{2} m \bar{v}_G \cdot \bar{v}_G + \tfrac{1}{2} (\tfrac{1}{12} mL^2) \, \dot{\theta}_2^2$$

$$= \tfrac{1}{2} mL^2 \, [\tfrac{4}{3} \dot{\theta}_1^2 + \tfrac{1}{3} \dot{\theta}_2^2 + \dot{\theta}_1 \dot{\theta}_2 \cos (\theta_2 - \theta_1)]$$

The elevation of pin A is a convenient datum for gravitational potential energy. The corresponding Lagrangian is

$$L = T - V = \tfrac{1}{2} mL^2 \, [\tfrac{4}{3} \dot{\theta}_1^2 + \tfrac{1}{3} \dot{\theta}_2^2 + \dot{\theta}_1 \dot{\theta}_2 \cos (\theta_2 - \theta_1)]$$

$$+ mg \frac{L}{2} \cos \theta_1 + mgL \, (\cos \theta_1 + \tfrac{1}{2} \cos \theta_2)$$

The generalized momenta are

$$p_1 = \frac{\partial L}{\partial \dot{\theta}_1} = mL^2 \, [\tfrac{4}{3} \dot{\theta}_1 + \tfrac{1}{2} \dot{\theta}_2 \cos (\theta_2 - \theta_1)]$$

$$p_2 = \frac{\partial L}{\partial \dot{\theta}_2} = mL^2 \, [\tfrac{1}{3} \dot{\theta}_2 + \tfrac{1}{2} \dot{\theta}_1 \cos (\theta_2 - \theta_1)]$$

We solve these relations for $\dot{\theta}_1$ and $\dot{\theta}_2$, which yields

$$\dot{\theta}_1 = \frac{12 p_1 - 18 p_2 \cos (\theta_2 - \theta_1)}{mL^2 [16 - 9 \cos^2 (\theta_2 - \theta_1)]}$$

$$\dot{\theta}_2 = \frac{-18 p_1 \cos (\theta_2 - \theta_1) + 48 p_2}{mL^2 [16 - 9 \cos^2 (\theta_2 - \theta_1)]}$$

We may now form the Hamiltonian. By definition,

$$H = p_1 \dot{\theta}_1 + p_2 \dot{\theta}_2 - L$$

from which we eliminate $\dot{\theta}_1$ and $\dot{\theta}_2$ by substituting the above expressions. Thus,

$$H = \frac{12 p_1^2 - 36 p_1 p_2 \cos (\theta_2 - \theta_1) + 48 p_2^2}{mL^2 [16 - 9 \cos^2 (\theta_2 - \theta_1)]} - \frac{1}{2mL^2} \, [16 - 9 \cos^2 (\theta_2 - \theta_1)]^{-2}$$

$$\times \{ \tfrac{4}{3} \, [12 p_1 - 18 p_2 \cos (\theta_2 - \theta_1)]^2 + \tfrac{1}{3} \, [-18 p_1 \cos (\theta_2 - \theta_1) + 48 p_2]^2$$

$$+ [12 p_1 - 18 p_2 \cos (\theta_2 - \theta_1)][-18 p_1 \cos (\theta_2 - \theta_1) + 48 p_2]$$

$$\times \cos (\theta_2 - \theta_1) \} - mgl \, (\tfrac{3}{2} \cos \theta_1 + \tfrac{1}{2} \cos \theta_2)$$

Collecting like coefficients enables us to rewrite H in a simpler form as

$$H = \frac{1}{2K(\theta_1, \theta_2)} \, [c_{11}(\theta_1, \theta_2) p_1^2 + 2 c_{12}(\theta_1, \theta_2) p_1 p_2 + c_{22}(\theta_1, \theta_2) p_2^2]$$

$$- mgL \, (\tfrac{3}{2} \cos \theta_1 + \tfrac{1}{2} \cos \theta_2)$$

where $K = mL^2[16 - 9\cos^2(\theta_2 - \theta_1)]$
$c_{11} = 192 - 108\cos^2(\theta_2 - \theta_1)$
$c_{12} = -[288 - 162\cos^2(\theta_2 - \theta_1)]\cos(\theta_2 - \theta_1)$
$c_{22} = 768 - 432\cos^2(\theta_2 - \theta_1)$

The last step preceding forming Hamilton's canonical equations is to identify that the generalized forces vanish, $Q_1 = Q_2 = 0$, because the constraints imposed by pins A and B are satisfied.

Differentiating H with respect to p_1 and p_2, in accord with the first of Eqs. (7.25), yields

$$\dot{\theta}_1 = \frac{1}{K}(c_{11}p_1 + c_{12}p_2)$$

$$\dot{\theta}_2 = \frac{1}{K}(c_{12}p_1 + c_{22}p_2)$$

The second part of Hamilton's equations (7.25) requires differentiation of H with respect to θ_1 and θ_2. However, K and the c_{ij} depend only on $\theta_2 - \theta_1$, so derivatives with respect to θ_2 are the negative of those with respect to θ_1. The resulting equations are

$$\dot{p}_1 = -\frac{1}{2K^2}(d_{11}p_1^2 + 2d_{12}p_1p_2 + d_{22}p_2^2) - \tfrac{3}{2}mgL\sin\theta_1$$

$$\dot{p}_2 = +\frac{1}{2K^2}(d_{11}p_1^2 + 2d_{12}p_1p_2 + d_{22}p_2^2) - \tfrac{1}{2}mgL\sin\theta_2$$

where

$$d_{ij} = K\frac{\partial c_{ij}}{\partial\theta_1} - \frac{\partial K}{\partial\theta_1}c_{ij} \qquad i, j = 1 \text{ or } 2$$

In the present context, little would be learned if we were to evaluate the coefficients d_{ij}. Indeed, we can see that the operations required to form Hamilton's canonical equations are far more tedious than those required for Lagrange's equations.

___

7.2.3 Ignorable Coordinates

The concept of generalized momenta leads to further understanding of the laws of mechanics in some special circumstances. The first arises when the Lagrangian function depends on a particular generalized velocity $\dot{q}_n$, but not on the generalized coordinate q_n itself. Since $\partial L/\partial q_n = 0$ in this case, Lagrange's equation for this generalized coordinate is

$$\frac{d}{dt}\left(\frac{\partial L}{\partial\dot{q}_n}\right) = Q_n \qquad (7.26)$$

By virtue of the definition of p_n, this equation may be integrated to obtain

$$p_n\big|_{t_2} = p_n\big|_{t_1} + \int_{t_1}^{t_2} Q_n(\tau)\, d\tau \tag{7.27}$$

This equation is a generalization of the impulse-momentum principles. Both linear and angular momentum are described by it, depending on the type of geometric quantity associated with q_n. It is interesting to note that if Q_n depends on any of the generalized coordinates q_n, we cannot obtain momentum principles in this manner. Such a situation corresponds to the Newtonian formulation of position-dependent forces, where momentum principles are not used because the impulse cannot be evaluated.

Consider the more restrictive situation, in which L does not depend on q_n *and* the corresponding generalized force Q_n vanishes, $Q_n = 0$. We find from the foregoing that p_n is constant, which corresponds to *conservation of generalized momentum*. When such a situation occurs, the corresponding generalized coordinate q_n is said to be *ignorable*. This term arises because the relation

$$p_n = \frac{\partial T}{\partial \dot{q}_n} = \frac{\partial L}{\partial \dot{q}_n} = \text{a constant} \tag{7.28}$$

may be solved for $\dot{q}_n$ in terms of the other generalized coordinates and velocities. That relation may be substituted into each of the remaining Lagrange's equations in order to obtain equations of motion in which neither q_n nor $\dot{q}_n$ appears.[†]

We used this procedure to simplify the equations of motion in Example 6.12. The procedure may be extended to treat a system in which several generalized coordinates are ignorable. However, it is important to recognize that all the Lagrange's equations must be formed prior to substitution for the ignorable coordinates. (Treating the conserved momenta as constants in the expression for T or L when Lagrange's equations are formulated is incorrect, because doing so does not allow for a description of the full effect of a variation of each generalized coordinate.)

It is possible, as an alternative, to use the relations for the ignorable coordinates to derive an equivalent system having a reduced number of degrees of freedom. The method by which this reduction may be achieved is *Routh's method for the ignoration of coordinates*. Suppose that from the original set of M generalized coordinates, there are $M - J$ ignorable coordinates, which we designate as $q_{J+1}, \ldots, q_M$. Then Eqs. (7.28) apply for $n = J + 1, \ldots, M$, that is, $p_{J+1}, \ldots, p_M$ are constants. Solving Eqs. (7.28) simultaneously allows us to evaluate the generalized velocities $\dot{q}_{J+1}, \ldots, \dot{q}_M$ as functions of the J generalized coordinates and velocities that are not ignorable, and, possibly, time. The *Routhian function* is defined in terms of the Lagrangian for the system, according to

$$R = L - \sum_{n=J+1}^{M} \frac{\partial L}{\partial \dot{q}_n}\, \dot{q}_n = L - \sum_{n=J+1}^{M} p_n \dot{q}_n \tag{7.29}$$

[†]Ignorable coordinates are sometimes called *cyclic coordinates*. This name stems from the observation that many cases where a generalized coordinate is ignorable involve rotation about an axis. They are also known as *kinosthenic coordinates*.

After substitution of the relations for the generalized velocities of the ignorable coordinates, the dependence of the Routhian is $R = R(q_1, \ldots, q_J, \dot{q}_1, \ldots, \dot{q}_J, p_{J+1}, \ldots, p_M, t)$.

We consider first the variation of R based on its functional dependence. Even though $p_{J+1}, \ldots, p_M$ are constants in the actual motion, they must be varied in deriving the equations of motion. Thus,

$$\delta R = \sum_{n=1}^{J} \left(\frac{\partial R}{\partial q_n} \delta q_n + \frac{\partial R}{\partial \dot{q}_n} \delta \dot{q}_n \right) + \sum_{n=J+1}^{M} \frac{\partial R}{\partial p_n} \delta p_n \tag{7.30a}$$

In comparison, a variation based on the definition of R, Eq. (7.29), yields

$$\delta R = \sum_{n=1}^{J} \frac{\partial L}{\partial q_n} \delta q_n + \sum_{n=1}^{M} \frac{\partial L}{\partial \dot{q}_n} \delta \dot{q}_n - \sum_{n=J+1}^{M} (p_n \delta \dot{q}_n + \delta p_n \dot{q}_n)$$

$$= \sum_{n=1}^{J} \left(\frac{\partial L}{\partial q_n} \delta q_n + \frac{\partial L}{\partial \dot{q}_n} \delta \dot{q}_n \right) - \sum_{n=J+1}^{M} \delta p_n \dot{q}_n \tag{7.30b}$$

The alternative forms of δR must be valid for arbitrary virtual increments, which means that like coefficients must match. Hence, we find from Eqs. (7.30) that

$$\frac{\partial R}{\partial q_n} = \frac{\partial L}{\partial q_n} \qquad \frac{\partial R}{\partial \dot{q}_n} = \frac{\partial L}{\partial \dot{q}_n} \qquad n = 1, \ldots, J \tag{7.31a}$$

$$\frac{\partial R}{\partial p_n} = -\dot{q}_n \qquad n = J + 1, \ldots, M \tag{7.31b}$$

According to Eqs. (7.31a), the original Lagrange's equations for the nonignorable generalized coordinates may be replaced by the equations of the same form that feature the Routhian.

$$\frac{d}{dt} \left(\frac{\partial R}{\partial \dot{q}_j} \right) - \frac{\partial R}{\partial q_j} = Q_j \qquad n = 1, \ldots, J \tag{7.32}$$

In other words, the Routhian may be considered to be the Lagrangian for a system described by J generalized coordinates. If the generalized coordinates are an unconstrained set, then the Routhian represents an equivalent system having J degrees of freedom.

EXAMPLE 7.6 ───

A particle slides on the interior of a smooth surface of revolution whose shape is defined in cylindrical coordinates as $r = f(z)$, where z is the vertical distance along the axis. Derive the differential equation of motion whose solution gives z as a function of time.

Solution. The azimuthal angle θ and elevation z are useful generalized coordinates, $q_1 = z$, $q_2 = \theta$. The radial distance r is subject to the configuration constraint, $r = f(z)$, so the radial velocity is

$$\dot{r} = f' \dot{z}$$

where a prime denotes differentiation with respect to z. The kinetic energy for a particle of mass m is therefore

$$T = \tfrac{1}{2} m[(f')^2 + 1]\dot{z}^2 + \tfrac{1}{2} mf^2\dot{\theta}^2$$

We let $z = 0$ be the gravitational datum. The corresponding Lagrangian is

$$L = \tfrac{1}{2} m [(f')^2 + 1]\dot{z}^2 + \tfrac{1}{2} mf^2\dot{\theta}^2 - mgz$$

We note that θ does not appear explicitly in L, and that the generalized force $Q_\theta = 0$ because the system is conservative. These are the conditions for which θ is ignorable. The corresponding generalized momentum p_θ is constant, where

$$p_\theta = \frac{\partial L}{\partial \dot{\theta}} = mf^2\dot{\theta}$$

We solve this expression for the generalized velocity,

$$\dot{\theta} = \frac{p_\theta}{mf^2}$$

and use this relation to replace $\dot{\theta}$ wherever it occurs in the Routhian. The result is

$$
\begin{aligned}
R &= L - p_\theta\dot{\theta} \\
&= \tfrac{1}{2} m[(f')^2 + 1]\dot{z}^2 + \tfrac{1}{2} mf^2 \left(\frac{p_\theta}{mf^2}\right)^2 - mgz - p_\theta \frac{p_\theta}{mf^2} \\
&= \tfrac{1}{2} m[(f')^2 + 1]\dot{z}^2 - \frac{p_\theta^2}{2mf^2} - mgz
\end{aligned}
$$

Since R depends only on z and $\dot{z}$, it represents an equivalent one-degree-of-freedom system. The first term in R, which contains the generalized velocity, is the equivalent kinetic energy, while the equivalent potential energy is the negative of the sum of the remaining terms.

The equation of motion is Lagrange's equation for $q_1 = z$, with R used instead of the Lagrangian. It is necessary to recognize that f and f' are functions of z, which is, in turn, a function of time. Thus, the required derivatives of R are

$$\frac{d}{dt}\left(\frac{\partial R}{\partial \dot{z}}\right) = m\frac{d}{dt}[(f')^2 + 1]\dot{z} = m[(f')^2 + 1]\ddot{z} + 2mf'f''\dot{z}^2$$

$$\frac{\partial R}{\partial z} = mf'f''\dot{z}^2 + \frac{p_\theta^2}{mf^3} f' - mg$$

Setting $Q_1 = 0$ for a conservative system leads to

$$[(f')^2 + 1]\ddot{z} + f'f''\dot{z}^2 - \frac{p_\theta^2}{mf^3} f' = -g$$

7.2.4 Conservation Theorems

The previous section derived an extended momentum conservation principle. Here we shall develop concepts that are related to conservation of energy. In the course of deriving Hamilton's canonical equations, we saw in Eq. (7.22) that the time derivative of the

Hamiltonian function H is given by

$$\dot{H} = \sum_{i=1}^{M} \left(\dot{p}_i \dot{q}_i - \frac{\partial L}{\partial q_i} \dot{q}_i \right) - \frac{\partial L}{\partial t} \tag{7.33}$$

Let us remove p_i from this relation by factoring out $\dot{q}_i$, and then recalling the definition of p_i and Lagrange's equations. This yields

$$\dot{H} = \sum_{i=1}^{M} \left[\frac{d}{dt} \left(\frac{\partial L}{\partial \dot{q}_n} \right) - \frac{\partial L}{\partial q_i} \right] \dot{q}_i - \frac{\partial L}{\partial t}$$

$$\dot{H} = \sum_{i=1}^{M} Q_i \dot{q}_i - \frac{\partial L}{\partial t} \tag{3.34}$$

A corollary of this relation is the result that H is constant for a conservative, time-independent system, which is *Jacobi's integral*.

Now recall that the virtual work is

$$\delta W = \sum_{i=1}^{M} Q_i \, \delta q_i \tag{7.35}$$

Changing the virtual increments in δW to actual time derivatives suggests that the summation on the right side of Eq. (7.35) might be related to the rate at which the forces acting on the system do work. If this were so, $\dot{H}$ would be related to the rate of change of the mechanical energy. Let us explore this question.

The generalized momenta are found from Eqs. (7.5) to (7.7), and (7.19) to be

$$p_n = \frac{\partial T}{\partial \dot{q}_n} = \sum_{j=1}^{M} M_{nj} \dot{q}_j + N_n \tag{7.36}$$

Substitution of this expression into the definition of H, Eq. (7.20), leads to

$$H = \sum_{n=1}^{M} \sum_{j=1}^{M} M_{nj} \dot{q}_j \dot{q}_n + \sum_{n=1}^{M} N_n \dot{q}_n - L = 2T_2 + T_1 - L$$

$$= 2T_2 + T_1 - (T_2 + T_1 + T_0 - V) = T_2 - T_0 + V \tag{7.37}$$

Now recall that the derivation of Eq. (6.48), which is the standard form of the kinetic energy, showed that if the system is catastatic, then the kinetic energy contains only quadratic terms in the generalized velocities, that is, $T = T_2$. Equation (7.37) then reduces to

$$H = T + V = E \tag{7.38}$$

where E is the total mechanical energy of the system. We therefore conclude that the Hamiltonian function reduces to the mechanical energy if the system is catastatic. Clearly, the important case of a scleronomic system fits this specification.

We observed another consequence of catastatic constraints in Chapter 6 when we developed the kinematical method for virtual displacement. By definition, the velocity of any point in a castastatic system is homogeneous in the generalized velocities. As a consequence, a velocity $\bar{v}$ and a virtual displacement $\delta \bar{r}$ have analogous forms.

$$\bar{v} = \sum_{j=1}^{M} \frac{\partial \bar{r}}{\partial q_j} \dot{q}_j \qquad \delta \bar{r} = \sum_{j=1}^{M} \frac{\partial \bar{r}}{\partial q_j} \delta q_j$$

Since $d\bar{r} = \bar{v}\, dt$, the virtual displacements are possible infinitesimal movements of the system. Therefore, expressions for virtual changes may be employed to treat actual increments, and vice versa. Replacing virtual quantities by real differential increments leads to the conclusion that the power supplied to a catastatic system is

$$\dot{W} = \sum_{i=1}^{M} Q_i \dot{q}_i \tag{7.39}$$

If all constraints are independent of time, rather than merely catastatic, we have $\partial L / \partial t = 0$, as well as $H = E$. In this case, Eq. (7.34) reduces to the power form of the work-energy principle,

$$\text{Power} = \dot{W} = \dot{E} \tag{7.40}$$

In contrast, if any of the constraints are not catastatic, $H \neq E$, from which it follows that Eq. (7.34) for $\dot{H}$ and Eq. (7.40) for $\dot{E}$ are independent principles. The work-energy principle is generally valid. However, the work to the system might come from reactions that impose the time-dependent motion, as well as from the applied forces. It is possible that Jacobi's integral applies, even if the mechanical energy is not conserved. When H is conserved, its constant value is determined from the initial conditions, which leads to a relation between the generalized velocities at different positions.

EXAMPLE 7.7 ⎯⎯⎯⎯⎯⎯⎯⎯⎯⎯⎯⎯⎯⎯⎯⎯⎯⎯⎯⎯⎯⎯⎯⎯⎯⎯⎯⎯⎯⎯⎯⎯⎯⎯⎯⎯

Bar AB, whose mass is m, is pinned to the vertical shaft. The assembly precesses about the vertical axis at the constant rate Ω due to the torque Γ. Consider the angle of nutation θ as the sole generalized coordinate. Compare the rate of change of the Hamiltonian in such a formulation with the rate of change of the mechanical energy of the system. The inertia of the vertical shaft is negligible.

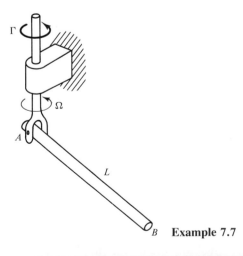

B **Example 7.7**

Solution. We consider the restriction that Ω is constant to be a time-dependent constraint on the precession angle ψ. Accordingly, we select $q_1 = \theta$ as the sole generalized coordinate. We must express the kinetic energy in terms of θ in order to form both H and E. The angular velocity of the bar is

$$\bar{\omega} = \Omega \bar{K} - \dot{\theta}\bar{j} = -\Omega \cos \theta \bar{i} - \dot{\theta}\bar{j} + \Omega \sin \theta \bar{k}$$

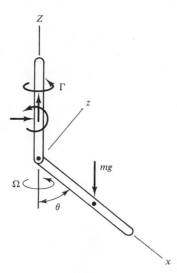

x Free body diagram.

The kinetic energy may be considered to be purely rotational relative to the stationary end A. The bar is slender, and xyz are principal axes, so we have

$$T = \tfrac{1}{2}\,(I_{yy}\omega_y^2 + I_{zz}\omega_z^2) = \tfrac{1}{6}\,mL^2(\dot{\theta}^2 + \Omega^2 \sin^2 \theta)$$

For a gravitational datum at the elevation of end A, the Lagrangian is

$$L = \tfrac{1}{6}\,mL^2(\dot{\theta}^2 + \Omega^2 \sin^2 \theta) + mg\,\frac{L}{2}\cos \theta$$

The generalized momentum corresponding to the only generalized coordinate is

$$p_1 = \frac{\partial L}{\partial \dot{\theta}} = \tfrac{1}{3}\,mL^2\dot{\theta}$$

We solve this expresion for $\dot{\theta}$, and use the result to eliminate $\dot{\theta}$ from the Hamiltonian. Thus,

$$H = p_1\dot{\theta} - L = \frac{3}{2mL^2}\,p_1^2 - \tfrac{1}{6}\,mL^2\Omega^2 \sin^2 \theta - mg\,\frac{L}{2}\cos \theta$$

No constraints are violated in a virtual movement resulting from incrementing θ by $\delta\theta$. Therefore, $Q_1 = 0$. Furthermore, L does not explicitly depend on time, so $\partial L/\partial t = 0$. (This would not be true if Ω was a given time-dependent function.)

According to Eq. (7.34), these conditions correspond to $\dot{H} = 0$, from which it follows that the Hamiltonian is conserved,

$$\frac{3}{2\,mL^2}\,p_1^2 - \tfrac{1}{6}\,mL^2\Omega^2\sin^2\theta - mg\,\frac{L}{2}\cos\theta = H_0$$

where H_0 is the value when the motion was initiated. This conservation equation may be expressed in terms of $\dot{\theta}$ by substituting $p_1 = mL^2\dot{\theta}/3$, which yields

$$\tfrac{1}{6}\,mL^2(\dot{\theta}^2 - \Omega^2\sin^2\theta) - mg\,\frac{L}{2}\cos\theta = H_0$$

The mechanical energy is

$$E = T + V = \tfrac{1}{6}\,mL^2(\dot{\theta}^2 + \Omega^2\sin^2\theta) - mg\,\frac{L}{2}\cos\theta$$

The only nonconservative force that does work is the couple Γ, so the power input is

$$\dot{W} = \Gamma\Omega$$

The work-energy principle, $\dot{E} = \dot{W}$, then yields

$$\dot{E} = \tfrac{1}{3}\,mL^2(\dot{\theta}\ddot{\theta} + \Omega^2\dot{\theta}\sin\theta\cos\theta) + mg\,\frac{L}{2}\dot{\theta}\sin\theta = \Gamma\Omega$$

For comparison, we now differentiate the equation for the constant Hamiltonian.

$$\dot{H} = \tfrac{1}{3}\,mL^2(\dot{\theta}\ddot{\theta} - \Omega^2\dot{\theta}\sin\theta\cos\theta) + mg\,\frac{L}{2}\dot{\theta}\sin\theta = 0$$

Aside from the presence of the common factor $\dot{\theta}$, this is the same equation of motion that we would obtain from the application of Lagrange's equations for $q_1 = \theta$. We also may obtain an expression for Γ by forming $\dot{E} - \dot{H}$, which yields

$$\Gamma = \tfrac{2}{3}\,mL^2\,\Omega\dot{\theta}\sin\theta\cos\theta$$

This relation could have been obtained from the Lagrangian formulation by using constrained generalized coordinates $q_1 = \theta$, $q_2 = \psi$, which must satisfy the constraint $\dot{\psi} = \Omega$.

7.3 GIBBS-APPELL EQUATIONS FOR QUASI-COORDINATES

The methods by which dynamic-system equations were reduced to first-order form in the preceding section are closely linked with the Lagrangian of the system. The formulation of first-order equations we shall develop here is out of this mainstream, in that it is not founded on the kinetic energy. It has other assets beyond leading to a first-order form, the primary one being that it yields equations for nonholonomic systems in which constraint forces and/or Lagrangian multipliers do not occur. We begin the development by extending the concept of generalized coordinates. Then we will derive the kinetics principle by considering a system of particles. After we have derived the basic relations for such a system, we will introduce the effect of constraint equations and the corresponding reactions. The last step in the derivation will be to specialize the equations to the case of systems of rigid bodies.

7.3.1 Quasi-Coordinates and Generalized Forces

By definition, we may uniquely describe the instantaneous position of any point in a system in terms of M generalized coordinates q_j, where $M \geqq N$ and N is the number of degrees of freedom of the system. We could correspondingly describe the instantaneous velocity of that point in terms of M generalized velocities $\dot{q}_j$. However, it might be more desirable in some cases to use a different set of parameters, called quasi-coordinates γ_j, whose derivatives are meaningful but whose values are not. For example, Euler's equations of motion for a rigid body are expressed in terms of the angular velocity components, but there is no angular orientation vector that may be used to form a corresponding set of generalized coordinates.

Consider a system of P particles. Let x_i, $i = 1, \ldots, 3P$ denote the set of (absolute) Cartesian coordinates for all particles in the system. The position of the system is known in terms of the generalized coordinates, but the value of the time t must also be specified if the physical constraints imposed on the system are time-dependent. We therefore have $x_i = x_i(q_j, t)$, where functional dependence on a set of variables will henceforth be indicated by a generic variable in that set. Differentiation of the position leads to expressions for the velocity components of each particle,

$$\dot{x}_i = \sum_{j=1}^{M} \frac{\partial x_i}{\partial q_j} \dot{q}_j + \frac{\partial x_i}{\partial t} \tag{7.41}$$

where all partial derivatives may be functions of the generalized coordinates and time. Clearly, $\dot{x}_i \, dt$ is an exact differential that may be integrated to return to the functional dependence of the position coordinates.

Let us generalize Eq. (7.41) such that the rate variable it defines is no longer a perfect differential. Specifically, we replace the physical velocity components $\dot{x}_i$ by parameters $\dot{\gamma}_i$, and also replace the partial derivatives by arbitrary functions. The γ_i are called *quasi-coordinates*. Thus, the *quasi-coordinate velocities* $\dot{\gamma}_i$ (or more briefly, the *quasi-velocities*) are related to the generalized coordinates by

$$\dot{\gamma}_i = \sum_{j=1}^{M} u_{ij}(q_k, t)\dot{q}_j + u_i(q_k, t) \qquad i = 1, \ldots, M \tag{7.42}$$

where the functions u_{ij} and u_i depend on the definitions of the quasi- and generalized coordinates. Note that we have defined only M quasi-coordinates, because that is the number of generalized velocities $\dot{q}_i$.

The differential displacement in a time interval dt corresponding to Eq. (7.42) is

$$d\gamma_i = \dot{\gamma}_i \, dt = \sum_{j=1}^{M} u_{ij}(q_k, t) \, dq_j + u_i(q_k, t) \, dt \qquad i = 1, \ldots, M \tag{7.43}$$

Because the coefficients u_{ij} and u_i are arbitrary, integrating $d\gamma_i$ in order to obtain a value of the quasi-coordinate γ_i is only possible if we express the dependence of the generalized coordinates and velocities on time. However, we do not know such time dependencies until we have solved the equations of motion, so the quasi-coordinates are not useful for specifying position. This is the reason for using the prefix "quasi" to describe the γ_i.

In the developments that follow, the only use of generalized coordinates will be to represent position. Quasi-coordinates will be used to represent velocities and accelerations, as well as virtual displacements. We may replace any dependence on the generalized velocities $\dot{q}_i$ by a comparable dependence on the quasi-velocities $\dot{\gamma}_i$. Equations (7.42) represent a set of M linear equations relating the $\dot{q}_i$ and the $\dot{\gamma}_i$. Solving them yields relations for the generalized velocities whose form is

$$\dot{q}_i = \sum_{j=1}^{M} v_{ij}(q_k, t)\dot{\gamma}_j + v_i(q_k, t) \qquad i = 1, \ldots, M \tag{7.44}$$

The linear form of the transformation between the $\dot{q}_i$ and the $\dot{\gamma}_i$ has an important implication for constraint equations. Consider the substitution of Eqs. (7.44) into Eqs. (7.1). The result is a set of constraints on the quasi-velocities, whose form is similar to Eq. (7.1).

$$\sum_{j=1}^{M} A_{ij}(q_k, t)\dot{\gamma}_j + B_i(q_k, t) = 0 \qquad i = 1, \ldots, M - N \tag{7.45}$$

Note that the coefficients A_{ij} appearing above are related to, but not identical to, the coefficients a_{ij} in Eq. (7.1).

In order to treat virtual displacements, we multiply Eq. (7.44) by dt, which leads to an expression relating the increments in the quasi- and generalized coordinate in an infinitesimal time interval

$$dq_i = \sum_{j=1}^{M} v_{ij}(q_k, t)\, d\gamma_j + v_i(q_k, t)\, dt$$

Since time is held constant in a virtual movement, the virtual increments imparted to the generalized coordinates are related to the corresponding increments in the quasi-coordinates by

$$\delta q_i = \sum_{j=1}^{M} v_{ij}(q_k, t)\, \delta\gamma_j \qquad i = 1, \ldots, M \tag{7.46}$$

As a consequence of Eq. (7.46), we may form a generalized force Γ_j corresponding to each quasi-coordinate γ_j. The definition of the generalized forces Q_i associated with a set of generalized coordinates is that the virtual work has the form $\delta W = \Sigma\, q_i\, \delta q_i$. Substitution of Eq. (7.46) into this definition yields an analogous form for the virtual work in terms of the Γ_j, specifically,

$$\delta W = \sum_{j=1}^{M} \Gamma_j\, \delta\gamma_j \tag{7.47}$$

where the generalized forces associated with the quasi-coordinates are

$$\Gamma_j = \sum_{i=1}^{M} Q_i v_{ij}(q_k, t) \tag{7.48}$$

We therefore may evaluate the generalized forces corresponding to the quasi-cooordinates directly, according to Eq. (7.47), or alternatively, by transforming according to Eq. (7.48) the generalized forces associated with the generalized coordinates. The latter is particularly useful for conservative forces. To do so, we merely use $Q_i = -\partial V/\partial q_i$, where $V(q_k, t)$ is the potential energy of the conservative forces.

The similarity of virtual increments in generalized and quasi-coordinates leads to an explicit statement of the manner in which the constraint forces affect the generalized forces. Since the constraint equations (7.45) have the same form as Eq. (7.1) for constrained generalized coordinates, the contribution of the reactions to the generalized forces must be as indicated in Eq. (7.3). Let λ_i be the Lagrangian multiplier for the ith constraint on the quasi-coordinates. Then the generalized force Γ_j is a superposition of $\Gamma_j^{(a)}$ due to active forces applied to the system, and the Lagrangian multiplier contributions $\lambda_i A_{ij}$ due to each constraint. The generalized forces are therefore given by

$$\Gamma_j = \Gamma_j^{(a)} + \sum_{i=1}^{M-N} \lambda_i A_{ij} \tag{7.49}$$

As we found in Section 7.1, using Lagrangian multipliers allows us to account for the reactions associated with constraints without actually formulating the virtual work they do. Even better, it is possible to employ a set of quasi-coordinates that entirely avoid the appearance of reaction forces in the equations of motion. Let us select a subset of N quasi-coordinates, which we denote as $\tilde{\gamma}_1, \tilde{\gamma}_2, \ldots, \tilde{\gamma}_N$. We correspondingly rewrite Eqs. (7.45) as

$$\sum_{j=1}^{M-N} C_{ij}(q_k, t)\dot{\gamma}_{(j+N)} = -\sum_{j=1}^{N} A_{ij}(q_k, t)\dot{\tilde{\gamma}}_j - B_i(q_k, t) \qquad i = 1, \ldots, M - N \tag{7.50a}$$

where

$$C_{ij} = A_{i(j+N)} \qquad i, j = 1, \ldots, M - N \tag{7.50b}$$

If the square array of coefficients C_{ij} is nonsingluar, that is, if $|C| \neq 0$, then the constraint equations may be solved for the remaining quasi-velocities in terms of the N parameters $\dot{\tilde{\gamma}}_i$. Clearly, any set of values of the $\dot{\tilde{\gamma}}_i$ may be selected without violating the constraint equations; they represent motions that are always consistent with the constraints. Hence, the $\dot{\tilde{\gamma}}_i$ are *unconstrained quasi-coordinates*.

As a corollary of Eq. (7.50a), the generalized velocities are uniquely related to the unconstrained quasi-velocities. Such relations have the same linear form as Eq. (7.44), but the coefficient functions are altered. The new expressions are

$$\dot{q}_i = \sum_{j=1}^{N} \tilde{v}_{ij}(q_k, t)\dot{\tilde{\gamma}}_j + \tilde{v}_i(q_k, t) \qquad i = 1, \ldots, M \tag{7.51}$$

Because the values of the $\dot{\tilde{\gamma}}_i$ are not constrained, virtual increments $\delta\tilde{\gamma}_i$ may be arbitrarily selected. Thus, the generalized force $\tilde{\Gamma}_i$ corresponding to each $\tilde{\gamma}_i$ will not contain any terms associated with reactions, that is, $\tilde{\Gamma}_i = \tilde{\Gamma}_i^{(a)}$. We may obtain these generalized forces according to either Eq. (7.47) or Eq. (7.48), which become

$$\delta W = \sum_{j=1}^{N} \tilde{\Gamma}_j^{(a)} \, \delta\tilde{\gamma}_j \tag{7.52a}$$

$$\tilde{\Gamma}_j = \sum_{i=1}^{N} Q_i \bar{v}_{ij}(q_k, t) \tag{7.52b}$$

By eliminating the reaction forces from the generalized forces, unconstrained quasi-coordinates simplify nonholonomic systems in the same way that unconstrained generalized coordinates simplify holonomic systems.

7.3.2 Gibbs-Appell Equations

It is possible to modify Lagrange's equations such that the term that depends on the generalized velocities $\dot{q}_j$ is replaced by terms that depend on the quasi-velocities $\dot{\gamma}_j$. However, the transformation may be conveniently carried out only for a scleronomic system. There is little to be gained from such a derivation, especially when an alternative principle that is more widely applicable is available.

We begin by formulating a function S for the system of particles. It resembles the kinetic energy, except that the speed is replaced by the magnitude of the acceleration. In particular, if the Cartesian position coordinates of the various particles in the system are denoted by the generic symbol x_i, then we have

$$S = \frac{1}{2} \sum_{i=1}^{3P} m_i \ddot{x}_i^2 \tag{7.53}$$

In some presentations, S is called the energy of acceleration, but it is more appropriate to call it the *Gibbs-Appell function*, after the researchers who developed these concepts (see Appell and Gibbs in the references at the end of this chapter).

The relationship between S and the quasi-coordinates is surprisingly straightforward to derive. As before, the initial derivation considers the situation in which we employ M generalized coordinates and M quasi-coordinates with $M > N$. Equation (7.44) indicates that all velocity components $\dot{x}_i$ may be expressed as

$$\dot{x}_i = \sum_{j=1}^{M} w_{ij}(q_k, t)\dot{\gamma}_j + w_i(q_k, t) \tag{7.54}$$

The acceleration obtained by differentiating this velocity component is

$$\ddot{x}_i = \sum_{j=1}^{M} w_{ij}\ddot{\gamma}_j + \sum_{j=1}^{M} \left(\frac{\partial w_{ij}}{\partial t} + \sum_{k=1}^{N} \frac{\partial w_{ij}}{\partial q_k} \dot{q}_k \right)\dot{\gamma}_j + \frac{\partial w_i}{\partial t} + \sum_{k=1}^{M} \frac{\partial w_i}{\partial q_k} \dot{q}_k \tag{7.55}$$

We will soon use two properties of this acceleration equation, both of which result from using Eq. (7.44) to express $\dot{q}_k$.

1. The accelerations depend on the values of the generalized coordinates q_i, the quasi-velocities $\dot{\gamma}_i$, the quasi-accelerations $\ddot{\gamma}_i$, and time t.
2. The quasi-accelerations $\ddot{\gamma}_i$ appear in the physical accelerations components only in the first summation, where the dependence is linear.

As a result of the first observation, we conclude that $S = S(q_i, \dot{\gamma}_i, \ddot{\gamma}_j, t)$. Let us consider the derivative of S with respect to a specific $\ddot{\gamma}_j$. Differentiating the definition, Eq. (7.53), yields

$$\frac{\partial S}{\partial \ddot{\gamma}_j} = \sum_{i=1}^{3P} m_i \ddot{x}_i \frac{\partial \ddot{x}_i}{\partial \ddot{\gamma}_j} \tag{7.56}$$

Newton's second law produces the first simplification of this expression, since we may replace $m_i \ddot{x}_i$ by the resultant force component f_i. Furthermore, from Eq. (7.55) and the second observation above, we have

$$\frac{\partial \ddot{x}_i}{\partial \ddot{\gamma}_j} = \frac{\partial \dot{x}_i}{\partial \dot{\gamma}_j} = w_{ij} \tag{7.57}$$

As a result, Eq. (7.56) becomes

$$\frac{\partial S}{\partial \ddot{\gamma}_j} = \sum_{i=1}^{3P} f_i w_{ij} \tag{7.58}$$

Finally, we note from Eq. (7.54) that the virtual displacements δx_i are related to the virtual increments in the quasi-coordinates by

$$\delta x_i = \sum_{j=1}^{M} w_{ij} \, \delta \gamma_j \tag{7.59}$$

so the virtual work is

$$\partial W = \sum_{i=1}^{3P} f_i \, \delta x_i = \sum_{i=1}^{3P} \sum_{j=1}^{M} f_i w_{ij} \, \delta \gamma_j \tag{7.60}$$

Comparison of this expression with Eq. (7.47) reveals that the right side of Eq. (7.58) is merely the generalized force Γ_j. When we employ Eq. (7.49) in order to display the role of the constraint forces, we find that

$$\frac{\partial S}{\partial \ddot{\gamma}_j} = \Gamma_j^{(a)} + \sum_{i=1}^{M-N} \lambda_i A_{ij} \qquad j = 1, \ldots, M \tag{7.61}$$

These are the *Gibbs-Appell equations*. In combination with the M equations for the generalized velocities given by Eqs. (7.44) and the $M - N$ constraint equations (7.45), we obtain a set of $3M - N$ differential equations for the M unknown generalized coordinates, the M unknown quasi-velocities, and the $M - N$ Lagrangian multipliers. The quasi-coordinates occur in these equations as velocities $\dot{\gamma}_i$ and first derivatives of those velocities, $\ddot{\gamma}_i$, while the generalized coordinates appear only as q_i and $\dot{q}_i$. Hence, the Gibbs-Appell equations are a set of first-order equations of motion.

The general situation addressed above may be simplified further if unconstrained quasi-coordinates are employed. Since those parameters do not violate any constraints, the corresponding Lagrangian multipliers vanish. Hence, the Gibbs-Appel equations then become

$$\frac{\partial S}{\partial \ddot{\tilde{\gamma}}_j} = \tilde{\Gamma}_j^{(a)} \qquad j = 1, \ldots, N \tag{7.62}$$

Also, there are no constraint equations that remain to be satisfied. Therefore, the full set of equations of motion in this case is obtained from the combination of Eqs. (7.62) and the M equations (7.51) relating the M unknown generalized coordinates to the N unknown quasi-velocities.

In some situations, especially those involving Coulomb friction forces, it is necessary that a constraint force appear in the equations of motion. We then need a set of quasi-coordinates that violate only the corresponding constraint, but not others. Without loss of generality, we may consider this constraint to be numbered 1. Since we wish to violate only one constraint, we employ $N + 1$ quasi-coordinates. We form this set from the N unconstrained parameters we would otherwise use, supplemented by a quasi-coordinate γ_{N+1} that is associated with movement in violation of the constraint. When we specialize the standard constraint equations (7.45) to the present situation, we find that

$$A_{1(N+1)} \dot{\gamma}_{N+1} + \sum_{j=1}^{N} A_{1j}(q_k, t)\dot{\tilde{\gamma}}_j + B_1(q_k, t) = 0 \tag{7.63}$$

(It is implicit in the discussion that $A_{1(N+1)} \neq 0$. If it were, then our choice for γ_{N+1} would not be suitable for the extra quasi-coordinate, because it does not represent a movement that violates the constraint.) The Gibbs-Appell equations consist of the set of N equations (7.62), and the one for the constrained quasi-coordinate,

$$\frac{\partial S}{\partial \ddot{\tilde{\gamma}}_{N+1}} = \Gamma_{(N+1)} \tag{7.64}$$

Note that the reaction associated with the constraint equation appears only in $\Gamma_{(N+1)}$, but not in the generalized forces $\tilde{\Gamma}_j^{(a)}$ for $j = 1$ to N. Thus, we gain the reaction and $\dot{\gamma}_{(N+1)}$ as unknowns, which are balanced by the addition of Eqs. (7.63) and (7.64).

As a closure to the development, we shall summarize the procedure required to formulate the Gibbs-Appell equations of motion when reaction forces are not of interest. The first step is to select a set of M generalized coordinates q_i that satisfy all holonomic constraints, and N unconstrained quasi-velocities $\dot{\tilde{\gamma}}_i$, where N is the number of degrees of freedom and $M - N$ is the number of nonholonomic constraints. The kinematical portion of the analysis requires derivation of the actual form of Eqs. (7.51) relating the $\dot{q}_i$ to the $\dot{\tilde{\gamma}}_i$. The kinematical analysis must also describe the acceleration parameters required to express S in terms of the quasi-velocities and accelerations, as well as the generalized coordinates. The kinetics analysis requires formation of the expression for S, and also evaluation of the generalized forces $\tilde{\Gamma}_i$. The latter may be evaluated by following Eq. (7.52a), which is based on the virtual work corresponding to increments in the quasi-coordinates. Alternatively, Eq. (7.52b) may be employed to transform the generalized forces Q_i corresponding to the generalized coordinates q_i. The latter alternative should be employed for conservative forces, for which $Q_i = -\partial V/\partial q_i$. In any case, all reactions may be omitted in the evaluation of generalized forces. The complete set of equations consists of the N Gibbs-Appell equations (7.62) and the M velocity relations in Eqs. (7.51).

EXAMPLE 7.8 _____

A small sphere is suspended by a spring of stiffness k from the collar, which rides on the horizontal guide bar. The collar and the sphere each have mass m. A force $\overline{F}$ acting transversely to the spring is applied to the sphere, with the result that the velocity of the sphere is always parallel to the spring. Derive Gibbs-Appell equations of motion for the system.

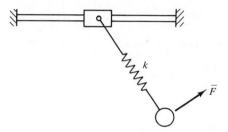

Example 7.8

Solution. The position of the system may be described by the length L and angle of orientation θ of the spring, and the horizontal position X_C of the collar, $q_1 = L$, $q_2 = \theta$, $q_3 = X_C$. A constraint is imposed on the direction of the velocity of the sphere, so the system has two degrees of freedom. Correspondingly, two quasi-velocities are unconstrained. A variety of definitions are possible; we shall employ

$$\dot{L} = \dot{\gamma}_1 \qquad \dot{\theta} = \dot{\gamma}_2 \qquad\qquad (1,2)$$

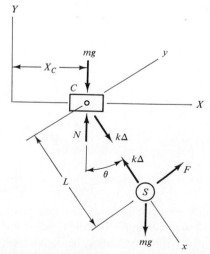

Free body diagram.

Note that these definitions are system equations relating two of the generalized coordinates to the quasi-velocities. The relation for $\dot{X}_C$ requires a kinematical analysis.

Let xyz be a moving reference frame whose origin coincides with collar C and whose x axis is always collinear with the spring. Then

$$\bar{v}_C = \dot{X}_C\bar{I} = \dot{X}_C(\sin\theta\,\bar{i} + \cos\theta\,\bar{j}) \qquad \bar{\omega} = \dot{\theta}\bar{k}$$
$$\bar{r}_{S/C} = L\bar{i} \qquad \bar{v}_S = \bar{v}_C + (\bar{v}_S)_{xyz} + \bar{\omega}\times\bar{r}_{S/C}$$

Using the definitions of the quasi-velocities, Eqs. (1) and (2), leads to

$$\bar{v}_S = (\dot{X}_C\sin\theta + \dot{\gamma}_1)\bar{i} + (\dot{X}_C\cos\theta + \dot{\gamma}_2L)\bar{j}$$

Since the y axis is normal to the spring in the plane of motion, the constraint equation is

$$\bar{v}_S\cdot\bar{j} = \dot{X}_C\cos\theta + \dot{\gamma}_2 L = 0$$

Thus, the relation between the third generalized coordinate and the $\dot{\gamma}_j$ is

$$\dot{X}_C = -\dot{\gamma}_2\frac{L}{\cos\theta} \tag{3}$$

from which it follows that

$$\bar{v}_S = (\dot{\gamma}_1 - \dot{\gamma}_2 L\tan\theta)\bar{i}$$

We will require an expression for the acceleration of each body in order to form the function S. Differentiating the expression for $\dot{X}_C$ yields

$$\ddot{X}_C = -\ddot{\gamma}_2\frac{L}{\cos\theta} - \dot{\gamma}_1\dot{\gamma}_2\frac{1}{\cos\theta} - \dot{\gamma}_2^2 L\frac{\sin\theta}{\cos^2\theta}$$

For the sphere, we have

$$\bar{a}_S = \bar{a}_C + (\bar{a}_S)_{xyz} + \bar{\alpha}\times\bar{r}_{S/C} + \bar{\omega}\times(\bar{\omega}\times\bar{r}_{S/C}) + 2\bar{\omega}\times(\bar{v}_S)_{xyz}$$
$$= \ddot{X}(\sin\theta\,\bar{i} + \cos\theta\,\bar{j}) + \ddot{L}\bar{i} + \ddot{\theta}\bar{k}\times L\bar{i} - \dot{\theta}^2 L\bar{i} + 2\dot{\theta}\bar{k}\times\dot{L}\bar{i}$$
$$= \left(\ddot{\gamma}_1 - \ddot{\gamma}_2 L\tan\theta - \dot{\gamma}_1\dot{\gamma}_2\tan\theta - \dot{\gamma}_2^2 L\frac{1}{\cos^2\theta}\right)\bar{i} + (\dot{\gamma}_1\dot{\gamma}_2 - \dot{\gamma}_2^2 L\tan\theta)\bar{j}$$

The Gibbs-Appell function is

$$S = \tfrac{1}{2}m\bar{a}_S\cdot\bar{a}_S + \tfrac{1}{2}m\bar{a}_C\cdot\bar{a}_C$$
$$= \tfrac{1}{2}m\left(\ddot{\gamma}_1 - \ddot{\gamma}_2 L\tan\theta - \dot{\gamma}_1\dot{\gamma}_2\tan\theta - \dot{\gamma}_2^2 L\frac{1}{\cos^2\theta}\right)^2$$
$$+ \tfrac{1}{2}m\left(\ddot{\gamma}_2\frac{L}{\cos\theta} + \dot{\gamma}_1\dot{\gamma}_2\frac{1}{\cos\theta} + \dot{\gamma}_2^2 L\frac{\sin\theta}{\cos^2\theta}\right)^2 + \cdots$$

The terms not listed in the above expression do not contain second derivatives, $\ddot{\gamma}_j$, so they will not contribute to the equations of motion. Also, note that there is no reason to collect like terms in S before evaluating its derivatives.

Let L_0 be the unstretched length of the spring, and let the datum for gravity be the elevation of the collar. Then

$$V = \tfrac{1}{2} k(L - L_0)^2 - mgL \cos \theta$$

We begin the evaluation of the Γ_j by considering the work done by the force $\overline{F}$. An expression for the virtual displacement of the sphere may be obtained by using the earlier expression to form $\overline{v}_s \, dt$, and then replace differentials by virtual increments. This yields

$$\delta \overline{r}_S = (\delta \gamma_1 - \delta \gamma_2 \, L \tan \theta) \overline{i}$$

Since $\delta \overline{r}_S$ is in the $\overline{i}$ direction and $\overline{F} = F\overline{j}$, we find that $\overline{F} \cdot \partial \overline{r}_S = 0$. We should note that the same result could have been obtained much more simply. The displacements resulting from virtual increments of unconstrained quasi-coordinates do not violate any constraint conditions. It is for this reason that $\overline{F}$, which is the reaction required to prevent the sphere from moving in the transverse direction, does no virtual work.

The forces doing virtual work are the spring and gravity, which are conservative. We shall illustrate the general procedure for identifying Γ_1 and Γ_2 based on the virtual work in Eq. (7.47), rather than by direct substitution into Eq. (7.48). The potential energy may depend on each of the three generalized coordinates, so the virtual work done by the conservative forces in the present system is

$$\delta W = -\frac{\partial V}{\partial L} \delta L - \frac{\partial V}{\partial \theta} \partial \theta - \frac{\partial V}{\partial X_c} \delta X_C$$

Since $\partial V / \partial X_C = 0$, there is no need to relate δX_C to the quasi-coordinates. For the other virtual increments, we use the theorems for partial differentiation to write

$$\delta L = \frac{\partial \dot{L}}{\partial \dot{\gamma}_1} \delta \gamma_1 + \frac{\partial \dot{L}}{\partial \dot{\gamma}_2} \delta \gamma_2 = \delta \gamma_1$$

$$\delta \theta = \frac{\partial \dot{\theta}}{\partial \dot{\gamma}_1} \delta \gamma_1 + \frac{\partial \dot{\theta}}{\partial \dot{\gamma}_2} \delta \gamma_2 = \delta \gamma_2$$

Thus, matching the virtual work in the system to the standard form, Eq. (7.47), leads to

$$\delta W = -\frac{\partial V}{\partial L} \delta \gamma_1 - \frac{\partial V}{\partial \theta} \delta \gamma_2 = \Gamma_1 \, \delta \gamma_1 + \Gamma_2 \, \delta \gamma_2$$

$$\Gamma_1 = -\frac{\partial V}{\partial L} = -k(L - L_0) + mg \cos \theta$$

$$\Gamma_2 = -\frac{\partial V}{\partial \theta} = -mgL \sin \theta$$

Next, we form the derivatives for the Gibbs-Appell equations.

$$\frac{\partial S}{\partial \ddot{\gamma}_1} = m \left(\ddot{\gamma}_1 - \ddot{\gamma}_2 L \tan \theta - \dot{\gamma}_1 \dot{\gamma}_2 \tan \theta - \dot{\gamma}_2^2 L \frac{1}{\cos^2 \theta} \right)$$

$$\frac{\partial S}{\partial \ddot{\gamma}_2} = m \left(\ddot{\gamma}_1 - \ddot{\gamma}_2 L \tan \theta - \dot{\gamma}_1 \dot{\gamma}_2 \tan \theta - \dot{\gamma}_2^2 L \frac{1}{\cos^2 \theta} \right)(-L \tan \theta)$$

$$+ m \left(\ddot{\gamma}_2 \frac{L}{\cos \theta} + \dot{\gamma}_1 \dot{\gamma}_2 \frac{1}{\cos \theta} + \dot{\gamma}_2^2 L \frac{\sin \theta}{\cos^2 \theta} \right) \frac{L}{\cos \theta}$$

Equating each derivative to the corresponding Γ_j, followed by canceling $\cos \theta$ from the denominators, yields

$$\ddot{\gamma}_1 \cos^2 \theta - (\ddot{\gamma}_2 L + \dot{\gamma}_1 \dot{\gamma}_2) \sin \theta \cos \theta - \dot{\gamma}_2^2 L$$

$$+ \frac{k}{m} (L - L_0) \cos^2 \theta - g \cos^3 \theta = 0 \quad (4)$$

$$-\ddot{\gamma}_1 L \sin \theta \cos^2 \theta + (\ddot{\gamma}_2 + \dot{\gamma}_1 \dot{\gamma}_2) L \cos \theta (1 + \sin^2 \theta)$$

$$+ 2\dot{\gamma}_2^2 L^2 \sin \theta + gL \sin \theta \cos^3 \theta = 0 \quad (5)$$

The full set of equations of motion are the first-order differential equations (1) to (5), in which the five dependent variables are L, θ, X_C, $\dot{\gamma}_1$, and $\dot{\gamma}_2$.

7.3.3 Gibbs-Appell Function for a Rigid Body

The definition of S, Eq. (7.53), is similar in form to that for kinetic energy, so the derivation of an expression for S in rigid-body motion will also follow a comparable approach. Indeed, we will find that S may be expressed in terms of the angular motion and angular momentum. As a preliminary to this development, it is appropriate to consider a basic aspect of the manner in which S is employed.

Since derivatives of S are taken only with respect to the quasi-coordinate accelerations, any terms in S that are independent of the $\ddot{\gamma}_j$ cannot be relevant to the equations of motion. (This is similar to noting that Lagrange's equations are unaffected by any constant that is added to either the potential or kinetic energy of a system.) However, the quasi-accelerations occur only in the physical acceleration components, as evidenced by a comparison of Eqs. (7.54) and (7.55). Therefore, any term in S that we find to depend solely on velocity parameters may be ignored. In general, this observation provides a shortcut, since it allows us to skip the evaluation of some terms.

We begin by expressing the magnitude of the acceleration as a dot product. The acceleration of a point P locating an element of mass dm in a rigid body is related to the motion of some reference point A in that body through the kinematical equation for acceleration. Thus, the infinitesimal contribution of this element to S is given by

$$dS = \tfrac{1}{2} [\bar{a}_A + \bar{\alpha} \times \bar{r} + \bar{\omega} \times (\bar{\omega} \times \bar{r})] \cdot [\bar{a}_A + \bar{\alpha} \times \bar{r} + \bar{\omega} \times (\bar{\omega} \times \bar{r})] \, dm$$

$$= \tfrac{1}{2} \{ \bar{a}_A \cdot \bar{a}_A + (\bar{\alpha} \times \bar{r}) \cdot (\bar{\alpha} \times \bar{r}) + 2\bar{a}_A \cdot (\bar{\alpha} \times \bar{r}) + 2\bar{a}_A \times [\bar{\omega} \times (\bar{\omega} \times \bar{r})]$$

$$+ 2(\bar{\alpha} \times \bar{r}) \cdot [\bar{\omega} \times (\bar{\omega} \times \bar{r})] \} \, dm + \text{negligible terms} \quad (7.65)$$

where $\bar{\omega}$ and $\bar{\alpha}$ are the angular velocity and angular acceleration of the body, and $\bar{r}$ is the position of dm relative to reference point A. Recall the opening remark regarding terms that do not contain derivatives of the quasi-velocities. On that basis, we dropped from Eq. (7.65) the contribution of the magnitude of the centripetal acceleration.

As we did for angular momentum, we now restrict point A to be either the center of mass or the fixed point when the body is in pure rotation. In the first case, the first moment of mass, which is the integral of $\bar{r}\, dm$, vanishes, while the second case gives $\bar{a}_A = \bar{0}$. In either case, the third and fourth terms in Eq. (7.65) vanish after integration. The first term requires no modification. The dependence of the other terms on the angular momentum may be displayed by using the following identities for the scalar and vector triple products:

$$\bar{a} \cdot (\bar{b} \times \bar{c}) = \bar{b} \cdot (\bar{c} \times \bar{a}) = \bar{c} \cdot (\bar{a} \times \bar{b}) \tag{a}$$

$$\bar{a} \times (\bar{b} \times \bar{c}) = \bar{b}(\bar{c} \cdot \bar{a}) - \bar{c}(\bar{a} \cdot \bar{b}) \tag{b}$$

Applying identity (a) to the second term in Eq. (7.65) leads to

$$(\bar{\alpha} \times \bar{r}) \cdot (\bar{\alpha} \times \bar{r}) = \bar{\alpha} \cdot [\bar{r} \times (\bar{\alpha} \times \bar{r})] \tag{7.66}$$

while the same step changes the fourth term in Eq. (7.65) to

$$(\bar{\alpha} \times \bar{r}) \cdot [\bar{\omega} \times (\bar{\omega} \times \bar{r})] = \bar{\alpha} \cdot \bar{r} \times [\bar{\omega} \times (\bar{\omega} \times \bar{r})]$$

We continue by applying identity (b) to this expression, which gives

$$(\bar{\alpha} \times \bar{r}) \cdot [\bar{\omega} \times (\bar{\omega} \times \bar{r})] = \bar{\alpha} \cdot \{\bar{\omega}[\bar{r} \cdot (\bar{\omega} \times \bar{r})] - (\bar{\omega} \times \bar{r})(\bar{r} \cdot \bar{\omega})\}$$

By definition, $\bar{r}$ is perpendicular to $\bar{\omega} \times \bar{r}$, so the first term vanishes. We next solve identity (b) for the first term on its right side, and cancel the dot product of orthogonal vectors. These operations lead to

$$(\bar{\alpha} \times \bar{r}) \cdot [\bar{\omega} \times (\bar{\omega} \times \bar{r})] = \bar{\alpha} \cdot \{-\bar{\omega} \times [(\bar{\omega} \times \bar{r}) \times \bar{r}] - [\bar{\omega} \cdot (\bar{\omega} \times \bar{r})]\bar{r}\}$$
$$= \bar{\alpha} \cdot \bar{\omega} \times [\bar{r} \times (\bar{\omega} \times \bar{r})] \tag{7.67}$$

The next step is to substitute Eqs. (7.66) and (7.67) into Eq. (7.65), and then to integrate over the entire body. Factoring out of the integrals terms that are functions of time only yields

$$S = \tfrac{1}{2}\bar{a}_A \cdot \bar{a}_A \iiint dm + \tfrac{1}{2}\bar{\alpha} \cdot \iiint \bar{r} \times (\bar{\alpha} \times \bar{r})\, dm$$
$$+ \bar{\alpha} \cdot \bar{\omega} \times \iiint \bar{r} \times (\bar{\omega} \times \bar{r})\, dm \tag{7.68}$$

The first integral is the mass of the body. Reference to Section 5.2.1 shows that the third integral is the angular momentum $\bar{H}_A$. Furthermore, the second integral is the same as the third, except that $\bar{\alpha}$ replaces $\bar{\omega}$. Hence, the second integral reduces to the rate of change of the components of the angular momentum, which we write as $\delta \bar{H}_A/\delta t$. The final form of S is therefore

$$S = \tfrac{1}{2} m\bar{a}_A \cdot \bar{a}_A + \tfrac{1}{2} \bar{\alpha} \cdot \frac{\delta \bar{H}_A}{\delta t} + \bar{\alpha} \cdot (\bar{\omega} \times \bar{H}_A) \tag{7.69}$$

where point A is the center of mass of the body, or a point in a purely rotating body that remains fixed. A feature of primary significance in this result is that the Gibbs-Appell function for a rigid body involves the same parameters (acceleration of the center of mass, angular velocity, angular acceleration, and the inertia properties) as those that arise in the Newton-Euler equations of motion.

Given the apparent simplicity of the Gibbs-Appell equations of motion, it might seem that there is no reason to use Lagrange's equations. However, each approach has certain merits. For a holonomic system, Lagrange's equations are easier to formulate. They do not require evaluation of accelerations, and the expression for the kinetic energy is simpler than that for the Gibbs-Appell function. Also, the potential energy may be employed directly in Lagrange's equations. When there are nonholonomic constraints, the Lagrangian multipliers must be eliminated algebraically from Lagrange's equations of motion. If the number of such constraints is large, this operation may be quite onerous. In that case, the Gibbs-Appell approach, which avoids the appearance of constraint forces and Lagrangian multipliers, becomes more attractive. A last feature is that the Gibbs-Appell first-order differential equations often are somewhat simpler to solve by numerical methods than are the state-space equations or Hamilton's canonical equations. This improvement may be attributed to the freedom to select quasi-coordinates that match the kinematical properties, independent of the choice of generalized coordinates, which describe the geometrical configuration.

EXAMPLE 7.9

Use the Gibbs-Appell equations to derive Euler's equations of motion for a rigid body.

Solution. Since we are concerned with the rotational motion, we define the quasi-velocities to be the components of angular velocity relative to a set of body-fixed axes; $\dot{\gamma}_1 = \omega_x$, $\dot{\gamma}_2 = \omega_y$, $\dot{\gamma}_3 = \omega_z$. It is sufficient to demonstrate the equivalence between a Gibbs-Appell equation for one γ_j and the corresponding Euler equation because the components of $\overline{\omega}$ and $\overline{H}_A$ are symbolic permutations. Hence, we shall evaluate the equation for $\partial S/\partial \ddot{\gamma}_1$.

We begin by expressing $\overline{\omega}$ and $\overline{\alpha}$ in terms of the $\dot{\gamma}_j$.

$$\overline{\omega} = \dot{\gamma}_1 \overline{i} + \dot{\gamma}_2 \overline{j} + \dot{\gamma}_3 \overline{k} \qquad \overline{\alpha} = \ddot{\gamma}_1 \overline{i} + \ddot{\gamma}_2 \overline{j} + \ddot{\gamma}_3 \overline{k}$$

Note that the expression for $\overline{\alpha}$ results from the identity $\overline{\alpha} = \delta\overline{\omega}/\delta t$. Since Euler's equations are based on xyz being principal axes, we have

$$\overline{H}_A = I_{xx}\dot{\gamma}_1 \overline{i} + I_{yy}\dot{\gamma}_2 \overline{j} + I_{zz}\dot{\gamma}_3 \overline{k}$$

$$\frac{\delta \overline{H}_A}{\delta t} = I_{xx}\ddot{\gamma}_1 + I_{yy}\ddot{\gamma}_2 \overline{j} + I_{zz}\ddot{\gamma}_3 \overline{k}$$

Substitution of the foregoing into Eq. (7.69) yields S in the required functional form.

A virtual rotation is the vector sum of the rotations $\delta\gamma_1$, $\delta\gamma_2$, and $\delta\gamma_3$ about the respective axes. Let $\overline{M}$ denote the resultant moment acting on the body. Then the virtual work is

$$\delta W = \overline{M} \cdot \overline{\delta\theta} = (\overline{M} \cdot \overline{i})\,\delta\gamma_1 + (\overline{M} \cdot \overline{j})\,\delta\gamma_2 + (\overline{M} \cdot \overline{k})\,\delta q_3$$

which corresponds to generalized forces that are the moments about the body-fixed axes.

$$\Gamma_1 = \overline{M} \cdot \overline{i} \qquad \Gamma_2 = \overline{M} \cdot \overline{j} \qquad \Gamma_3 = \overline{M} \cdot \overline{k}$$

The next step is to evaluate $\partial S / \partial \ddot{\gamma}_1$. We assume that $\overline{a}_A$ is independent of $\dot{\gamma}_1$, $\dot{\gamma}_2$, and $\dot{\gamma}_3$, in which case differentiating Eq. (7.69) yields

$$\frac{\partial S}{\partial \ddot{\gamma}_1} = \frac{1}{2} \frac{\partial \overline{\alpha}}{\partial \ddot{\gamma}_1} \cdot \frac{\delta \overline{H}_A}{\delta t} + \frac{1}{2} \overline{\alpha} \cdot \left[\frac{\partial}{\partial \ddot{\gamma}_1} \left(\frac{\delta \overline{H}_A}{\delta t} \right) \right] + \frac{\partial \overline{\alpha}}{\partial \ddot{\gamma}_1} \cdot (\overline{\omega} \times \overline{H}_A)$$

In view of the earlier expressions, we have

$$\frac{\partial \overline{\alpha}}{\partial \ddot{\gamma}_1} = \overline{i} \qquad \frac{\partial}{\partial \ddot{\gamma}_1} \left(\frac{\delta \overline{H}_A}{\delta t} \right) = I_{xx} \overline{i}$$

Thus

$$\frac{\partial S}{\partial \ddot{\gamma}_1} = \tfrac{1}{2} \overline{i} \cdot \frac{\delta \overline{H}_A}{\delta t} + \tfrac{1}{2} \overline{\alpha} \cdot (I_{xx} \overline{i}) + \overline{i} \cdot (\overline{\omega} \times \overline{H}_A) = \Gamma_1$$

$$I_{xx} \ddot{\gamma}_1 + (I_{zz} - I_{yy}) \dot{\gamma}_2 \dot{\gamma}_3 = \overline{M} \cdot \overline{i}$$

Permutations replacing the various $\dot{\gamma}_j$ by the appropriate term ω_x, ω_y or ω_z demonstrates the equivalence.

EXAMPLE 7.10 _____

Consider the bar in Example 7.7 under the conditions stated there. Use the Gibbs-Appell equations to derive the differential equation of motion for the angle θ, and an equation for the torque Γ.

Solution. In the Gibbs-Appell formulation, generalized and quasi-coordinates are defined individually. Since the precession angle at any instant is $\psi = \Omega t$, the system has only one degree of freedom. The nutation angle θ is suitable as the generalized coordinate, and a useful quasi-velocity is

$$\dot{\theta} = \dot{\gamma}_1 \tag{1}$$

Also, in order to obtain an equation for Γ, we must violate the constraint that Ω is constant. Hence, we let $\dot{\gamma}_2 = \dot{\psi}$ be the precession rate, which is subject to the constraint equation

$$\dot{\gamma}_2 = \Omega \tag{2}$$

Note that we must derive the Gibbs-Appell equation associated with γ_2 by considering $\ddot{\gamma}_2 \ne 0$.

Since end A is stationary, we may formulate the angular momentum of the bar relative to point A. This enables us to avoid formulating the acceleration of the center of mass. The angular motion (for nonconstant $\dot{\gamma}_2$) is

$$\overline{\omega} = \Omega \overline{K} + \dot{\theta}(-\overline{j}) = \dot{\gamma}_2 \overline{K} - \dot{\gamma}_1 \overline{j}$$
$$= -\dot{\gamma}_2 \cos \theta \overline{i} - \dot{\gamma}_1 \overline{j} + \dot{\gamma}_2 \sin \theta \overline{k}$$
$$\overline{\alpha} = \ddot{\gamma}_2 \overline{K} - \ddot{\gamma}_1 \overline{j} - \dot{\gamma}_1 (\overline{\omega} \times \overline{j})$$
$$= (-\ddot{\gamma}_2 \cos \theta + \dot{\gamma}_1 \dot{\gamma}_2 \sin \theta) \overline{i} - \ddot{\gamma}_1 \overline{j} + (\ddot{\gamma}_2 \sin \theta + \dot{\gamma}_1 \dot{\gamma}_2 \cos \theta) \overline{k}$$

Setting $I_{xx} = 0$ for a slender bar then gives

$$\overline{H}_A = I_{yy}\omega_y \overline{j} + I_{zz}\omega_z \overline{k} = \tfrac{1}{3} mL^2(-\dot{\gamma}_1 \overline{j} + \dot{\gamma}_2 \sin \theta \overline{k})$$

$$\frac{\delta \overline{H}_A}{\delta t} = I_{yy}\alpha_y \overline{j} + I_{zz}\alpha_z \overline{k}$$

$$= \tfrac{1}{3} mL^2[-\ddot{\gamma}_1 \overline{j} + (\ddot{\gamma}_2 \sin \theta + \dot{\gamma}_1 \dot{\gamma}_2 \cos \theta)\overline{k}]$$

From these expressions, we obtain

$$S = \frac{1}{2}\overline{\alpha} \cdot \frac{\delta \overline{H}_A}{\delta t} + \overline{\alpha} \cdot (\overline{\omega} \times \overline{H}_A)$$

$$= \tfrac{1}{2} (\tfrac{1}{3} mL^2) [\ddot{\gamma}_1^2 + (\ddot{\gamma}_2 \sin \theta + \dot{\gamma}_1 \dot{\gamma}_2 \cos \theta)^2]$$

$$+ (\tfrac{1}{3} mL^2) [-\ddot{\gamma}_1 \dot{\gamma}_2^2 \sin \theta \cos \theta + (\ddot{\gamma}_2 \sin \theta)(\dot{\gamma}_1 \dot{\gamma}_2 \cos \theta)] + \text{first derivatives}$$

The constraints imposed by the bearing force $\overline{F}$ and couple $\overline{M}$ are not violated, so virtual work is done by the applied torque and gravity. For a datum at the elevation of pin A, the potential energy as a function of the generalized coordinate is

$$V = -mg \frac{L}{2} \cos \theta$$

Hence, the virtual work is

$$\delta W = \Gamma \, \delta\gamma_2 - \frac{\partial V}{\partial \theta} \frac{\partial \dot{\theta}}{\partial \dot{\gamma}_1} \, \delta\gamma_1 = \Gamma \, \delta\gamma_2 - mg \frac{L}{2} \sin \theta \, \delta\gamma_1$$

so the generalized forces are

$$\Gamma_1 = -mg \frac{L}{2} \sin \theta \qquad \Gamma_2 = \Gamma$$

The Gibbs-Appell equations are

$$\frac{\partial S}{\partial \ddot{\gamma}_1} = \tfrac{1}{3} mL^2(\ddot{\gamma}_1 - \dot{\gamma}_2^2 \sin \theta \cos \theta) = -mg \frac{L}{2} \sin \theta \tag{3}$$

$$\frac{\partial S}{\partial \ddot{\gamma}_2} = \tfrac{1}{3} mL^2(\ddot{\gamma}_2 \sin \theta + 2\dot{\gamma}_1 \dot{\gamma}_2 \sin \theta \cos \theta) = \Gamma \tag{4}$$

In combination with Eqs. (1) and (2), these equations are identical to the comparable results in Example 7.7

EXAMPLE 7.11 ───

The wheelbarrow is pushed in the horizontal plane by forces $\overline{F}_A$ and $\overline{F}_B$ acting at each handle. The body of the wheelbarrow has mass m_1 with its center of mass at point G; the corresponding centroidal moment of inertia about a vertical axis is I. The wheel, which rolls without slipping, has mass m_2, moment of inertia J about its axle, and radius r. Derive the Gibbs-Appell equations of motion.

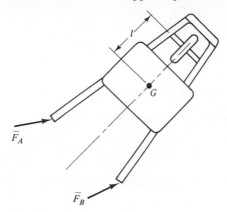

Example 7.11

Solution. We assume that the wheelbarrow remains upright. Then, because the wheel rolls without slipping, we know that the angular velocity component of the wheel about the horizontal axle is $\omega_1 = v/r$, where v is the speed of the center C of the axle. Also, because there is no slippage, the velocity of point C must be along the longitudinal axis, x. Convenient generalized coordinates for the wheelbarrow are the absolute position coordinates of point C, $q_1 = X_C$, $q_2 = Y_C$, and the angle θ locating the x axis. For the wheel, we let $q_4 = \phi$, the angle of rotation about the axle.

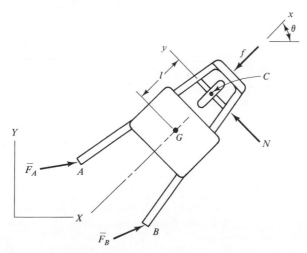

Free body diagram.

We let v be a quasi-velocity, $\dot{\gamma}_1 = v$, because that expedites the description of the constraint on the movement of point C. The other quasi-velocity is $\dot{\gamma}_2 = \dot{\theta}$. Note that although four generalized coordinates have been defined, there are two constraint conditions due to rolling without slipping. Therefore, the system has only two degrees of freedom, and only two quasi-coordinates are unconstrained.

We relate the $\dot{q}_i$ to the $\dot{\gamma}_i$ by describing the constraint conditions. By definition

$$\dot{\theta} = \dot{\gamma}_2 \qquad (1)$$

The constraint on the velocity of point C is

$$\bar{v}_C = \dot{X}_C \bar{I} + \dot{Y}_C \bar{J} = v\bar{i} = v(\cos\theta\bar{I} + \sin\theta\bar{J})$$

so

$$\dot{X}_C = \dot{\gamma}_1 \cos\theta \qquad \dot{Y}_C = \dot{\gamma}_1 \sin\theta \qquad (2,3)$$

Also, for rolling without slipping, $v = r\dot{\phi}$, which yields

$$\dot{\phi} = \frac{\dot{\gamma}_1}{r} \qquad (4)$$

The next task is to form the acceleration variables for S. For the center of mass G, we use

$$\bar{a}_G = \bar{a}_C + \bar{\alpha} \times \bar{r}_{G/C} + \bar{\omega} \times (\bar{\omega} \times \bar{r}_{G/C})$$

where $\bar{\omega}$ and $\bar{\alpha}$ are for the xyz reference frame. The individual terms are

$$\bar{a}_C = \ddot{X}_C \bar{I} + \ddot{Y}_C \bar{J}$$
$$= (\ddot{\gamma}_1 \cos\theta - \dot{\gamma}_1\dot{\gamma}_2 \sin\theta)(\cos\theta\bar{i} - \sin\theta\bar{j})$$
$$+ (\ddot{\gamma}_1 \sin\theta + \dot{\gamma}_1\dot{\gamma}_2 \cos\theta)(\sin\theta\bar{i} + \cos\theta\bar{j})$$
$$= \ddot{\gamma}_1 \bar{i} + \dot{\gamma}_1\dot{\gamma}_2 \bar{j}$$
$$\bar{\omega} = \dot{\theta}\bar{k} = \dot{\gamma}_2\bar{k} \qquad \bar{\alpha} = \ddot{\gamma}_2\bar{k} \qquad \bar{r}_{G/C} = -l\bar{i}$$

so that

$$\bar{a}_G = (\ddot{\gamma}_1 + \dot{\gamma}_2^2 l)\bar{i} + (-\ddot{\gamma}_2 l + \dot{\gamma}_1\dot{\gamma}_2)\bar{j}$$

In order to express the contribution of the wheel to S, we must describe its angular motion. If the wheelbarrow does not tilt, then the wheel precesses at $\dot{\theta}$ about the vertical axis, while it spins about its axle at $\dot{\phi}$. Thus

$$\bar{\omega}_w = \frac{1}{r}\dot{\gamma}_1\bar{j} + \dot{\gamma}_2\bar{K} = \frac{1}{r}\dot{\gamma}_1\bar{j} + \dot{\gamma}_2\bar{k}$$

$$\bar{\alpha}_w = \frac{1}{r}\ddot{\gamma}_1\bar{j} + \ddot{\gamma}_2\bar{K} + \frac{1}{r}\dot{\gamma}_1\bar{\omega} \times \bar{j} = -\frac{1}{r}\dot{\gamma}_1\dot{\gamma}_2\bar{i} + \frac{1}{r}\ddot{\gamma}_1\bar{j} + \ddot{\gamma}_2\bar{k}$$

Now that the kinematical parameters have been characterized, we form the angular momentum properties. For the wheelbarrow, we have

$$\bar{H}_G = I\dot{\gamma}_2\bar{k} \qquad \frac{\delta\bar{H}_G}{\delta t} = I\ddot{\gamma}_2\bar{k}$$

We consider the wheel to be very thin, so its moment of inertia about a centroidal in-plane axis is $\frac{1}{2}J$. Thus,

$$\bar{H}_C = \frac{J}{r}\dot{\gamma}_1\bar{j} + \frac{J}{2}\dot{\gamma}_2\bar{k} \qquad \frac{\delta\bar{H}_C}{\delta t} = -\frac{J}{2r}\dot{\gamma}_1\dot{\gamma}_2\bar{i} + \frac{J}{r}\ddot{\gamma}_1\bar{j} + \frac{J}{2}\ddot{\gamma}_2\bar{k}$$

The Gibbs-Appell function for the system is the sum of the contributions of each body.

$$S = \tfrac{1}{2} m \bar{a}_G \cdot \bar{a}_G + \tfrac{1}{2} \bar{\alpha} \cdot \frac{\delta \bar{H}_G}{\delta t} + \bar{\alpha} \cdot (\bar{\omega} \times \bar{H}_G)$$

$$+ \tfrac{1}{2} m \bar{a}_C \cdot \bar{a}_C + \tfrac{1}{2} \bar{\alpha}_w \cdot \frac{\delta \bar{H}_C}{\delta t} + \bar{\alpha}_w \cdot (\bar{\omega}_w \times \bar{H}_C)$$

$$= \tfrac{1}{2} m_1 [(\ddot{\gamma}_1 + \dot{\gamma}_2^2 l)^2 + (-\ddot{\gamma}_2 l + \dot{\gamma}_1 \dot{\gamma}_2)^2]$$

$$+ \tfrac{1}{2} I \ddot{\gamma}_2^2 + \tfrac{1}{2} m_2 \ddot{\gamma}_1^2 + \tfrac{1}{2} J \left(\frac{1}{r^2} \ddot{\gamma}_1^2 + \frac{1}{2} \ddot{\gamma}_2^2 \right) + \text{first derivatives}$$

Note that the terms associated with a gyroscopic moment for the wheel do not appear in S because the system has been constrained to move in the horizontal plane. Momentum effects tending to cause the wheelbarrow to tilt are balanced by reactions, which are the out-of-plane components of the forces $\bar{F}_A$ and $\bar{F}_B$.

The next step is to evaluate the virtual work. Since the velocity of point C has already been described, we replace the applied forces $\bar{F}_A$ and $\bar{F}_B$ by a force-couple system, $\bar{R}$ and $\bar{M}$, acting at point C.

$$\bar{R} = \bar{F}_A + \bar{F}_B \qquad \bar{M} = \bar{r}_{A/C} \times \bar{F}_A + \bar{r}_{B/C} \times \bar{F}_B$$

The analogy between differential and virtual displacements leads to

$$d\bar{r}_C = \bar{v}_C \, dt = d\gamma_1 \bar{i} \Rightarrow \delta \bar{r}_C = \delta \gamma_1 \bar{i}$$

so the virtual work is

$$\delta W = \bar{R} \cdot \delta \bar{r}_C + \bar{M} \cdot \delta \bar{\theta} = \Gamma_1 \, \delta \gamma_1 + \Gamma_2 \, \delta \gamma_2$$

Matching the two descriptions of δW yields the generalized forces,

$$\Gamma_1 = \bar{R} \cdot \bar{i} \qquad \Gamma_2 = \bar{M} \cdot \bar{k}$$

It should be noted that the reaction forces $\bar{f}$ and $\bar{N}$ exerted on the wheel by the ground do no virtual work. Their role is to prevent the wheel from slipping, and that constraint is not violated.

The Gibbs-Appell equations resulting from the expression for S are

$$\frac{\partial S}{\partial \ddot{\gamma}_1} = \left(m_1 + m_2 + \frac{J}{r^2} \right) \ddot{\gamma}_1 + m_1 l \dot{\gamma}_2^2 = \bar{R} \cdot \bar{i} \tag{5}$$

$$\frac{\partial S}{\partial \ddot{\gamma}_2} = \left(m_1 l^2 + I + \frac{1}{2} J \right) \ddot{\gamma}_2 - m_1 l \dot{\gamma}_1 \dot{\gamma}_2 = \bar{M} \cdot \bar{k} \tag{6}$$

In combination with Eqs. (1) to (4), which relate the generalized and quasi-velocities, we have derived six first-order differential equations for the unknown X_C, Y_C, θ, ϕ, $\dot{\gamma}_1 = v$, and $\dot{\gamma}_2 = \dot{\phi}$.

7.4 LINEARIZATION

Lagrange's equations provide a highly versatile approach by which we may obtain differential equations of motion. The next task after such equations are formulated is to solve them for the dynamic response. Two complications might arise, associated with our ability to solve the resulting equations. Clearly, if the system contains many independently moving pieces, the number of equations will be rather large. In addition, the equations might be rather complicated, particularly if the geometrical details of the system are intricate. The combination of these features often makes it necessary to devote a major effort to the process of solving the equations of motion.

In most cases, little is known about the response when a system is studied for the first time. It might then be adequate to simplify the equations of motion. A broad class of systems feature vibratory responses in which displacements oscillate about a reference state in which the generalized coordinates have constant values. We often are concerned with the case where such displacements are a small fraction of the overall dimensions of the system. In that condition we may simplify the equations of motion through the process of linearization.

For a system whose physical constraints are independent of time, a suitable reference state would be the static equilibrium position. When there are time-dependent constraints, it might be possible to define a constant reference state, such as steady rotation about an axis, even if there are no constant values of the generalized coordinates that actually satisfy the equations of motion. Alternatively, suppose the time-dependent features are a result of small imposed motion of a support, such as a displacement of a pin. Then we could consider the reference state to be the static equilibrium position that the system would occupy if the support were fixed.

Let us denote as q_i^* the constant values of the generalized coordinates at the reference state. For the sake of simplicity, we treat only holonomic systems here. Applying Lagrange's equations to the reference state then yields

$$\frac{d}{dt}\left(\frac{\partial T}{\partial \dot{q}_i}\right)^* - \left(\frac{\partial T}{\partial q_i}\right)^* + \left(\frac{\partial V}{\partial q_i}\right)^* = Q_i^* \qquad i = 1, \ldots, M \qquad (7.70)$$

where an asterisk is used to mark terms that are evaluated at the reference state, where $q_i = q_i^*$ and $\dot{q}_i = 0$.[†] Solution of these equations yields the reference values q_i^*. When the generalized coordinates are a constrained set ($M > N$), Eqs. (7.70) also describe the static reactions associated with the constraint conditions that are not identically satisfied.

In a linearized analysis, we are only concerned with very small displacements relative to the reference configuration. We measure displacement relative to this position by a set of *relative generalized coordinates* ξ_i, such that

$$\xi_i = q_i - q_i^* \Rightarrow q_i = \xi_i + q_i^* \qquad \text{and} \qquad \dot{q}_i = \dot{\xi}_i \qquad (7.71)$$

There are essentially two approaches whereby linear equations of motion for the ξ_i may be obtained. The first, and most reliable, operates directly on the equations of

[†]In the case of a scleronomic system, the kinetic energy vanishes when the generalized velocities are zero. Then Eq. (7.70) reduces to the principle of virtual work and potential energy, Eq. (6.42).

motion. Let F be an arbitrary function of the q_i, $\dot{q}_i$, and t that appears as a term in an equation of motion. We employ Eqs. (7.71) to replace this dependence by an equivalent dependence on the ξ_i. Assuming that the function F is analytic, a Taylor series yields

$$F(q_i, \dot{q}_i, t) = F(q_i^* + \xi_i, \dot{\xi}_i, t)$$

$$= F^* + \sum_{j=1}^{M} \left[\left(\frac{\partial F}{\partial q_j}\right)^* \xi_j + \left(\frac{\partial F}{\partial \dot{q}_j}\right)^* \dot{\xi}_j \right] + \cdots \qquad (7.72)$$

As before, an asterisk marking a term indicates that the respective quantities are to be evaluated at the static equilibrium values $q_i = q_i^*$, $\dot{q}_i^* = 0$, and $\ddot{q}_i^* = 0$. Thus,

> One method by which the equations of motion may be linearized is to substitute $q_i = q_i^* + \xi_i$ into those equations, and then to truncate at linear terms in ξ_i or $\dot{\xi}_i$ a Taylor series of each term.

Truncation of a series in this manner implies that quadratic products of the ξ_i are negligible (in a nondimensional sense) in comparison with the variables to the first power. Technically, this is only true if all ξ_j are infinitesimal. Indeed, linearized equations of motion are often said to constitute an *infinitesimal displacement theory*. There are numerous situations, however, where linearized equations of motion accurately describe the system response for very substantial displacements.

In order to illustrate this process, consider the pendulum in Figure 7.3, whose cable is elastic with a linear stiffness k and an unstretched length L_0. The nonlinear equations of motion for the system are

$$m\ddot{R} - mR\dot{\theta}^2 + k(R - L_0) - mg \cos \theta = 0$$
$$mR\ddot{\theta} + 2m\dot{R}\dot{\theta} + mg \sin \theta = 0$$

The equilibrium positions of the system are obtained by solving these equations with all time derivatives set to zero. The stable position is found to be

$$R^* = L_0 + \frac{mg}{k} \qquad \theta^* = 0$$

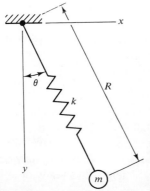

Figure 7.3 Spring-supported pendulum.

For the linearization, let $R = R^* + \xi_1$, $\theta = \xi_2$. Since $\xi_2 \ll 1$ rad, the series for the trigonometric functions may be trucated as

$$\sin \theta = \sin \xi_2 \approx \xi_2 \qquad \cos \theta = \cos \xi_2 \approx 1 - \tfrac{1}{2} \xi_2^2 \approx 1$$

In addition to the above, nonlinear terms arise in the accelerations in each of the equations of motion. When all terms that contain products of the ξ_i are dropped, the result is

$$m\ddot{\xi}_1 + k\xi_1 = 0 \qquad R^*\ddot{\xi}_2 + g\xi_2 = 0$$

It is clear that the linearized equations of motion for this system are substantially easier to solve than the original nonlinear equations.

The process of linearizing the equations of motion directly is straightforward. However, it introduces the simplifications of linearization subsequent to the application of Lagrange's equations. This approach therefore involves the derivation of energy expressions that are more accurate than necessary. Let us therefore consider the alternative of deriving simplified energy expressions based on a kinematical restriction to small displacements.

The kinetic and potential energies appear in Lagrange's equations as first derivatives with respect to the generalized coordinates. Consequently, quadratic and higher terms in the equations of motion may be avoided by neglecting cubic and higher terms in the energies. In other words:

A linearized set of equations of motion may be obtained by retaining in the kinetic and potential energy expressions only those terms that are either quadratic or linear in the generalized coordinates measured relative to the static equilibrium configuration.

The implications of this statement for the potential energy are readily identified because V can only depend on the generalized coordinates, and possibly time. A Taylor-series expansion relative to the static equilibrium configuration leads to

$$V(q_1^* + \xi_1, q_2^* + \xi_2, \ldots, t) = V^* + \sum_{i=1}^{M} \left(\frac{\partial V}{\partial q_i}\right)^* \xi_i + \frac{1}{2} \sum_{i=1}^{M} \sum_{j=1}^{M} \left(\frac{\partial^2 V}{\partial q_i \, \partial q_j}\right)^* \xi_i \xi_j \tag{7.73}$$

The constants derived from the second derivative are commonly known as *stiffness coefficients*,

$$K_{ij} = \left(\frac{\partial^2 V}{\partial q_i \, \partial q_j}\right)^* \qquad K_{ij} = K_{ji} \qquad i, j = 1, \ldots, M \tag{7.74}$$

The corresponding expression for the potential energy in a linearized analysis is

$$V = V^* + \sum_{i=1}^{M} \left(\frac{\partial V}{\partial q_i}\right)^* \xi_i + \frac{1}{2} \sum_{i=1}^{M} \sum_{j=1}^{M} K_{ij} \xi_i \xi_j \tag{7.75}$$

The reason for the term "stiffness" is obvious when one considers the analogy between the quadratic terms in Eq. (7.75) and the potential energy for a spring, $V = \tfrac{1}{2} k \Delta^2$.

A comparable treatment of the kinetic energy may be derived by recalling the general form of T, Eqs. (7.5) and (7.6). The terms M_{ij}, N_j, and T_0 may be functions of time and of the generalized coordinates. The coefficients M_{ij} already describe terms that are quadratic in the generalized velocities. These coefficients should be evaluated at the equilibrium position, because their Taylor-series expansions in powers of the ξ_i would otherwise lead to cubic and higher-order terms. Similarly, the coefficients N_j are associated with terms in T that are linear in the generalized coordinates, so their Taylor-series expansions should be truncated at linear terms. Terms up to quadratic need to be retained for T_0, which represents the kinetic energy that is present even when the generalized velocities are zero.

The result of substituting the series expansions of the coefficients into Eqs. (7.5) and (7.6) is therefore found to be

$$T = \frac{1}{2} \sum_{i=1}^{M} \sum_{j=1}^{M} \left[M_{ij}^* \dot{\xi}_i \dot{\xi}_j + 2 \left(\frac{\partial N_i}{\partial q_j} \right)^* \dot{\xi}_i \dot{\xi}_j + \left(\frac{\partial^2 T_0}{\partial q_i \, \partial q_j} \right)^* \xi_i \xi_j \right]$$
$$+ \sum_{i=1}^{M} \left[N_i^* \dot{\xi}_i + \left(\frac{\partial T_0}{\partial q_j} \right)^* \xi_i \right] + T_0^* \quad (7.76)$$

The quantities marked by an asterisk are to be evaluated at the reference position. Hence, they are not functions of the generalized coordinates and velocities. The coefficients M_{ij}^* are known as *inertia coefficients*, analogous to the terminology for the K_{ij}. It must be noted, however, that all coefficients might be functions of time if the physical constraints imposed on the system are time-dependent.

The primary purpose in writing Eqs. (7.75) and (7.76) is to emphasize the general form of the energy expressions that can be expected to arise in the linearization process. It also leads to simplifications in the kinematical analysis of velocity and displacement in the special case of a scleronomic system. In that situation, the coefficients N_j and T_0 are identically zero and the coefficients M_{ij} are independent of time; see the discussion of Eq. (6.48). When these coefficients are evaluated at the reference state in order to form the kinetic energy for the linearized analysis, the resulting M_{ij}^* reduce to constants. This observation leads to the conclusion that

When a system is scleronomic, the kinematical relationship between the linear and angular velocities and the generalized velocities may be evaluated using the geometrical configuration at the static equilibrium position.

A useful corollary of the ability to consider only the geometry of the static equilibrium position is substantial simplification in the kinematical analysis. For example, an instant-center analysis may be performed by locating points relative to the position of the instant center when the system passes the static equilibrium position. This saves the additional and sometimes difficult task of evaluating geometrical relations for an arbitrary position.

The linearization procedures for the potential energy usually do not admit the kinematical simplifications available for the kinetic energy of a system whose constraints are independent of time. Let us refer back to Eq. (7.73) in order to identify the reason for this difference. The parameter V^* is the potential energy at the equilibrium position. That quantity is unimportant, because only derivatives of V occur in Lagrange's equations. Also, the term $(\partial V/\partial q_i)^*$ must have been evaluated already if the reference state is known; see Eq. (7.70). Therefore, we need only evaluate the stiffness coefficients K_{ij} in order to characterize the potential energy function for a linearized analysis.

The difficulty arises in the description of the geometric parameters required to form these coefficients. Consider a linear spring, whose potential energy is $V = \frac{1}{2}k\Delta^2$, where Δ is the total elongation. For convenience in this discussion, let us restrict our attention to the case where the system has only one degree of freedom. Suppose that the spring is stretched by an amount Δ^* when the system is at its static equilibrium position. Then expanding the spring length in a Taylor series in the relative displacement ξ leads to

$$\Delta = \Delta^* + c_1\xi + c_2\xi^2$$

The potential energy corresponding to this expression is

$$V = \frac{1}{2}\, k(c_1^2 + 2\Delta^*c_2)\xi^2 + \text{nonquadratic terms}$$

Note that c_2 is the coefficient of the quadratic term in the series expansion for Δ. This simple example leads us to the general conclusion that

For the derivation of linearized equations of motion, the elongation Δ of each spring may be carried out by expanding Δ in a Taylor series that retains quadratic terms in the relative displacements ξ_i. The only situation where Δ may be truncated at linear terms in the ξ_i arises if the spring is unstretched at the static equilibrium position.

We should mention that we will see in Example 7.12 that there are general situations where an approximate expression for the potential energy may be obtained from only the linear term in the spring deformation.

It is equally important to be careful in dropping higher-order terms when treating other types of conservative forces. For example, consider the potential energy of the gravitational attraction on the bar in Figure 7.4. Let θ^* denote the angle of inclination of the bar at the static equilibrium position. (A nonzero value of θ^* may be obtained by attaching springs to the bar, or if the bar is part of a linkage.) For a datum at the elevation of the pivot, the potential energy is

$$V = -mg\frac{L}{2}\cos(\theta^* + \xi) = -mg\frac{L}{2}(\cos\theta^* \cos\xi - \sin\theta^* \sin\xi)$$

We expand the cosine and sine terms in powers of ξ, and drop nonquadratic terms, with the result that

$$V = \frac{1}{4}\,(mgL \cos\theta^*)\,\xi^2$$

In this case, it was the height of the center of mass G relative to the datum that must be expanded in a Taylor series including quadratic terms. Only when the bar is horizontal in the static equilibrium position, $\theta^* = 90°$, does gravity have a purely static effect.

A different aid is available when it is necessary to account for the effects of energy dissipation in a linearized analysis. The dissipation mechanisms may be modeled by dashpots, such as the one depicted schematically in Figure 7.5. The most common actual use of this device is as a shock absorber in an automobile. The magnitude of the force exerted by a dashpot depends on the rate of change of its elongation, as well as on the

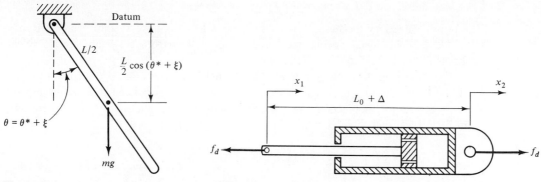

Figure 7.4

Figure 7.5   Dashpot.

elongation itself in some designs, $f_d = f_d(\Delta, \dot{\Delta})$. The direction of the forces exerted at each end is tensile if the elongation is increasing, that is, if $\dot{\Delta} > 0$.

When the dashpot force is linearized, the force it exerts is taken to be proportional to its elongation rate,

$$f_d = \mu \dot{\Delta} \tag{7.77}$$

where the coefficient μ is the *dashpot constant,* whose dimensions are FT/L. This relation has obvious similarities to that for a linear spring, since the only difference is the proportionality with the elongation rate, rather than the elongation itself.

The analogy between the forces exerted by a spring and a dashpot is exploited to define the *Rayleigh dissipation function D.* Just as the potential energy in a spring is $\frac{1}{2}k\Delta^2$, we define

$$D = \tfrac{1}{2}\,\mu\dot{\Delta}^2 \tag{7.78}$$

Since the relations for a linear dashpot differ from those for a linear spring only by the presence of a time derivative of the elongation, the generalized force associated with the dashpot is obtained by a differentiation with respect to the generalized velocity

$$(Q_d)_i = -\frac{\partial D}{\partial \dot{q}_i} \tag{7.79}$$

When the system contains several dashpots, the dissipation function is the sum of the effects associated with each,

$$D = \frac{1}{2}\sum_i \mu_i \dot{\Delta}_i^2 \tag{7.80}$$

The dissipation function is related to the instantaneous power entering the dashpot. In the situation in Figure 7.5 the elongation is $\Delta = x_2 - x_1$. The dashpot force is tensile because the elongation is increasing, $\dot{x}_2 > \dot{x}_1$. Thus, power is input to the dashpot at the right end at the rate $f_d\dot{x}_2$, and it is expended at the left at the rate $f_d\dot{x}_1$. The net power input is

$$P_d = f_d \dot{x}_2 - f_d \dot{x}_1 = \mu \, (\dot{x}_2 - \dot{x}_1)^2 = 2D \tag{7.81}$$

The power gained by the dashpot is energy that is lost from the system. Thus, we may conclude that the instantaneous mechanical energy loss from a system due to linear dashpots is twice the Rayleigh dissipation function.

The Rayleigh dissipation function may be used to write Lagrange's equations in a modified form. Let $Q_i^{(a)}$ now denote the generalized force associated with given applied forces that are not included in the potential energy or dissipation function. It follows from Eq. (7.79) that the equations of motion may be written as

$$\frac{d}{dt}\left(\frac{\partial T}{\partial \dot{q}_i}\right) - \frac{\partial T}{\partial q_i} + \frac{\partial D}{\partial \dot{q}_i} + \frac{\partial V}{\partial q_i} = Q_i^{(a)} \qquad i = 1, 2, \ldots, M \tag{7.82}$$

It must be emphasized that *the Rayleigh dissipation function is valid solely for linear dashpot models.* In systems where the dashpot force does not depend linearly on the rate of elongation, the generalized forces associated with the dashpot must be evaluated from the virtual work in the usual manner. The above form of Lagrange's equations is not valid in that case.

When a system is scleronomic, the dissipation function will be a quadratic sum of the generalized velocities,

$$D = \frac{1}{2}\sum_{i=1}^{M}\sum_{j=1}^{M} D_{ij}\dot{\xi}_i\dot{\xi}_j \qquad D_{ij} = D_{ji} \tag{7.83}$$

where the coefficients D_{ij} are constants. This is the same form as the quadratic terms in Eq. (7.75) for the potential energy, and in Eq. (7.76) for the kinetic energy. Let us examine the corresponding form of the linearized equations of motion. Since the ξ_i differ from the original generalized coordinates only by the constant q_i^*, it must be that $\delta q_i = \delta \xi_i$. The consequence of the equivalence of the virtual increments of the two sets of generalized coordinates is that the generalized forces are the same $Q_i^{(a)}$ for either set.

Let ξ_n be a specified generalized coordinate. All coefficients $N_j = 0$ for a scleronomic system. The details of the derivative of a quadratic sum were treated earlier; see Eq. (7.7a). When we apply that result here to Eqs. (7.75), (7.76), and (7.83), we find that

$$\frac{\partial T}{\partial \dot{\xi}_n} = \sum_{j=1}^{M} M_{nj}^*\dot{\xi}_j \qquad \frac{\partial D}{\partial \dot{\xi}_n} = \sum_{j=1}^{M} D_{nj}\dot{\xi}_j \qquad \frac{\partial V}{\partial \xi_n} = \sum_{j=1}^{M} K_{nj}\xi_j + \left(\frac{\partial V}{\partial q_n}\right)^* \tag{7.84}$$

Also, because the linearized kinetic energy in Eq. (7.76) does not depend on the generalized coordinates when the constraints are independent of time, we find that $\partial T/\partial \xi_n = 0$.

The derivatives in Eqs. (7.84) may be substituted into the modified Lagrange's equations, Eq. (7.82), to obtain standard equations of motion.

$$\sum_{j=1}^{M} M_{nj}^*\ddot{\xi}_j + \sum_{j=1}^{M} D_{nj}\dot{\xi}_j + \sum_{j=1}^{M} K_{nj}\xi_j + \left(\frac{\partial V}{\partial q_n}\right)^* = Q_n^{(a)} \qquad n = 1, \ldots, M \tag{7.85}$$

These equations may be written more compactly in matrix form. Let ξ_i be the ith element of $\{\xi\}$, and let M_{ij}^*, D_{ij}, and K_{ij} be the elements in the ith row and jth column of the mass, dissipation, and stiffness arrays, $[M]$, $[D]$, and $[K]$, respectively. Then Eqs. (7.85) are equivalent to

$$[M]\{\ddot{\xi}\} + [D]\{\dot{\xi}\} + [K]\{\xi\} = \{Q^{(a)}\} - \left\{\frac{\partial V}{\partial q}\right\}^{*} \tag{7.86}$$

In this expression, $\{Q^{(a)}\}$ is the set of generalized forces, excluding the conservative and dissipative forces, whose effects appear in the potential energy and dissipation function, respectively. Also, $\{\partial V/\partial q\}^{*}$ is the set of static external forces that establish the equilibrium position. Thus, the right side of Eq. (7.86) represents the set of nondissipative generalized forces that tend to move the system away from its static equilibrium position, at which $\{\xi\} = \{0\}$.

Many methods are available for solving these standard equations of motion for a scleronomic system, for they have constant coefficients. Study of the responses that can be obtained from these equations forms a major portion of texts on linear vibrations. It must be emphasized, however, that Eq. (7.86) is valid only for analysis of a scleronomic system. If the physical constraints of a system are time-dependent, there will be additional terms due to the more general form of the kinetic energy in Eq. (7.76). Also, if there are nonholonomic constraint conditions to be satisfied, the differential equations of motion must be modified to include either the constraint forces or the Lagrangian multipliers. Also, the system of equations must be supplemented by constraint equations, which also may be linearized.

EXAMPLE 7.12 _____

The bar is in static equilibrium in the upright position shown. The unstretched length of the identical springs is $l_0 < (a^2 + b^2)^{1/2}$, in order that the springs may remain taut throughout any small displacement away from this position. Derive the corresponding equation of motion.

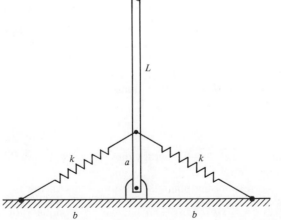

Example 7.12

Solution. We select as the generalized coordinate the angle θ relative to the vertical. Since $\theta = 0$ at the static equilibrium position, this variable serves directly

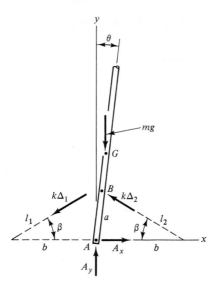

Free body diagram.

as a relative generalized coordinate. The bar is in pure rotation about end A, so the kinetic energy is

$$T = \tfrac{1}{2} (\tfrac{1}{3} mL^2)\dot\theta^2$$

This expression needs no simplification, because it is quadratic in $\dot\theta$.

We select the elevation of the pin as the datum for the gravitational potential energy. The springs also are conservative, so

$$V = \tfrac{1}{2} k\Delta_1^2 + \tfrac{1}{2} k\Delta_2^2 + mg\frac{L}{2}\cos\theta$$

In order to obtain some general properties regarding the potential energy of a spring, let us consider the left spring first. We use the law of cosines to describe the length of this spring at an arbitrary position in terms of the generalized coordinate.

$$l_1 = [a^2 + b^2 - 2ab\cos(90° + \theta)]^{1/2} = (a^2 + b^2 + 2ab\sin\theta)^{1/2}$$

We need only terms up to quadratic in the elongations, so we drop cubic and higher-degree terms in the expansion of the sine, as well as in the binomial expansion of the square root. The resulting elongation is

$$\Delta_1 = l_1 - l_0 \approx [(a^2 + b^2)^{1/2} - l_0] + ab\theta(a^2 + b^2)^{-1/2}$$
$$- \tfrac{1}{2}(ab\theta)^2(a^2 + b^2)^{-3/2}$$

Since the linear terms in the potential energy merely define the equilibrium position, we retain only the quadratic portion of Δ_1^2. Some terms cancel in this product, with the result that the corresponding potential energy of this spring is

$$V_1 = \tfrac{1}{2} k\Delta_1^2 = \tfrac{1}{2} ka^2b^2l_0(a^2 + b^2)^{-3/2}\theta^2 + \text{nonquadratic terms}$$

A more interesting form of this result comes from expressing the horizontal distance

b in terms of the angle of elevation β, from which we find that

$$V_1 = \frac{1}{2} k \frac{l_0}{(a^2 + b^2)^{1/2}} (a\theta \cos \beta)^2$$

In the small displacement approximation $a\theta$ is the horizontal movement of point B. Then $a\theta \cos \beta$ is the component of this displacement parallel to the spring, which is the same as the change in the length of the spring relative to the length at the equilibrium position. The above result for V_1 is illustrative of a general feature. If a spring is stretched very little at the static equilibrium position [$l_0 \approx (a^2 + b^2)^{1/2}$ for V_1], the potential energy of that spring may be obtained with reasonable accuracy by following the displacements of the ends of the spring.[†] Note that, if the stiffness k is large, this condition can be obtained even though the spring force at the equilibrium position is large.

We now return to the derivation of the equation of motion by noting that the quadratic term in the energy of the other spring will be identical to the above expression. For the gravitational potential energy, we expand $\cos \theta$ in a series for small θ, and truncate that series at the second degree, which yields

$$\cos \theta \approx 1 - \tfrac{1}{2} \theta^2$$

The total potential energy that results is

$$V = \left[k\frac{l_0}{(a^2 + b^2)^{1/2}} (a \cos \beta)^2 - \tfrac{1}{4} mg \, L \right] \theta^2 + \text{nonquadratic terms}$$

It is interesting to note that the appearance of l_0 in this expression is surprising for some who have learned linearization concepts as ad hoc procedures.

The only forces remaining are the reactions, whose constraints are not violated. Therefore, $Q_1 = 0$. Substitution of T and V into Lagrange's equations therefore leads to

$$\tfrac{1}{3} mL^2\ddot{\theta} + \left[2k \frac{l_0}{(a^2 + b^2)^{1/2}} a^2 \cos^2 \beta - \tfrac{1}{2} mg \, L \right] \theta = 0$$

EXAMPLE 7.13 _____

Bar AB, whose mass is m, is in static equilibrium in the position shown owing to the presence of the counterweight. The spring and dashpot restraining the collar have stiffness k and viscosity μ, respectively, and the cable joining the two bodies has ideal properties. The horizontal force $F(t)$ is not present when the system is in static equilibrium. Determine the equations of motion for small displacements relative to the equilibrium position.

[†]This derivation of the potential energy of a spring is not valid when $\beta = 0$. However, it is not difficult to show in that case that the static elongation is *never* significant to the quadratic terms in potential energy.

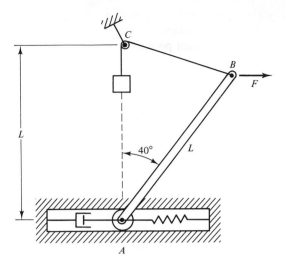

Example 7.13

Solution. The mass m_2 of the counterweight may be evaluated from a moment sum about pin A. The horizontal position ξ_1 and the angle ξ_2 are defined to be zero at the static equilibrium position, and the cable force $P = m_2 g$ in that case. Thus, the moment sum for the static system is

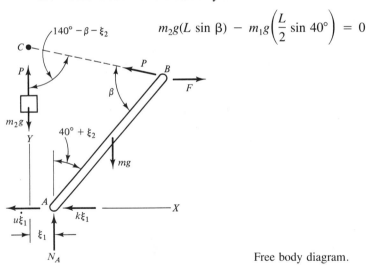

$$m_2 g(L \sin \beta) - m_1 g\left(\frac{L}{2} \sin 40°\right) = 0$$

Free body diagram.

Since triangle BAC is isosceles,

$$2\beta + (40° - \xi_2) = 180° \qquad \beta = 70° - \frac{\xi_2}{2}$$

which for $\xi_2 = 0$ yields

$$m_2 = m_1 \frac{\sin 40°}{2 \sin 70°} = 0.3420 \, m_1$$

As suggested above, we use ξ_1 and ξ_2 as generalized coordinates for movement relative to the equilibrium position. The angular velocity of the bar is $\bar{\omega} = -\dot{\zeta}_2 \bar{K}$, so the velocity of the relevant points on the bar is

$$\bar{v}_A = \dot{\zeta}_1 \bar{I} \qquad \bar{v}_B = \bar{v}_A + \bar{\omega} \times \bar{r}_{B/A} \qquad \bar{v}_G = \bar{v}_A + \bar{\omega} \times \bar{r}_{G/A}$$

Because the analysis is to be linearized, it is permissible to describe the position of points at static equilibrium. Hence

$$\bar{v}_B = \dot{\xi}\bar{I} + (-\dot{\xi}_2\bar{K}) \times (L \sin 40°\bar{I} + L \cos 40°\bar{J})$$
$$= (\dot{\xi}_1 + 0.7660L\dot{\xi}_2)\bar{I} - 0.6428L\dot{\xi}_2\bar{J}$$
$$\bar{v}_G = \tfrac{1}{2}(\bar{v}_A + \bar{v}_B) = (\dot{\xi}_1 + 0.3830L\dot{\xi}_2)\bar{I} - 0.3214L\dot{\xi}_2\bar{J}$$

It is reasonable to assume that the counterweight only moves vertically. We may determine its speed v_C from the condition that the cable is inextensible. This condition may be satisfied in the linearized analysis by letting v_C equal the component of $\bar{v}_B$ parallel to the cable in the equilibrium position. Setting $\beta = 70°$ and $\xi_2 = 0$ to describe $\bar{e}_{B/C}$ leads to

$$v_C = \bar{v}_B \cdot \bar{e}_{B/C} = \bar{v}_B \cdot [\sin(140° - \beta)\bar{I} - \cos(140° - \beta)\bar{J}]$$
$$= 0.9397(\dot{\xi}_1 + L\dot{\xi}_2)$$

We may now form the kinetic energy. Since the bar is in general motion, we have

$$T = \tfrac{1}{2} m_1 \bar{v}_G \cdot \bar{v}_G + \tfrac{1}{2}(\tfrac{1}{12} m_1 L^2)\omega^2 + \tfrac{1}{2} m_2 v_C^2$$
$$= \tfrac{1}{2} m_1[(\dot{\xi}_1 + 0.3830L\dot{\xi}_2)^2 + (0.3214L\dot{\xi}_2)^2 + (\tfrac{1}{12}L^2)\dot{\xi}_2^2]$$
$$+ \frac{1}{2}(0.3420m_1)(0.9397)^2(\dot{\xi}_1 + L\dot{\xi}_2)^2$$
$$= \tfrac{1}{2} m_1(1.3020\dot{\xi}_1^2 + 0.6353L^2\dot{\xi}_2^2 + 1.3700L\dot{\xi}_1\dot{\xi}_2)$$

The potential energy is much easier to obtain because the elongation of the spring relative to the static equilibrium position is ζ_1. Selecting end A as the datum for gravitational potential energy then leads to

$$V = \tfrac{1}{2} k\xi_1^2 + mg\frac{L}{2}\cos(40° + \xi_2)$$

In keeping with the linearization process, we expand V in a series in ξ_2 and drop nonquadratic terms. Thus

$$V = \tfrac{1}{2} k\xi_1^2 + mg\frac{L}{2}(\cos 40° \cos \xi_2 - \sin 40° \sin \xi_2)$$
$$= \tfrac{1}{2} k\xi_1^2 + \tfrac{1}{2} mgL \cos 40°(-\tfrac{1}{2}\xi_2^2) + \cdots$$

We account for the nonconservative dashpot force by means of the dissipation function,

$$D = \tfrac{1}{2} \mu\dot{\xi}_1^2$$

It is possible to oversimplify the virtual work in the linearization process. Let us formulate the virtual work done by the horizontal force $\overline{F}$ for arbitrarily large displacements. The position of point B in that case is

$$\overline{r}_{B/O} = [\xi_1 + L \sin (40° + \xi_2)]\overline{I} + L \cos (40° + \xi_2)\overline{J}$$

so

$$\delta \overline{r}_B = \frac{\partial \overline{r}_B}{\partial \xi_1} \delta \xi_1 + \frac{\partial \overline{r}_B}{\partial \xi_2} \delta \xi_2$$
$$= \overline{I}\delta \xi_1 + L[\cos (40° + \xi_2)\overline{I} - \sin (40° + \xi_2)\overline{J}] \delta \xi_2$$

The corresponding virtual work is

$$\delta W = F\overline{I} \cdot \delta \overline{r}_B = F \delta \xi_1 + FL \cos (40° + \xi_2) \delta \xi_2$$

according to which the generalized forces are

$$Q_1 = F \qquad Q_2 = FL \cos (40° + \xi_2)$$

Note that the Q_j occur linearly in the equations of motion. Correspondingly, we drop all quadratic and higher terms from a series expansion of Q_2.

$$Q_2 = FL(\cos 40° \cos \xi_2 - \sin 40° \sin \xi_2)$$
$$= FL(\cos 40° - \xi_2 \sin 40°)$$

It could be argued that the smallness of ξ_2 makes the second term negligible in comparison with the first. However, the terms affect the equations of motion in different ways. The first term corresponds to an inhomogeneous excitation, but the second will occur in an equation of motion as a time-dependent coefficient. There are some situations where the presence of such a term might be important. In general, if we are not concerned with that possibility, it is permissible to obtain the virtual displacements from the kinematical relationships at the reference position.

We obtain the equations of motion from Eq. (7.82), which is the version of Lagrange's equations that includes the dissipation function. The results are

$$1.3020\ddot{\xi}_1 + 0.6850l\ddot{\xi}_2 + \mu\dot{\xi}_1 + k\xi_1 = F$$
$$0.6353L^2\ddot{\xi}_2 + 0.6850L\ddot{\xi}_1 + (0.6428F - 0.3830mg)L\xi_2 = 0.7660FL$$

EXAMPLE 7.14 _____

Servomotors maintain the spin rate $\dot{\phi}$ and the precession rate $\dot{\psi}$ at constant values. When $\dot{\psi}$ exceeds a minimum value, a steady precessional motion is possible, in which θ is a constant nonzero value. Determine this minimum value of $\dot{\psi}$. Then, for the case where $\dot{\psi}$ exceeds this minimum, derive the equation of motion for small changes of θ from its steady precession value.

Solution. The reference state in this system is a steady-state precession. In order to establish the conditions for such motion from the same equations as those for

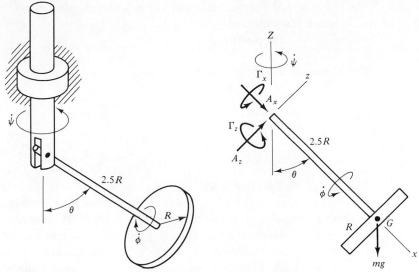

Example 7.14   Free body diagram.

small deviations away from steady state, we begin by considering the nutation angle θ to be arbitrary, $q_1 = \theta$. The precession and spin rates are constrained to be constant, so the system is rheonomic with one degree of freedom.

The flywheel is in pure rotation about pin A. The angular motion of the disk is

$$\overline{\omega} = \dot{\psi}\overline{K} + \dot{\theta}(-\overline{j}') + \dot{\phi}\overline{i} = (\dot{\phi} - \dot{\psi}\cos\theta)\overline{i} - \dot{\theta}\overline{j} + \dot{\psi}\sin\theta\overline{k}$$

$$\overline{\alpha} = -\ddot{\theta}\overline{j}' - \dot{\theta}(\dot{\psi}\overline{K} \times \overline{j}') + \dot{\phi}\overline{\omega} \times \overline{i}$$
$$= \dot{\psi}\dot{\theta}\sin\theta\overline{i} + (-\ddot{\theta} + \dot{\psi}\dot{\phi}\sin\theta)\overline{j} + (\dot{\psi}\dot{\theta}\cos\theta + \dot{\theta}\dot{\phi})\overline{k}$$

We assume that the mass of the flywheel in much greater than that of either shaft. Because xyz are principal axes for the flywheel, we may employ Euler's equations. Considering the flywheel to be a thin disk leads to $I_{xx} = 0.5mR^2$, $I_{yy} = I_{zz} = 6.50mR^2$. The pin exerts no moment about its own axis, so the equation of motion is

$$\sum M_{Ay} = mg(2.5R\sin\theta)$$
$$= I_{yy}\alpha_y - (I_{zz} - I_{xx})\omega_x\omega_y$$
$$= mR^2[6.5(-\ddot{\theta} + \dot{\psi}\dot{\phi}\sin\theta) - 6.0(\dot{\phi} - \dot{\psi}\cos\theta)(\dot{\psi}\sin\theta)]$$

which simplifies to

$$13\ddot{\theta} + \left(5\frac{g}{R} - \dot{\psi}\dot{\phi}\right)\sin\theta - 12\dot{\psi}^2\sin\theta\cos\theta = 0 \qquad (1)$$

For steady precession, we set $\dot{\theta} = \ddot{\theta} = 0$. In that case, the above equation of motion yields

$$\sin\theta^* = 0 \qquad \text{or} \qquad \cos\theta^* = \frac{5g - R\dot{\psi}\dot{\phi}}{12R\dot{\psi}^2} \qquad (2)$$

The first case is the vertical position. The other solution is possible only if $|\cos \theta*|$ < 1, which leads to

$$-12R\dot{\psi}^2 < 5g - R\dot{\psi}\dot{\phi} < 12R\dot{\psi}^2$$

This condition may be rewritten as

$$\dot{\phi}_{min} < \dot{\phi} < \dot{\phi}_{max} \tag{3a}$$

where

$$\dot{\phi}_{min} = 5\frac{g}{R\dot{\psi}} - 12\dot{\psi} \qquad \dot{\phi}_{max} = 5\frac{g}{R\dot{\psi}} + 12\dot{\psi} \tag{3b}$$

When the spin rate is less than $\dot{\phi}_{min}$, only $\theta* = 0$ is possible. In that case, the dynamic tendency to swing outward is inadequate to overcome the tendency of gravity to return the system to the upright position. In contrast, a spin rate that exceeds $\dot{\phi}_{max}$, which is not physically possible because of interference with the vertical shaft, corresponds to $\theta* = 180°$. Then the tendency of gravity to move the flywheel away from the inverted position is inadequate to balance the angular momentum change that would result if the system were to move away from upright.

In order to study the effect of small changes in θ away from a steady-state position, we let

$$\theta = \theta* + \xi \qquad |\xi| << 1$$

Linearization of the equation of motion requires dropping quadratic and higher-order terms in ξ from series expansions. Thus, we write

$$\sin \theta = \sin(\theta* + \xi) = \sin \theta* \cos \xi + \cos \theta* \sin \xi$$
$$= \sin \theta* + \xi \cos \theta* + \cdots$$
$$\cos \theta = \cos(\theta + \xi) = \cos \theta* - \xi \sin \theta*$$
$$\sin \theta \cos \theta = \tfrac{1}{2} \sin 2\theta* + \xi \cos 2\theta*$$

Substitution of these expressions into the nonlinear equation of motion (1) gives

$$13\ddot{\xi} + \left[\left(5\frac{g}{R} - \dot{\psi}\dot{\phi} \right) \sin \theta* - 6\dot{\psi}^2 \sin 2\theta* \right]$$
$$+ \left[\left(5\frac{g}{R} - \dot{\psi}\dot{\phi} \right) \cos \theta* - 12\dot{\psi}^2 \cos 2\theta* \right] \xi = 0$$

The first bracketed term vanishes identically, by virtue of Eq. (2) for $\theta*$, so we obtain

$$13\ddot{\xi} + \left[\left(5 - \frac{g}{R} \dot{\psi}\dot{\phi} \right) \cos \theta* - 12\dot{\psi}^2 \cos 2\theta* \right] \xi = 0$$

This is the desired equation for motion relative to steady precession. It has several interesting features, all of which result from the observation that $\xi(t)$ is oscillatory if the coefficient of ξ in the differential equation is positive. Similarly, a negative coefficient corresponds to exponentially increasing motions. Define β to be this coefficient,

$$\beta = \left(5\frac{g}{R} - \dot{\psi}\dot{\phi}\right)\cos\theta^* - 12\dot{\psi}^2\cos 2\theta^* \tag{4}$$

If $\beta > 0$, a small disturbance away from $\theta = \theta^*$ will not grow in time. Thus, $\beta > 0$ means that $\theta = \theta^*$ is a stable steady precession. In contrast, the steady precession is unstable if $\beta < 0$.

Consider first the case $\theta^* = 0$, which is always a possibility. In that case,

$$\beta = 5\frac{g}{R} - \dot{\psi}\dot{\phi} - 12\dot{\psi}^2 \tag{5}$$

Instability of $\theta^* = 0$ occurs when $\beta < 0$, which is satisfied if $\dot{\phi} > 5g/R\dot{\psi} - 12\dot{\psi}$. This is identical to the condition $\dot{\phi} > \dot{\phi}_{min}$ for which the nonzero value of θ^* is possible.

Suppose now that $\dot{\phi}$ satisfies the inequality (3a), corresponding to $\theta^* > 0$. We substitute Eq. (2) and

$$\cos 2\theta^* = 2\cos^2\theta^* - 1$$

into the definition of β. After simplification, the result is

$$\beta = \frac{(12R\dot{\psi}^2)^2 - (5g - R\dot{\psi}\dot{\phi})^2}{12R^2\dot{\psi}^2}$$

The conditions for $\beta > 0$ are identical to the inequality condition (3a).

In summary, the flywheel remains below the pivot if the spin rate is below the lower limit

$$\dot{\phi}_{min} < \frac{5g}{R\dot{\psi}^2} - 12\dot{\psi}$$

If $\dot{\phi}$ exceeds this value, a stable steady precession results with

$$\theta^* = \cos^{-1}\left(\frac{5g - R\dot{\psi}\dot{\phi}}{12\dot{\psi}^2}\right)$$

Such a precessional motion ceases to be possible if the spin rate exceeds the upper limit

$$\dot{\phi}_{max} > \frac{5g}{R\dot{\psi}^2} + 12\dot{\psi}$$

If the upright position $\theta^* = 180°$ were physically admissible, we could prove by examining β that it is unstable if $\dot{\phi} < \dot{\phi}_{max}$, and stable if $\dot{\phi}$ exceeds that value.

REFERENCES

P. Appell, "Sur les Mouvements de Roulement; Equations du Mouvement Analogues à celles de Lagrange," *Comptes Rendus,* vol. 129, pp. 317–320, 1899.

E. A. Desloge, *Classical Mechanics,* vol. 2, John Wiley & Sons, New York, 1982.

J. W. Gibbs, "On the Fundamental Formulae of Dynamics," *American Journal of Mathematics,* vol. 2, pp. 49–64, 1879.

H. Goldstein, *Classical Mechanics,* 2d ed., Addison-Wesley, Reading, Mass., 1980.

L. Meirovitch, *Methods of Analytical Dynamics,* McGraw-Hill, New York, 1970.

L. A. Pars, *A Treatise on Analytical Dynamics,* William Heinemann, Ltd., London, 1965.

R. M. Rosenberg, *Analytical Dynamics of Discrete Systems,* Plenum Press, New York, 1977.

K. R. Symon, *Mechanics,* 3d ed., Addison-Wesley, Reading, Mass., 1971.

E. T. Whittaker, *A Treatise on the Analytical Dynamics of Particles and Rigid Bodies,* 4th ed., Cambridge University Press, Cambridge, 1937.

HOMEWORK PROBLEMS

7.1 The torque Γ acting about the vertical shaft is such that the rotation rate Ω is constant. The blocks, having masses m_A and m_B, are tied together by an inextensible cable. Derive the equation of motion for the radial distance R, and also obtain an expression for Γ.

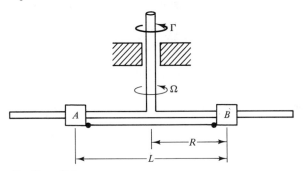

Problem 7.1

7.2 Derive the Lagrange equations of motion for the system in Example 7.7, and also obtain a relation for Γ. Compare the results with those obtained in the solution to the example.

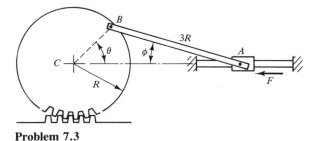

Problem 7.3

7.3 Force $F(t)$ pushes piston A, whose mass is m, to the left. This causes the gear to roll over the horizontal rack. The mass of the gear is $2m$ and its radius of gyration about center point C is κ. Use the angles θ and ϕ as constrained generalized coordinates to derive the equations of motion for the system.

7.4 The bar is slipping relative to the ground as it falls. The coefficient of kinetic friction is μ. Use Lagrange's equations to derive the equations of motion for the bar.

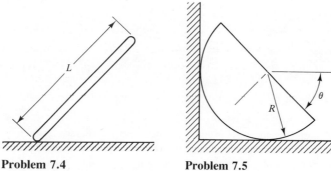

Problem 7.4   **Problem 7.5**

7.5 The semicylinder, whose mass is m, is released from rest at an initial orientation $\theta > 0$. The floor is smooth, and the coefficient of kinetic friction μ between the cylinder and the wall is not adequate to prevent sliding. Use Lagrange's equations to derive the equations of motion for the semicylinder.

7.6 The bar is supported by a ball-and-socket joint at end A and the rough wall at end B; the coefficient of sliding friction is μ. Use Lagrange's equations to derive the equation of motion governing the angle of inclination θ.

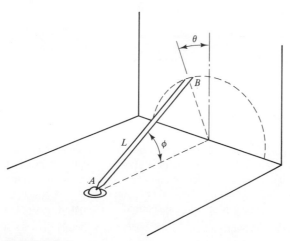

Problem 7.6

7.7 The cylinder of mass m is free to rotate by angle β relative to the gimbal, which rotates about the horizontal axis. The precessional rate Ω is held constant by varying the torque Γ. Use Lagrange's equations to derive the equation of motion governing β, as well as an expression for Γ.

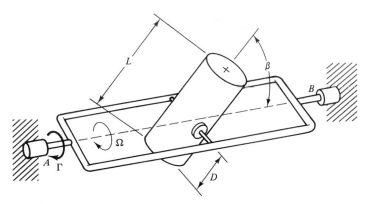

Problem 7.7

7.8 A known couple $\overline{M}(t)$ is applied to the upper bar. Force $\overline{F}$, which is applied perpendicularly to the lower bar, acts to make the velocity of end C always be collinear with the line from joint B to end C. The bars have equal mass m, and the system lies in the horizontal plane. Use the method of Lagrangian multipliers to derive the equations of motion.

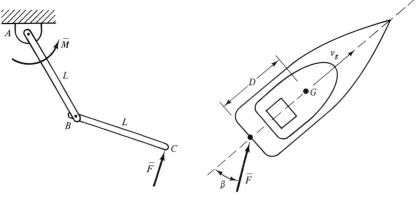

Problem 7.8   **Problem 7.9**

7.9 The thrust of an outboard motor on a boat may be represented as a force $\overline{F}$ acting at an angle β relative to the axis of the boat. The hydrodynamic properties of the boat are such that the velocity of the center of mass G is constrained to be parallel to the longitudinal axis of the boat. Derive the equations of motion for the boat using Lagrangian multipliers. The mass of the boat is m, and the centroidal moment of inertia is I.

7.10 Write the equations of motion for the system in Example 7.1 in state-space form.

7.11 Moment $Q(t)$ about the vertical axis causes the gyroscope to rotate such that the precession angle $\psi = c\beta$, where c is a constant. The spin rate ω_s is held constant by a servomotor. The mass of this motor and the gimbal is negligible. The mass of the flywheel is m and its principal radii of gyration for centroidal axes are κ_1 about its spin axis and κ_2 normal to that axis. Derive the state-space form of the equations of motion. Also, derive an expression for Q.

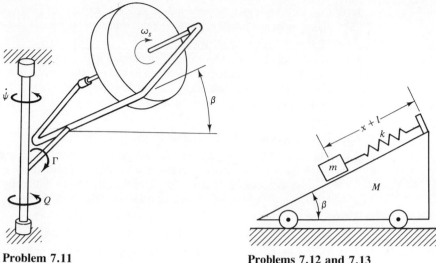

Problem 7.11   **Problems 7.12 and 7.13**

7.12 The small mass m is supported by a spring as it moves along the smooth incline on the cart, whose mass is M. The spring has stiffness k and its unstretched length is l. Derive Hamilton's canonical equations for the system.

7.13 Consider the system in Problem 7.12. Derive a single differential equation of motion for the relative distance x using Routh's method for the ignoration of coordinates.

7.14 Derive Hamilton's canonical equations for the system in Example 6.14.

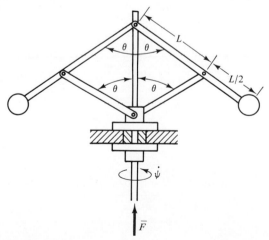

Problem 7.15

7.15 Angle θ for the flyball governor is controlled by applying force F, which moves the vertical shaft up or down. The system rotates freely about the vertical axis. The mass of each sphere is m and the mass of the linkage is negligible. Use Routh's method for the ignoration of coordinates to derive a single differential equation governing the nutation angle θ.

7.16 Consider the system in Problem 7.1. Use conservation of the Hamiltonian and the work-energy principle to derive the differential equation for the radial distance R, and to obtain an expression for the couple Γ.

7.17 Consider the system in Problem 7.7 in the case where the torque $\Gamma = 0$, so that the rotation rate Ω about the horizontal axis is unknown. Let $D = L/2$, so that the center of mass is coincident with the horizontal axis. Use Routh's method for the ignoration of coordinates to derive a single differential equation governing the nutation angle β. Can such a formulation be used when $D \neq L/2$? Explain your answer.

7.18 Use conservation of the Hamiltonian and the work-energy principle to solve Problem 7.7.

7.19 The absolute velocity of a particle may be represented by its components v_x, v_y, and v_z along the axes of a moving reference system xyz. Suppose that the angular velocity $\bar{\omega}$ of xyz and the velocity $\bar{v}_0$ of the origin of xyz are known as functions of time. Derive the Gibbs-Appell equations of motion corresponding to the quasi-velocities $\dot{\gamma}_1 = v_x$, $\dot{\gamma}_2 = v_y$, and $\dot{\gamma}_3 = v_z$.

7.20 Derive the Gibbs-Appell equations of motion for the flyball governor in Problem 7.15.

7.21 Derive the Gibbs-Appell equations of motion for the boat in Problem 7.9.

7.22 Friction between the rod and the surfaces it contacts is negligible. Determine the Gibbs-Appell equations of motion for the system. Assume that the rod remains in contact with the wall.

7.23 The coefficient of kinetic friction between the rod and corner A is μ, while frictional resistance at the wall is negligible. Determine the Gibbs-Appell equations of motion for the system. Assume that the rod remains in contact with the wall.

7.24 Derive the Gibbs-Appell equations of motion for the system in Example 7.1.

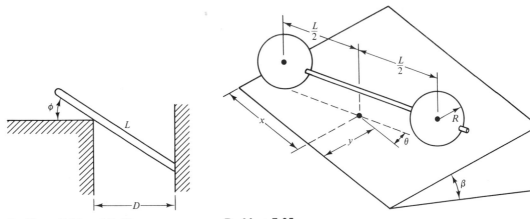

Problems 7.22 and 7.23   **Problem 7.25**

7.25 The identical spheres of mass m spin freely relative to the massless shaft, such that their centers are at constant distance L. Derive equations of motion for position coordinates x and y and angle θ of the shaft relative to the incline.

7.26 A square plate pivoted about corner A is supported by two springs of stiffness k, such that the inverted position shown is a static equilibrium position. Derive the equation of motion for small rotation away from this position. From that equation determine the minimum allowable value of k for which the equilibrium position is stable.

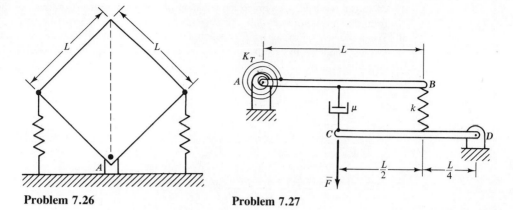

Problem 7.26   **Problem 7.27**

7.27 Bars AB and CD, each of whose mass per unit length is m/L, are connected by a spring whose stiffness is k and a dashpot whose constant is μ. In addition a torsional spring whose stiffness is K_T restrains rotation at pivot A. The system is in static equilibrium in the horizontal position shown when the force $\overline{F}$ is not present. Determine the equations of motion for small displacements relative to this position.

7.28 The linkage, which lies in the vertical plane, is loaded by a horizontal force $\overline{F}(t)$. The mass of each bar is m, and the spring, whose unstretched length is $L/2$, has stiffness k. The hydraulic cylinder at end C, which only permits vertical movement, acts like a dashpot whose constant is μ. The static equilibrium position of the system is $\theta = 36.87°$ when $F = 0$. Derive an expression for k. Then derive the equation of motion for small displacements away from the static equilibrium position.

7.29 When the vertical force $\overline{F}$ is not present, the system is in static equilibrium in the position shown. The mass of the bar and of the block, which is attached to the bar by the inextensible

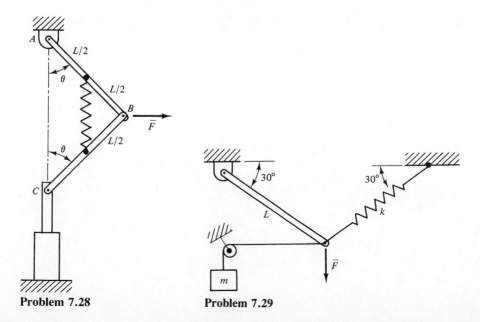

Problem 7.28   **Problem 7.29**

cable, are each m. Determine the equation of motion for small rotations of the bar away from this position.

7.30 When the vertical excitation $\bar{F}(t)$ is not present, the linkage is in equilibrium in the position shown. The mass of each bar is m and the stiffness of the spring is k. Derive the linearized equation of motion.

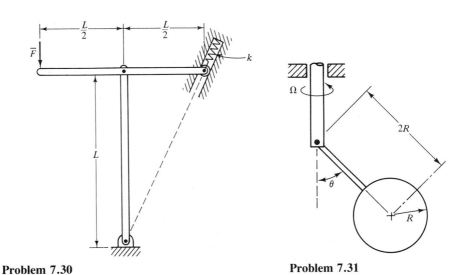

Problem 7.30   **Problem 7.31**

7.31 The sphere, whose mass is m, is connected to the vertical shaft by a bar whose mass is negligible. The rotation rate Ω about the vertical axis is constant. Depending on the value of Ω, the system may precess in a steady manner with either the vertical orientation, $\theta = 0$, or in a nutated orientation, in which θ is nonzero. Derive the minimum value Ω_{cr} for which the nutated position is possible. By considering small changes of θ from its steady value, evaluate the stability of the vertical and nutated positions as a function of Ω.

7.32 Consider the system in Problem 7.1 in the case where $m_A = m$ and $m_B = m/2$, where m is a basic mass unit. It is possible for the system to rotate such that the radial distance R to collar A is constant. Derive an expression for this constant distance. Then evaluate its stability by considering a small displacement away from the steady position.

Chapter 8

Gyroscopic Effects

The emphasis thus far has been on the development of basic principles for treating the kinematics and kinetics of rigid-body motion. Regardless of whether we employed a Newtonian or Lagrangian formulation, it was usually necessary to account for constraints associated with the way in which the system is supported. Sometimes the goal was to characterize the force system required to produce a specified motion, as is the case when the reactions must be evaluated. Other situations required the determination of conditions that are satisfied during the motion, as typified by the task of deriving differential equations of motion. This chapter is devoted to a fuller investigation of the manner in which rigid bodies move through space. Because the angular momentum in spatial motion is usually not aligned with the instantaneous axis of rotation, a portion of the rotational effect does not coincide with that axis. Such phenomena are exploited in gyroscopes, whose theory will be introduced here. However, we may learn much about the nature of dynamical responses by beginning with studies of simpler, yet more common, systems that display comparable effects.

8.1 FREE MOTION

One of the first types of spatial motion treated in basic physics and engineering courses on mechanics is projectile motion, whose study is devoted to the determination of the motion of the center of mass. In contrast, the manner in which the body rotates about its center of mass is seldom discussed in a fundamental course. Our study of this *free motion* will be based on the assumption that the external force system resolves to a single equivalent force acting at the center of mass. In that case the rotational motion is uncoupled from that of the mass center. In reality, aerodynamic forces acting on a body usually may be represented as a force-couple system acting at the center of pressure, which does not necessarily coincide with the mass center. Such forces depend on the orientation (angle of attack) as well as the overall velocity. Hence, accurate models of the motion of objects through the air might require consideration of coupling between the translational and rotational motions.

8.1.1 Arbitrary Bodies

A direct consequence of the assumption that the force system acts through the center of mass is that the angular momentum $\overline{H}_G$ about the center of mass is constant. This simple fact provides the foundation for the entire development. The constant magnitude of $\overline{H}_G$ leads to a constant that relates the rates of rotation. The direction of $\overline{H}_G$ is also invariable, so it provides a fixed direction in space that may be employed as a reference.

Eulerian angles are useful for the kinematical formulation of free motion. Suppose that we know the orientation of the body at the instant it is released, and that the angular velocity at that instant is also known. It is a simple matter to use such information to form $\overline{H}_G$. Let us *define the precession axis to coincide with the direction of the angular momentum*. It is convenient for the development to use as the body-fixed reference frame a set of axes xyz that are principal, about which the principal moments of inertia are I_1, I_2, and I_3, respectively. (The evaluation of the orientation of these axes for an arbitrary body is discussed in Section 5.2.2) The definition of the Eulerian angles associated with these axes is depicted in Figure 8.1. The precession rate $\dot{\psi}$ is about the Z axis, which is parallel to $\overline{H}_G$. The spin rate $\dot{\phi}$ is about the z axis, which may be any of the principal axes. The nutation rate $\dot{\theta}$ is about the line of nodes, which is the y' axis perpendicular to the plane formed by the Z and z axes.

The angular velocity at any instant is

$$\overline{\omega} = (-\dot{\psi}\sin\theta\cos\phi + \dot{\theta}\sin\phi)\overline{i} + (\dot{\psi}\sin\theta\sin\phi + \dot{\theta}\cos\phi)\overline{j} + (\dot{\psi}\cos\theta + \dot{\phi})\overline{k}$$
(8.1)

An expression for the angular momentum may be obtained by combining these angular velocity components with the respective moments of inertia.

$$\overline{H}_G = I_1\omega_x\overline{i} + I_2\omega_y\overline{j} + I_3\omega_z\overline{k}$$
(8.2)

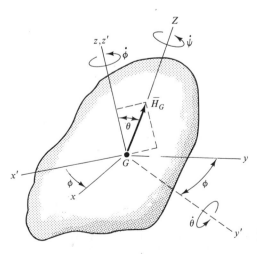

Figure 8.1 Eulerian angles for free motion.

However, projecting the $\overline{H}_G$ vector in Figure 8.1 onto the respective axes leads to a different expression.

$$\overline{H}_G = -H_G \sin \theta \cos \phi \bar{i} + H_G \sin \theta \sin \phi \bar{j} + H_G \cos \theta \bar{k} \qquad (8.3)$$

Substitution of the angular velocity components in Eq. (8.1) into Eq. (8.2), followed by matching like components of that expression to Eq. (8.3), yields

$$I_1(\dot{\psi} \sin \theta \cos \phi - \dot{\theta} \sin \phi) = H_G \sin \theta \cos \phi$$
$$I_2(\dot{\psi} \sin \theta \sin \phi + \dot{\theta} \cos \phi) = H_G \sin \theta \sin \phi \qquad (8.4)$$
$$I_3(\dot{\psi} \cos \theta + \dot{\phi}) = H_G \cos \theta$$

Since the value of H_G is known from the motion at the instant the body was released, these relations constitute a set of first-order, coupled differential equations for the Eulerian angles. Their initial conditions are the values of the Eulerian angles at the instant of release.

It is interesting to note that we could have used Lagrange's equations to obtain an alternative set of differential equations of motion. Such equations would have been second-order. In essence, the constancy of $\overline{H}_G$ led us in Eqs. (8.4) to first integrals of the Lagrangian equations.

These equations of motion are highly nonlinear. If one is interested in developing solutions for specified initial conditions, integration schemes are easier to implement if the equations are not coupled in the derivatives. We may obtain equations for the precession and nutation rates from the first two of Eqs. (8.4), after which we find an expression for the spin rate from the last equation. The result is

$$\dot{\psi} = H_G \left(\frac{\cos^2 \phi}{I_1} + \frac{\sin^2 \phi}{I_2} \right)$$

$$\dot{\theta} = H_G \left(\frac{1}{I_2} - \frac{1}{I_1} \right) \sin \theta \sin \phi \cos \phi \qquad (8.5)$$

$$\dot{\phi} = H_G \left(\frac{1}{I_3} - \frac{\cos^2 \phi}{I_1} - \frac{\sin^2 \phi}{I_2} \right) \cos \theta$$

We shall not pursue solutions of these differential equations, although analytical solutions in the form of elliptic functions are possible; see Synge and Griffith in the references at the end of this chapter. Also, numerical solutions may be readily implemented. We will study a graphical way of understanding the rotation in a later section on the Poinsot construction, after we treat the special case of an axisymmetric body in the next section. Before we continue to those topics, we can make a few general observations.

First, note that $\dot{\psi}$ is always positive, which means that a body in free motion never changes the direction in which it precesses. However, the signs of $\dot{\theta}$ and $\dot{\phi}$ depend on the relative magnitudes of the moments of inertia, and on the current quadrant in which θ and ϕ reside. The latter observation suggests that there might be free motions in which the nutation and spin rates, and therefore the corresponding angles, oscillate. This, in turn, leads us to be concerned with the stability of a rotational motion that has been established. We may obtain specific results regarding stability in the important case where we attempt to make a body rotate at rate Ω about a principal axis. Without loss of generality, we consider such a motion to consist solely of a spin, with the precession axis aligned with the spin axis; both the precession and nutation rates are zero in such a motion.

The question of stability for this motion arises from the recognition that we are not likely to impart an initial rotation to a body in which the axis of rotation is *exactly* aligned with one of the principal axes. A more realistic expectation is that because of a small error the initial motion will feature nutation and precession rates that are much smaller than Ω, and that the nutation angle will be small. If an evaluation of the response confirms that these initial *perturbations* remain small for the overall response, we may conclude that the rotation is stable.

Accordingly, let

$$\dot{\phi} = \Omega + \Delta\dot{\phi} \qquad \theta = \Delta\theta \qquad \dot{\psi} = \Delta\dot{\psi} \tag{8.6}$$

where $\Delta\psi$, $\Delta\theta$, and $\Delta\phi$ represent small deviations from the ideal. We may derive solvable solutions for the perturbations in situations where they are sufficiently small to linearize the equations of motion. It is easier to investigate the stability by returning to the original set of equations of motion, Eqs. (8.4). The linearized differential equations resulting from substitution of Eqs. (8.6) are

$$-I_1 \Delta\dot{\theta} \sin\phi = H_G \Delta\theta \cos\phi$$
$$I_2 \Delta\dot{\theta} \cos\phi = H_G \Delta\theta \sin\phi \tag{8.7}$$
$$I_3 \Omega = H_G$$

In order to solve these equations, let us consider $\overline{\Delta\theta}$ to be a vector aligned along the line of nodes in the right-hand sense. Then $\Delta\theta \cos\phi$ and $\Delta\theta \sin\phi$ are the projections of the small nutation angle onto the x and y axes. Let u and v denote these components,

$$u = \Delta\theta \cos\phi \qquad v = \Delta\theta \sin\phi \tag{8.8}$$

Then, because $\dot{\phi} \approx \Omega$ when higher-order perturbation terms are neglected, differentiation of Eqs. (8.8) produces

$$\dot{u} = \Delta\dot{\theta} \cos\phi - \Omega \Delta\theta \sin\phi = \Delta\dot{\theta} \cos\phi - \Omega v$$
$$\dot{v} = \Delta\dot{\theta} \sin\phi + \Omega \Delta\theta \cos\phi = \Delta\dot{\theta} \sin\phi + \Omega u \tag{8.9}$$

Substitution of Eqs (8.8) and (8.9) into the differential equations (8.7) yields

$$I_1(\dot{u} - \Omega v) = -I_3 \Omega v \qquad I_2(\dot{v} + \Omega u) = I_3 \Omega u \tag{8.10}$$

These are a pair of coupled, homogeneous, linear differential equations with constant coefficients. Their solution must be exponential in time,

$$u = A \exp(\lambda t) \qquad v = B \exp(\lambda t) \tag{8.11}$$

Substitution of these forms into Eq. (8.10) leads to

$$\begin{bmatrix} I_1\lambda & (I_3 - I_1)\Omega \\ (I_2 - I_3)\Omega & I_2\lambda \end{bmatrix} \begin{Bmatrix} A \\ B \end{Bmatrix} = \begin{Bmatrix} 0 \\ 0 \end{Bmatrix} \tag{8.12}$$

In order for there to be a nontrivial solution, the determinant of this set of equations must vanish, which leads to the following characteristic equation:

$$I_1 I_2 \lambda^2 + (I_3 - I_1)(I_3 - I_2)\Omega^2 = 0 \qquad (8.13)$$

Because the moments of inertia are positive values, the roots of this quadratic equation occur either as a pair of conjugate imaginary values or as real values with alternating sign, depending on the sign of the last term. A positive value of λ corresponds to exponential growth of u and v, and therefore of $\Delta\theta$. In this case, a small perturbation from the ideal condition grows with time. Hence, the free rotation is unstable if $(I_3 - I_1)(I_3 - I_2) < 0$. In contrast, the case where the values of λ are imaginary, $(I_3 - I_1)(I_3 - I_2) > 0$, corresponds to an oscillation. The perturbation in the nutation angle $\Delta\theta$ in that case never exceeds a bounded value that is determined by the initial conditions. This means that the spin axis will remain close to the precession axis, which corresponds to stability of the initial rotation.

The stability condition is obtained if I_3 is either the largest moment of inertia, $I_3 > I_2$ and $I_3 > I_1$, or the smallest, $I_3 < I_2$ and $I_3 < I_1$. In other words, a body that is released with an initial angular velocity that is essentially a spin about the principal axis for which the moment of inertia is either the largest or smallest value will continue with that type of rotation. An initial spin about the principal axis for which the moment of inertia is the intermediate value will show a growth in the nutation angle, such that the eventual rotation does not resemble the attempted initial state. Note in this regard that Eqs. (8.10) merely describe the onset of instability. They may not be employed to study the unstable response, because the assumption of a small nutation angle would not be valid in such a case.

If you wish, you may test these stability properties by throwing a homogeneous rectangular object, such as a wooden block or a board eraser. Try to impart to it an initial spin about an axis parallel to one of its edges. It is fairly easy to obtain a motion in which the object spins about an axis parallel to the shortest or longest edge. However, a comparable attempt for rotation about the intermediate edge does not produce the desired steady spin.

8.1.2 Axisymmetric Bodies

The preceding discussion of stability of an initial spinning rotation was predicated on the assumption that the principal moments of inertia about the center of mass are three distinct values. An axisymmetric body, that is, any body whose mass is distributed symmetrically about an axis, has identical moments of inertia for all centroidal axes that perpendicularly intersect the axis of symmetry. Also, any set of axes containing the axis of symmetry are principal axes. Without loss of generality, the z axis may be selected to be the axis of symmetry. Further, let $I_3 = I$ and $I_1 = I_2 = I'$.†

Equations (8.5) remain valid for an axisymmetric body. Substitution for the moments of inertia converts those relations to

---

†Any body having two equal principal moments of inertia behaves as though it were axisymmetric. The present analysis is valid for the free motion of such an object, provided that the z axis is aligned with the axis that has the distinct moment of inertia.

$$\dot{\psi} = \frac{H_G}{I'} \qquad \dot{\theta} = 0 \qquad \dot{\phi} = H_G \left(\frac{1}{I} - \frac{1}{I'} \right) \cos \theta \qquad (8.14)$$

This shows that the free rotation of an axisymmetric body is characterized by a steady spinning rotation about the axis of symmetry accompanied by a steady precession about an axis that is parallel to the angular momentum; the nutation angle between these axes is constant.

Although Eqs. (8.14) fully characterize the motion, further examination will greatly enhance our qualitative understanding of free motion. First, we evaluate the angular velocity by substituting Eqs. (8.14) into Eq. (8.1). This yields

$$\bar{\omega} = -\frac{H_G}{I'} \sin \theta \cos \phi \bar{i} + \frac{H_G}{I'} \sin \theta \sin \phi \bar{j} + \frac{H_G}{I} \cos \theta \bar{k} \qquad (8.15)$$

Recall that one of the reference frames used in Chapter 4 to define the Eulerian angles was $x'y'z'$, which undergoes only the precessional and nutational rotations. As shown in Figure 8.1, the y' axis is the line of nodes, whereas the z' axis coincides with the z axis. The unit vector along the x' axis is

$$\bar{i}' = \cos \phi \bar{i} - \sin \phi \bar{j} \qquad (8.16)$$

which means that the angular velocity in Eq. (8.15) may be written as

$$\bar{\omega} = -\frac{H_G}{I'} \sin \theta \bar{i}' + \frac{H_G}{I} \cos \theta \bar{k}' \qquad (8.17)$$

One check on the correctness of this expression comes from using its components to reconstitute the angular momentum,

$$\bar{H}_G = I' \omega_{x'} \bar{i}' + I \omega_z \bar{k}' = -H_G \sin \theta \bar{i}' + H_G \cos \theta \bar{k}' = H_G \bar{K} \qquad (8.18)$$

This confirms that the angular momentum is aligned along the precession axis.

Because H_G and θ are constant, we find from Eq. (8.17) that the angular velocity is formed from two orthogonal components having constant magnitude. One component is parallel to the axis of symmetry and the other lies in the plane formed by the axis of symmetry and the fixed direction characterized by the angular momentum. This representation of $\bar{\omega}$ is shown in Figure 8.2, as is the representation obtained by vectorially adding the precession and spin rates.

Suppose that the initial motion of the body is specified, which is equivalent to specifying the initial value of $\bar{\omega}$ and the initial orientation of the body. Such conditions mean that we know the initial angular speed $\omega = |\bar{\omega}|$, as well as the angle β between $\bar{\omega}$ and the axis of symmetry at the instant of release. (In order to avoid ambiguity, and without loss of generality, we consider the angle to be acute, $\beta < 0$.) We could use the relations we have already established to express the other parameters in terms of these initial conditions. Let us instead develop the appropriate relations by referring to Figure 8.2, which displays three methods for constructing the components of $\bar{\omega}$. Specifically,

$$\omega_{z'} = \omega \cos \beta = \frac{H_G}{I} \cos \theta = \dot{\phi} + \dot{\psi} \cos \theta$$

$$(8.19)$$

$$-\omega_{x'} = \omega \sin \beta = \frac{H_G}{I'} \sin \theta = \dot{\psi} \sin \theta$$

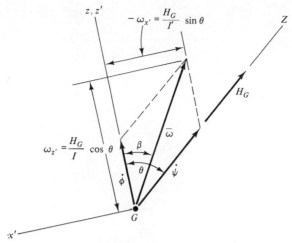

Figure 8.2 Construction of the angular velocity in free motion of an axisymmetric body.

Eliminating all kinematical parameters except ω and β from these relations yields

$$\tan \theta = \frac{I'}{I} \tan \beta$$

$$H_G = (I'^2 \sin^2 \beta + I^2 \cos^2 \beta)^{1/2} \, \omega$$

$$\dot{\psi} = \left[\sin^2 \beta + \left(\frac{I}{I'} \right)^2 \cos^2 \beta \right]^{1/2} \omega \qquad (8.20)$$

$$\dot{\phi} = \left(1 - \frac{I}{I'} \right) \omega \cos \beta$$

The picture provided by Figure 8.2 may be considered to be general, except that we must remember that the entire system precesses at a constant rate about the Z axis. As the motion evolves, the angular velocity sweeps out a cone in space whose axis is parallel to $\overline{H}_G$ and whose vertex half angle is $\theta - \beta$. Similarly, the axis of symmetry precesses such that it is coplanar with the $\overline{\omega}$ and $\overline{H}_G$ vectors, at a constant angle θ relative to the precession axis.

Such a motion may be represented by a conceptual model formed from (right circular) cones. The *body cone,* which is fixed to the body, rolls without slipping over the stationary *space cone*. The angular velocity of the body cone must be parallel to the line of contact between the cones, because the instantaneous axis of rotation is the locus of points in the body having zero velocity. Hence, $\overline{H}_G$ defines the axis of the space cone whose semivertex angle is $\beta - \theta$, whereas the axis of symmetry is the axis of the body cone whose semivertex angle is β. The sides of the space and body cone contact at $\overline{\omega}$.

Two types of rotation are recognizable from this model. *Regular precession,* which is shown in Figure 8.3, corresponds to rotations in which $\beta < \theta$. The body cone in this

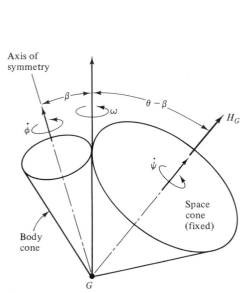

Figure 8.3 Body and space cones for regular precession.

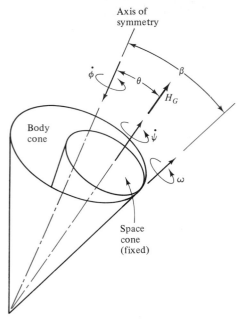

Figure 8.4 Body and space cones for retrograde precession.

case rolls over the exterior of the space cone. Note from the first of Eqs. (8.20) that regular precession is obtained if $I' > I$, which is characteristic of a slender body such as a football. We see from the last of Eqs. (8.20) that whenever $I' > I$, the spin rate is positive; the third equation shows that the precession rate is always positive. Therefore, the angle between the precession and spin rate vectors is acute in a regular precession.

Figure 8.4 depicts a *retrograde precession*, which is the case where $\beta > \theta$. The body cone in this case rolls over the interior of the space cone. Such a rotation arises when $I' < I$, which corresponds to a squat body such as a disk. In this case the spin rate is negative, so the spin-rate vector is oriented along the negative z axis. Because the precession rate is always positive, the angle between the two rotation-rate vectors is now $180° - \theta$, so we perceive the precession to be generally opposite to the sense of the spin. This counterrotation is the source of the term ''retrograde.''

EXAMPLE 8.1 _____

A football has an instantaneous velocity of 80 ft/s parallel to its longitudinal axis, z, and it is spinning about that axis at 5 rev/s. At that instant, the ball is deflected by a transverse force $\overline{F}$ at the forward tip. As a result of the action of $\overline{F}$, whose duration is very short, the ensuing motion relative to the center of mass is such that the longitudinal axis always lies on the surface of a cone whose apex angle is 60°. The radii of gyration about centroidal axes are 1.8 and 2.4 in. along and transversely to the longitudinal axis, respectively. Determine

1. The angular velocity and the velocity of the center of mass immediately after the application of $\bar{F}$.
2. The orientation of the precession axis for the subsequent rotation relative to the orientation of the longitudinal axis prior to the application of $\bar{F}$.
3. The precession and spin rates for the rotational motion.

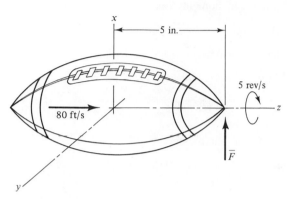

Example 8.1

Solution. The force $\bar{F}$ fits the impulsive model because it induces a substantial change in the motion over a short time interval. We neglect position changes during this interval, which means that the orientation of the body-fixed xyz reference frame changes negligibly during the interval of the impulse. The moments of inertia are

$$I = I_{zz} = m(1.8^2) = 3.24m \text{ lb-s}^2\text{-in.}$$
$$I' = I_{xx} = I_{yy} = m(2.4)^2 = 5.76m \text{ lb-s}^2\text{-in.}$$

where m is the mass in units of lb-s²-in.

The initial linear and angular momenta are

$$\bar{P}_1 = m(\bar{v}_G)_1 = m(80)(12)\bar{k} \text{ lb-s}$$
$$(\bar{H}_G)_1 = I\omega_z\bar{k} = I(-5)(2\pi)\bar{k} = -101.79m\bar{k} \text{ lb-s-in.}$$

Because the change in the position of any point on the football is negligible during the impulse interval, the point of application of $\bar{F}$ is essentially constant at $\bar{r}_{P/G} = 5\bar{k}$ in. The corresponding impulse-momentum principles are

$$m(\bar{v}_G)_2 = 960m\bar{k} + F\,\Delta t\bar{i}$$
$$(\bar{H}_G)_2 = (\bar{H}_G)_1 + \bar{r}_{P/G} \times F\,\Delta t\bar{i} = -101.79m\bar{k} + 5F\,\Delta t\bar{j}$$

The given information that the z axis sweeps out a 60° cone in the subsequent rotation means that the nutation angle is $\theta = 30°$, with the precession axis coincident with the axis of that cone. We find the corresponding angle β between the angular velocity and the axis of symmetry from Eqs. (8.20).

$$\beta = \tan^{-1}\left(\frac{I}{I'}\tan\theta\right) = 17.992°$$

Since $\beta < \theta$, this is a case of regular precession.

The preceding expression for $(\overline{H}_G)_2$ indicates that $\omega_x = 0$ at the instant when $\overline{F}$ terminates, while $\omega_y > 0$ and $\omega_z < 0$ at that instant. Hence, the angular velocity at that instant must be

$$\overline{\omega}_2 = -\omega\cos\beta\overline{k} + \omega\sin\beta\overline{j} = \omega(-0.9511\overline{k} + 0.3089\overline{j})$$

The corresponding angular momentum is

$$(\overline{H}_G)_2 = I\omega_z\overline{k} + I'\omega_y\overline{j} = m\omega(-3.082\overline{k} + 1.7791\overline{j})$$

Matching this to the first expression for $\overline{H}_G$ yields

$$3.082m\omega = 101.79m \qquad 1.7791m\omega = 5F\,\Delta t$$

$$\omega = 33.03 \text{ rad/s} \qquad F\frac{\Delta t}{m} = 11.752 \text{ in./s}$$

The corresponding motion parameters are

$$\overline{\omega}_2 = -31.41\overline{k} + 10.202\overline{j} \text{ rad/s}$$
$$(\overline{v}_G)_2 = 960\overline{k} + 11.762\overline{j} \text{ in./s} = 80\overline{k} + 0.979\overline{j} \text{ ft/s}$$

The precession axis is parallel to $(\overline{H}_G)_2$, which is the angular momentum for the subsequent free-motion axis. Thus

$$\overline{K} = \frac{(\overline{H}_G)_2}{|(\overline{H}_G)_2|} = \frac{-3.082\overline{k} + 1.7791\overline{j}}{(3.082^2 + 1.7791^2)^{1/2}} = -0.8660\overline{k} + 0.500\overline{j}$$

Note that the angle between $\overline{K}$ and $\overline{k}$, that is, between the symmetry and precession axes, is $\cos^{-1}(-0.8660) = 150°$, in agreement with the stated conditions.

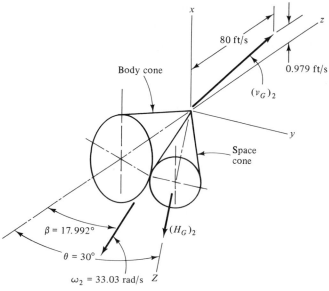

Motion at the end of the impulse.

From these results we may draw a sketch of the position of the body cone relative to the space cone at the initiation of the free motion. We also show $(\bar{v}_G)_2$ in that sketch. The corresponding precession and spin rates are given by Eqs. (8.20).

$$\dot{\psi} = 20.40 \text{ rad/s} \qquad \dot{\phi} = 13.74 \text{ rad/s}$$

8.1.3 Poinsot's Construction for Arbitrary Bodies

When the principal moments of inertia are unequal, the nutation angle will generally not be constant. As was mentioned earlier, one approach in this case is to seek analytical or numerical solutions of the first-order equations of motion, Eqs. (8.5). Here, we develop a pictorial representation of the motion that considerably enhances our qualitative understanding of free rotation. The framework for this development is the ellipsoid of inertia, which was described in Section 5.2.2.

We begin by noting that the general expression for the rotational kinetic energy, which employs the angular velocity components and inertia properties $[I]$ relative to xyz, is more simply represented in terms of the moment of inertia I about the instantaneous axis of rotation. Specifically,

$$T_{\text{rot}} = \tfrac{1}{2} \bar{\omega} \cdot \bar{H}_G = \tfrac{1}{2} \{\omega\}^T [I]\{\omega\} = \tfrac{1}{2} I \omega^2 \tag{8.21}$$

The ellipsoid of inertia is a fictitious body that executes the free motion in unison with the actual body. The major, minor, and intermediate axes of the ellipsoid coincide with the principal axes of the body, which are xyz in the current situation. The distance from the center of mass G to any point P on the inertia ellipsoid is defined to be the rate at which the body should rotate about axis GP in order that the rotational kinetic energy be one-half. Setting $T_{\text{rot}} = \tfrac{1}{2}$ in Eq. (8.21) shows that the required rotation rate is $1/\sqrt{I}$, where I is the moment of inertia about axis GP.

If (x, y, z) are the coordinates of point P relative to the body-fixed reference frame whose origin is the center of mass, then the position $\bar{\rho}$ may be written in vector and matrix notation as

$$\bar{\rho} = x\bar{i} + y\bar{j} + z\bar{k} \qquad \{\rho\} = \begin{Bmatrix} x \\ y \\ z \end{Bmatrix} \tag{8.22}$$

By definition, we have $\rho = 1/\sqrt{I}$. This does not represent an explicit relation among the (x, y, z) values because I depends on the location of point P. We obtain such a relation in a different manner. In view of the definition of $\bar{\rho}$, it represents the angular velocity required for $T_{\text{rot}} = \tfrac{1}{2}$ when the axis of rotation is parallel to line GP. Substitution of $\bar{\omega} = \bar{\rho}$ and $\omega = 1/\sqrt{I}$ into Eq. (8.21) yields

$$\{\rho\}^T [I]\{\rho\} = 1 \tag{8.23a}$$

This product may be expanded for arbitrary $[I]$. However, we defined x, y, and z to be principal axes, with I_1, I_2, and I_3 the respective principal values. Thus, the expanded form of Eq. (8.23a) is

$$I_1x^2 + I_2y^2 + I_3z^2 = 1 \tag{8.23b}$$

In order to describe the motion of the inertia ellipsoid in a free motion, we draw a line through the center of mass parallel to the angular velocity $\bar{\omega}$ at an arbitrary instant; this construction appears in Figure 8.5. The point P we shall follow is the intersection of this line with the surface of the ellipsoid. If we know $\bar{\omega}$, the corresponding $\bar{\rho}$ is a vector parallel to $\bar{\omega}$ with magnitude $1/\sqrt{I}$, that is,

$$\bar{\rho} = \frac{1}{\sqrt{I}} \frac{\bar{\omega}}{\omega} \tag{8.24a}$$

Thus, the coordinates of point P are

$$x = \frac{\omega_x}{\sqrt{I}\,\omega} \qquad y = \frac{\omega_y}{\sqrt{I}\,\omega} \qquad z = \frac{\omega_z}{\sqrt{I}\,\omega} \tag{8.24b}$$

Recall that the angular momentum of a body is constant in a free motion. Since we may construct this vector from the rotational motion at the instant the body was released, $\overline{H}_G$ defines a convenient reference direction.

The first significant aspect of the motion of the inertia ellipsoid comes from an evaluation of the component of $\bar{\rho}$ parallel to the angular momentum. As shown in Figure 8.5, this component, which we denote ρ_H, may be obtained from a dot product of $\bar{\rho}$ with a unit vector parallel to $\overline{H}_G$. In view of Eq. (8.24), we obtain

$$\rho_H = \bar{\rho} \cdot \frac{\overline{H}_G}{H_G} = \frac{1}{\sqrt{I}} \frac{\bar{\omega}}{\omega} \cdot \frac{\overline{H}_G}{H_G} = \frac{(2\,T_{\mathrm{rot}})^{1/2}}{H_G} \tag{8.25}$$

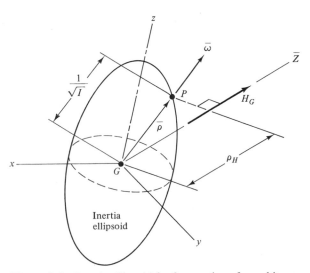

Figure 8.5 Inertia ellipsoid for free motion of an arbitrary body.

Now note that because the forces acting on a body in free motion exert no moment about the center of mass G, no work is done in the rotation. Consequently, the rotational kinetic energy is constant. Because both T_{rot} and H_G are constant, the distance ρ_H remains constant as the body and its ellipsoid of inertia rotate.

A general property of a plane is that all points on it are at the same normal distance to a point above the plane. It follows that point P always lies on a plane that is at the constant distance ρ_H from the center of mass, with $\overline{H}_G$ being normal to the plane. If we ignore the movement of the center of mass, this plane appears to be stationary; it is the *invariable plane*.

Knowledge of the invariable plane does not fully prescribe the motion of the ellipsoid of inertia. We have not established how point P moves along the plane, nor do we know how the ellipsoid is oriented relative to the plane. In order to address these questions, we derive another property of the ellipsoid of inertia.

Let us define a family of concurrent ellipsoids having the same shape as the inertia ellipsoid. Let C be a constant that scales the magnitude of $\overline{\rho}$. In other words, if point P' is a point on a different ellipsoid that is collinear with center of mass G and point P on the inertia ellipsoid, then the distance from point G to point P' is defined to be $C/\sqrt{I}$. It follows from Eq. (8.23b) that the coordinates of point P' satisfy

$$F(x, y, z, T_{\text{rot}}) = I_1 x^2 + I_2 y^2 + I_3 z^2 = C^2 \tag{8.26a}$$

where $C = 1$ corresponds to the ellipsoid of inertia.

The gradient operator applied to Eq. (8.25) indicates the direction in which the value of C changes most rapidly in going from one surface to another. Therefore, the gradient of F, which is

$$\nabla F = 2I_1 x \overline{i} + 2I_2 y \overline{j} + 2i_z z \overline{k} \tag{8.26b}$$

defines the normal to the ellipsoid associated with that value of C. This expression describes the normal direction to any of the ellipsoids. The coordinates of point P, at which the rotation axis intersects the inertia ellipsoid, are given by Eqs. (8.24b), which when substituted into Eq. (8.26) yield

$$\nabla F = \frac{2}{\sqrt{I}\,\omega}(I_1\omega_x\overline{i} + I_2\omega_y\overline{j} + I_3\omega_z\overline{k}) = \frac{2}{\sqrt{I}\,\omega}\overline{H}_G \tag{8.27}$$

We see now that the normal to the inertia ellipsoid at point P is parallel to the angular momentum. However, the normal to the invariable plane is also parallel to $\overline{H}_G$. As shown in Figure 8.6, these two statements means that the ellipsoid of inertia is always tangent at point P to the invariable plane. Furthermore, the velocity of point P is zero, because it is on the instantaneous axis of rotation. These observations lead us to the Poinsot construction, which states that

The ellipsoid of inertia of a body in free motion rotates about the center of mass such that it rolls without slipping over the invariable plane. The point where the ellipsoid (tangentially) contacts the invariable plane is on the instantaneous axis of rotation passing through the center of mass, and the constant distance from the center of mass to the invariable plane is set by the rotational motion at the instant the body is released.

At each instant, a different point on the inertia ellipsoid contacts the invariable

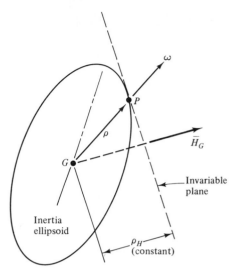

Figure 8.6 Motion of the inertia ellipsoid for free motion of an arbitrary body.

plane. The locus of points of contact is a curve on the inertia ellipsoid called the *polhode*. This locus also forms a curve on the invariable plane called the *herpolhode*. The herpolhode is generally an open curve, which means that the rotation does not repeat, but the polhode is a closed curve.

The closure of the polhode, as well as the overall nature of these curves, may be established by noting that the inertia ellipsoid represents the constancy of the kinetic energy in the rotational motion. However, the angular momentum is also constant, which means that

$$H_G^2 = \overline{H}_G \cdot \overline{H}_G = (I_1\omega_x)^2 + (I_2\omega_y)^2 + (I_3\omega_z)^2 = \text{constant} \tag{8.28}$$

Let us use Eqs. (8.24b) to express this relation in terms of the coordinates of point P at which the inertia ellipsoid contacts the invariable plane. Thus,

$$H_G^2 = I\omega^2[(I_1x)^2 + (I_2y)^2 + (I_3z)^2] \tag{8.29}$$

However, at each instant the body is rotating at rate ω about the instantaneous axis through origin G and point P, and I is the moment of inertia about that axis. Thus, $I\omega^2 = 2T_{\text{rot}}$, so the foregoing becomes

$$I_1^2x^2 + I_2^2y^2 + I_3^2z^2 = \frac{H_G^2}{2T_{\text{rot}}} = D \tag{8.30}$$

This relation characterizes another ellipsoid that is fixed to the body. The intersection of the ellipsoid given by the foregoing with the ellipsoid of inertia, Eq. (8.23b), is the polhode. The closure of the polhode is a direct consequence of the fact that both ellipsoids, and therefore their intersection, rotate with the body.

The value of the constant D is determined by the initial motion. In view of Eqs. (8.24b) and (8.30), initial rotations about each of the principal axes correspond to $D = I_1, I_2,$ and I_3, respectively. Without loss of generality, we now specify the labeling of the xyz axes to be such that I_1 is the smallest value and I_3 is the largest. Then, $I_1 \leqq D \leqq I_3$. The polhode for a specified value of D may be constructed by picking a value of one coordinate and then solving Eqs. (8.23b) and (8.30) simultaneously for the other two. Alternatively, the projection of a polhode curve onto any of the principal coordinate planes may be derived by eliminating the coordinate normal to that plane from the two relations. These projection equations are

xy plane: $I_1(I_3 - I_1)x^2 + I_2(I_3 - I_2)y^2 = I_3 - D$

yz plane: $I_2(I_2 - I_1)y^2 + I_3(I_3 - I_1)z^2 = D - I_1$ (8.31)

xz plane: $I_1(I_2 - I_1)x^2 - I_3(I_3 - I_2)z^2 = I_2 - D$

We wrote each of these equations such that the coefficients are positive for the assigned sequence $I_1 < I_2 < I_3$. For this ordering, the projections onto the xy and yz planes are ellipses, and the projections onto the xz plane are hyperbolas. These projections and the outline of the inertia ellipsoid are illustrated in Figure 8.7 for the positive quadrants. The curve corresponding to $D = I_2$ is the separatrix between the hyperbolas in the xz plane, but it appears as an ellipse in the other coordinate planes. The corresponding polhode curves on the ellipsoid of inertia are shown in Figure 8.8.

We concluded in Section 2.1 that an attempt to impart a rotation about the principal axis of smallest or largest moment of inertia would produce a stable rotation. This is further demonstrated here. Recall that the instantaneous angular velocity $\overline{\omega}$ is parallel to a line from the origin (that is, the center of mass) to the point where the polhode curve contacts the invariable plane. If the initial rotation is approximately about the z axis, then D is slightly smaller than I_3. In that case the projection of the polhode curve onto the xy plane is a small ellipse, corresponding to an angular velocity that always is nearly parallel to the z axis. Similarly, an initial rotation approximately about the x axis, which gives a value of D slightly larger than I_1, leads to a polhode curve projection on the yz

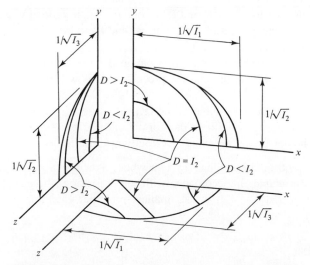

Figure 8.7 Projection of the polhode curves on the principal-axis planes.

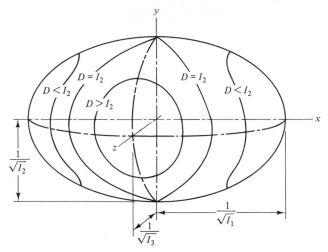

Figure 8.8   Polhode curves.

plane that is a small ellipse. This corresponds to an angular velocity that is always nearly parallel to the x axis. In either case, the axis of rotation remains close to the respective principal axis. In contrast, if the initial motion is approximately about the y axis, then $D \approx I_2$. Then the polhode curves are close to the separatrices. Depending on whether D is greater than or less than I_2, the closed polhode curve is centered about either the z axis or the x axis, respectively. In either case, the angle between $\bar{\omega}$ and any of the coordinate axes varies greatly in the motion. This explains why the rotation of an arbitrary body is often difficult to observe.

EXAMPLE 8.2

The rectangular plate is released with an initial angular velocity $\bar{\omega}$ such that edge AB is horizontal, and the normal to the plate and $\bar{\omega}$ lie in a common vertical plane.

(a) Determine the angle γ between the normal and the vertical direction for which the precession axis is vertical.

(b) Determine the maximum value of the angle β between the normal and $\bar{\omega}$ for which the angle between the normal and the precession axis will not exceed $90°$ in the rotation after release.

(c) For the case where β is one-half the critical value in part (b), determine the maximum and minimum angles between the normal and the precession axis during the rotation. Evaluate the corresponding angular velocity at these limits.

Solution.   A centroidal coordinate system whose axes are aligned with the edges of the plate is principal. In accordance with the derivation, the axes in the sketch are labeled such that $I_{xx} = I_1$ is the smallest principal value and $I_{zz} = I_3$ is the largest. These values are

$$I_1 = \tfrac{1}{12} m(6^2) = 3m \qquad I_2 = \tfrac{1}{12} m(12^2) = 12m$$
$$I_3 = \tfrac{1}{12} m(6^2 + 12^2) = 15m \text{ lb-s}^2\text{-in.}$$

where m is the mass in units of lb-s^2/in.

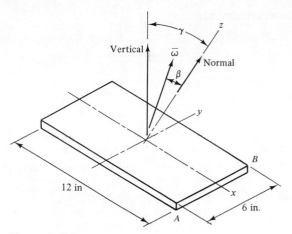

Example 8.2

The initial angular velocity is given as

$$\bar{\omega} = \omega(-\sin\beta\bar{i} + \cos\beta\bar{k})$$

so the constant angular momentum is

$$\bar{H}_G = I_1\omega_x\bar{i} + I_2\omega_y\bar{j} + I_3\omega_z\bar{k} = m\omega(-3\sin\beta\bar{i} + 15\cos\beta\bar{k})$$

In contrast, the condition that the precession axis be vertical requires that the angular momentum be parallel to the Z axis. Thus

$$\bar{H}_G = H_G(-\sin\gamma\bar{i} + \cos\gamma\bar{k})$$

Matching the two descriptions of $\bar{H}_G$ leads to

$$m\omega(3\sin\beta) = H_G\sin\gamma \qquad m\omega(15\cos\beta) = H_G\cos\gamma$$

which we solve to find

$$H_G = m\omega(9\sin^2\beta + 225\cos^2\beta)^{1/2} \qquad \gamma = \tan^{-1}(\tfrac{1}{5}\tan\beta)$$

In order to address the second question, we examine Figure 8.8. The angle β corresponds to the angle between the z axis and a line from the center to a point on the polhode curve. If the polhode curve lies inside the separatrix defined by $D = I_2$, this angle will always be acute. Thus, the maximum allowable value of β is that which gives $D = I_2$.

The earlier expressions for $\bar{\omega}$ and $\bar{H}_G$ in terms of β give

$$2T_{\text{rot}} = \bar{\omega} \cdot \bar{H}_G = m\omega^2(3\sin^2\beta + 15\cos^2\beta)$$

Hence, the critical condition is

$$D = \frac{H_G^2}{2T_{\text{rot}}} = m\frac{9\sin^2\beta + 225\cos^2\beta}{3\sin^2\beta + 15\cos^2\beta} = I_2 = 12m$$

which becomes

$$225 - 216 \sin^2 \beta = 12(15 - 12 \sin^2 \beta) \Rightarrow \sin^2 \beta = \tfrac{45}{72}$$
$$\text{Critical } \beta = 52.24°$$

For part (c), we set $\beta = 26.12°$. The corresponding value of D is readily obtained from the general formula above.

$$D = 14.449m$$

Since $D > I_2$, the polhode curve surrounds the z axis. Now examine the polhode curve for this case in Figure 8.8. The maximum and minimum angles between the z axis and the normal to the tangent plane occur when the polhode curve intersects the yz plane and xz plane, respectively. The polhode curves, in general, correspond to values of (x, y, z) that simultaneously satisfy Eqs. (8.23b) and (8.30). For the present values of the parameters, these equations are

$$3x^2 + 12y^2 + 15z^2 = \frac{1}{m}$$

$$9x^2 + 144y^2 + 225z^2 = \frac{14.449}{m}$$

At the intersection with the yz plane, $x = 0$. The corresponding values of y and z satisfying the above polhode equations are

$$y = \frac{0.12372}{\sqrt{m}} \qquad z = \frac{0.2333}{\sqrt{m}}$$

According to Eqs. (8.24), the corresponding angular velocity is

$$\bar{\omega}_i = \left(\frac{I_i}{m}\right)^{1/2} \omega_i(0.12372\bar{j} + 0.2333\bar{k})$$

where the subscript i identifies the quantities as instantaneous values. Taking $|\bar{\omega}|$ in this relation yields

$$I_i = \frac{m}{0.12372^2 + 0.2333^2} = 14.341m$$

which, when substituted back into the expression for $\bar{\omega}_i$, leads to

$$\bar{\omega}_i = \omega_i(0.4685\bar{j} + 0.8835\bar{k})$$

The angular momentum corresponding to this expression for $\bar{\omega}_i$ is

$$\bar{H}_G = I_2\omega_y\bar{j} + I_3\omega_z\bar{k} = m\omega_i[12(0.4685)\bar{j} + 15(0.8835)\bar{k}]$$
$$= m\omega_i(5.622\bar{j} + 13.252\bar{k})$$

We may evaluate ω_i by equating $|\bar{H}_G|$ to the (constant) initial value corresponding to the present value of β.

$$H_G^2 = m^2\omega^2(9 \sin^2 \beta + 225 \cos^2 \beta) = 183.13m^2\omega^2$$

Thus

$$\omega_i^2(5.622^2 + 13.252^2) = 183.13\omega^2 \Rightarrow \omega_i = 0.9401\omega$$

Substituting this value into the expression for $\bar{\omega}_i$ yields

$$\bar{\omega}_i = \omega(0.4404\bar{j} + 0.8306\bar{k})$$

The corresponding angle between the z axis and $\bar{H}_G$ may be found from a dot product,

$$\cos\theta_i = \frac{\bar{H}_G \cdot \bar{k}}{|H_G|} = \frac{13.252}{(5.622^2 + 13.252^2)^{1/2}}$$

$$\theta_i = 22.99°$$

and the angle between the angular velocity and the z axis is

$$\beta_i = \cos^{-1}\left(\frac{\bar{\omega}_i \cdot \bar{k}}{\omega_i}\right) = 27.93°$$

The orientation of the principal axes and the angular velocity at this instant are depicted in the sketch on the next page.

We follow the same analysis in order to evaluate the minimum angle condition, which we denote by subscript ii. The only difference is that now we set $y = 0$ for the intersection of the polhode curve with the xz plane. The solution of the polhode equations in this case is

$$x = \frac{0.12372}{\sqrt{m}} \qquad z = \frac{0.2522}{\sqrt{m}}$$

Hence,

$$\bar{\omega}_{ii} = \left(\frac{I_{ii}}{m}\right)^{1/2} \omega_{ii}(0.12372\bar{i} + 0.2522\bar{k})$$

from which we obtain

$$I_{ii} = \frac{m}{0.12372^2 + 0.2522^2} = 12.672m$$

$$\bar{\omega}_{ii} = \omega_{ii}(0.4404\bar{i} + 0.8978\bar{k})$$

$$\bar{H}_G = I_1\omega_x\bar{i} + I_3\omega_z\bar{k} = m\omega_{ii}(1.3212\bar{i} + 13.467\bar{k})$$

Matching $|\bar{H}_G|$ to the initial value yields

$$\omega_{ii}^2(1.3212^2 + 13.467^2) = 183.13\omega^2 \Rightarrow \omega_{ii} = 1.001\omega$$

$$\bar{\omega}_{ii} = \omega(0.4404\bar{i} + 0.8979\bar{k})$$

Then

$$\theta_{ii} = \cos^{-1}\left(\frac{\bar{H}_G \cdot \bar{k}}{H_G}\right) = 5.60° \qquad \beta_{ii} = \cos^{-1}\left(\frac{\bar{\omega}_{ii} \cdot \bar{k}}{\omega_{ii}}\right) = 26.13°$$

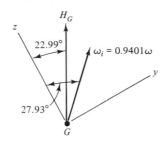

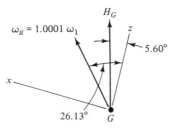

Extreme conditions of the free motion.

These conditions are also shown in the sketch. At an arbitrary instant, the angles between the z axis and the precession axis, and between the angular velocity and the precession axis, will be intermediate to the illustrated conditions.

___

8.2 SPINNING TOP

The toy known as a spinning top consists of an axially symmetric body that executes a pure rotation about an apex situated on the axis of symmetry. (We shall not worry here about the drift that occurs when the apex is not anchored, primarily because such effects are complicated by minor irregularities in the surface over which the apex would move.) The study of a spinning top leads to many insights regarding the interplay between rotation, angular momentum, and the moment exerted by forces. The results for its motion may be extended to other bodies that rotate about a reference point due to the moment of the gravitational force; such systems include certain types of gyroscopes.

In Figure 8.9, point O is held stationary by a reaction force having components F_j in the horizontal and vertical directions. The gravity force acts through the center of mass G, so it exerts a moment of magnitude $mgL \sin \theta$. This moment is about the horizontal axis through point O that is perpendicular to the axis of symmetry. Because such an axis is the line of nodes (nutation axis) for a set of Eulerian angles, it is natural to formulate the equations of motion in terms of those parameters. Note that the reactions exert no moments about the precession, spin, and nutation axes, so the generalized nonconservative force associated with each angle is identically zero.

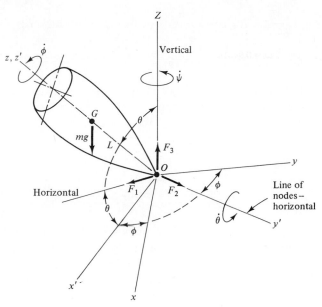

Figure 8.9 Free body diagram for a spinning top.

Let I be the moment of inertia about the axis of symmetry and let I' be the moment of inertia about any axis perpendicular to the axis of symmetry and intersecting point O. Lagrange's equations are useful for formulating the equations of motion, although the same conclusions could be obtained from the Newtonian formulation. In either case the primary difference between a top and an axisymmetric body in free motion is the moment exerted by the gravity force about the reference point for the rotation, which must be balanced by an angular momentum that varies with time.

When resolved into components relative to the $x'y'z'$ axes for the Eulerian angles, the angular velocity of the body is

$$\overline{\omega} = -\dot{\psi} \sin\theta \,\overline{i}' + \dot{\theta}\overline{j}' + (\dot{\psi}\cos\theta + \dot{\phi})\overline{k}' \qquad (8.32)$$

The moment of inertia is the same for any axis through the apex O and perpendicular to the axis of symmetry, so the kinetic energy corresponding to this expression for $\overline{\omega}$ is

$$T = \tfrac{1}{2} I(\dot{\psi}\cos\theta + \dot{\phi})^2 + \tfrac{1}{2} I'(\dot{\psi}^2 \sin^2\theta + \dot{\theta}^2) \qquad (8.33)$$

The elevation of the apex is a convenient reference for the gravitational potential energy, so

$$V = mg\,L\cos\theta \qquad (8.34)$$

We noted earlier that the generalized forces are all zero in our idealized model. Furthermore, the Lagrangian, $L = T - V$, in this case does not depend explicitly on either the precession or spin angles. As a result the precession and spin angles are ignorable coordinates, corresponding to conservation of the generalized momenta associated with these variables. These momenta are

$$p_\psi = \frac{\partial T}{\partial \dot\psi} = I(\dot\psi \cos\theta + \dot\phi)\cos\theta + I'\dot\psi \sin^2\theta = I'\beta_\psi$$

$$p_\phi = \frac{\partial T}{\partial \dot\phi} = I(\dot\psi \cos\theta + \dot\phi) = I'\beta_\phi$$

(8.35)

where β_ψ and β_ϕ are constants having the units of angular speed.

The values of β_ψ and β_ϕ are specified by the initial conditions, so Eqs. (8.35) yield the following first-order differential equations for the precession and spin angles:

$$\dot\psi = \frac{\beta_\psi - \beta_\phi \cos\theta}{\sin^2\theta}$$

$$\dot\phi = \frac{\beta_\phi(I' \sin^2\theta + I \cos^2\theta) - \beta_\psi I \cos\theta}{I \sin^2\theta}$$

(8.36)

In both differential equations the as yet undetermined nutation is the excitation. A constant value of β_ϕ corresponds to constancy of the total rotation rate about the axis of symmetry, $\omega_z = \dot\psi \cos\theta + \dot\phi$. The above expressions reveal that the precession and spin rates are individually constant only when the nutation angle is constant. Otherwise, each rate must change with changes in θ in order to maintain the constant value of the component of angular velocity about the spin axis.

Constant values of p_ψ and p_ϕ satisfy the Lagrange's equations associated with ψ and ϕ. In the derivation of the third Lagrange's equation, which governs θ, it is necessary to evaluate the derivatives of the energy expressions before Eqs. (8.36) are used to eliminate the ignorable coordinates. We find from Eq. (8.33) that

$$\frac{\partial T}{\partial \theta} = \dot\psi \sin\theta[I'\dot\psi \cos\theta - I(\dot\psi \cos\theta + \dot\phi)]$$

(8.37)

The next step is to substitute Eqs. (8.36) into this expression in order to remove the dependence on the precession and spin rates, and then to use that result to form the Lagrange equation. These operations yield

$$\ddot\theta + \frac{1}{\sin^3\theta}(\beta_\phi - \beta_\psi \cos\theta)(\beta_\psi - \beta_\phi \cos\theta) - \frac{mg\,L}{I'}\sin\theta = 0$$

(8.38)

We shall employ this equation of motion later. A first integral of it could be obtained by separating variables using the chain-rule identity

$$\ddot\theta = \frac{d\dot\theta}{d\theta}\dot\theta = \frac{1}{2}\frac{d}{d\theta}(\dot\theta)^2$$

However, it is much simpler to observe that mechanical energy, $E = T + V$, is conserved. Expressions for the kinetic and potential energy are given by Eqs. (8.33) and (8.34). Eliminating the precession and spin rates with the aid of Eqs. (8.36) yields

$$E = \tfrac{1}{2}I'\dot\theta^2 + \frac{1}{2}I'\frac{(\beta_\psi - \beta_\phi \cos\theta)^2}{\sin^2\theta} + \frac{(I')^2}{2I}\beta_\phi^2 + mg\,L\cos\theta$$

(8.39)

The value of the energy E, just like the generalized momenta, is known from the initial conditions.

When we multiply this equation by $\sin^2 \theta$, we see that the derivative of θ appears in the combination $\dot\theta \sin \theta$, whereas the terms that do not contain a derivative depend on $\cos \theta$, because $\sin^2 \theta = 1 - \cos^2 \theta$. This suggests that it would be useful to define a new variable such that

$$u = \cos \theta \qquad \dot u = -\dot\theta \sin \theta \tag{8.40}$$

Also, it is convenient to define the following combination of parameters:

$$\varepsilon = \frac{2E}{I'} - \frac{I'}{I}\beta_\phi^2 \qquad \gamma = \frac{2mg\,L}{I'} \tag{8.41}$$

Substitution of Eqs. (8.40) and (8.41) converts the energy expression in Eq. (8.39) to

$$\dot u^2 = (\varepsilon - \gamma u)(1 - u^2) - (\beta_\psi - \beta_\phi u^2) \tag{8.42}$$

It is possible to separate variables in this differential equation, which would lead to an expression for the time t required to attain a certain value of θ. Such a relation would have the form of an elliptic integral. Numerical methods provide another approach by which the differential equation, Eq. (8.42), may be solved for a relation between θ and t. However, we may determine much qualitative information about the motion by merely studying the roots of the cubic polynomial in the right side of Eq. (8.42). These roots describe the conditions for which $\dot\theta$ is zero, corresponding to either extrema or constant values of the nutation angle.

The polynomial is

$$f(u) = (\varepsilon - \gamma u)(1 - u^2) - (\beta_\psi - \beta_\phi u)^2 \tag{8.43}$$

In view of the definition of u, Eq. (8.40), the physically meaningful values of u must lie in the range $-1 \le u \le 1$, subject to the requirement that $f(u) \ge 0$ in order that $\dot\theta$ be real. (For an actual toy top on the ground, $\theta > 0$ is the only realistic case, but $\theta < 0$ is possible by placing the apex O on an elevated pivot.)

Let us study the nature of the roots of $f(u)$. When u is very large, we find that $f(u) \approx \gamma u^3 > 0$ because γ is a positive parameter. Furthermore, $f(1) < 0$ because the first term vanishes. It follows that one root of $f(u)$ must be in the range $u > 1$, so it is of no interest. A comparable evaluation for large negative values of θ shows that $f(u)$ is asymptotically negative and $f(-1) < 0$. Since there must be some range of u over which there is a real value of the nutation rate, we may conclude that $f(u)$ must have two roots in the range $-1 \le u \le 1$. One possible situation for the significant roots u_1 and u_2 is shown in Figure 8.10, although it might be that both roots are positive or negative.

The variable u may be interpreted geometrically as being the elevation above the apex of a point P on the axis of symmetry (i.e., the z axis) at a unit distance from the apex. In this interpretation the precession angle ψ and the nutation angle θ are spherical coordinates for point P, whose path lies on a sphere of unit radius.

Note that the highest elevation of point P corresponds to the largest value $u = u_2$, for which the nutation angle is the smallest. Similarly, the lowest elevation attained in the motion is $u = u_1$, corresponding to the largest nutation angle. Hence, the nutational

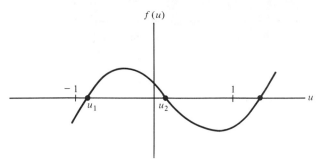

Figure 8.10   Roots of $f(u) = 0$ for a spinning top.

motion is such that the symmetry axis oscillates between high and low positions. In the exceptional situation where the roots are repeated, $u = u_1 = u_2$ throughout the motion, corresponding to a constant nutation angle. This is an important possibility, because we saw in Eqs. (8.38) that the precession and spin rate are constant when θ is constant. Thus, the case of repeated roots corresponds to *steady precession,* which we treat later.

The values of the parameters β_ψ, β_ϕ, ε, and γ are set by the initial conditions. The relation between the elevation u and the precession rate for specified initial conditions is found from Eqs. (8.36) and (8.40) to be

$$\dot\psi = \frac{\beta_\phi(u_0 - u)}{1 - u^2} \qquad u_0 = \frac{\beta_\psi}{\beta_\phi} \tag{8.44}$$

Since $|u| \leqq 1$, we observe from this relation that the sense of the precession, which is defined by the sign of $\dot\psi$, is determined by the parametric combination $u - u_0$. Indeed, $\dot\psi$ vanishes at $u = u_0$. Whether $\dot\psi$ actually goes to zero in a motion depends on whether the value of u_0 lies in the range $u_1 \leqq u \leqq u_2$, which is the physically meaningful range of values.

There are three ranges in which the value of u_0 may be situated relative to u_1 and u_2. Understanding each requires recognition of the interplay between the alteration in the rotational motions necessary to conserve angular momentum and energy. The second of Eqs. (8.35) shows that the total rate of rotation, $\omega_z = \dot\phi + \dot\psi \cos\theta$, about the axis of symmetry remains constant in order to conserve momentum about that axis. Thus, a decrease in the precession rate or an increase in the nutation angle must be compensated by a rise in the spin rate. The effect of the nutation on the precession rate may be seen from the first of Eqs. (8.35). The precessional momentum originates from two sources: the projection of the spin momentum onto the precession axis, and the angular momentum associated with the precession itself. The equivalent moment of inertia for the latter effect is $I' \sin^2\theta$. Increasing the nutation angle decreases this moment of inertia, while it simultaneously decreases the projection of p_ϕ. Hence, an increase in the nutation angle has competing effects on the precession rate, depending on the value of β_ϕ relative to β_ψ.

In regard to the energies, Eq. (8.39) indicates that the portion of the mechanical energy E attributable to the precession and spin increases when θ increases (assuming that β_ψ and β_ϕ have the same sign). This is accompanied by a decrease in the potential energy with increasing θ. The nutational portion of the mechanical energy must maintain the balance between kinetic and potential energy. At the extremes of the nutational motion, the change in potential energy is exactly compensated by the change in the precession and spin kinetic energy, so the nutational energy vanishes at those locations.

Unidirectional Precession: $u_0 < u_1$ or $u_0 > u_2$ In this situation the precession rate is never zero. Whatever sense it has at the initial instant is what it retains throughout the motion. The nutation angle has its maximum and minimum values at $u = u_1$ and u_2, but the precession continues at those locations. As shown in Figure 8.11, the path of point P is tangent to the circles on the unit sphere that mark the highest and lowest elevations of point P. One way in which we may initiate a unidirectional precession is to release the top at the highest elevation of point P, $u = u_2$, with the appropriate angular velocity. The initial nutation rate $\dot{\theta}$ at this location must be zero, corresponding to $\dot{u} = 0$ and $\theta \neq 0$, and the initial precession rate should be relatively large, sufficient to make $u_0 = \beta_\psi / \beta_\phi$ exceed u_2. Then Eq. (8.44) yields a value of $\dot{\psi}$ that does not change sign as the elevation u decreases.

Looping Precession: $u_1 < u_0 < u_2$ In this situation the precession rate is zero at the elevation u_0 intermediate to the extreme values u_1 and u_2 that mark the limits of the nutation. This null corresponds to a change in the sense of the precession as the elevation rises and falls. In contrast, the nutation rate vanishes at the lowest and highest elevations. In other words, point P moves tangent to circles of maximum and minimum elevation in opposite senses, as shown in Figure 8.12. The vertical tangencies in the loops correspond to positions where $u = u_0$.

A looping precession may be attained by releasing a top at the highest elevation, $u = u_2$, with a comparatively small precession rate. The nutation rate at release must be zero in order for u_2 to be the maximum elevation. As the top falls, the portion of the precession associated with β_ψ is eventually overwhelmed by the countereffect associated with β_ϕ. Thus, the overall precession comes to rest at elevation u_0 and then proceeds

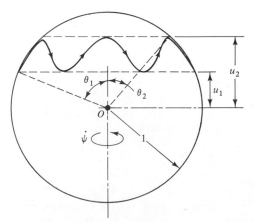

Figure 8.11 Path of the spin axis of a top-unidirectional precession.

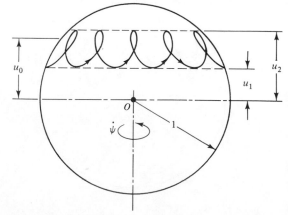

Figure 8.12 Path of the spin axis of a top-looping precession.

opposite to the initial sense down to u_1. The process repeats with the return to elevation u_2. As shown in Figure 8.12, the overall precessional motion matches the sense of the precession at the minimum elevations, $u = u_1$, even though the precession rate oscillates.

Cuspidial Motion: $u_0 = u_2$ This case is a transition between the unidirectional and looping precessions discussed above. In it, the precession comes to rest at the highest elevation, $u = u_2$. Point P approaches the circle of highest elevation perpendicularly, which results in the appearance of cusps in the path of point P at these locations. As shown in Figure 8.13, the path of point P resembles a cycloidal path that is wrapped around the unit sphere.

Cuspidial motion may be attained by releasing the top at the highest elevation, $u = u_2$, with no initial precessional or nutational motion. The precessional motion that arises as the top falls is therefore only due to the spin momentum β_ϕ. As the top falls, it gains kinetic energy and loses potential energy, until the changes in the precession and spin rates result in an increase in the kinetic energy that equals the decrease in the potential energy. Incidentally, we may prove by this reasoning that the cusps cannot arise at the largest nutation angle, where $u = u_1$. Such a motion would lead to kinetic and potential energies that are both maximum values at $u = u_1$, in violation of energy conservation.

Cuspidial motion shares many of the characteristics of the unidirectional and looping precessions. The coincidence of the values of u_0 and u_2 in this case leads to several simplifications that allow us to derive analytical expressions for the Eulerian angles. We have seen that suitable initial conditions leading to cuspidial motion are that $\dot{\psi} = \dot{\theta} = 0$ when $u_0 = u_2 = \cos\theta_2$, with $\dot{\phi}_2$ nonzero. The corresponding momentum parameters are given by Eqs. (8.35) to be

$$\beta_\phi = \frac{I}{I'}\,\dot{\phi}_2 \qquad \beta_\psi = \beta_\phi u_0 \tag{8.45}$$

The energy-level parameters obtained from Eqs. (8.39) and (8.41) in this case are

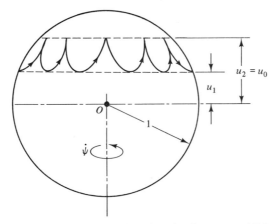

Figure 8.13   Path of the spin axis of a top-cuspidial motion.

$$\gamma = \frac{2mg\,L}{I'} \qquad \varepsilon = \gamma u_0 \tag{8.46}$$

Upon substitution of these parameters, the energy function $f(u)$ defined in Eq. (8.43) factorizes as

$$f(u) = (u_0 - u)[\gamma(1 - u^2) - \beta_\phi^2(u_0 - u)] \tag{8.47}$$

The roots of $f(u)$ for cuspidial motion are readily found to be

$$\begin{aligned}
u_1 &= U - (U^2 - 2u_0 U + 1)^{1/2} \\
u_2 &= u_0 \\
u_3 &= U + (U^2 - 2u_0 U + 1)^{1/2}
\end{aligned} \tag{8.48}$$

where

$$U = \frac{\beta_\phi^2}{2\gamma} = \frac{I^2 \dot{\phi}_2^2}{4I'mg\,L} \tag{8.49}$$

Because $|u_0| < 1$ when the top is released away from the vertical, we have $U^2 - 2u_0 U + 1 > (U - 1)^2$. It follows that $u_3 > 1$, which means that it is the meaningless root, while $-1 < u_1 < 1$. This, of course, agrees with our earlier assessment of the nature of the roots of $f(u)$ in the general case.

The limits of the nutation in cuspidial motion are given explicitly by Eqs. (8.48) in terms of the initial-condition parameters in Eq. (8.49). Further simplifications are possible when we consider the situation that usually arises: a *fast top*, in which the spin rate imparted in the initial motion is large. We quantify this restriction by specifying that $U \gg 1$. The corresponding minimum elevation obtained from the leading terms in a series expansion of the first of Eqs. (8.48) is

$$u_1 = u_0 - \frac{1 - u_0^2}{2U} \tag{8.50}$$

In view of Eq. (8.49), we may conclude from this expression that, when U is large, the difference between the maximum and minimum elevations decreases as the inverse square of the initial spin rate.

The smallness of the cuspidial motion at high spin rates allows us to evaluate the precessional and nutational rotations as explicit functions of time. The technique for such an investigation is *perturbation analysis*. The value of u_1 in Eq. (8.50) suggests that in general the elevation u may be expressed in a series as

$$u = u_0 - \frac{1}{U} v_1(t) - \frac{1}{U^2} v_2(t) + \cdots \tag{8.51}$$

where the $v_j(t)$ are unknown functions of time that are independent of the parameter U. Many terms would be required to make the series converge when the value of U is arbitrary. In contrast, the error that arises from truncating the series becomes smaller and smaller as the value of U increases. We say that Eq. (8.51) is an *asymptotic series* for the variable u in terms of the perturbation parameter $1/U$.

We obtain differential equations for the unknown functions v_j by requiring that the asymptotic series satisfy the equation of motion at each level of approximation, associated with increasing powers of $1/U$. The equation of motion we shall employ is the energy-conservation relation, Eq. (8.42), with the function $f(u)$ for cuspidial motion given by Eq. (8.47). The parameter γ may be removed from the expression by applying the definition of U, Eq. (8.49), which leads to

$$
\left[\frac{1}{U} \dot{v}_1(t) + \frac{1}{U^2} \dot{v}_2(t) + \cdots \right]^2
$$

$$
= \left(\frac{1}{U} v_1 + \frac{1}{U^2} v_2 + \cdots \right) \left\{ \frac{\beta_\phi^2}{2U} \left[1 - \left(u_0 - \frac{1}{U} v_1 + \cdots \right)^2 \right] \right.
$$

$$
\left. - \beta_\phi^2 \left(\frac{1}{U} v_1 + \frac{1}{U^2} v_2 + \cdots \right) \right\} \quad (8.52)
$$

Although we truncated the asymptotic series for u at two terms beyond the initial approximation $u = u_0$, we shall consider only the first approximation here. In other words, we shall only evaluate v_1. Matching the coefficients of $1/U^2$ on each side of Eq. (8.52) yields

$$
\dot{v}_1^2 = \beta_\phi^2 \left[\tfrac{1}{2} (1 - u_0^2) v_1 - v_1^2 \right] \quad (8.53)
$$

Taking the square root of this nonlinear, first-order differential equation for v_1, in order to form an equation whose variables may be separated, leads to an ambiguity in sign that can only be resolved by addressing the initial conditions. A simpler technique is to convert the equation to a second-order differential equation by differentiating it once with respect to time. This operation leads to a common factor of $\dot{v}_1$, which when canceled, results in

$$
\ddot{v}_1 + \beta_\phi^2 v_1 = \tfrac{1}{4} \beta_\phi^2 (1 - u_0^2) \quad (8.54)
$$

The solution of this differential equation must satisfy the initial conditions for cuspidial motion, which we have taken to be that $u = u_0$ and $\dot{u} = 0$ at the instant of release. The leading term in Eq. (8.51) satisfies these conditions, so the next order of approximation must satisfy rest conditions, that is,

$$
v_1 = \dot{v}_1 = 0 \qquad \text{when } t = 0
$$

The sum of the complementary and particular solutions satisfying the initial conditions is

$$
v_1 = \tfrac{1}{4} (1 - u_0^2)(1 - \cos \beta_\phi t)
$$

$$
u \approx u_0 - \frac{1}{U} v_1 = u_0 - \frac{\gamma}{2\beta_\phi^2} (1 - u_0^2)(1 - \cos \beta_\phi t) \quad (8.55)
$$

The rate of change of the elevation obtained from the foregoing leads to an expression for the nutation rate. Differentiating Eq. (8.55) gives

$$
\dot{u} = - \frac{\gamma}{2\beta_\phi} (1 - u_0^2) \sin \beta_\phi t \quad (8.56)
$$

The expression for $\dot{u}$ in Eq. (8.40) may be simplified for the present situation because the value of θ remains close to the initial value θ_2, so

$$\sin \theta \approx \sin \theta_2 = (1 - u_0^2)^{1/2}$$
$$\dot{u} = -\sin \theta \; \dot{\theta} \approx -(1 - u_0^2)^{1/2} \; \dot{\theta} \tag{8.57}$$

The result of equating Eq. (8.56) and (8.57) is

$$\dot{\theta} \approx \frac{\gamma}{2\beta_\phi} \sin \theta_2 \sin \beta_\phi t \tag{8.58}$$

We find an expression for the precession rate by using $u \approx u_0$ to simplify the denominator of Eq. (8.44). Substitution of Eqs. (8.45) and (8.55) then yields

$$\dot{\psi} \approx \frac{\gamma}{2\beta_\phi} (1 - \cos \beta_\phi t) \tag{8.59}$$

The interpretation of these results is that the cuspidial precession rate of a fast top varies harmonically about the mean value $\gamma/2\beta_\phi$, with an amplitude equal to the mean value. When the precession rate is zero ($\cos \beta_\phi t = 1$), the nutation rate is zero and the top is at its highest elevation. In contrast, at the instant when the precession rate is maximum ($\cos \beta_\phi t = -1$), the nutation rate is also zero, corresponding to the lowest elevation.

Steady Precession   It is possible to obtain a rotation in which the nutation angle is constant if the appropriate initial motion is imparted to the top. The corresponding spin and precession rates in that case will not vary from their initial values. The most direct approach leading to this response employs the equation of motion for the nutation angle, Eq. (8.38). We see there that if the nutation angle is constant, then

$$(\beta_\phi - \beta_\psi \cos \theta)(\beta_\psi - \beta_\phi \cos \theta) - \frac{mg\,L}{I'} \sin^4 \theta = 0 \tag{8.60}$$

We could consider this relation to govern the nutation angle for specified values of the momentum parameters. However, it is more meaningful to use it to derive an expression for the precession rate corresponding to a specified nutation angle. The definitions of the momentum parameters in Eqs. (8.35) are

$$\beta_\psi = \dot{\psi} \sin^2 \theta + \beta_\phi \cos \theta \qquad \beta_\phi = \frac{I}{I'} (\dot{\psi} \cos \theta + \dot{\phi}) \tag{8.61}$$

Because the spin momentum β_ϕ is proportional to the component of angular velocity parallel to the axis of symmetry, we shall retain that parameter, rather than the spin rate. We therefore substitute only the above relation for β_ψ into Eq. (8.60), and cancel a common factor of $\sin^4\theta$, which leads to

$$(\beta_\phi - \dot{\psi} \cos \theta)\dot{\psi} - \frac{\gamma}{2} = 0 \tag{8.62}$$

The solution of this quadratic equation is

$$\dot{\psi} = \frac{\beta_\phi \pm (\beta_\phi^2 - 2\gamma \cos \theta)^{1/2}}{2 \cos \theta} \tag{8.63}$$

An interesting corollary of this result is that, for a specified nutation angle, there is a minimum spin momentum for which steady precession is possible, specifically

$$(\beta_\phi)_{min} = (2\gamma \cos \theta)^{1/2} \qquad (8.64)$$

Equation (8.63) seems to be fairly straightforward. However, a complication arises if one desires to determine the steady precession rate for a specified spin rate. This is due to the dependence of the spin momentum on the value of $\dot\psi$, according to the second of Eqs. (8.61). Example 8.3 describes an accurate evaluation of the relation between $\dot\phi$ and $\dot\psi$. Here, we derive simple formulas for the case of a fast top, for which $\beta_\phi^2 >> 2\gamma$. It is permissible in that case to truncate a binomial expansion of the square root in Eq. (8.63) at the first two terms. The corresponding roots are

$$\dot\psi_1 = \frac{\gamma}{2\beta_\phi} \qquad \dot\psi_2 = \frac{\beta_\phi}{\cos \theta} \qquad (8.65)$$

The first value is comparatively small, because it varies inversely with β_ϕ; similarly, the second value is large. It follows that we may neglect the contribution of the precession rate to β_ϕ in the first case, but not in the second. Specifically, we find from the second of Eqs. (8.61) that

$$(\beta_\phi)_1 = \frac{I}{I'} \dot\phi \Rightarrow \dot\psi_1 = \frac{I'\gamma}{2I\dot\phi} = \frac{mg\,L}{I\dot\phi}$$

$$(\beta_\phi)_2 = \frac{I}{I'} [(\beta_\phi)_2 + \dot\phi] \Rightarrow \dot\psi_2 = \frac{I\dot\phi}{(I' - I) \cos \theta} \qquad (8.66)$$

The fast precession rate $\dot\psi_2$ matches the value obtained from Eqs. (8.19) for a symmetric body in free motion. In essence, the spin and precession rates in the fast case are so high that the gravitational moment is negligible in comparison with the moments required to alter the angular momentum of the top. Steady precession of a top usually occurs at the slow precession rate, because the kinetic energy required to attain $\dot\psi_2$ is prohibitive.

A special case of steady rotation is the *sleeping top*, which is the term used when the axis of symmetry of the top is vertical, $\cos \theta = 1$. The precession and spin are indistinguishable in a sleeping top, because both rotations are about concurrent axes. (The name ''sleeping top'' stems from the merger of spin and precession, which causes a polished body of revolution without markings to appear to be stationary.) Because of the similarity of precession and spin in such a rotation, some of the relations for steady precession become trivial. For example, because $\beta_\psi = \beta_\phi$ when $\theta = 90°$, Eq. (8.60) is satisfied identically. However, all relations for steady precession remain valid in the limit as $\theta \to 0$. We treat this degenerate case by noting that the angular velocity of a sleeping top is merely

$$\omega = \dot\phi + \dot\psi$$

so the definition of the spin momentum in Eq. (8.61) reduces to

$$\beta_\phi = \frac{I}{I'} \omega$$

Hence, we find from Eq. (8.64) that the minimum rotation rate required for a top to sleep is

$$\omega_{cr} = \frac{I'}{I}\,(2\gamma)^{1/2} = \left(\frac{4\,mg\,L\,I'}{I^2}\right)^{1/2} \tag{8.67}$$

If $\omega < \omega_{cr}$, the axis of symmetry cannot remain vertical.[†] In actuality, the effect of friction at the apex O is to slow the rate of rotation. When the value of ω for a sleeping top falls below ω_{cr}, the top begins to nutate. Because the nutational velocity is zero at the instant when the rotation rate falls below critical, the ensuing motion is a cuspidial precession. If the spin rate decreases slowly, the amplitude of the nutation will slowly increase until the top hits the ground or falls from its support.

EXAMPLE 8.3

A 2-kg top is in a state of steady slow precession at a spin rate of 500 rev/min with its axis at $\theta = 120°$. A vertical impulsive force suddenly induces an upward nutation, such that the ensuing motion is observed to be cuspidial. The radii of gyration of the top about its pivot A are 360 and 480 mm parallel and transverse to the axis of symmetry, respectively. Determine:

(a) The nutation rate induced by the impulsive force.

(b) The largest and smallest values of the nutation angle in the cuspidial precession.

(c) The number of cusps in the path of the axis of symmetry for one revolution of the top about the vertical axis.

(d) The maximum, minimum, and average precession rates in the cuspidial motion.

Solution. We begin by evaluating the steady precession preceding the application of the impulse force. We could employ Eqs. (8.66) for this purpose, provided that we verify the condition $\beta_\phi^2 \gg 2\gamma$ for a fast top. However, an alternative relation for the steady precession rate that does not require preconditions is available. In the present situation, we know the spin rate, which is only one contribution to β_ϕ. Therefore, we substitute the second of Eqs. (8.35) into Eq. (8.62) in order to obtain a relation between the precession and spin rates, with the result that

$$\left[\left(\frac{I}{I'} - 1\right)\cos\theta\right]\dot\psi^2 + \frac{I}{I'}\,\dot\phi\dot\psi - \frac{\gamma}{2} = 0$$

Solving this quadratic equation shows the fast and slow precession rates to be[‡]

$$\dot\psi = \frac{I\dot\phi \pm [I^2\dot\phi^2 - 2(I' - I)I'\gamma\cos\theta]^{1/2}}{2(I' - I)\cos\theta}$$

[†]The vertical position, $\theta = 0$, is a solution to the equations of motion for any value of ω, because the angular momentum is constant and the moment of the gravity force is zero. In essence, our analysis of the sleeping top demonstrates that the vertical position is unstable if $\omega < \omega_{cr}$.

[‡]Setting the discriminant of this equation to zero shows that the minimum spin rate for which steady precession is possible is

$$\dot\phi_{min} = \left[2\left(\frac{I'}{I} - 1\right)\frac{I'}{I}\,\gamma\cos\theta\right]^{1/2}$$

This value is smaller than the spin rate corresponding to Eq. (8.64), which is the condition for the minimum angular velocity component parallel to the axis of symmetry.

The parameters for the present system are

$$m = 2 \text{ kg} \qquad I = m\kappa^2 = 0.2592 \text{ kg-m}^2 \qquad I' = m(\kappa')^2 = 0.4608 \text{ kg-m}^2$$

$$L = 0.2 \text{ m} \qquad \gamma = \frac{2mgL}{I'} = 17.026 \text{ rad}^2/\text{s}^2 \qquad \dot{\phi} = 52.36 \text{ rad/s}$$

which leads to the two roots

$$\dot{\psi}_1 = 0.2884 \text{ rad/s} \qquad \dot{\psi}_2 = -134.93 \text{ rad/s}$$

Both values are extremely close to the approximations in Eqs. (8.66). Because it is stated that the initial precession is slow, we use $\dot{\psi}_1$ as the initial rate.

The impulsive force induces an unknown nutation rate $\dot{\theta}$, because it exerts a moment about the horizontal axis through the pivot. However, the spin and precession rates are not altered during the impulse interval. We find $\dot{\theta}$ from the fact that the subsequent precession is cuspidial. We need the values of the precession and spin momentum parameters to evaluate cuspidial motion. Equations (8.61) for the slow precession rate found above and the given spin rate yield

$$\beta_\phi = 29.371 \text{ rad/s} \qquad \beta_\psi = -14.469 \text{ rad/s}$$

Then the highest elevation u_0 is

$$u_0 = \frac{\beta_\psi}{\beta_\phi} = -0.4926$$

which corresponds to the position where the nutation angle is a minimum.

$$\theta_{\min} = \theta_2 = \cos^{-1} u_0 = 119.514°$$

Before we may employ the results of the perturbation analysis of cuspidial motion, we must check the value of the parameter U in Eq. (8.49). We calculate

$$U = \frac{\beta_\phi^2}{2\gamma} = 25.33$$

which is sufficiently large. Because $u = \cos\theta$, Eq. (8.55) represents an expression for time dependence of the nutation angle. We find that

$$\theta = \cos^{-1}\left[\cos\theta_2 - \frac{1}{4U}\sin\theta_2(1 - \cos\beta_\phi t)\right]$$

$$= \cos^{-1}\{-0.4926 - 0.008588[1 - \sin(29.37t)]\} \text{ rad}$$

We obtain the corresponding nutation and precession rates by direct substitutions into Eqs. (8.58) and (8.59), which yield

$$\dot{\theta} \approx 0.2522 \sin(29.37t) \text{ rad/s}$$
$$\dot{\psi} \approx 0.2898[1 - \cos(29.37t)] \text{ rad/s}$$

With the above relations, we have fully evaluated the cuspidial response. In order to answer the question regarding the nutation rate induced by the impulse, we first determine the initial time t_0. Since $\theta = 120°$ initially we have

$$-0.4926 - 0.008588[1 - \sin (29.37t_0)] = -0.50$$
$$\sin (29.37t_0) = 0.1424$$

The corresponding initial value of the nutation rate is

$$\dot{\theta} = 0.0426 \text{ rad/s}$$

The average of a sinusoidal function is zero, so the expression for the precession rate indicates that

$$\text{Average } \dot{\psi} = 0.2898 \text{ rad/s}$$
$$\text{Minimum } \dot{\psi} = 0 \qquad \text{Maximum } \dot{\psi} = 0.5796 \text{ rad/s}$$

Finally, we note that cusps occur when $\theta = \theta_2$, corresponding to the minimum precession angle. This condition occurs whenever $\sin (29.37t) = 1$. Therefore, the time interval between two adjacent cusps is

$$\Delta t = \frac{2\pi}{29.37} = 0.2139 \text{ s}$$

At the average precession rate, the time interval for one revolution about the vertical axis is

$$T = \frac{2\pi}{\dot{\psi}_{\text{average}}} = 21.68 \text{ s}$$

The ratio $T/\Delta t$ is the number of cusps per precessional revolution. Thus

$$\frac{T}{\Delta t} = 101.36 \Rightarrow N = 101 \text{ or } 102 \text{ cusps}$$

8.3 GYROSCOPES FOR INERTIAL GUIDANCE

We know from our studies thus far that the moment required to change the orientation of the rotation axes of a body is directly correlated to the change in the state of motion. We explore here a number of ways in which this effect has been employed in devices that direct moving vehicles without reference to the frame of reference provided by the earth. These devices are called *inertial guidance systems* because they provide an inertial reference system that moves with the vehicle. Our conceptual pictures will be quite crude. In practice, the various pieces of equipment are manufactured with exceptionally high accuracy and with the finest bearings, in order to match as closely as possible the ideal conditions that we shall treat.

8.3.1 Free Gyroscope

The gyroscope in Figure 8.14 is said to be *free* because the rotation of the rotor is unconstrained. The outer gimbal permits precessional rotation, the inner gimbal permits nutation, and the rotor shaft permits spin. For our introductory study we ignore the effect

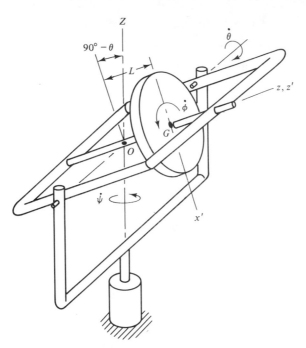

Figure 8.14　Free gyroscope.

of the motion of the vehicle supporting the outer gimbal. In that case the center point O is stationary, because the three rotation axes are concurrent at the center O.

When the center of mass G of the rotor does not coincide with the fixed point O, the gimbals must rotate. The excitation is the moment of the gravity force about the line of nodes, which is the axis about which the inner gimbal rotates relative to the outer gimbal. Aside from the fact that the nutation angle may exceed $90°$, the configuration of the system, as represented by the position of the center of mass relative to the fixed point, is identical to that of a spinning top. It follows that the two systems behave in the same manner. In the special case in which the center of mass coincides with the fixed point O, the free gyroscope behaves like a body in free motion, section A, because there are no external moments. We shall employ the results in the previous sections, as necessary.

Suppose a steady precession, in which the nutation angle is constant, has been established. The relation between the precession rate, the spin rate, and the nutation angle is given by Eq. (8.63). In order for there to be a steady precession, the value of β_ψ for the gyroscope must exceed the minimum rotation rate about the symmetry axis, given by Eq. (8.64). Note that if the center of mass coincides with the fixed point O, then $\gamma = 0$. The steady slow precession rate in that case is zero, which means that the axis of symmetry has a constant orientation.

An important question that must be addressed is whether the steady precession is a stable response. If it is not, such motion would not be observed in reality. One technique by which *dynamic stability* may be studied is to disturb the nutation angle from the steady value it has when a steady precession has been established. Thus, let

$$\theta = \theta^* + \Delta \qquad (8.68)$$

where θ^* denotes the constant value for steady precession and Δ is a small disturbance that may vary with time.

A linearized equation of motion governing Δ may be obtained from the general equation, Eq. (8.38), which we multiply by $\sin^3 \theta$. Then, we substitute Eq. (8.68), and expand in a Taylor series that we truncate at terms that contain quadratic and higher powers of Δ. For example,

$$\cos \theta \approx \cos \theta^* - \Delta \sin \theta^*$$
$$(\sin \theta)^4 \approx (\sin \theta^*)^4 + 4\Delta (\sin \theta^*)^3 \cos \theta^* \qquad (8.69)$$

By definition, θ^* is a solution of Eq. (8.60). Consequently, the zero-order terms (i.e., those that are independent of Δ) cancel. The first-order equations that results from the foregoing procedure is

$$\sin^3 \theta^* \ddot{\Delta} + [\beta_\phi(\beta_\phi - \beta_\psi \cos \theta^*) + \beta_\psi(\beta_\psi - \beta_\phi \cos \theta^*)] \Delta \sin \theta^*$$
$$- 2\gamma\Delta \sin^3 \theta^* \cos \theta^* = 0 \quad (8.70)$$

We may simplify this expression further by eliminating β_ψ with the aid of Eq. (8.61). This yields

$$\ddot{\Delta} + \omega^2 \Delta = 0 \qquad (8.71)$$

where

$$\omega^2 = \beta_\phi^2 + \dot{\psi}^2 \sin^2 \theta^* - 2\gamma \cos \theta^* \qquad (8.72)$$

The steady precession is stable to small disturbances if the value of Δ remains bounded. Such a condition is obtained if $\omega^2 > 0$, since the corresponding solution of Eq. (8.71) is oscillatory. However, Eq. (8.64) states that a steady precession can exist only if the spin momentum is sufficiently large, $\beta_\phi^2 > 2\gamma \cos \theta$. It follows that $\omega^2 > 0$ for any free precession. Therefore, if the spin momentum is sufficiently large to establish a steady precession, an attempt to change the nutation angle by a small amount will result in an oscillatory nutational motion whose mean value is θ^*.

The balanced free gyroscope, for which $L = 0$, has its primary application in inertial navigation systems that track vehicle motion. The concept is remarkably simple, although its actual implementation for aircraft and rockets requires the highest degree of technological sophistication. The invariability of the direction of the balanced gyroscope provides a translating reference frame. Measurements of the vehicle's rotation relative to this reference frame are used to drive servomotors, which maintain a platform in a horizontal orientation (relative to the earth's surface). The movement of the vehicle is tracked by accelerometers on the horizontal platform, which measure the acceleration of the platform relative to the earth's surface. The displacement relative to the earth may then be determined by electronically integrating the accelerations twice in time.

EXAMPLE 8.4 _____

In order to overcome the effects of friction, a servomotor applies a torque about the spin axis of a symmetric gyroscope, with the result that the spin rate is constant. The initial conditions are such that the initial precession rate $\dot{\psi}^*$ and nutation angle θ^* correspond to steady precession. Determine whether the action of the servomotor can cause the gyroscope to be unstable to small disturbances.

Solution. The primary difference between the present system and a free gyroscope is that there are only two degrees of freedom, because the spin rate is constrained to be constant. In this case, $\dot{\phi}$, rather than the spin momentum, p_ϕ, is constant. We commence to derive the equations of motion for the servogyroscope by forming the Lagrangian using Eqs. (8.33) and (8.34).

$$L = \tfrac{1}{2} I(\dot{\psi} \cos \theta + \dot{\phi})^2 + \tfrac{1}{2} I'(\dot{\psi}^2 \sin^2 \theta + \dot{\theta}^2) - mg L \cos \theta$$

The precession angle is an ignorable generalized coordinate, because only its derivative appears in L. The corresponding conservation-of-momentum equation is identical to the first of Eqs. (8.35), which leads to

$$\beta_\psi = \dot{\psi} \left(\frac{I}{I'} \cos^2 \theta + \sin^2 \theta \right) + \frac{I}{I'} \dot{\phi} \cos \theta \tag{1}$$

The Lagrange's equation for θ may be written as

$$\ddot{\theta} + g(\dot{\psi}, \theta) = 0 \tag{2}$$

where the function g is found from Eq. (8.37) to be

$$g(\dot{\psi}, \theta) = \left(\frac{I}{I'} - 1 \right) \dot{\psi}^2 \sin \theta \cos \theta + \frac{I}{I'} \dot{\psi}\dot{\phi} \sin \theta - \tfrac{1}{2} \gamma \sin \theta \tag{3}$$

Because the initial conditions are those appropriate to a steady precession, we have

$$g(\dot{\psi}^*, \theta^*) = 0 \tag{4a}$$

Since $\sin \theta \neq 0$ for steady precession in the nonvertical position, Eq. (4a) is satisfied when

$$\left[\left(\frac{I}{I'} - 1 \right) \cos \theta^* \right] (\dot{\psi}^*)^2 + \frac{I}{I'} \dot{\phi}\dot{\psi}^* - \tfrac{1}{2} \gamma = 0 \tag{4b}$$

which matches the expression established in Example 8.3 for the free gyroscope.

Although the angular motion in steady precession is identical to that of a free gyroscope, the stability situation is different. Constancy of β_ψ now requires that any fluctuation in the nutation angle will also be manifested by a change in the precession rate. We only consider very small changes $\Delta\theta$ for a stability analysis, so the corresponding increment in the precession rate, $\Delta\dot{\psi}$, is also small. Therefore, a Taylor-series expansion of Eq. (1) may be truncated at first-order terms. This leads to

$$\Delta\beta_\psi = 0 = \frac{\partial\beta_\psi}{\partial\dot{\psi}} \Delta\dot{\psi} + \frac{\partial\beta_\psi}{\partial\theta} \Delta\theta \Rightarrow \Delta\dot{\psi} = - \frac{(\partial\beta_\psi/\partial\theta)^*}{(\partial\beta_\psi/\partial\dot{\psi})^*} \Delta\theta \tag{5}$$

where asterisks indicate that derivatives are evaluated at the steady precession condition.

According to Eq. (5), a small increment in the nutation angle produces a proportional change in the precession rate. We now must determine the equation governing $\Delta\theta$. We let

$$\theta = \theta^* + \Delta\theta \qquad \dot\psi = \dot\psi^* + \Delta\dot\psi$$

and substitute this expression into Eq. (2). Truncation of a Taylor-series expansion at first-order terms leads to

$$\Delta\ddot\theta + g(\dot\psi^* + \Delta\dot\psi, \theta^* + \Delta\theta) = \Delta\ddot\theta + g(\dot\psi^*, \theta^*) + \Delta\dot\psi \frac{\partial g^*}{\partial\dot\psi} + \Delta\theta \frac{\partial g^*}{\partial\theta} = 0$$

In view of Eq. (4) for the steady precession condition, and Eq. (5) for the fluctuation in the precession rate, this simplifies to

$$\Delta\ddot\theta + \omega^2\,\Delta\theta = 0 \tag{6}$$

where

$$\omega^2 = \left(\frac{\partial g}{\partial\theta}\right)^* - \frac{(\partial\beta_\psi/\partial\theta)^*}{(\partial\beta_\psi/\partial\dot\psi)^*}\left(\frac{\partial g}{\partial\dot\psi}\right)^*$$

The specific result obtained from actually differentiating β_ψ in Eq. (1) and g in Eq. (3) is

$$\omega^2 = \left[\left(\frac{I}{I'} - 1\right)(\dot\psi^*)^2 \cos 2\theta^* + \left(\frac{I}{I'}\dot\phi\dot\psi^* - \frac{1}{2}\gamma\right)\cos\theta^*\right]$$
$$- \frac{2(I - I')\dot\psi^* \cos\theta^* + I\dot\phi}{I \cos^2\theta^* + I' \sin^2\theta^*}\left[2\left(\frac{I}{I'} - 1\right)\dot\psi^* \cos\theta^* + \frac{I}{I'}\dot\phi\right]\sin^2\theta^* \tag{7}$$

As was true for the free gyroscope, whose stability was described by Eqs. (8.71) and (8.72), the servo-driven gyroscope is unstable to small disturbances $\Delta\theta$ when $\omega^2 < 0$. The differences between Eq. (7) and Eq. (8.72) are apparent. It is not a trivial matter to identify the sign of ω^2 now, because $\dot\psi^*$ is related to the other parameters through the quadratic equation (4b). However, even without further manipulation, it seems likely that $\omega^2 < 0$ is possible. This is indeed the case. For example, a gyroscope having the inertial and geometrical parameters in Example 8.3 nutating at $\theta = 120°$ is unstable if the spin rate is less than 1.521 rad/s. (The slow precession rate in that condition is 4.582 rad/s.)

The occurrence of the instability could have been anticipated on the basis of physical arguments. The free gyroscope is a conservative system, whereas the servo-driven gyroscope in the present case is not, since the servomotor does work to hold the spin rate constant. The energy provided to the system from this source might have the effect of driving the nutational motion away from steady precession. However, the situation in most cases of practical interest is like the numerical example above, since the instability arises only at spin rates that are sufficiently low to be of no concern.

8.3.2 Gyrocompass

A fundamental requirement for earthbound navigation is knowledge of the orientation of true north. As the earth rotates, the balanced free gyroscope maintains a fixed orientation as the earth rotates; an observer on the earth would perceive the gyroscope to be rotating. The gyrocompass has the feature that its steady precession can be adjusted such that its precession always matches the earth's rotation, so that the rotor axis is always oriented in the northerly (or southerly) direction.

The gyrocompass bears much similarity to an unbalanced free gyroscope, except for the placement of the mass causing the imbalance. As shown in Figure 8.15, the center of mass of the rotor is situated on the intersection of the precession and spin axes, but a small additional mass m is attached to the inner gimbal. This arrangement is selected in order that the gravitational moment may be like that for a pendulum. It is necessary to include the rotation of the earth in the analysis, for which the earth-based reference frame we defined in Chapter 3 is useful. Thus, in Figure 8.15 the Z axis is oriented in the direction perceived as vertical to an observer on the earth and the X axis is situated in the northerly direction. The angular velocity of the earth is therefore

$$\bar{\omega}_e = \omega_e(\cos \lambda \bar{I} + \sin \lambda \bar{K}) \qquad (8.73)$$

where $\omega_e = 7.292(10^{-5})$ rad/s $\approx 2\pi$ rad/24 h is the rotation rate.

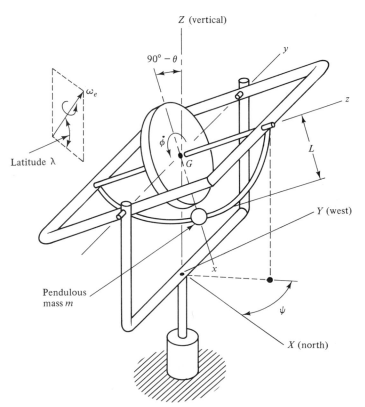

Figure 8.15 Gyrocompass.

The gyroscope may assume any position. We use the Eulerian angles to describe the orientation of the rotor *relative to the earth fixed reference frame XYZ*. Toward that end, we introduce an intermediate reference frame *xyz* that is fixed to the inner gimbal. We align the *z* axis with the axis of symmetry of the rotor and the *y* axis is the line of nodes formed by the bearings of the inner gimbal. Our goal here is to determine whether there is any set of precession and nutation angles for which the axis of the rotor remains stationary relative to the earth. For this reason we consider the values of ψ and θ to be constant, and we also assume that $\dot{\phi}$ remains constant. The corresponding angular velocity of the rotor relative to *XYZ* is $\dot{\phi}\bar{k}$. In order to combine this term with the rotation of the earth, we transform the unit vectors according to

$$\begin{aligned}
\bar{I} &= \cos\psi(\cos\theta\bar{i} + \sin\theta\bar{k}) - \sin\psi\bar{j} \\
\bar{J} &= \sin\psi(\cos\theta\bar{i} + \sin\theta\bar{k}) + \cos\psi\bar{j} \\
\bar{K} &= -\sin\theta\bar{i} + \cos\theta\bar{k}
\end{aligned} \tag{8.74}$$

Adding the earth's rotation to the rotor spin then leads to the following absolute angular velocity of the rotor:

$$\bar{\omega} = \omega_e(\cos\lambda\cos\psi\cos\theta - \sin\lambda\sin\theta)\bar{i} - \omega_e\cos\lambda\sin\psi\bar{j}$$
$$+ (\omega_e\cos\lambda\cos\psi\sin\theta + \omega_e\sin\lambda\cos\theta + \dot{\phi})\bar{k} \tag{8.75}$$

Terms containing ω_e have a very small value, so we may simplify the kinetic energy of the system by neglecting effects that are of the order of ω_e^2. The corresponding kinetic energy for the system is

$$\begin{aligned}
T &= \tfrac{1}{2}(I'\omega_x^2 + I'\omega_y^2 + I\omega_z^2) \\
&= \tfrac{1}{2}I[\dot{\phi}^2 + 2\omega_e(\cos\lambda\cos\psi\sin\theta + \sin\lambda\cos\theta)\dot{\phi}] \tag{8.76}
\end{aligned}$$

The corresponding potential energy is associated with the unbalanced mass on the inner gimbal. When the datum is set at the fixed point O, we find that

$$V = mgL\cos\left(\frac{\pi}{2} + \theta\right) = -mgL\sin\theta \tag{8.77}$$

There are no nonconservative forces in this ideal model, so the Lagrange's equations for the generalized coordinates ψ, θ, and ϕ in this case of steady precession are

$$\begin{aligned}
\omega_e\dot{\phi}\cos\lambda\sin\psi\sin\theta &= 0 \\
I\omega_e\dot{\phi}(-\cos\lambda\cos\psi\cos\theta + \sin\lambda\sin\theta) - mgL\cos\theta &= 0 \\
\dot{\phi} + \omega_e(\cos\lambda\cos\psi\sin\theta + \sin\lambda\cos\theta) &= \beta_\phi
\end{aligned} \tag{8.78}$$

where $\beta_\phi = p_\phi/I$ is the spin-momentum parameter associated with the ignorable coordinate ϕ.

Because of the smallness of ω_e, the last of Eqs. (8.78) indicates that the spin momentum is primarily associated with the spin itself. The first equation is satisfied when $\sin\theta = 0$ or $\sin\psi = 0$. The first possibility is not useful, because then the rotor does not provide any directional information. The second case is the one we seek, because it is satisfied when $\psi = 0$ or π, so that the spin axis is aligned along the north-south meridian. Setting $\cos\psi = \pm 1$ in the second of Eqs. (8.78) yields

$$(mgL \pm I\omega_e\dot{\phi}\cos\lambda)\cos\theta = I\omega_e\dot{\phi}\sin\lambda\sin\theta \tag{8.79}$$

We observe that the smallness of ω_e means that the value of $\tan \theta$ obtained from the foregoing is much larger than unity, which corresponds to $\theta \approx \pi/2$. Thus, set

$$\theta = \frac{\pi}{2} - \Delta \qquad \Delta << 1$$

Furthermore, we may neglect $I\omega_e\dot{\phi}$ in comparison with $mg\ l$. We therefore find from Eq. (8.79) that

$$\Delta = \frac{I\omega_e\dot{\phi} \sin \lambda}{mg\ L} \qquad \theta = \frac{\pi}{2} - \Delta \qquad \psi = 0 \text{ or } \pi \qquad (8.80)$$

As a summary of these results, recall that the analysis treated a balanced gyroscope having a pendulous mass mg attached to the inner gimbal. We have established that if the rotor is released with its spin axis tilted at an angle Δ above the north-south horizontal, the precession of the gyroscope will match the component of the earth's angular velocity in the direction of the local vertical. Thus, the plane containing the rotor and the bearings of the outer gimbal will indicate the northward direction.

Our analysis of the gyrocompass has established the conditions for dynamic equilibrium at a specified latitude λ. We shall see in the next example that it is stable to small disturbances. The primary limitation on the use of the gyrocompass is its loss of accuracy due to rapid movement of the vehicle in which it is mounted. Linear motion relative to the earth is actually motion along a great circle. This represents a base rotation that appears from the reference frame of the outer gimbal to be a precession equivalent to the earth's rotation. Consider the situation in Figure 8.16, where the velocity $\bar{v}$ of the base of the gyrocompass is oriented at angle β west of north. This velocity may be represented as the result of a movement relative to the center O of the earth associated with a relative angular speed $\bar{\omega}_{rel}$, where

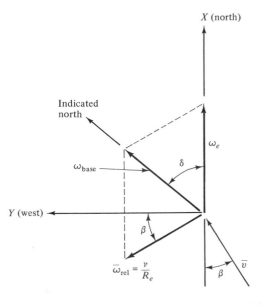

Figure 8.16 Directional error in a gyrocompass due to movement of the vehicle in a great circle.

$$\bar{v} = \bar{\omega}_{\mathrm{rel}} \times \bar{r}_{O'/O} \qquad \bar{r}_{O'/O} = R_e \bar{K}$$

$$\bar{\omega}_{\mathrm{rel}} = -\frac{v}{R_e} \sin \beta \bar{I} + \frac{v}{R_e} \cos \beta \bar{J} \qquad (8.81)$$

Then the total angular velocity of the base of the gyroscope is the sum of the earth's rotation and the foregoing rotation of the base relative to the earth.

$$\bar{\omega}_{\mathrm{base}} = \left(\omega_e \cos \lambda - \frac{v}{R_e} \sin \beta \right) \bar{I} + \frac{v}{R_e} \cos \beta \bar{J} + \omega_e \sin \lambda \bar{K} \qquad (8.82)$$

The component of this angular velocity parallel to the earth's surface is deviated from the true northerly direction, as shown in Figure 8.16.

Since the direction of the component of $\bar{\omega}_{\mathrm{base}}$ parallel to the earth's surface is the reference for our derivation of Eq. (8.80), we may conclude that the gyrocompass in steady precession will align with this component. The angle δ, indicating the error, is found from either Figure 8.16 or Eq. (8.82) to be

$$\delta = \tan^{-1} \left(\frac{v \cos \beta}{\omega_e R_e \cos \lambda - v \sin \beta} \right) \qquad (8.83)$$

If $\omega_e R_e \cos \lambda \gg v$, this error is quite small. However, if the gyrocompass is mounted on a moving vehicle near either the north or south poles, the error will be quite substantial, even if the speed is quite low. In practice, since δ is known from the above, it is possible to compensate readings accordingly. However, this does not entirely remove the difficulty near the poles, since the manner in which a gyrocompass responds to disturbances at the poles introduces additional errors, as is discussed in Example 8.5.

Another source of error arises from acceleration of the pendulous mass, which introduces inertial forces, in addition to the weight of that body. In effect, this alters the apparent magnitude and direction of the gravitational force. For all these reasons, the gyrocompass is used primarily as a navigational aid for slowly moving vehicles, such as ships.

EXAMPLE 8.5 _____

A gyrocompass tracking the northerly direction in a steady precession is given a small initial nutational disturbance $\Delta\theta$, causing it to deviate from its proper direction. Determine the response to this initial disturbance. Then, from that result, assess the stability of the gyrocompass.

Solution. The precessional and nutational motions resulting from the disturbance are time-dependent, so Eqs. (8.78) are not adequate for the stability analysis. In order to derive the equations of motion for this case, we form the angular velocity of the flywheel as a superposition of the spin and rotation of the earth, given by Eq. (8.75), the nutational motion $\dot{\theta}$, and the precession rate $\dot{\psi}$ relative to the earth. Thus,

$$\begin{aligned}
\bar{\omega} = &[-\dot{\psi} \sin \theta + \omega_e (\cos \lambda \cos \psi \cos \theta - \sin \lambda \sin \theta)]\bar{i} \\
&+ (\dot{\theta} - \omega_e \cos \lambda \sin \psi)\bar{j} \\
&\qquad + (\dot{\psi} \cos \theta + \omega_e \cos \lambda \cos \psi \sin \theta + \omega_e \sin \lambda \cos \theta + \dot{\phi})\bar{k}
\end{aligned}$$

Since the disturbance is small, the nutation angle θ will remain close to $\pi/2$, provided that the system is stable. We therefore define

$$\gamma = \frac{\pi}{2} - \theta \qquad \dot{\gamma} = -\dot{\theta}$$

The corresponding expression for the Lagrangian is

$$L = \tfrac{1}{2} I'[-(\dot{\psi} + \omega_e \sin \lambda) \cos \gamma + \omega_e \cos \lambda \cos \psi \sin \gamma]^2$$
$$+ \tfrac{1}{2} I'(\dot{\gamma} + \omega_e \cos \lambda \sin \psi)^2$$
$$+ \tfrac{1}{2} I[\dot{\phi} + (\dot{\psi} + \omega_e \sin \lambda) \sin \gamma + \omega_e \cos \lambda \cos \psi \cos \gamma]^2 + mg L \cos \gamma$$

The spin angle ϕ is ignorable, as it was in the case of steady precession. The corresponding generalized momentum parameter β_ϕ is constant, where now

$$\beta_\phi = \dot{\phi} + (\dot{\psi} + \omega_e \sin \lambda) \sin \gamma + \omega_e \cos \lambda \cos \psi \cos \gamma$$

This expression may be substituted into the Lagrange's equations for γ and ψ, after the derivatives with respect to the generalized coordinates and velocities have been evaluated. The result is

$$I'(\ddot{\gamma} + \dot{\psi}\omega_e \cos \lambda \cos \psi)$$
$$- [I\beta_\phi - I'\dot{\psi} \sin \gamma - I'\omega_e(\sin \lambda \sin \gamma + \cos \lambda \cos \psi \cos \gamma)]$$
$$\times [(\dot{\psi} + \omega_e \sin \lambda) \cos \gamma - \omega_e \cos \lambda \cos \psi \sin \gamma] + mg L \sin \gamma = 0 \qquad (1)$$

$$I'[\ddot{\psi} \cos^2 \gamma - (\dot{\psi} + \omega_e \sin \lambda)(\dot{\gamma} + \tfrac{1}{2} \omega_e \cos \lambda \sin \psi) \sin 2\gamma$$
$$+ \omega_e \cos \lambda \,(\tfrac{1}{2} \dot{\psi} \sin \psi \sin 2\gamma - 2\dot{\gamma} \cos \psi \cos^2 \gamma)$$
$$- \tfrac{1}{2} \omega_e^2 \cos^2 \lambda \sin 2\psi \cos^2 \gamma] + I[\beta_\phi(\dot{\gamma} + \omega_e \cos \lambda \sin \psi) \cos \gamma] = 0 \qquad (2)$$

In the case of steady precession, $\psi = 0$ and γ is the constant value Δ given by Eq. (8.80), which is a very small value. Stable responses $\gamma(t)$ and $\psi(t)$ resulting from a small disturbance will remain small. Hence, we may linearize equations of motion (1) and (2). In this process, we also ignore terms that are quadratic in ω_e, and use $mg L >> I\beta_\phi\omega_e$, $\beta_\phi >> \dot{\psi}$, $\beta_\phi >> \omega_e$. The linearized equations of motion simplify to

$$I'\ddot{\gamma} - I\beta_\phi\dot{\psi} + mg L \,\gamma = I\beta_\phi\omega_e \sin \lambda \qquad (3)$$
$$I'\ddot{\psi} + (I\beta_\phi - 2I'\omega_e \cos \lambda)\dot{\gamma} + (I\beta_\phi\omega_e \cos \lambda)\psi = 0 \qquad (4)$$

We form the solution of these differential equations by adding complementary and particular solutions. The latter are the values for steady precession,

$$\psi_p = 0 \qquad \gamma_p = \frac{I\beta_\phi\omega_e \sin \lambda}{mg L}$$

Since $\beta_\phi \approx \dot{\phi}$, the foregoing is equivalent to $\gamma_p = \Delta$.

For the complementary solution, let us assume that the system is stable and that γ and θ vary sinusoidally. We note that each homogeneous equation relates a generalized coordinate and its second derivative to the first derivative of the other generalized coordinate. Consequently, one generalized coordinate must be 90° out of phase relative to the other. A suitable trial form for the complementary solution therefore is

$$\gamma_c = A \sin(\sigma t - \nu) \qquad \psi_c = B \cos(\sigma t - \nu) \qquad (5)$$

where A, B, σ, and ν are constants. Requiring that these expressions be solutions of the homogeneous portions of Eqs. (3) and (4) leads to

$$(mg\,L - I'\sigma^2)A + I\beta_\phi\sigma B = 0$$
$$(I\beta_\phi - 2I'\omega_e\cos\lambda)\sigma A + (I\beta_\phi\omega_e\cos\lambda - I'\sigma^2)B = 0 \qquad (6)$$

In order for there to be a nontrivial solution, the determinant of the coefficients of A and B must vanish, which leads to the characteristic equation,

$$(I')^2\sigma^4 + [I'(I\beta_\phi\omega_e\cos\lambda - mg\,L) - I^2\beta_\phi^2]\sigma^2 + mg\,L\,I\beta_\phi\omega_e\cos\lambda = 0 \quad (7)$$

For practical applications, the spin rate is sufficiently large that $\beta_\phi^2 \gg mg\,L/I$. Then the two values of $\sigma \gtreqqless 0$ obtained from this quadratic equation are well approximated as

$$\sigma_1 \approx \left(\frac{mg\,L\,\omega_e\cos\lambda}{I\beta_\phi}\right)^{1/2} \qquad \sigma_2 \approx \frac{I\beta_\phi}{I'} \qquad (8)$$

For each frequency σ_j, there is a corresponding ratio A/B. The first of Eqs. (6) indicates that

$$B_j = \mu_j A_j \qquad \mu_j = \frac{I'\sigma_j^2 - mg\,L}{I\beta_\phi\sigma_j} \qquad (9)$$

For the assumed orders of magnitude of β_ϕ, $mg\,L/I$, and ω_e, substitution of each of Eqs. (8) leads to

$$\mu_1 \approx \left(\frac{mg\,L}{I\beta_\phi\omega_e\cos\lambda}\right)^{1/2} \qquad \mu_2 \approx 1$$

We conclude from the foregoing that the complementary solution, which is a free vibration, occurs as either of two modes. The first is a low-frequency mode at σ_1, in which the amplitude of the nutation is much smaller than that of the precession ($\mu_1 \gg 1$), while the second is a high-frequency mode at σ_2, in which the amplitudes of the nutation and the precession are approximately equal.

The most general solution is a sum of the two modes, and of the particular solution. Thus, we find that the response to the disturbance is

$$\gamma = \Delta + A_1 \sin(\sigma_1 t - \nu_1) + A_2 \sin(\sigma_2 t - \nu_2)$$
$$\psi = \mu_1 A_1 \cos(\sigma_1 t - \nu_1) + \mu_2 A_2 \cos(\sigma_2 t - \nu_2)$$

The actual values of the amplitudes A_j and phase angles ν_j depend on the initial conditions, which are not stated specifically. In most actual situations, dissipation effects damp the high-frequency mode much more than the low-frequency mode, in which case the oscillation at frequency σ_1 is most likely to be observed.

In regard to the question of stability, we note that the values of σ_1 and σ_2 are always real. Hence, disturbances of the gyrocompass always result in bounded oscillations, corresponding to a stable steady motion. However, the value of σ_1

becomes very small if $\lambda \approx 0$ or π, corresponding to locations near the north or south poles. Hence, at those locations the gyrocompass executes very slow oscillations when disturbed, which make it difficult to make accurate readings.

8.3.3 Single-Axis Gyroscope

An important aspect to the navigation of vehicles, particularly those that move rapidly, is the measurement of rotation of the vehicle. The single-axis gyroscope, which only has an inner gimbal, provides such information because its nutation is directly correlated to the precession rate.

A conceptual model of a single-axis gyro appears in Figure 8.17, where the platform is assumed to undergo arbitrary rotations about the axes $\xi\eta\zeta$. These axes are defined such that the $\xi\zeta$ plane is parallel to the platform, with ξ aligned parallel to the bearings of the gimbal. The *xyz* reference frame is attached to the gimbal. The gimbal is mounted on the platform by a shaft that is loaded by a linear torsional spring of stiffness K and a torsional damper whose constant is C. We require that the spring be mounted such

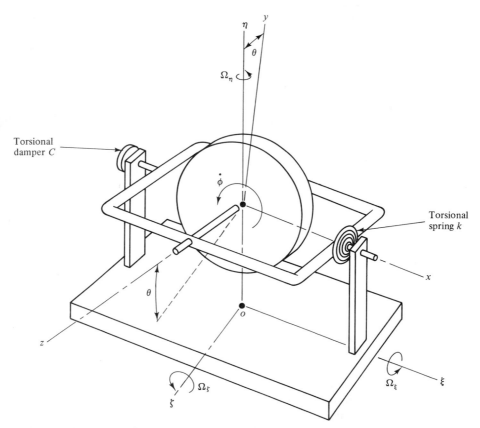

Figure 8.17   Single-axis gyroscope.

that, in the absence of movement of the platform, the rotor axis aligns parallel to the platform.

The presence of only one gimbal means that the rotor has only two degrees of freedom relative to the platform. (We are assuming that the rotor spins freely.) Several alternative definitions of the Eulerian angles are possible in this case. We shall consider the x axis to be the nutation axis and set $\psi = 0$. The nutation angle θ is the angle from the ξ axis to the rotor axis, and the spin angle ϕ is the rotation of the rotor about its axis. According to this definition, $\theta = 0$ represents the undeformed position of the spring.

We describe the rotation of the platform in terms of the rotation rates Ω_ξ, Ω_η Ω_ζ of a set of axes that are fixed to the platform. The angular velocity of the rotor relative to the platform will be described in terms of the Eulerian angles. Adding this relative quantity to the angular velocity of the platform yields the absolute angular velocity of the rotor.

$$\bar{\omega} = \Omega_\xi \bar{i}_\xi + \Omega_\eta \bar{i}_\eta + \Omega_\zeta \bar{i}_\zeta + \dot{\theta}\bar{i} + \dot{\phi}\bar{k} \tag{8.84}$$

Let I denote the moment of inertia of the rotor about the z axis. Owing to the axial symmetry, the moments of inertia of the rotor about the x and y axes are both I', regardless of the angle of spin of the rotor. Correspondingly, we express $\bar{\omega}$ in terms of components relative to xyz in order to form the kinetic energy. The result is

$$\bar{\omega} = (\Omega_\xi + \dot{\theta})\bar{i} + (\Omega_\eta \cos\theta - \Omega_\zeta \sin\theta)\bar{j} + (\Omega_\eta \sin\theta + \Omega_\zeta \cos\theta + \dot{\phi})\bar{k} \tag{8.85}$$

The corresponding general expression for the kinetic energy is

$$T = \tfrac{1}{2}I'[(\Omega_\xi + \dot{\theta})^2 + (\Omega_\eta \cos\theta - \Omega_\zeta \sin\theta)^2] + \tfrac{1}{2}I(\Omega_\eta \sin\theta + \Omega_\zeta \cos\theta + \dot{\theta})^2 \tag{8.86}$$

The position where the nutation angle is zero corresponds to the unstretched position of the torsional spring, so the potential energy is

$$V = \tfrac{1}{2}K\theta^2 \tag{8.87}$$

We shall describe the effect of the linear torsional dashpot by employing the Rayleigh dissipation function, which treats damping forces analogously to a linear spring. The dashpot constant is C, so we have

$$D = \tfrac{1}{2}C\dot{\theta}^2 \tag{8.88}$$

It is important to recognize that the base rotations are specified, so they are not generalized coordinates. Also, the spin angle appears in the formulation only as a time derivative. Hence, ϕ is an ignorable coordinate; the corresponding Lagrange equation reduces to $\partial T/\partial\dot{\phi} = I\beta_\phi$, where the spin momentum parameter β_ϕ is a constant. For the kinetic energy in Eq. (8.86), this reduces to

$$\beta_\phi = \dot{\phi} + \Omega_\eta \sin\theta + \Omega_\zeta \cos\theta \tag{8.89}$$

The equation of motion for the nutation angle is the full Lagrange's equation, including the term $\partial D/\partial\dot{\theta}$ for the dashpot. The result is

$$I'(\dot{\Omega}_\xi + \ddot{\theta}) + I'(\Omega_\eta \cos\theta - \Omega_\zeta \sin\theta)(\Omega_\eta \sin\theta + \Omega_\zeta \cos\theta)$$
$$-I(\Omega_\eta \sin\theta + \Omega_\zeta \cos\theta + \dot{\phi})(\Omega_\eta \cos\theta - \Omega_\zeta \sin\theta) + C\dot{\theta} + K\theta = 0 \quad (8.90)$$

We remove the spin rate from this relation by substituting the spin momentum given by Eq. (8.89). When all terms containing the rotation rates of the platform are moved to the right side, the result is

$$I'\ddot{\theta} + C\dot{\theta} + K\theta = -I'[\tfrac{1}{2}(\Omega_\eta^2 - \Omega_\zeta^2)\sin 2\theta + \Omega_\zeta \Omega_\eta \cos 2\theta]$$
$$+ I\beta_\phi(\Omega_\eta \cos\theta - \Omega_\zeta \sin\theta) - I'\dot{\Omega}_\xi \quad (8.91)$$

The rotor in a practical single-axis gyroscope is made to spin much more rapidly than the highest anticipated rate of rotation of the platform, so that $\beta_\phi \gg \Omega_\eta$ or Ω_ζ. Also, the stiffness and damping parameters are usually selected to restrict the nutation angle to a small magnitude. Under these assumptions, the right side is dominated by two terms. The main effect of the rotation rates is contained in $I\beta_\phi\Omega_\eta \cos\theta$, which may be linearized by setting $\cos\theta \approx 1$, and the effect of unsteadiness in the rates appears only in $I'\dot{\Omega}_\xi$. The equation of motion then reduces to

$$I'\ddot{\theta} + C\dot{\theta} + K\theta = I\beta_\phi\Omega_\eta - I'\dot{\Omega}_\xi \quad (8.92)$$

This is the equation of motion for a damped, one-degree-of-freedom linear oscillator. Its natural frequency, ω, and ratio of critical damping, σ, are

$$\omega = \left(\frac{K}{I'}\right)^{1/2} \qquad \sigma = \frac{1}{2\omega}\frac{C}{I'} = \frac{C}{2(KI')^{1/2}} \quad (8.93)$$

Let us begin by evaluating the nutation when Ω_η is a constant, nonzero value. The largeness of β_ϕ in that case makes it permissible to ignore the term containing $\dot{\Omega}_\xi$. The corresponding response may be obtained by adding the complementary and particular solutions. In the absence of rotations of the platform, the gimbal will be at rest at its equilibrium position $\theta = 0$, so we set $\theta = \dot{\theta} = 0$ when $t = 0$ as initial conditions. If the damping is light, $\sigma < 1$, the corresponding response is

$$\theta = \frac{I\beta_\phi\Omega_\eta}{K}\left\{1 - \exp(-\sigma\omega t)\left[\cos\omega_d t + \frac{\sigma}{(1-\sigma)^{1/2}}\sin\omega_d t\right]\right\} \quad (8.94)$$

where $\omega_d = \omega(1 - \sigma^2)^{1/2}$ is the damped natural frequency. The important aspect of Eq. (8.94) is that it indicates that the steady-state response, which is obtained as $t \to \infty$, is a constant nutation angle that is proportional to the rate at which the platform is rotating about the η axis,

$$\theta_{ss} = \frac{I\beta_\phi}{K}\Omega_\eta \quad (8.95)$$

Thus, the nutation angle may be measured and compared with a scale that is calibrated in units of the rotation rate Ω_η. We should note that the foregoing steady-state response would also be obtained if Ω_η were time-dependent, provided that the free vibration response decays in a much smaller time than the interval required to observe substantial changes in Ω_η. This condition may be achieved by designing the system to have a high natural frequency and to be highly damped, subject to $\sigma < 1$. A single-axis gyro that is constructed with a spring and a dashpot that are both stiff is called a *rate gyroscope*. Since a rate gyroscope indicates the rotation about only one axis, three, mounted about orthogonal axes, are employed in inertial guidance systems for aerospace applications.

There is an alternative configuration for a single-axis gyro that is employed frequently. Suppose the torsional spring is not present. Setting $K = 0$ in the equation of motion for the nutation leads to

$$I'\ddot{\theta} + C\dot{\theta} = I\beta_\phi\Omega_\eta \qquad (8.96)$$

where we have assumed that $\beta_\phi\Omega_\eta$ is sufficiently large to justify neglecting the term in Eq. (8.92) that contains $\dot{\Omega}_\xi$. It is possible to obtain a solution valid for arbitrary Ω_η, not necessarily constant. Such a result consists of a convolution integral that may be derived either from a Laplace transform or from a Duhamel integral using the impulse response of a second-order linear oscillator that has no spring. The result is

$$\theta = \frac{I\beta_\phi}{C} \int_0^t \Omega_\eta(\tau) \left\{ 1 - \exp\left[-\frac{C}{I'}(t - \tau)\right]\right\} d\tau \qquad (8.97)$$

It is desirable that the damping rate be large, in order to make the exponential term in the integrand quickly decay. Then, after an initial startup interval, the nutation angle will be well approximated by

$$\theta = \frac{I\beta_\phi}{C} \int_0^t \Omega_\eta(\tau)\, d\tau \qquad (8.98)$$

We see that the nutation angle in this case is proportional to the integral of the rotation rate about the η axis, which represents the cumulative rotation. For this reason, a well-damped single-axis gyro that is not restrained by a spring is called an *integrating gyroscope*. As is true for rate gyroscopes, a complete guidance system would require three integrating gyroscopes whose nutation axes are aligned with mutually orthogonal axes.

We must note in closing that our discussions of inertial guidance systems have been drastically simplified, through both the models we created and the assumptions used to obtain responses. For example, we generally idealized systems by neglecting the inertia of the gimbals. In some cases this merely affects oscillation frequencies. However, the additional inertial resistance can lead to qualitative differences. Such is the case for a free gyroscope that is subjected to a small disturbance. The inertia of the outer gimbal can cause a precession that drifts away from the initial orientation, rather than merely oscillating about it. In regard to the analysis of responses, linearization often avoids some important questions, such as loss of dynamic stability. Practical usage of the gyroscope as a tool for navigation over long ranges requires more sophisticated analyses, accounting for gimbal inertia and bearing friction, than those we have presented here. However, the features of such an investigation would show many similarities to the steps we have pursued.

EXAMPLE 8.6 ──

An airplane in level flight executes a body-fixed rotation about an axis that lies in the $\xi\eta$ plane in Figure 8.17, at angle γ from the η axis. The rotation rate Ω about this axis is a sinusoidal pulse over a time interval τ,

$$\Omega = \Omega_0 \sin\left(\frac{\pi t}{\tau}\right) \qquad \text{for } 0 \leqq t \leqq \tau$$

$$\Omega = 0 \qquad \text{for } t \geqq \tau$$

The rotor was spinning in its reference position when the aircraft began its rotation. Determine the nutational response $\theta(t)$ of the gyroscope for the case where damping is less than critical. From that solution determine the conditions for which the value of $K\theta/I\beta_\phi$ closely matches the nominal response in Eq. (8.95).

Solution. Because the rotation is about a body-fixed axis, the components of the angular velocity $\overline{\Omega}$ of the aircraft relative to the (body-fixed) $\xi\eta\zeta$ axes are constant at

$$\Omega_\xi = \Omega \sin \gamma \qquad \Omega_\eta = \Omega \cos \gamma \qquad \Omega_\zeta = 0$$

The response we seek is the solution to Eq. (8.92) for the above rotation of the base, subject to the initial conditions that $\theta = \dot{\theta} = 0$ when $t = 0$. There are several methods for determining this response, including the method of undetermined coefficients, Laplace transforms, and the convolution integral. We shall exploit the similarity of the problem to that encountered in conventional transient vibrations.

Substitution of the given functional form of Ω for $t < \tau$ leads to

$$I'\ddot{\theta} + C\dot{\theta} + K\theta = \Omega_0 \left[I\beta_\phi \cos \gamma \sin\left(\frac{\pi t}{\tau}\right) - \frac{I'\pi}{\tau} \sin \gamma \cos\left(\frac{\pi t}{t}\right) \right]$$

This represents a one-degree-of-freedom system that is being subjected to a sum of two sinusoidal excitations at frequency π/τ. We may construct the particular solution, known as the steady-state response in vibration theory, by multiplying the quasistatic response (based on $\tau >> \pi/\omega$) by a dynamic magnification factor M. Also, the sinusoidal response to each term is delayed by a phase lag δ. Hence, we construct the particular solution for the present response as

$$\theta_p = \Omega_0 \frac{M}{K} \left[I\beta_\phi \cos \gamma \sin\left(\frac{\pi t}{\tau} - \delta\right) - \frac{I'\pi}{\tau} \sin \gamma \cos\left(\frac{\pi t}{\tau} - \delta\right) \right]$$

where

$$M = \left\{ \left[1 - \left(\frac{\pi}{\omega\tau}\right)^2 \right]^2 + 4\sigma^2 \left(\frac{\pi}{\omega\tau}\right)^2 \right\}^{-1/2}$$

$$\delta = \tan^{-1} \left[\frac{2\sigma\pi/\omega\tau}{1 - (\tau/\omega\tau)^2} \right] \qquad 0 \leqq \delta \leqq \pi$$

We rewrite θ_p in a nondimensional form by substituting $K = I'\omega^2$, which leads to

$$\theta_p = \frac{\Omega_0}{\omega} M \left[\frac{I}{I'} \frac{\beta_\phi}{\omega} \cos \gamma \sin\left(\frac{\pi t}{\tau} - \delta\right) - \frac{\pi}{\omega\tau} \sin \gamma \cos\left(\frac{\pi t}{\tau} - \delta\right) \right]$$

The initial conditions must be satisfied by the combination of the particular and complementary solutions. We note for the latter that damping is less than critical, so $\sigma < 1$. The complementary solution, which is equivalent to the free vibration response, therefore consists of an oscillatory response that decays exponentially in time according to

$$\theta_h = \exp{(-\sigma\omega t)}[B \sin{(\omega_d t)} + C \cos{(\omega_d t)}] \qquad \omega_d = \omega(1 - \sigma^2)^{1/2}$$

Setting $\theta_p + \theta_h = 0$ at $t = 0$ yields

$$C = \frac{\Omega_0}{\omega} M \left(\frac{I}{I'} \frac{\beta_\phi}{\omega} \cos{\gamma} \sin{\delta} + \frac{\pi}{\omega\tau} \sin{\gamma} \cos{\delta} \right)$$

Similarly, the condition that $\dot{\theta}_p + \dot{\theta}_h = 0$ at $t = 0$ requires that

$$\omega_d B - \sigma\omega C = -\frac{\Omega_0}{\omega} M \frac{\pi}{\tau} \left(\frac{I}{I'} \frac{\beta_\phi}{\omega} \cos{\gamma} \cos{\delta} - \frac{\pi}{\omega\tau} \sin{\gamma} \sin{\delta} \right)$$

When we substitute the above expression for C into this relation, we obtain the value of B. The total response then consists of

$$\theta = \frac{\Omega_0}{\omega} M \left[\frac{I}{I'} \frac{\beta_\phi}{\omega} \cos{\gamma} \sin{\left(\frac{\pi t}{\tau} - \delta \right)} - \frac{\pi}{\omega\tau} \sin{\gamma} \cos{\left(\frac{\pi t}{\tau} - \delta \right)} \right]$$
$$+ \exp{(-\sigma\omega t)}[B \sin{(\omega_d t)} + C \cos{(\omega_d t)}] \qquad t < \tau \quad (1)$$

As we have noted, this is the response only as long as the body-fixed rotation is active. At $t = \tau$, the airplane's rotation ceases. However, since the rotor was nutating at that instant, the nutation angle and angular velocity must change continuously at $t = \tau$. It follows that the response for $t > \tau$ consists of a free vibration. Let θ^* and $\dot{\theta}^*$ be the angle and rotation rate, respectively, at $t = \tau$. (We find these quantities by solving the response in the interval $t < \tau$.) We may use the complementary solution, which we found earlier, to form the free-vibration response. The task of satisfying the continuity conditions at $t = \tau$ is expedited by measuring time by the variable $t - \tau$. Hence, we let

$$\theta = \exp{[-\sigma\omega(t - \tau)]}\{B^* \sin{[\omega_d(t - \tau)]} + C^* \cos{[\omega_d(t - \tau)]}\} \qquad t > \tau$$
$$(2)$$

where satisfying the continuity conditions yields

$$C^* = \theta^* \qquad \omega_d B^* - \sigma\omega C^* = \dot{\theta}^*$$

Given values of the system parameters, it would be a simple matter to use Eqs. (1) and (2) to determine the value of θ at any instant. For comparison, the nominal response given by Eq. (8.95) would be

$$\theta = \frac{I\beta_\phi}{I'\omega^2} \Omega_\eta$$

which for the given rotation becomes

$$\theta = \frac{\Omega_0}{\omega} \frac{I}{I'} \frac{\beta_\phi}{\omega} \cos\gamma \sin\left(\frac{\pi t}{\tau}\right) \qquad t < \tau \tag{3}$$

$$\theta = 0 \qquad t > \tau \tag{4}$$

We desire that the expression in Eq. (1) match closely Eq. (3). We observe that for any rotor $I'/I \gtrsim 0.5$, with the lower bound corresponding to a thin disk. Also, the values of the coefficients B and C and the amplitudes of the sinusoidal terms have comparable magnitudes. Therefore, if the two expressions are to match, the following conditions must apply:

(a) The exponential factor, representing the decay of the homogeneous solution, must become very small in a time interval much shorter than τ.

(b) The magnification factor M must be close to unity.

(c) The phase lag δ must be close to zero.

(d) The first term in Eq. (1) must be much larger than the second.

These, in turn, require that

$$\sigma\omega\tau \gg 1 \qquad \frac{\pi}{\omega\tau} \ll 1 \qquad \beta_\phi \gg \frac{\pi}{\tau}$$

In general, the spin rate will be much larger than the rate at which the aircraft rotates, so $\beta_\phi \approx \dot\phi$. The above conditions lead to the following requirements. The natural period of free vibration, $2\pi/\omega$, should be much smaller than the time interval τ over which the pulse occurs, the damping should be reasonably close to critical, and the spin rate should be much larger than the frequency at which the airplane's rotation fluctuates.

These requirements are not difficult to meet, because the spin, roll, and yaw motions of even a very high performance aircraft are moderate from a mechanical standpoint. For example, a very violent maneuver might consist of several rolls in a few seconds, for which τ might be of the order of 1 s. In contrast, a natural frequency of 1000 rad/s and a spin rate of 20,000 rev/min are readily attainable.

When the above criteria are met, differences between the responses in Eqs. (1) and (3), and between Eqs. (2) and (4), are significant only for early elapsed times. For example, if $\sigma\omega t$ in the initial phase, or $\sigma\omega(t - \tau)$ after the cessation of rotation, equals unity, the homogeneous solution will only have decayed to 37 percent of its initial magnitude. Clearly, designing the damping to be as close as possible to critical, $\sigma = 1$, can substantially reduce observable discrepancies to a negligible amount.

REFERENCES

R. N. Arnold and M. Maunder, *Gyrodynamics and Its Engineering Applications*, Academic Press, New York, 1961.

H. Goldstein, *Classical Mechanics*, 2d ed., Addison-Wesley, Reading, Mass., 1980.

A. Gray, *A Treatise on Gyrostatics and Rotational Motion*, Macmillan, London, 1918.

D. T. Greenwood, *Principles of Dynamics*, Prentice-Hall, Englewood Cliffs, N.J., 1965.

L. Meirovitch, *Methods of Analytical Dynamics*, McGraw-Hill, New York, 1970.

E. J. Routh, *Dynamics of a System of Rigid Bodies*, Part I, *Elementary Part*, 7th ed., Macmillan, New York, 1905.

E. J. Routh, *Dynamics of a System of Rigid Bodies*, Part II, *Advanced Part,* 6th ed., Macmillan, New York, 1905.

J. L. Synge and B. A. Griffith, *Principles of Mechanics,* 3d ed., McGraw-Hill, 1959.

HOMEWORK PROBLEMS

8.1 Prove that the polhode description of free motion for an arbitrary body reduces to the space and body cone analogy when the body is axisymmetric.

8.2 An axially symmetric earth satellite, whose ratio of principal moments of inertia is $I/I' = 1.6$, precesses about its axis once every 2 s. The spin rate in this state is 0.1 rad/s. Determine the overall rate of rotation and the angle from the axis of symmetry to the precession axis. Then determine the minimum angular impulse that a set of control rockets fastened to the satellite must exert in order to bring the precession axis into coincidence with the axis of symmetry. What is the rotation rate of the satellite at the conclusion of such a maneuver? (Assume that the control rockets act impulsively.)

8.3 The cylinder, whose mass is 2 kg, translates downward such that its axis of symmetry remains horizontal. The spin rate about that axis is 50 rad/s. The cylinder has a speed of 40 m/s when it collides with the ledge. Immediately after the impact, the center of mass of the cylinder has a downward velocity of 10 m/s. Describe the rotational motion of the cylinder after impact.

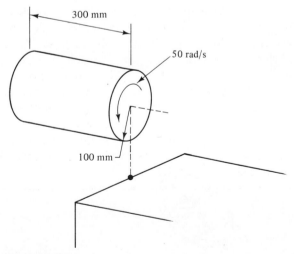

300 mm

50 rad/s

100 mm

Problem 8.3

8.4 The angular velocity of a wooden block at the instant it is released is as shown. Which body-fixed axis is surrounded by the polhode curve for the free rotation? What are the maximum and minimum angles between this axis and the constant direction of the angular

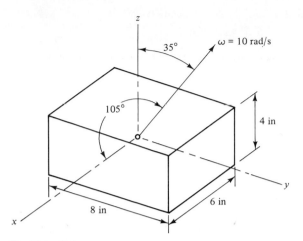

Problem 8.4

momentum? What are the angular velocities of the block at these maximum and minimum conditions?

8.5 The thin disk of mass m is welded to bar AB, which is fastened to the vertical shaft by a pin. The rotation rate of this shaft is $\dot{\psi}$. The mass of bar AB is negligible. Evaluate the stability of a steady precession about the vertical orientation of bar AB, $\theta = 0$, as a function of the precession rate $\dot{\psi}$ and the length ratio R/l.

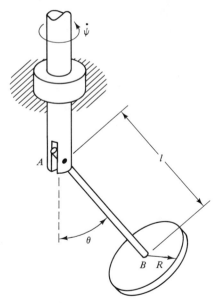

Problem 8.5

8.6 A free gyroscope is in a state of slow, steady precession at a nutation angle of 53.13°, and the spin rate at 10,000 rev/min. The rotor weighs 10 lb, its radii of gyration about its pivot are $\kappa = 5$ in., $\kappa' = 8$ in., and its center of mass is 2 in. from the pivot. A person accidentally touches the outer gimbal, causing the precession rate to decrease suddenly by 0.6 rad/s. Determine whether the ensuing motion is unidirectional, looping, cuspidial, or steady precession. What are the maximum and minimum nutation angles in that motion?

8.7 A free symmetric gyroscope initially in a state of steady slow precession is subjected to a small disturbing torque $\varepsilon mgL \sin \Omega t$ acting about the fixed vertical shaft supporting the outer gimbal. Use a perturbation analysis for $\varepsilon << 1$ to determine the frequency, if any, at which the system resonates.

8.8 The device shown is a *gyropendulum*, which is a system used in some inertial guidance applications to locate the vertical direction. The spin rate $\dot{\phi}$ is held constant by a servomotor. Let m be the mass, and let I and I' be the centroidal principal moments of inertia of the flywheel parallel and transverse to the spin axis; ignore the inertia of the gimbals. Evaluate the nutation response $\theta(t)$ and the precessional response $\psi(t)$ of the flywheel to a disturbance that causes it to rotate by very small angles away from the vertical reference position, at which $\psi = 0$ and $\theta = 90°$. Compare the frequency of these responses with that of a simple pendulum, and use that result to discuss an advantage of the gyropendulum.

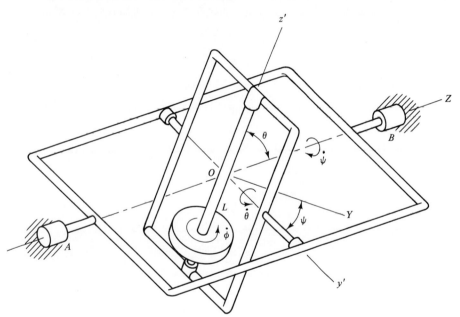

Problem 8.8

8.9 Consider the gyropendulum in Problem 8.8. Because of movement of the vehicle, the center point O has a constant acceleration $\dot{v}$ directed parallel to the axis of the outer gimbal (that is, $\bar{a}_0 = \dot{v}\bar{e}_{B/A}$). Let this acceleration be directed at angle β north of east. Derive equations of motion for the Eulerian angles including the effect of the earth's rotation and the movement of the vehicle in a great circle.

8.10 The platform of an integrating gyroscope is rotated about the η axis in a time-dependent manner. Consider an angular speed that consists of an average value Ω_0, over which is superposed a harmonic fluctuation at amplitude Ω_1 and frequency λ, that is, $\Omega_\eta = \Omega_0 + \Omega_1 \sin (\lambda t)$. What conditions must be true if the nutational response $\theta(t)$ following the initial transient phase is to be proportional to the mean rotation $\Omega_0 t$?

8.11 A top is initially in a state of steady precession at a precession rate $\dot\psi^*$ and a nutation angle θ^*. The precession rate is suddenly increased by the amount $\Delta\dot\psi$, owing to the application of an impulsive force. Determine the precessional and nutational responses after cessation of the impulsive force. Use a perturbation analysis of the basic equations for a top, Eqs. (8.35) to (8.38), in which the small parameter is $\epsilon = \Delta\dot\psi/\dot\psi^*$.

8.12 *Whirling* is a phenomenon in turbomachinery in which a rotating shaft undergoes deformation as a beam. In order to study this effect, consider the shaft supporting the disk of mass m and radius R to be flexible in bending, rigid in extension and torsion, and massless. The disk is welded to the shaft, such that its center C is on the centerline of the shaft. The rotation rate of the shaft about its bearing is constant at Ω. Let $x'y'z'$ be a reference frame that rotates at this rate, with its z axis concurrent with the line connecting the bearings. Let the deflected position of the center of the disk relative to $x'y'z'$ be $\bar{r}_{C/O'} = \xi\bar{i}' + \eta\bar{j}'$. Furthermore, let xyz be principal centroidal axes for the disk, and let β and γ be the rotations about the x' and y' axes, respectively, of xyz relative to $x'y'z'$. For a shaft of symmetric cross section and a rotor situated at the midpoint of the shaft, each deformation is resisted solely by a corresponding proportional elastic force or moment; the elastic constants are k_ξ, k_η, k_β, and k_γ. Derive the corresponding linearized equations of motion.

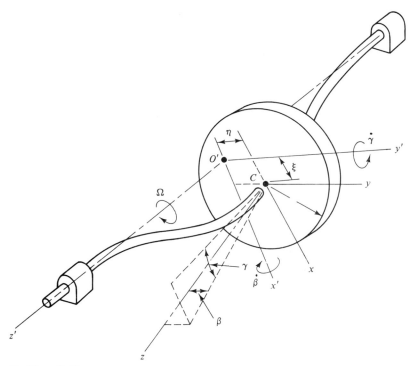

Problem 8.12

8.13 Consider the effects of the inertia of the gimbals in a balanced free gyroscope ($L = 0$). Let A, B, and C denote the (principal) moments of inertia of the inner gimbal about the xyz axes, where y is the line of nodes and z is the spin axis of the rotor. Also, let A' denote the moment of inertia of the outer gimbal about the precession axis. Derive the equations of motion for the gyroscope in this case.

8.14 Suppose that the gyroscope in Problem 8.13 is initially spinning at $\dot{\phi}$ and the nutation angle is constant at θ_0; there is no precession in this initial motion. At $t = 0$, a nutational velocity $\varepsilon << \dot{\phi}$ is imparted to the inner gimbal. Use a perturbation analysis in which $\theta = \theta_0 + \varepsilon\theta_1 + \varepsilon^2\theta_2$ and $\dot{\psi} = \varepsilon\dot{\psi}_1 + \varepsilon^2\dot{\psi}_1$ to determine the nutational and precessional fluctuations induced by the disturbance. Show that, because of gimbal inertia, the response exhibits *gimbal walk*, in which there is a nonzero average precessional rotation rate, even though the gyroscope is balanced.

Centroidal Inertia Properties

Slender Bar

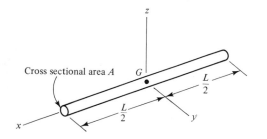

$$V = LA$$
$$I_{xx} = 0$$
$$I_{yy} = I_{zz} = \tfrac{1}{12}\, m\, L^2$$

Rectangular Prism

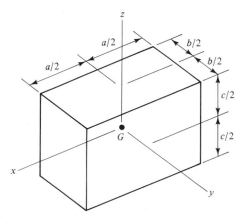

$$V = abc$$
$$I_{xx} = \tfrac{1}{12}\, m(b^2 + c^2)$$
$$I_{yy} = \tfrac{1}{12}\, m(a^2 + c^2)$$
$$I_{zz} = \tfrac{1}{12}\, m(a^2 + b^2)$$

Cylinder

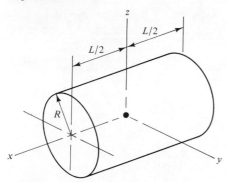

$$V = \pi R^2 L$$
$$I_{xx} = \tfrac{1}{2} mR^2$$
$$I_{yy} = I_{zz} = \tfrac{1}{12} m(3R^2 + L^2)$$

Thin Disk

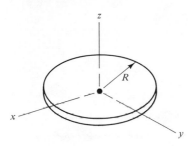

$$V = \pi R^2 h$$
$$I_{xx} = I_{yy} = \tfrac{1}{4} mR^2$$
$$I_{zz} = \tfrac{1}{2} mR^2$$

Semicylinder

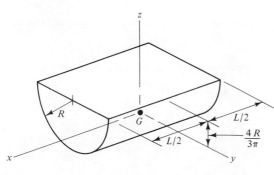

$$V = \tfrac{1}{2} \pi R^2 L$$
$$I_{xx} = \frac{9\pi^2 - 32}{18\pi^2} mR^2$$
$$I_{yy} = \frac{9\pi^2 - 64}{36\pi^2} mR^2 + \tfrac{1}{12} mL^2$$
$$I_{zz} = \tfrac{1}{12} m(3R^2 + L^2)$$

Sphere

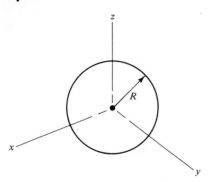

$$V = \tfrac{4}{3}\,\pi R^2$$
$$I_{xx} = I_{yy} = I_{zz} = \tfrac{2}{5}\,mR^2$$

Hemisphere

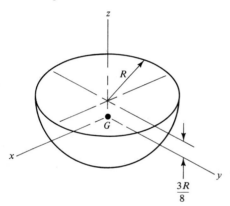

$$V = \tfrac{2}{3}\,\pi R^2$$
$$I_{xx} = I_{yy} = \tfrac{83}{320}\,mR^2$$
$$I_{zz} = \tfrac{2}{5}\,mR^2$$

Ellipsoid

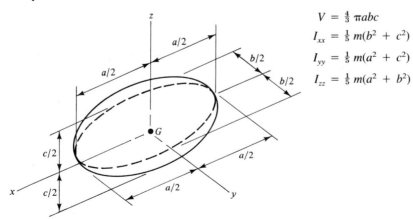

$$V = \tfrac{4}{3}\,\pi abc$$
$$I_{xx} = \tfrac{1}{5}\,m(b^2 + c^2)$$
$$I_{yy} = \tfrac{1}{5}\,m(a^2 + c^2)$$
$$I_{zz} = \tfrac{1}{5}\,m(a^2 + b^2)$$

Cone

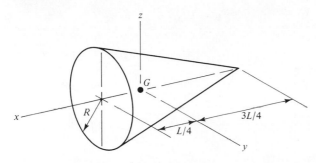

$$V = \tfrac{1}{3}\,\pi R^2 L$$
$$I_{xx} = \tfrac{3}{10}\,mR^2$$
$$I_{yy} = I_{zz} = \tfrac{3}{80}\,m(4R^2 + L^2)$$

Semicone

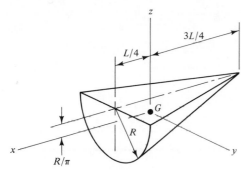

$$V = \tfrac{1}{6}\,\pi R^2 L$$
$$I_{xx} = \left(\frac{3}{10} - \frac{1}{\pi^2}\right) mR^2$$
$$I_{yy} = \left(\frac{3}{20} - \frac{1}{\pi^2}\right) mR^2 + \tfrac{3}{80}\,m\,L^2$$
$$I_{zz} = \tfrac{3}{80}\,m(4R^2 + L^2)$$
$$I_{xz} = -\frac{1}{20\pi}\,mRL \qquad I_{xy} = I_{yz} = 0$$

Right Triangular Prism

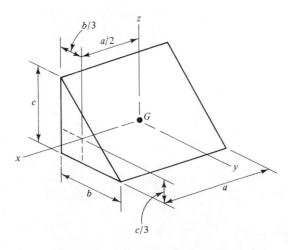

$$V = \tfrac{1}{2}\,abc$$
$$I_{xx} = \tfrac{1}{18}\,m(b^2 + c^2)$$
$$I_{yy} = \tfrac{1}{36}\,m(3a^2 + 2c^2)$$
$$I_{zz} = \tfrac{1}{36}\,m(3a^2 + 2b^2)$$
$$I_{yz} = -\tfrac{1}{36}\,mbc \qquad I_{xy} = I_{xz} = 0$$

Orthogonal Tetrahedron

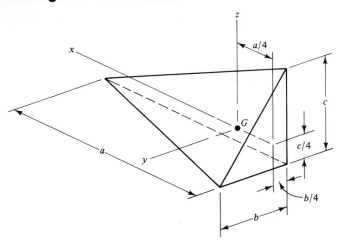

$$V = \tfrac{1}{6}\,abc$$
$$I_{xx} = \tfrac{3}{80}\,m(b^2 + c^2)$$
$$I_{yy} = \tfrac{3}{80}\,m(a^2 + c^2)$$
$$I_{zz} = \tfrac{3}{80}\,m(a^2 + b^2)$$
$$I_{xy} = -\tfrac{1}{80}\,mab$$
$$I_{xz} = -\tfrac{1}{80}\,mac$$
$$I_{yz} = -\tfrac{1}{80}\,mbc$$

Conversion Factors from U.S. Customary Units to SI Units

CONVERSION FACTORS FROM U.S. CUSTOMARY UNITS TO SI UNITS

To convert from U.S. unit	To SI unit	Multiply by
Length		
foot (ft)	meter (m)	0.3048*
inch (in.)	meter (m)	0.0254*
statute mile (mi)	meter (m)	1609.3
Speed		
foot/second (ft/s)	meter/second (m/s)	0.3048*
mile/hour (mi/h)	meter/second (m/s)	0.44704
mile/hour (mi/h)	kilometer/hour (km/h)	1.6093
knot (nautical mi/h)	meter/second (m/s)	0.51444
Acceleration		
foot/second² (ft/s²)	meter/second² (m/s²)	0.3048*
inch/second² (in./s²)	meter/second² (m/s²)	0.0254*
Force (1 N = 1 kg-m/s²)		
pound (lb)	newton (N)	4.4482

To convert from U.S. unit	To SI unit	Multiply by
Mass		
slug (lb-s^2/ft)	kilogram (kg)	14.594
pound-mass (lbm)	kilogram (kg)	0.45359
Moment of Inertia		
slug-ft^2 (lb-s^2-ft)	kilogram-meter^2 (kg-m^2)	1.3558
Density		
slug/foot^3 (lb-s^2/ft^4)	kilogram/meter^3 (kg/m^3)	515.38
pound-mass/foot^3 (lbm/ft^3)	kilogram/meter^3 (kg/m^3)	16.018
Work and Energy (1 joule = meter-newton)		
foot-pound (ft-lb)	joule (J)	1.3558
British thermal unit (Btu)	joule (J)	1055.1
Power (1 watt = 1 joule/second)		
foot-pound/second (ft-lb/s)	watt (W)	1.3558
horsepower (hp)	watt (W)	745.70
Pressure and Stress (1 pascal = 1 newton/meter^2)		
pound/foot^2 (lb/ft^2)	pascal (Pa)	47.880
pound/inch^2 (psi)	pascal (Pa)	6894.8
atmosphere (= 14.7 psi)	pascal (Pa)	1.0133(10^5)

*Denotes an exact factor.

Answers to Even-Numbered Problems

CHAPTER 2

2.2. $\bar{v} = \sqrt{2gh}\,\bar{e}_t$, $\bar{a} = g \cos \dfrac{s}{R}\bar{e}_t + 2g \sin \dfrac{s}{R}\bar{e}_n$

2.4. $\bar{v} = v_0 (1 - k \sinh x)\,\bar{e}_t$

$\bar{a} = v_0^2 (1 - k \sinh x) \left[-k\bar{e}_t + \dfrac{1 - k \sinh x}{\cosh^2 x}\bar{e}_n \right]$

2.6. $\bar{v} = k(-0.890\bar{i} + 4.839\bar{j} + 0.890\bar{k})$

$\bar{a} = k^2(0.771\bar{i} - 7.611\bar{j} - 0.727\bar{k})$

2.8. $\bar{v} = 2(kt \sin \omega t^2 + \omega kt^3 \cos \omega t^2)\bar{i} + 4kt^3\bar{j}$
$\qquad + 2 (kt \cos \omega t^2 - \omega kt^3 \sin \omega t^2)\bar{k}$

$\bar{a} = [10\omega kt^2 \cos \omega t^2 - (4\omega^2 kt^4 - 2k) \sin \omega t^2]\bar{i}$
$\qquad + 12kt^2\bar{j} + [-10\omega kt^2 \sin \omega t^2 - (4\omega^2 kt^4 - 2k) \cos \omega t^2]\bar{k}$

2.10. $D = \dfrac{2u^2 \sin (\theta + \beta) \cos \beta}{g \cos^2 \theta}$, $H = \dfrac{u^2 \sin^2 (\theta + \beta)}{2g \cos \theta}$

2.12. $\bar{r} = v_0\tau \cos \beta\, [1 - \exp(-t/\tau)]\,\bar{i}$
$\qquad + \tau\{(v_0 \sin \beta + g\tau)[1 - \exp(-t/\tau)] - gt\}\,\bar{j}$

$\bar{v} = v_0 \cos \beta \exp(-t/\tau)\bar{i} + [(v_0 \sin \beta + g\tau) \exp(-t/\tau)] - g\tau]\bar{j}$
where $\tau = m/c$

2.14. $\bar{v} = 1.773\omega\sqrt{bc}\,\bar{e}_R + 33.41\omega\sqrt{bc}\,\bar{e}_\theta + 10.883\omega b\bar{k}$

$\bar{a} = -366.7\omega^2\sqrt{bc}\,\bar{e}_R - 12.556\omega^2 b\bar{k}$

2.16. $\bar{v} = \dot{\theta}(R'\bar{e}_R + R\bar{e}_\theta)$

$\bar{a} = [(R'' - R)\bar{e}_R + 2R'\bar{e}_\theta]\,\dot{\theta}^2$

$\rho = \dfrac{[(R')^2 + R^2]^{3/2}}{|R''R - 2(R')^2 - R^2|}$

2.18. $\dot{r} = -404.1$ m/s, $\dot{\phi} = 0.3157$ rad/s, $\dot{\theta} = 0.18940$ rad/s

$\ddot{r} = 10.358$ m/s^2, $\ddot{\phi} = 0.3189$ rad/s^2, $\ddot{\theta} = 0.1913$ rad/s^2

2.20. $F_\phi = 11.517$ newtons, $F_\theta = 3.770$ newtons

2.22. $\bar{v} = \dot{\rho}\bar{e}_\rho + (r + \rho \cos \psi) \dot{\theta}\bar{e}_\theta + \rho\dot{\psi}e_\psi$

$\quad \bar{a} = [\ddot{\rho} - (r + \rho \cos \psi) \dot{\theta}^2 \cos \psi - \rho\dot{\psi}^2]\bar{e}_\rho$

$\qquad + [(r + \rho \cos \psi) \ddot{\theta} + 2\dot{\rho}\dot{\theta} \cos \psi - 2\rho\dot{\theta}\dot{\psi} \sin \psi]\bar{e}_\theta$

$\qquad + [\rho\ddot{\psi} + 2\dot{\rho}\dot{\psi} + (r + \rho \cos \psi) \dot{\theta}^2 \sin \psi]\bar{e}_\psi$

2.24. $\dot{v} = 43.96$ ft/s^2, $\rho = 12903$ ft,

$\quad \bar{r}_{C/P} = -11902\bar{i} + 1704\bar{j} - 4683\bar{k}$ ft

2.26. $\dot{\phi} = -0.7673$ rad/s, $\ddot{\phi} = -0.1838$ rad/s^2

CHAPTER 3

3.2. $\Delta\bar{r} = -12.853\bar{I} - 5.208\bar{J} + 5.583\bar{K}$ in.

3.4. $\bar{r}_C = -0.579\bar{I} + 4.449\bar{J} - 0.010\bar{K}$ in.

3.6. $\bar{r}_D = 76.60\bar{I} + 155.67\bar{J} - 141.06\bar{K}$ mm

3.8. $\bar{\omega} = -104.72\bar{j} + 0.660\bar{k}$ rad/s

$\quad \bar{\alpha} = 69.12\bar{i}$ rad/s^2, $\bar{j} = \bar{e}_t$

3.10. $\bar{\omega} = 8.66\bar{i} - 6\bar{j} + 623.3\bar{k}$ rad/s

$\quad \bar{\alpha} = -3713\bar{i} - 5351\bar{j} - 102\bar{k}$ rad/s^2; $\bar{k} = \bar{e}_{\omega2}$

3.12. $\bar{\omega} = -10\bar{i} + 1.2566\bar{j} + 123.99\bar{k}$ rad/s

$\quad \bar{\alpha} = 207.9\bar{i} + 1273.4\bar{j} + 12.6\bar{k}$ rad/s^2; $\bar{k} =$ axis of disk

3.14. $\bar{v} = [u \cos \theta - \omega R(1 - \cos \theta)]\bar{i} + (u + \omega R) \sin \theta\bar{j}$

$\quad \bar{a} = -\left(\dfrac{u^2}{R} + 2\omega u + \omega^2 R\right) \sin \theta\bar{i} + \left[\left(\dfrac{u^2}{R} + 2\omega u\right) \cos \theta - \omega^2 R(1 - \cos \theta)\right]\bar{j}$

3.16. $\bar{v}_B = \omega_2 L \sin \theta\bar{i} + \omega_1 L \cos \theta\bar{j} - \omega_1 L \sin \theta\bar{k}$

$\quad \bar{a}_B = 2\omega_1\omega_2 L \cos \theta\bar{i} - (\omega_1^2 + \omega_2^2) L \sin \theta\bar{j} - \omega_1^2 L \cos \theta\bar{k}$

3.18. $\bar{v} = -3.371\bar{i} - 2.451\bar{j} - 10.828\bar{k}$ m/s

$\quad \bar{a} = -158.2\bar{i} - 193.8\bar{j} + 113.4\bar{k}$ m/s^2; $\bar{j} = \bar{e}_{B/C}$, $\bar{k} = \bar{K}$

3.20. $\bar{v}_C = -0.4\dot{\theta}\bar{i} + 2\dot{\theta}\bar{j} + \dot{\psi} (0.4 \sin \theta - 2 \cos \theta)\bar{k}$ m/s

$\quad \bar{v}_D = 0.4\dot{\psi} \cos \theta\bar{i} + (2\dot{\theta} - 0.4\dot{\psi} \sin \theta)\bar{j} - 2\dot{\psi} \cos \theta\bar{k}$

$\quad \bar{a}_C = (-0.4\ddot{\theta} - 2\dot{\theta}^2 + 0.4\dot{\psi}^2 \sin \theta \cos \theta - 2\dot{\psi}^2 \cos^2\theta)\bar{i}$

$\qquad + (2\ddot{\theta} - 0.4\dot{\theta}^2 - 0.4\dot{\psi}^2 \sin^2\theta + 2\dot{\psi}^2 \sin \theta \cos \theta)\bar{j}$

$\qquad + \dot{\psi}\dot{\theta} (4 \sin \theta + 0.8 \cos \theta)\bar{k}$ m/s^2

$\quad \bar{a}_D = (-2\dot{\theta}^2 - 2\dot{\psi}^2 \cos^2\theta)\bar{i} + (2\ddot{\theta} + 2\dot{\psi}^2 \sin \theta \cos \theta)\bar{j}$

$\qquad + (4\dot{\psi}\dot{\theta} \sin \theta - 0.4\dot{\psi}^2)\bar{k}$

$\quad$ where $\dot{\psi} = \dfrac{\pi}{450}$ rad/s, $\theta = \dfrac{\pi}{3} \sin\left(\dfrac{\pi t}{7200}\right)$ rad and $\bar{i} =$ axis of the scope

3.22. $\bar{N} = -3304\bar{i} + 779\bar{k}$ lb, $\bar{a} = -3543\bar{i} - 6\bar{j} + 804\bar{k}$ ft/s^2; $\bar{k} = \bar{e}_\theta$

3.24. $\bar{v} = -2.87L\bar{i} + 17.84L\bar{j} - 0.03L\bar{k}$

$\bar{a} = -165.09L\bar{i} - 24.91L\bar{j} + 16.13L\bar{k}; \bar{i} = \bar{e}_{B/A}$

3.26. $\bar{v}_{P/G} = -\bar{i} + 3\bar{j} - 2.5\bar{k}$ m/s

$\bar{a}_{P/G} = -0.65\bar{i} + 5.2\bar{j} + 6.5\bar{k}$ m/s^2, $\bar{i}$ = longitudinal axis

3.28. $(\bar{a}_C)_{xyz} = -\omega_1^2 s \sin\theta\bar{j} - 2\omega_1 v_C \sin\theta\bar{k}$, $\bar{i}$ = axis of cylinder

3.30. $d^2 + \left(s - \dfrac{u}{2\omega_e \sin\lambda}\right)^2 = \left(\dfrac{u}{2\omega_e \sin\lambda}\right)^2$

3.32. (a) $x = 0$, $y = -0.9428\omega_e(H^3/g)^{1/2}\cos\lambda$ (east of O')

(b) $x = 0$, $y = 0.4714\omega_e(H^3/g)^{1/2}(1 + 3R_e/H)\cos\lambda$ (west of O')

CHAPTER 4

4.2. $[R] = \begin{bmatrix} -0.5797 & 0.4731 & -0.6636 \\ -0.0400 & -0.8297 & -0.5567 \\ -0.8138 & -0.2962 & 0.50 \end{bmatrix}$

4.4. $\omega_{AB} = 0.866$ rad/s (cw), $\omega_{CD} = 0.6928$ rad/s (ccw)

$\alpha_{AB} = 21.22$ rad/s^2 (ccw), $\alpha_{CD} = 23.00$ rad/s^2 (cw)

4.6. $v_A = 3$ m/s (down), $\dot{\beta} = 1.3416$ rad/s

4.8. $\bar{v}_C = 25.72\bar{i} - 60.32\bar{j} - 4.29\bar{k}$ m/s; $\bar{k}$ is vertical

4.10. $\bar{v}_C = u\bar{i} - 0.5774u\bar{j} - 0.5\Omega R\bar{k}$

$\bar{a}_C = -0.5\Omega^2 R\bar{i} - 1.5396\dfrac{u^2}{R}\bar{j} - 2\Omega u\bar{k}$; $\bar{k}$ is vertical

4.12. $\bar{\omega}_{BC} = -100\bar{k}$ rad/s, $\bar{\alpha}_{BC} = 10^4(3.464\bar{I} - 4\bar{J} + 2.309\bar{K})$ rad/s^2; $\bar{K}$ is vertical

4.14. $\bar{\omega}_{BC} = 2\dfrac{v}{R}\sin^2(\theta/2)$ cw, $\bar{v}_E = v(1 + \cos\theta)$ to right

4.16. $\bar{\omega} = \dfrac{v}{\rho}\cos\theta\bar{i} + \dfrac{v}{R}\bar{k}$

$\bar{\alpha} = -\dfrac{v^2}{\rho R}\left(1 + \dfrac{R}{\rho}\sin\theta\right)\bar{j}$; $\bar{k}$ is the axis of the disk

4.18. $\bar{\omega}_A = \omega_1 \sin(\beta + \gamma)\bar{i} - \left[\omega_1 \cos(\beta + \gamma) + (\omega_1 - \omega_2)\dfrac{\sin\gamma}{\sin\beta}\right]\bar{k}$

$\bar{\alpha}_A = \dot{\omega}_1 \sin(\beta + \gamma)\bar{i} + \omega_1(\omega_1 - \omega_2)\dfrac{\sin\gamma}{\sin\beta}\sin(\beta + \gamma)\bar{j}$

$- \left[\dot{\omega}_1 \cos(\beta + \gamma) + (\dot{\omega}_1 - \dot{\omega}_2)\dfrac{\sin\gamma}{\sin\beta}\right]\bar{k}$; $\bar{k}$ = axis of gear A

4.20. $\bar{\omega} = -12.116\bar{i} + 20.99\bar{k}$ rad/s

$\bar{\alpha} = 79.31\bar{j}$ rad/s^2; $\bar{k}$ = axis of body cone

4.22. $\dot{\psi} = -16.400$ rad/s and $\dot{\theta} = -5.282$ rad/s

or

$\dot{\psi} = -5.90$ rad/s and $\dot{\theta} = -14.680$ rad/s

4.24. $\bar{\omega} = \omega_1 \cos \beta \bar{i} - \dfrac{u}{R} \dfrac{1}{\sin \beta - 2 \cos \beta} \bar{j} - 2\omega_1(1 + \cos \beta)\bar{k}$

$\bar{\alpha} = 2 \dfrac{u\omega_1}{R} \dfrac{(1 + \cos \beta - \sin \beta)}{\sin \beta - 2 \cos \beta} \bar{i} + \left[\dfrac{u^2}{R^2} \dfrac{\cos \beta + 2 \sin \beta}{(\sin \beta - \cos \beta)^3} \right.$

$\left. - \omega_1^2(2 + 2 \cos \beta - \sin \beta) \right] \bar{j} + \dfrac{u\omega_1}{R} \left(\dfrac{2 \sin \beta}{\sin \beta - 2 \cos \beta} \right) \bar{k}$

where $\bar{k}$ is the axis of the disk

CHAPTER 5

5.2. $I_{xx} = \frac{1}{6}m\left[a^2 + b^2 + \dfrac{a + 3b}{a + b} c^2 \right]$, $I_{xy} = \frac{1}{6}md \dfrac{a^2 + ab + b^2}{a + b}$

5.4. $[I] = \begin{bmatrix} 0.0388 & 0.0208 & 0 \\ 0.0208 & 0.3812 & 0 \\ 0 & 0 & 0.420 \end{bmatrix}$

5.6. $I_1 = 0.1136$, $I_2 = 1.3864$, $I_3 = 1.5$ kg-m^2

5.8. $I_1 = 0.2919$, $I_2 = 2.9817$, $I_3 = 3.1715$ lb-s^2-ft

$[R] = \begin{bmatrix} 0.8819 & 0.2667 & -0.3888 \\ 0.4647 & -0.6311 & 0.6211 \\ 0.0797 & 0.7287 & 0.6805 \end{bmatrix}$

5.10. $[I] = \begin{bmatrix} 0.1520 & 0.0540 & 0 \\ 0.0540 & 0.2330 & 0 \\ 0 & 0 & 0.225 \end{bmatrix}$ kg-m^2

$\bar{A} = \omega^2(-0.0675\bar{j} + 0.60\bar{k})$, $\bar{B} = \omega^2(0.0675\bar{j} + 0.60\bar{k})$ newtons

5.12. $\bar{M}_A = \frac{1}{16}mR^2[-6\omega_1\omega_2 + \sqrt{3}(\omega_1^2 - \omega_2^2)]\bar{i}$; $\bar{i}$ is outward

5.14. (a) $13.19°$ between $\bar{\omega}$ and AB, $4.87°$ between $\bar{H}_G$ and AB

(b) $\dot{\bar{H}}_G = -0.2242mR^2\omega_1^2\bar{k}$; $\bar{k}$ is outward

(c) $\bar{A} = -\bar{B} = 0.561mR\omega_1^2\bar{j}$

5.16. $(\frac{1}{4}R^2 + L^2)\ddot{\theta} + \frac{1}{2}R^2\Omega\dot{\psi} \sin \theta + (\frac{1}{4}R^2 - L^2)\dot{\psi}^2 \sin \theta \cos \theta + gL \sin \theta = 0$

$C = m[\frac{1}{2}R^2 \cos^2 \theta + (\frac{1}{4}R^2 + L^2) \sin^2 \theta]\ddot{\psi} + 2m(L^2 - \frac{1}{4}R^2)\dot{\theta}\dot{\psi} \sin \theta \cos \theta$
$- \frac{1}{2}mR^2 \Omega\dot{\theta} \sin \theta$

5.18. $\frac{1}{3}L^2\ddot{\theta} + L\Omega^2 \sin \theta(\frac{1}{2} + \frac{1}{3} \cos \theta) = \frac{1}{2}gL \cos \theta$

5.20. $N = mg \dfrac{2(1 + \cos \gamma)}{2 + 2 \cos \gamma - \sin \gamma}$

$+ \frac{1}{4}mR\Omega^2 \left[\dfrac{4 \cos \gamma + 4 \cos^2 \gamma - 16 \sin \gamma - 17 \sin \gamma \cos \gamma}{2 + 2 \cos \gamma - \sin \gamma} \right]$

5.22. $\omega_1^2 > \dfrac{5}{7}\dfrac{g}{R}$

5.24. $(\frac{1}{12} + \frac{1}{4}\cos^2\theta)\,\ddot{\theta} - \frac{1}{4}\dot{\theta}^2\sin\theta\cos\theta + \frac{1}{2}\dfrac{g}{L}\cos\theta = 0$

$y_G = \dfrac{L}{2}\sin\theta,\ \ddot{x}_G = 0,\ N = mg + \frac{1}{2}mL(\ddot{\theta}\cos\theta - \dot{\theta}^2\sin\theta)$

5.26. $\mu_{min} = 0.1136;\ \dot{\omega} = 56.3\ \mathrm{rad/s^2}$

5.28. $\dot{v} = 0.0585g$

5.30. $\dot{\theta} = 6.549\ \mathrm{rad/s}$

5.32. $\theta_{max} = 48.8°$

5.34. $\overline{\omega} = 3.98\bar{i} + 7.824\bar{k}\ \mathrm{rad/s};\ \bar{k}$ is vertical

5.36. $\Omega = 0.949(g/L)^{1/2},\ \beta = 46.6°$

$(\bar{v}_G)_2 = 1.032(gL)^{1/2},\ \overline{\omega}_2 = \overline{0},\ \overline{F} = 0.1149m(gL)^{1/2}/\Delta t$

5.38. $\omega_B = \omega_A - \dfrac{2gL}{3R^2\dot{\psi}}$

5.40. $\Delta N = 2mv^2\,\dfrac{\kappa^2}{w\rho r}$

CHAPTER 6

6.2. $\dot{X} = \dot{x} + u,\ \dot{Y} = 2\beta x\dot{x}$

6.4. $\dot{R} = \dot{l}\sin\theta + l\dot{\theta}\cos\theta,\ \dot{\phi} = -\Omega,\ \dot{z} = \dot{l}\cos\theta - l\dot{\theta}\sin\theta$

6.6. $\dot{X} = \dot{D}\sin\theta + \dot{\theta}(D\cos\theta - R\sin\theta)$

$\dot{Y} = -\dot{D}\cos\theta + \dot{\theta}(D\sin\theta + R\cos\theta),\ \omega = \dot{\theta} - \dot{D}/R$

6.8. $m\ddot{x} + kx\left[1 - \dfrac{0.8L}{(L^2 + x^2)^{1/2}}\right] = 0$

6.10. $\frac{1}{3}mL^2\ddot{\theta} + 2kb^2\sin\dfrac{\theta}{2} - mg\dfrac{L}{2}\cos\theta = 0$

6.12. $3mL^2\ddot{\theta} + 2kL^2(\sqrt{2}\sin\theta - \sin 2\theta) + (\frac{5}{2}mg + 2P)L\cos\theta = 0$

6.14. $\frac{1}{3}L\ddot{\theta} + \frac{1}{2}(\ddot{y} + g)\sin\theta = 0$

6.16. $6\ddot{s}_1 + \ddot{s}_2\cos\theta + 8\dfrac{k}{m}s_1 = 0,\ \ddot{s}_2 + \ddot{s}_1\cos\theta + 3\dfrac{k}{m}s_2 = \frac{1}{2}mg$

$s_1 = $ displacement of cart

6.18. $\ddot{\theta}U + \frac{1}{2}\dot{\theta}^2\dfrac{dU}{d\theta} + \dfrac{5g}{2L}\cos\theta + \dfrac{P}{\sigma L^2}\sin\theta\left[1 + \dfrac{\cos\theta}{(16 - \sin^2\theta)^{1/2}}\right] = 0$

where $U = \frac{4}{3} + 4\sin^2\theta + 6\dfrac{\sin^2\theta\cos\theta}{(16 - \sin^2\theta)^{1/2}}$

$+ (2\sin^2\theta + \frac{16}{3})\dfrac{\cos^2\theta}{(16 - \sin^2\theta)}$

6.20. $m_1(1 + 4\sigma_1^2/9R^2)\ddot{x}_1 + 2k(2x_1 - 3x_2) = 0$

$m_2(1 + \sigma_2^2/R^2)\ddot{x}_2 + 3k(3x_2 - 2x_1) = 0$

6.22. $\frac{1}{2}\ddot{\theta}[(3m_A - m_B)\sin^2\theta + m_B(3 - 2\sin\theta)]$

$+ \frac{1}{2}\dot{\theta}^2\cos\theta[(3m_A + m_B)\sin\theta - m_B] + \dfrac{m_Bg}{R_A + R_B}\cos\theta = 0$

where θ is the angle of elevation of the line of centers

6.24. $mR[4\ddot{\theta} + 2\omega^2\sin 2\theta + g\sin(2\theta + \omega t)] = 2F\sin\theta$

where θ is the angle between the cable and diameter AO

6.26. $3m\ddot{l} - 2ml\dot{\psi}^2\sin^2\beta + mg(1 - 2\cos\beta) = 0, \ 2ml^2\sin^2\beta\ \ddot{\psi} = M$

where l is the distance to a block on the inclined arm

6.28. $I_2\ddot{\theta} + \frac{1}{2}(I_1 - I_2)\omega_1^2\sin 2\theta + I_1\omega_1\omega_2\sin\theta = 0$

6.30. $\frac{1}{4}mR^2[(2\cos^2\theta + \sin^2\theta)\ \dot{\Omega} + 2\dot{\omega}_1\cos\theta] = C$

6.32. $(1 + 8\cos^2\theta)\ddot{\theta} - (8\dot{\theta}^2 + 9\Omega^2)\sin\theta\cos\theta + \dfrac{4F - 2mg}{mL}\sin\theta = 0$

6.34. $\frac{7}{3}[\ddot{\theta} - \dot{\psi}^2\sin\theta\cos\theta + \dfrac{3g}{L}\sin\theta = 0$

$[\frac{7}{3}mL^2\sin^2\theta + I]\ \ddot{\psi} + \frac{14}{3}mL^2\dot{\psi}\dot{\theta}\sin\theta\cos\theta = \Gamma$

where θ is the angle of bar AB from the vertical

CHAPTER 7

7.2. $\frac{1}{3}(\ddot{\theta} - \dot{\psi}^2\sin\theta\cos\theta) + \dfrac{g}{2L}\sin\theta = 0$

$\frac{1}{3}mL^2[\ddot{\psi}\sin^2\theta + 2\dot{\psi}\dot{\theta}\sin\theta\cos\theta] = \Gamma, \ \dot{\psi} = \Omega$

7.4. $\ddot{x}_G = -\mu\left(g + \dfrac{L}{2}\ddot{\theta}\cos\theta - \dfrac{L}{2}\dot{\theta}^2\sin\theta\right)\mathrm{sgn}\left(\dot{x}_G + \dfrac{L}{2}\dot{\theta}\sin\theta\right)$

$\frac{1}{6}L\ddot{\theta} + \left(g + \dfrac{L}{2}\ddot{\theta}\cos\theta - \dfrac{L}{2}\dot{\theta}^2\sin\theta\right)\left[\cos\theta + \mu\sin\theta\ \mathrm{sgn}\left(\dot{x}_G + \dfrac{L}{2}\dot{\theta}\sin\theta\right)\right] = 0$

7.6. $\frac{1}{3}L\ddot{\theta}\sin\phi - \frac{1}{2}g\sin\theta = -\mu[\frac{1}{2}g\cot\phi\cos\theta - \frac{1}{3}L\dot{\theta}^2\cos\phi]\ \mathrm{sgn}\ \dot{\theta}$

7.8. $mL^2[\frac{4}{3}\ddot{\theta}_1 + \frac{1}{2}\ddot{\theta}_2\cos(\theta_1 - \theta_2) + \frac{1}{2}\dot{\theta}_2^2\sin(\theta_1 - \theta_2)] + \frac{3}{2}mgL\sin\theta_1$

$= \lambda_1\cos(\theta_1 - \theta_2) + M$

$mL^2[\frac{1}{3}\ddot{\theta}_2 + \frac{1}{2}\ddot{\theta}_1\cos(\theta_1 - \theta_2) - \frac{1}{2}\dot{\theta}_1^2\sin(\theta_1 - \theta_2)] + \frac{1}{2}mgL\sin\theta_2 = \lambda_1$

$\dot{\theta}_1\cos(\theta_1 - \theta_2) + \dot{\theta}_2 = 0$ where θ_j are measured from vertical

7.10. $\dot{x}_1 = x_3, \; \dot{x}_2 = x_4$

$(m_1 L^2 + I_2 \sin x_2 + \tfrac{1}{3} mL^2 \sin^2 x_2)\dot{x}_3 = \Gamma - m_1 L^2 (1 + \tfrac{2}{3} \cos x_2)(\sin x_2) x_3 x_4$

$\tfrac{1}{3}\dot{x}_4 = (\tfrac{1}{2} + \tfrac{1}{3} \cos x_2)(\sin x_2) x_3^2 - \dfrac{g}{2L} \sin x_2$

7.12. $\dot{s} = \dfrac{m}{D}(p_1 + p_2 \cos \beta), \; \dot{x} = \dfrac{m}{D} p_1 \cos \beta + \dfrac{M + m}{D} p_2$

$\dot{p}_1 = 0, \; \dot{p}_2 = mg \sin \beta - kx$

where s is the displacement of the cart and $D = m(M + m \sin^2 \beta)$

7.14. $\dot{\psi} = \dfrac{p_1 - I_1 \omega_1 \cos \theta}{I_1 \cos^2 \theta + I_2 \sin^2 \theta}, \; \dot{\theta} = p_2/I_2, \; \dot{p}_1 = 0$

$\dot{p}_2 = -\dfrac{[p_1(I_1 - I_2) \cos \theta + I_1 I_2 \omega_1]}{[I_1 \cos^2 \theta + I_2 \sin^2 \theta]^2} (p_1 - I_1 \omega_1 \cos \theta) \sin \theta$
$\qquad - (m_1 + \tfrac{1}{2}m_2)gL \sin \theta$

7.16. $H = \tfrac{1}{2}(m_A + m_B)\dot{R}^2 - \tfrac{1}{2}[m_A R^2 + m_B(L - R)^2] \, \Omega^2$

$\Gamma = 2[(m_A + m_B)R - m_B L] \, \dot{R}\Omega$

7.18. $H = \tfrac{1}{2}I_{xx}\dot{\beta}^2 - \tfrac{1}{2}(I_{xx} \sin^2 \beta + I_{zz} \cos^2 \beta)\Omega^2 + mg\left(\dfrac{L}{2} - D\right) \sin \beta \cos \Omega t$

$\Gamma = (I_{xx} - I_{zz}) \, \Omega \, \dot{\beta} \sin 2\beta - mg\left(\dfrac{L}{2} - D\right) \sin \beta \sin \Omega t$

where $I_{xx} = m(\tfrac{1}{4}R^2 + \tfrac{1}{3}L^2 - LD + D^2), \; I_{zz} = \tfrac{1}{2}mR^2$

7.20. $\dot{\gamma}_1 = \dot{\psi}, \; \dot{\gamma}_2 = \dot{\theta}$

$\ddot{\gamma}_1 \sin^2 \theta + \dot{\gamma}_1 \dot{\gamma}_2 \sin 2\theta = 0$

$\ddot{\gamma}_2 (9 \cos^2 \theta + \sin^2 \theta) - (\tfrac{9}{2}\dot{\gamma}_1^2 + 4\dot{\gamma}_2^2) \sin 2\theta = \dfrac{2(mg - F)}{mL} \sin \theta$

7.22. $\left(\tfrac{1}{3}L^2 - \dfrac{DL}{\cos \theta} + \dfrac{D^2}{\cos^4 \theta}\right) \ddot{\gamma}_1 + \left(2D^2 \dfrac{\sin \theta}{\cos^5 \theta} - \tfrac{1}{2}DL \dfrac{\sin \theta}{\cos^2 \theta}\right) \dot{\gamma}_1^2$

$\qquad + g\left(\dfrac{D}{\cos^2 \theta} - \dfrac{L}{2} \cos \theta\right) = 0; \; \dot{\gamma}_1 = \dot{\theta}$

7.24. $\dot{\gamma}_1 = \dot{\theta}, \; \dot{\gamma}_2 = \dot{\psi} = c\dot{\theta}$

$\tfrac{1}{3}\ddot{\gamma}_1 - \dot{\gamma}_2^2 \cos \theta(\tfrac{1}{2} + \tfrac{1}{3} \sin \theta) + \dfrac{g}{2L} \sin \theta = 0$

$[mL^2(1 + \sin \theta + \tfrac{1}{3} \sin^2 \theta) + I_2]\ddot{\gamma}_2$
$\qquad + mL^2(1 + \tfrac{2}{3} \sin \theta)(\cos \theta)\dot{\gamma}_1 \dot{\gamma}_2 = \Gamma$

7.26. $\tfrac{2}{3}mL\ddot{\xi} + (kL - \tfrac{1}{2}\sqrt{2}mg) \, \xi = 0, \; k > mg/\sqrt{2}L$

7.28. $1.3687mL\ddot{\xi} + 1.44\mu L\dot{\xi} + 2.4mg\xi = F(0.8 - 0.6\xi)$

7.30. $\tfrac{8}{3}mL\ddot{\theta} + (5kL - \tfrac{3}{2}mg) \, \theta = F(2 - 14\theta)$

7.32. $R^* = \dfrac{m_B}{m_A + m_B} L, \; \ddot{\xi} - \Omega^2 \xi = 0$, always unstable

CHAPTER 8

8.2. $\omega = 3.135$ rad/s at $\beta = 93.05°$, $\theta = 94.87°$

$\Delta \overline{H}_G = -3.130 I' \overline{i}$, $\omega_2 = -0.1667$ rad/s

8.4. rotation about z axis

maximum angle: $\theta_1 = 38.56°$, $\overline{\omega}_1 = 6.902 \overline{i} + 6.927 \overline{k}$ rad/s

minimum angle: $\theta_2 = 18.93°$, $\overline{\omega}_2 = 5.526 \overline{j} + 8.379 \overline{k}$ rad/s

8.6. looping precession: $\theta_{min} = 53.13°$, $\theta_{max} = 53.26°$

8.8. $\theta = \dfrac{\pi}{2} + C_1 \sin\left(\dfrac{\omega^2}{\sigma} t + v_1\right) - C_2 \sin(\sigma t + v_2)$

$\psi = C_1 \cos\left(\dfrac{\omega^2}{\sigma} t + v_1\right) + C_2 \cos(\sigma t + v_2)$

where $\omega^2 = mgL/(I' + mL^2)$, $\sigma = \dot{\phi} I/(I' + mL^2)$

8.10. $\Omega_0 \gg \Omega_1$, $I'\lambda/c \gg 1$, and $ct/I' \gg 1$

8.12. $\ddot{\xi} + 2\Omega\dot{\eta} + (k_\xi/m - \Omega^2)\xi = g \cos \Omega t$

$\ddot{\eta} - 2\Omega\dot{\xi} + (k_\eta/m - \Omega^2)\eta = g \sin \Omega t$

$\ddot{\beta} - 2\Omega\dot{\gamma} + (k_\beta/I_2 - \Omega^2)\beta = 0$

$\ddot{\gamma} + 2\Omega\dot{\beta} + (k_\gamma/I_2 - \Omega^2)\gamma = 0$

8.14. $\theta = \theta_0 + \dfrac{\epsilon}{\lambda} \sin \lambda t + \dfrac{\epsilon^2}{4\lambda^2}\left[\dfrac{(I' + A - C)}{J} \sin 2\theta_0 - \cot \theta_0\right][3 - 4\cos \lambda t + \cos 2\lambda t]$

$\dot{\psi} J/p_\phi = \dfrac{\epsilon}{\lambda} \sin \theta_0 \sin \lambda t + \dfrac{\epsilon^2}{2\lambda^2} \cos \theta_0 \left[1 - 4 \dfrac{I' + A - C}{J} \sin^2 \theta_0\right] \sin^2 \lambda t$

$\qquad + \dfrac{\epsilon^2}{4\lambda^2}\left[2\dfrac{(I' + A - C)}{J} \sin^2 \theta_0 - 1\right] \cos \theta_0 (3 - 4\cos \lambda t + \cos 2\lambda t)$

where $J = A' + (I' + A) \sin^2 \theta_0 + C \cos^2 \theta_0$ and $\lambda^2 = \dfrac{p_\phi^2 \sin^2 \theta_0}{J(I' + B)}$

Index